NASA's First Space Shuttle Astronaut Selection

Redefining the Right Stuff

David J. Shayler and Colin Burgess

NASA's First Space Shuttle Astronaut Selection

Redefining the Right Stuff

 Springer

Published in association with
Praxis Publishing
Chichester, UK

David J. Shayler
Astronautical Historian
Astro Info Service Ltd
Halesowen, West Midlands, UK

Colin Burgess
Spaceflight Historian
Bangor, New South Wales, Australia

SPRINGER-PRAXIS BOOKS IN SPACE EXPLORATION

Springer Praxis Books
Space Exploration
ISBN 978-3-030-45741-9   ISBN 978-3-030-45742-6   (eBook)
https://doi.org/10.1007/978-3-030-45742-6

Project Editor: Michael D. Shayler
Cover design: Jim Wilkie

CAPTIONS FOR COVER IMAGES:

Front: A stunning bird's-eye view of the Space Shuttle *Atlantis*, with its External Tank and twin Solid Rocket Boosters, during its rollout to Pad 39A at the Kennedy Space Center, Florida, as STS-79 on August 20, 1996. During this mission, John Blaha traded places with TFNG Shannon Lucid at the end of her six-month tour on the Russian Mir space station.

Back, top left: The crew of STS-7 on the flight deck of *Challenger*, June 1983. Under the command of Robert 'Crip' Crippen [rear left], the first four members of the TFNG to fly on the Space Shuttle were [l to r] Mission Specialist Sally Ride, the first American woman to fly in space, Pilot Rick Hauck, and Mission Specialists Norman Thagard and John Fabian.

Back, top right: Sixteen years later, in July 1999, Mission Specialist Steve Hawley – seen here on the middeck of *Columbia* during his fifth mission, STS-93 – became the last member of the Class of 1978 to fly in space, ending an era of over 100 missions in Earth orbit.

Back, lower right: The official emblem for the Class of 1978, designed by Robert McCall.

This Springer imprint is published by the registered company Springer Nature Switzerland AG
The registered company address is: Gewerbestrasse 11, 6330 Cham, Switzerland

Contents

Authors' Preface

At first, this book was never going to happen. Previously, either individually or working as co-authors, we had researched and put together four books covering the early NASA astronaut selections involving Groups 1 through to 7 between April 9, 1959 and August 14, 1969. During that time, 73 men were chosen to crew and support America's pioneering manned space missions under the Mercury, Gemini and Apollo programs. As mentioned in our previous work, *The Last of NASA's Original Pilot Astronauts*, several of these astronauts remained active long enough to crew missions in the first decade of Shuttle flight operations between 1981 and 1990.

We knew that 2018 would mark the 40th anniversary of the first astronaut selection of the new Space Shuttle era, and even though we had intended that our previous book on the Group 5 and 7 astronauts would end our collaboration on the subject of astronaut selections, we decided this anniversary should not go unmarked. We therefore determined to research and write the history of NASA's Group 8 astronauts, unofficially known as The Thirty-Five New Guys, or by the acronym TFNG (which is further explained later in this book).

Although there have been biographies and memoirs published on several of the 29 men and six women selected in this 1978 group, in addition to countless magazine and newspaper articles, there has never been a book published which fully details the lives, accomplishments and sometimes tragedies associated with each member of this group. That is all presented here, along with the chronology of the selection process begun in 1976, and the personal recollections of some who were successful in their application – and others who were not.

Unlike previous astronaut intakes, NASA opened the way for women and minorities with this selection process, not only to apply for the position of Pilot astronaut and the new category of Mission Specialist, but actually to be selected. History records that the civilian space agency announced the names of the 35 newly-selected astronaut candidates in January 1978, including four minority selectees and six women. The names of those six women became instantly renowned across the world.

This *will* be the final group selection book we will tackle jointly, but it is a tale which history has dictated should be recorded. We are proud to have compiled and presented this fascinating and truly involving story.

Colin Burgess

This new collaborative effort between Dave Shayler and I never started out as part of a series on the selection of NASA's astronauts leading up to the era of the Space Shuttle program. In fact, it had its origins in a friendship going back many years, as well as a shared interest in the history of human space exploration – and in recording that history. I cannot recall who first came up with the idea all those years ago (most likely Dave) of combining our talents, enthusiasm and research in putting together a book on NASA's fourth and sixth astronaut groups, known collectively as the Scientist-Astronauts, but we found we worked well together, even though we happened to live on opposite sides of the globe. The result, published in 2007, was our first co-authored book for Springer-Praxis, *NASA's Scientist-Astronauts*.

That book may have ended our cooperative work on the subject of astronaut groups, were it not for one occasion when I was visiting England and the two of us were sharing a meal in a restaurant. During our conversation, we were discussing the selection of the space agency's very first astronauts for the Mercury program. We agreed that we shared a mutual frustration, in that the names of five candidates for that role had eluded both of us for many years.

Then, in a stroke of remarkably good fortune, I was contacted by former military man, Walter ("Sully") Sullivan, who had acted as the liaison officer for the 32 Mercury astronaut finalists back in 1959 at Wright-Patterson Air Force Base, Ohio. He had read a post on the subject I placed in the collectSPACE.com forum, and kindly offered to assist me if I wanted to write a book about this selection group. Most importantly, he had a complete list of the 32 Mercury finalists. I quickly agreed, and he subsequently sent me the names of the missing five candidates. Sadly he is no longer with us, but with Sully working as a contact liaison and guarantor for me, I managed to locate all of the Mercury candidates or their surviving family members, and the result was the 2011 book, *Selecting the Mercury Seven: The Search for America's First Astronauts*.

Following the publication of this book, I had no plans to put together a follow-on title delving into the selection of the second NASA astronaut group. That is until one propitious day when Dave contacted me in excitement. He said that in his latest search of NASA's treasure trove of historical records he had unearthed documents giving the names of all the finalists for not only the second, but also the third astronaut groups. With his own research for a number of books meaning he could not even look at this new project, he kindly offered those lists to me, and a whole new (and successful) hunt for these men or their family members began, culminating in a 2013 book on both groups for Springer-Praxis called *Moon Bound: Choosing and Preparing NASA's Lunar Astronauts.*

That same year, with the selection of NASA groups 1, 2, 3, 4 and 6 now covered in books, Dave and I began serious discussions on the possibility of combining resources once again to produce a new book detailing the process of choosing astronauts who became part of the two remaining pre-Shuttle classes: NASA Groups 5 and 7 (the latter also known as the MOL group). A contract was signed and work began, albeit slowly at first as we were both engaged in putting together other contracted books for Springer-Praxis over the next couple of years. Initially, our research and writing on the two groups was sporadic, but it began to pick up steam once those other obligations had been cleared. In 2017 this new book was published, and we felt we had finally recorded, separately and in partnership, the selection of seven groups of amazingly talented and bold men who pioneered America's dynamic space program before the advent of the reusable Space Shuttle.

The lure of writing about the 1978 Group 8 selection inexorably led to an email correspondence on the subject, and once we had agreed to tackle this new book the work began.

I know that we are both proud to have been able to record the stories of these very remarkable men and women, some of whom are no longer with us, and we are grateful to everyone who assisted in putting together this wonderful book, which we hope will be read and enjoyed by many generations to come.

David J. Shayler

From the late 1960s, the arrival of a familiar brown envelope – emblazoned with the NASA logo – was eagerly anticipated, answering my latest request and filled with documents or photos from the public affairs departments of the different NASA field centers. In the summer of 1978, I awaited the envelope containing background information and images of the recently selected 35 individuals of the eighth NASA astronaut class. When I finally received the package and looked at the images, I recall thinking how youthful they looked compared to the seasoned veterans chosen in the 1960s. This first intake of NASA astronauts for a decade was certainly a different composition, featuring the first female and minority astronaut candidates chosen by the space agency. Even the phrase 'astronaut candidate' was different. No longer were these new astronauts considered to be fully-fledged 'astronauts' from their first day at NASA. They were going to have to earn that title by completing a training and evaluation program first. For me that did not matter, as they had been named as the next astronaut class by NASA and, in my system, qualified for individual archive folders that expanded over the coming years as their biographical details were researched, assignments recorded and space flights logged.

I had been collecting information about each NASA astronaut group and their members since the 1960s, subsequently expanded to include details on the Soviet cosmonauts and other international space explorers. From the late 1970s, this had led to articles in the publications of the British Interplanetary Society – most notably *Spaceflight* – assisting other authors in their research, and self-publishing a range of titles on the early NASA astronaut groups and the Shuttle program through my company Astro Info Service.

Over the years, I had established contacts at NASA's Johnson Space Flight Center in Houston, and at the NASA History Office in Washington D.C., who had kindly replied to my numerous queries and questions. By the 1980s, I had in my collection a growing number of letters from former and current astronauts which gradually filled the gaps in my research into the structure and workings of the NASA Astronaut Office, the experiences of those lucky few who had flown in space, and the assignments that the astronauts held in between their missions. Aided by a growing network of co-researchers across the globe, including a certain Colin Burgess in Australia, my archive of information on the world's space explorers has continued to expand to this day.

One area in which I was especially interested was researching the extensive support roles the astronauts fulfilled in between missions, and with it the story about what astronauts *really* do when not flying in space. Starting with the first crewed flights of Apollo, and in addition to detailing the astronauts' roles and accomplishments on each space flight, I became intrigued by the roles fulfilled by those who supported the missions, securing that little snippet of detail, obscure piece of information, or item of little-known fact. Simply reporting that an astronaut had 'supported' a mission was not enough for me. I wanted to know what they had supported and what that role involved. By the 1980s, that became more convoluted when trying to piece together the careers of the astronauts during the far more complex Shuttle program. To my mind, understanding the evolution of the astronauts' roles and their assignments is as important as the missions they fly.

While compiling this current title, we were able to access a number of official documents pertaining to the various support assignments across the 135 Shuttle missions. These have been used to track the support roles of both the TFNG and other members of the Astronaut Office. However, there are gaps in the records and I would appreciate any additional information (via the contact details on my website, www.astroinfoservice.co.uk) from readers who may be able to fill in these gaps to document the complete picture of astronaut support roles during the Shuttle program.

For me, a significant change to the system of NASA astronaut selection, training and assignments occurred with the naming of the Class of 1978. These "Thirty-Five New Guys" represented a different era from those chosen a decade earlier. Even the selection process had matured to reflect the changes in the program, the experiences from past selections, and the hopes for the future. With the selection of the eighth group of astronauts, there were new elements to this process: pilots with a broader range of flying experiences and the introduction of multi-skilled Mission Specialists instead of Scientist-Astronauts. Unlike their scientific predecessors, the Mission Specialists were not required to qualify from an intensive military jet pilot course before continuing with their astronaut career. Even in the late 1970s, the prospect of flying regular, more complex missions in the Shuttle program hinted at a team of astronauts who would almost be earning 'frequent-flyer' status.

By the early 1980s, many of the legendary pioneering NASA astronauts – those who had the so-called '*Right Stuff*' qualities born in the 1950s test-pilot fraternity, as coined by author Tom Wolfe in his 1979 book – had long since retired. They had blazed the trail into space and on to the Moon, and had put the untried Shuttle system through its paces. From now on, a new era of astronauts would take up the mantle and define a new, different era of '*Right Stuff*', as team players in a more bureaucratic NASA and a far more complex program than it had been just decade or so before.

When Colin first suggested a third collaborative venture on the Class of 1978, I was as intrigued by their story as I had been with those from the earlier selections. The TFNG were indeed pioneers in their own right, creating an era that would last longer than that of the original astronauts they had followed. They, in turn, unwittingly molded a new image of a multi-talented crewmember, which would morph again in the new millennium and yet another generation of post-Shuttle NASA astronauts. For me, these 35 individuals of the eighth selection represented the next step, after the Moon, in American space flight history. Almost immediately, the youthful-looking 35 took up the mantle of the 'original 73' astronauts chosen between 1959 and 1969, remolded it to fit a new program and a new NASA, and fulfilled their remit of active participation in the Shuttle program, either in support roles, as crewmembers, or latterly as managers across the 30-year program. This, then, is their story, one which ranks alongside that of their pioneering colleagues, who together earned the right to be known as an *American astronaut*.

Acknowledgements

Dave Shayler

A project such as this requires a significant amount of research, and by the very nature of such research, the assistance of a number of key individuals to guide and aid you in that task.

Prominent for me in this 'support team' over the past four decades have been the various members of the NASA History and Public Affairs Offices at NASA HQ and JSC. Without their help and support, access to the information presented in this and many of my past and upcoming publications would not have been possible. The oral histories of the TFNG within the NASA JSC Oral History Project have been especially useful in writing this book. In addition, the archived documents of former NASA Flight Director Cliff Charlesworth, originally located at JSC when researched by the authors, have also proven to be a valuable asset in detailing the Astronaut Office assignments for each Shuttle mission quoted. Thanks also to Ed Gibson, Group 4 Scientist-Astronaut and Group 8 selection board member, for his generous Foreword.

Considerable assistance for this project has been provided by Group 8 astronauts Hoot Gibson, Steve Hawley and Rhea Seddon. Special mention must also be made of fellow TFNG Mike Coats, Dick Covey, Jeff Hoffman, George "Pinky" Nelson, Norman Thagard and Kathy Sullivan, who were all interviewed over the years – primarily for other writing projects – but whose answers and information provided have also proven valuable for this book.

Special thanks are extended to Michael Cassutt, esteemed author of the trilogy of *Who's Who in Space* compendiums and authoritative biographies on Deke Slayton, Tom Stafford and George Abbey, for his guidance through the challenging and changing history of the NASA Astronaut Office and its various branches. Thanks also go to Bert Vis for his support and suggestions. Special thanks are due to Mario and Susan Runco for sharing with us their memories of their own applications to Group 8.

A special thank you also to Tim Gagnon for permission to replicate his design for the official TFNG 40th anniversary emblem.

Thanks also to the staff and Council of the British Interplanetary Society for access to their library and archive, which has been of immense value in compiling this book. Once again, the extensive resources of our co-author and friend, the late Rex Hall MBE, former President of the BIS, have proven invaluable during our research. We also express our continued thanks to Clive Horwood of Praxis Publishers for his support and encouragement of our numerous book proposals over the past two decades and to the team at Springer Nature, including Maury Solomon and Hannah Kaufman in New York. Appreciation also goes to Jim Wilkie for his skills in turning our cover ideas into the final design. Finally, on the production side, many thanks to my brother Mike Shayler who, during very difficult times, once again applied his professional and dedicated editing and wordsmith skills to turn our original draft into the finished product you see here.

Thanks also to my mother, Jean Shayler, who at 91 continues to be involved in reading each draft page of my latest book project, and to my wife Bel for her continued love, support and understanding in that I will always wish to write just one more book… and maybe another after that… and then another. Also, to our beloved German Shepherd Shado, our wolf in dog's clothing, who still cannot quite work out why I spend so much time away from more serious pursuits such as throwing a ball or running around a field.

To all a very large and appreciative thank you.
Dave Shayler

Colin Burgess
I can really only echo Dave Shayler's thanks to and comments on so many people who participated to varying degrees in the research and compilation of this book, which we regard as an important insight into the selection of the first group of astronauts specifically chosen to operate on NASA's fleet of Shuttle Orbiters. From the outset, we have received a great deal of support, not only from those astronauts named by Dave, but from a whole range of people connected in so many ways to the selection of the TFNG group, their training, missions and myriad achievements.

There are so many helpers to whom Dave and I owe our sincere thanks. The mere listing of names can in no way suggest our tremendous feelings of gratitude towards these individuals. Without their interest and cooperation it would have been literally impossible to collect, transcribe, organize or publish the information and stories gathered together in this book. Dave, as chief motivator for this project, has covered much of this territory in listing his acknowledgements, but I have a few more names to add. Each has added immeasurably to this publication, and both of us are extremely grateful.

For their kind assistance, I would like to thank these good folks who were involved in the selection of the TFNG group, and responded to our queries. Special personal thanks go to Carolyn Huntoon, the first American woman to serve as a Director of the Johnson Space Center (JSC) in Houston, and a member of the Group 8 selection panel. Collective thanks also to a representative few of the unsuccessful candidates for relating their experiences while undergoing the selection process, namely: Frank R. Harnden, Jr., William D. Heacox, Major Gary W. Matthes (USAF, Ret.), Dr. Joseph K. (Ken) Ortega, Dr. Lawrence (Larry) Pinsky, and Wilton T. Sanders. III. Additional information was provided by retired NASA astronaut Jerry Ross and former Payload Specialist Charlie Walker.

I would also like to thank a long-time supporter of our projects, namely Ed Hengeveld, for his assistance in tracking down some lesser-known photographs for use in this book, and for the same reason our appreciation goes to Joachim Becker of Spacefacts.de. The definitive source for many answers to our questions is space forum Collectspace.com, and its founder and chief motivator, Robert Pearlman. Other helpers include Francis French and David Shomper. I thank them for their wisdom and guidance.

As Dave has expressed his love and thanks to his family, so I also shower my wife of more than five decades, Pat, with love and gratitude for putting up with me and my countless hours spent hunched over my laptop, surrounded on our study floor by piles of books, magazines, photographs, scrap books, folders and old, yellowing newspapers.

If I have forgotten anyone, my sincere apologies. You are all champions.

Colin Burgess

Foreword

On January 16, 1978, NASA announced the selection of 35 new astronauts, the eighth group selected, who chose to call themselves the Thirty-Five New Guys (TFNG). There had been no additions to the Corps since August of 1969. This delay resulted from the long gap between the end of the Apollo-type missions in 1975 and the first Shuttle flight in 1981. But in the late 1970s, the need for highly qualified and motivated professionals became immediate as well as long term. The capabilities of the 35 new guys greatly exceeded requirements and expectations.

I was given the opportunity to be a member of their selection board and thereby got to know them on paper and in person rather quickly. Coming into the Astronaut Office with little previous experience in this type of high-powered, informal operation required some time to adjust. However, every one of the TFNG were hard chargers and there to contribute to the advancement of America in space to the best of their abilities. Any intimidation was quickly replaced with their drive to perform.

It was a diverse group: 15 military pilots and 20 scientists composed of 29 males and NASA's first six females. Other firsts included three African-Americans and one Asian-American. Despite their pride in their diversity, they quickly put their team identity and its contributions above all considerations. In addition, just like the pilots and scientists since the fourth group of astronauts, pilots also contributed to the onboard science and scientists also performed many space operational duties.

Unlike previous groups, except for the Apollo-Soyuz mission with Russia, the TFNG would soon encounter working relationships with many international astronauts and experimenters from Europe, Canada, Mexico, Japan, the Middle East and Russia. Once the International Space Station was in operation, working in an international environment became routine for all astronauts, especially the TFNG.

Over the next two decades, the TFNG contributed a total of 103 individual flights for an average of three flights per member. They performed:

- Retrieval and repair of satellites and other facilities and equipment in orbit
- The first untethered spacewalks 'flying' the MMU
- Spacelab operations
- Shuttle-Russian Mir station dockings
- Russian Mir station operations using Soyuz and Shuttle transportation systems

Skylab 4 Science Pilot Ed Gibson

A brief mention of a few of the TFNG illustrate the capabilities, drive and accomplishments of this group.

During STS-61 commanded by Richard Covey, Jeff Hoffman, partnered by veteran astronaut Story Musgrave, performed three of the five very difficult space walks to restore the designed-in level of visual acuity to the Hubble Space Telescope, which then revolutionized the field of Astronomy.

After earning BSc and MSc degrees in engineering, Norm Thagard flew 163 combat missions in Vietnam as a U.S. Marine Corps aviator, then earned his MD before flying four Shuttle missions as a NASA Mission Specialist. Lastly, he was the first American to fly a mission on the Russian Mir space station and paved the way for six of his fellow countrymen. On his 115-day mission, he also broke the American record for time in space that had lasted for 21 years. Another member of the group, Shannon Lucid, later broke this record with a flight of slightly over six months, also on the Mir space station.

Regrettably, five members of the group paid the ultimate price for their participation. On the morning of January 28, 1986, STS-51L lifted off with ice hanging from the launch gantry. It had been approximately 20 degrees below the stated SRB-lower-temperature limit overnight. Thus, at 73 seconds into the mission, the booster exploded, killing all seven crew members including four members of the TFNG: Dick Scobee (Commander), Judy Resnik, Ellison Onizuka and Ron McNair.

In the selection process, I became quite familiar with Ron McNair's background. As he grew up in poverty, he became skilled in football and interested in science. While earning a degree in a small college, he was given the opportunity to work at MIT in the summers and later was accepted there for graduate school where he earned a PhD. Along the way, he had also become a black belt in karate, which he used to help other children growing up in the same situation that he had. Each time he moved into a new environment, he found a church that had a basement that he could use to gather local youth and teach them the discipline of the sport. He got many young men moving in a constructive direction. He was an admirable black youth from poverty who made all the right decisions.

Additionally, Dave Griggs, who flew on Discovery in April 1985, when he made NASAs first unrehearsed spacewalk to rescue a stranded satellite, died in an aircraft accident on June 17, 1989.

The many significant contributions of the TFNG did not end when their flying days concluded. Many went on to perform exceptionally well in management. For example, Mike Coats became the Director of the Johnson Space Center and Kathy Sullivan went on to become the head of the National Oceanic and Atmospheric Administration. Many others stepped into leadership positions in industry or other government organizations.

Also, although not quantifiable but certainly important and real, the performance and diversity of the TFNG inspired many students in science, technology, engineering and math and many women and minorities to pursue their dreams where they might not have been so inclined without the leadership examples in the TFNG.

The strength and diversity of the numerous ways in which the TFNG have displayed their own unique '*Right Stuff*' has significantly enhanced America's performance in space and science. It flowed directly from the group's 35 individual skill sets, drives and perseverance.

The TFNG are a tough act to follow!

Edward G. Gibson (PhD)
Former NASA Astronaut (Group 4, 1965)
Science Pilot Skylab 4 (November 1973 – February 1974)
Member, Group 8 Selection Board

NASA Astronaut Gold Pin

This book is dedicated with gratitude to the lives and many accomplishments of the Group 8 astronauts who no longer share our days, but will live on in our memories.

Dale A. Gardner
S. David Griggs
Ronald E. McNair
Steven R. Nagel
Ellison S. Onizuka
Judith A. Resnik
Sally K. Ride
Francis R. Scobee
David M. Walker
Donald E. Williams

A blade of grass is commonplace on Earth; it would be a miracle on Mars.
Our descendants on Mars will know the value of a patch of green.
And if a blade of grass is priceless, what is the value of a human being?

Carl Sagan (1934–1996)

Abbreviations and Acronyms

Informal Military Designations

USAF
"FS" stands for Fighter Squadron; "RS" for Reconnaissance Squadron; "BS" for Bomber Squadron

U.S. Navy
"V" stands for fixed wing; "F" for fighter wing; "A" for attack; "Q" for electronic; "R" stands for Reserve but can also stand for Reconnaissance; "W" for early Warning; "T" for training; "X" for test and evaluation (as in eXperimental)

USMC
Marine air units use the suffix "M" within the U.S. Naval designation coding, such as VMFA for fixed, with AW added for "All Weather" squadrons

AB	Air Base
AFB	Air Force Base
AFTPS	Air Force Test Pilot School (Edwards AFB, California)
AFIT	Air Force Institute of Technology (Wright-Patterson AFB, Ohio)
AFSC	Air Force Systems Command
ATO	Abort To Orbit (abort mode)
ALT	Approach and Landing Tests (Space Shuttle)
AOA	Abort Once Around (abort mode)
ARPS	Aerospace Research Pilot School (Edwards AFB, California)
ASCAN	Astronaut Candidate
BIS	British Interplanetary Society (London, England)
BSc	Bachelor of Science degree
BUp	Back Up (crewmember)

CapCom	Capsule Communicator (Mission Control)
C^2	Cape Crusader (KSC astronaut support team)
CB	Astronaut Office, LBJ Space Center (JSC) mail code
CDR	Commander
CEIT	Crew Equipment Interface Test
CERN/EP	European Organization for Nuclear Research/Experimental Physics Division
CfA	Center for Astrophysics (Cambridge, Massachusetts)
CO	Commanding Officer (DOD)
CU Boulder	University of Colorado (Boulder, Colorado)
DOD	Department of Defense
DOTF	Deployments and Operations Task Force
DSO	Detailed Supplementary Objectives
DTO	Detailed Test Objectives
DVM	Doctor of Veterinary Medicine
EAFB	Edwards Air Force Base (California)
EDT	Eastern Daylight Time
EMU	Extravehicular Mobility Unit (Shuttle spacesuit)
EOM	End Of Mission
ER	Emergency Room
ESA	European Space Agency
EST	Eastern Standard Time
ET	External (fuel) Tank (Space Shuttle)
EV	EVA astronaut 1, 2, 3 or 4
EVA	Extra-Vehicular Activity (spacewalk)
FBI	Federal Bureau of Investigation
FBIS	Fellow of the British Interplanetary Society
FDF	Flight Data File
FCOD	Flight Crew Operations Directorate
FPO	Fleet Post Office
FRF	Flight Readiness Firing
g	Gravity force
GAS	Get Away Special
GET	Ground (Mission) Elapsed Time
GSFC	(Dr. Robert H.) Goddard Space Flight Center, Greenbelt, Maryland
HF/DF	High Definition/Direction Finding (also known as Huff Duff)
HQ	Headquarters
HST	Hubble Space Telescope

ISS	International Space Station
IUS	Interim Upper Stage
IV	IVA astronaut (EVA support crewmember)
IVA	Intra-Vehicular Activity
JSC	(Lyndon B.) Johnson Space Center (from 1973), Houston, Texas (formerly MSC)
KSC	(John F.) Kennedy Space Center (Florida)
LANTIRN	Low Altitude Navigation and Targeting Infrared for Night
LC	Launch Complex
LCDR	Lieutenant Commander
LST	Large Space Telescope (renamed Hubble Space Telescope)
MAW	Marine Air Wing
MCAS	Marine Corps Air Station
MCC	Mission Control Center (MSC/JSC, Houston, Texas)
MD	Doctor of Medicine
MECO	Main Engine Cut-Off
MET	Mission (Ground) Elapsed Time
MIT	Massachusetts Institute of Technology (Cambridge, Massachusetts)
MLP	Mobile Launcher Platform
MMU	Manned Maneuvering Unit (Space Shuttle)
MOCR	Mission Operations Control Room (MSC/JSC)
MOL	Manned Orbiting Laboratory
MPS	Main Propulsion System (SSME/SRB/ET)
MS	Mission Specialist (Space Shuttle)
MSc	Master of Science degree
MSC	Manned Spacecraft Center, Houston (Texas); renamed Lyndon B. Johnson Space Center (JSC) in 1973
MSE	Manned Spaceflight Engineer (military Payload Specialists)
MSFC	(George C.) Marshall Space Flight Center, Huntsville (Alabama)
NAS	Naval Air Station
NAS/NRC	National Academy of Sciences/National Research Council
NASA	National Aeronautics and Space Administration
NATC	Naval Air Training Command
NAVELEX	Naval Electronics System Command
NBL	Neutral Buoyancy Laboratory (Sonny Carter Facility, Houston)
NBS	Neutral Buoyancy Simulator (Marshall Space Flight Center, Huntsville)
NORAD	North American Aerospace Defense Command (Colorado Springs, Colorado)

NRC	National Research Council (Washington D.C.)
OFT	Orbital Flight Tests
OMS	Orbital Maneuvering System (Space Shuttle)
OPF	Orbiter Processing Facility
OV	Orbital Vehicle (Space Shuttle)
PAM	Payload Assist Module (upper stage)
PAO	Public Affairs Office/Officer
PC	Payload Commander
PDRS	Payload Deployment and Retrieval System
PhD	Doctorate degree
PLT	Pilot
PS	Payload Specialist
RADM	Rear Admiral
RAF	Royal Air Force
RCA	Radio Corporation of America
RMS	Remote Manipulator System (Space Shuttle)
ROTC	Reserve Officer Training Corps
RSLS	Redundant Set Launch Sequence (abort mode)
RTLS	Return To Launch Site (abort mode)
SAIL	Shuttle Avionics Integration Laboratory
SAMSO	Space And Military Systems Organization
SCA	Shuttle Carrier Aircraft (modified Boeing 747)
SL	Spacelab
SLS	Spacelab Life Sciences
SMM	Shuttle Mir (docking) Mission
SMS	Shuttle Mission Simulator
SRB	Solid Rocket Booster (Space Shuttle)
SRM	Solid Rocket Motor (Space Shuttle)
SSME	Space Shuttle Main Engine
STA	Shuttle Training Aircraft
STS	Space Transportation System (Space Shuttle)
TACT	Transonic Aircraft Technology
TAL	Transoceanic Abort Landing (abort mode)
TDRS(S)	Tracking and Data Relay Satellite (System)
TFNG	Thirty-Five New Guys (Group 8 astronaut candidates)
TPS	Test Pilot School
UCLA	University of California, Los Angeles (California)
USA	United Space Alliance or United States Army

USAF	United States Air Force
USAFA	United States Air Force Academy (Colorado Springs, Colorado)
USC	University of Southern California (California)
USGS	United States Geological Survey
USMA	United States Military Academy (West Point, New York State)
USMC	United States Marine Corps
USMCR	United States Marine Corps Reserve
USN	United States Navy
USNA	United States Naval Academy (Annapolis. Maryland)
USSR	Union of Soviet Socialist Republics (1917–1991) now Russia
VAB	Vehicle Assembly Building (KSC, Florida)
VAFB	Vandenberg Air Force Base (California)
WETF	Weightless Environment Training Facility (JSC)

Prologue

Some 40 years ago, in the intermediate days of the post-Apollo and pre-Space Shuttle era, the roads south east of Houston, Texas were a lot less congested than one would expect to find today. For 29-year-old Barbara Anne (Tracey) Sauerland MD, this made for an easy 19-mile (31-km) drive to the Johnson Space Center (JSC) from her home in Galveston.

Born in New England, Connecticut, Dr. Sauerland had attended the nearby St. Thomas Aquinas High School, later earning a biology degree in Boston and then a doctorate in Neurochemistry and general practitioner's degree from the University of Texas Medical Branch. By January 1978, she was working in JSC's Space and Life Sciences Directorate, and was a regular, familiar sight to the center's guard, who would promptly smile and wave her through.

Like her then husband, prominent psychiatrist Professor Eberhardt Sauerland MD, Tracey was keen on taking her profession and experience into space, completing an application for the 1978 astronaut selection program by the due date of June 30, 1977. In the previous decade, her husband had applied to become an Apollo-era astronaut, but had not made it through to the final selection. Tracey was determined to go further, and even held out hopes – however remote – of one day becoming the first American woman to venture into space.

As she told one reporter, there was an odd childhood pointer that could work in her favor, namely that she did not get sick in machines – handy enough when being whirled around at 30 rpm by a dizzying device designed to measure motion sickness. "I was the only one of our gang that used to go to the fairground and try everything without a murmur," she said[1].

---

[1] *Sydney Sun* newspaper, unaccredited article, "Going Up – A Real Bionic Woman," December 9, 1977

By midnight on the due date, a total of 8,079 applications had been received by NASA for consideration, which included 1,544 from women. There were 1,261 pilot applications and 6,818 for the position of Mission Specialist. Eventually, the number of suitably qualified applications was processed down to just 208. To her delight, Tracey Sauerland's name was still on this reduced list.

Beginning with the first 20 in August 1977, the 208 remaining astronaut applicants began arriving at JSC to undertake a week-long series of personal interviews, coupled with physical and psychiatric examinations. The final selection procedure was expected to be completed by November. The following month, the selection board would meet to determine which of the applicants best suited the requirements for future Shuttle pilots as well as the new category of Mission Specialist, with up to 20 to be chosen in each category. Those selected would report for duty at JSC in July 1978. After two years of training and evaluation – and assuming there were no unforeseen problems with any candidate – they would officially be named as NASA astronauts and become eligible for flights on future Space Shuttle missions. The training/evaluation time was later decreased to just one year.

 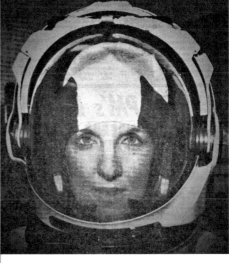

Dr. Tracy Sauerland and (right) trying on an Apollo spacesuit.

Dr. Sauerland was notified that she would be assessed in the third group of applicants, reporting to JSC on Monday, August 29 for her week of individual interviews and physical examinations. All 20 of the applicants in her group held doctorates or medical degrees, or both, and one (Michael Kastello) even had a degree in veterinary science. Eight of their number were women.

On January 16, 1978, NASA Administrator Dr. Robert A. Frosch announced the selection of the 35 candidates who would begin astronaut training with America's space corps later in the year. Included in that number were six women, plus the first three African-Americans, and the first Asian-American (who later became the first person of Japanese ancestry to fly into space).

Collectively, they were introduced to the world in February 1978 as NASA's Group 8 astronauts. However, they also came to be known by the less formal acronym TFNG, which (in polite usage) stood for 'the Thirty-Five New Guys'. Two decades earlier, America's first astronauts had been selected for the Project Mercury series of human space flights, and NASA was now preparing for the upcoming test flights of the spacecraft known as the Shuttle Transportation System (STS), or more simply, the Space Shuttle.

While pilots would still be required to fly the innovative winged vehicle, this group marked a time of operational and cultural changes within the nation's space program, which came with the emergence of a new breed of astronaut known as the Mission Specialist. Among their intended tasks was the operation of critical systems aboard the Shuttle, conducting intricate tests and experiments, and assisting in the deployment of satellites from the Orbiter's capacious payload bay through the manual manipulation of a multi-jointed robotic arm.

In the previous seven all-male astronaut selections, the task of crewing Mercury, Gemini and Apollo spacecraft had fallen to men with previous test-piloting experience, or scientists willing to be trained as jet pilots. Now, for the first time, six of the 35 newly-appointed Astronaut Candidates (Ascans) were women. Minorities were also included.

A breakdown of the Group 8 list demonstrates an end to the archaic and controversial gender and racial barriers that had existed since the space agency named its Group 1 Mercury astronauts back in 1959. Looking at statistics, 15 of those named fell into the Pilot category and 20 were potential Mission Specialists. Furthermore, the group consisted of 14 civilians and 21 military officers.

Pressed as to why there had been no ethnic minorities or women in the astronaut corps until that time, JSC Director Christopher Columbus Kraft boldly stated at a press conference in Washington that there had been few qualified minorities and women when the last group of astronauts had been chosen in August 1967. "The most rewarding thing [about this astronaut group] is that there were large numbers of qualified women and minorities this time around," a diplomatic Kraft said. "We had no problem finding women and minorities who were qualified and highly motivated as to what they wanted to do[2]."

---

[2] *Washington Post* newspaper, article, "NASA announces selection of 35 Shuttle astronauts", January 17, 1978

With her own qualifications and experience, Tracey Sauerland had been hoping to make the final cut, but she would be disappointed. Her name was not one of those chosen by NASA for their Group 8 astronaut cadre, and she received the dreaded phone call from a NASA aide, who said that her application had been unsuccessful on this occasion. Shaken but undeterred, she would try again for the May 1980 selection, but once again missed out. That particular dream was now at an end and she went into a lifetime career in radiology.

The six women chosen were identified by Dr. Frosch as Anna Fisher of Rancho Palos Verdes, California; Shannon Lucid of Oklahoma City; Judith Resnik of Redondo Beach, California; Sally Ride of Stanford, California; Margaret Rhea Seddon of Memphis, Tennessee; and Kathryn Sullivan of Halifax, Nova Scotia. All would serve as Mission Specialists – a combination flight engineer and scientist – on Space Shuttle crews.

The African-Americans (referred to as "blacks" in the 1978 announcement) were Air Force Maj. Guion Bluford of Dayton, Ohio, picked as a Mission Specialist; Air Force Maj. Frederick Gregory of Hampton, Virginia, chosen as a full-fledged Shuttle pilot; and Ronald McNair of Marina Del Rey, California, a Mission Specialist. Named as the "oriental" candidate, from Hawaii, was Air Force Capt. Ellison Onizuka of Edwards Air Force Base, California, also a Mission Specialist.

The announcement further stated that the astronaut candidates would report to JSC in Houston on July 1, joining the 27 existing astronauts on active duty, for an anticipated two years of training to fly the Space Shuttle in the 1980s.

With as many as 60 flights a year then being proposed for the Shuttle program, it was reasonably expected that the 35 candidates would quickly become seasoned space travelers, given far more flight opportunities than the pioneering pilot astronauts of the preceding two decades.

"We really did look for people who had an open view of what was expected of them," said Dr. Stuart A. Bergman, Jr., then chief of the medical operations branch at JSC. "Some of the people who came in for interviews voluntarily dropped out after they learned what they would have to do." Still, he and others did expect there would be some teething problems as the non-pilot candidates moved through their training and evaluation period. "Scientists are not the most cooperative boy scouts you will find," Bergmann added. "They tend to be individualists who want to know why they are being asked to perform a certain task. The Space Shuttle program is designed to give the scientist-astronauts a broad role in managing the scientific payloads to be carried into space in the Shuttle's cargo hold[3]."

---

[3] Lane, Earl, *Los Angeles Times* article, "The Space Shuttlers – A New Generation of 'Career' Astronauts," November 12, 1978, pg. F1

Dr. Stuart Alonzo (Lon) Bergman Jr.

Although they missed out on selection in the Group 8 intake, 13 of the unsuccessful Pilot/Mission Specialist candidates would become NASA astronauts over the next two intakes. Twelve of these were selected in the Group 9 cadre in May 1980 (James Bagian, John Blaha, Roy Bridges, Bonnie Dunbar, William Fisher, Guy Gardner, Ronald Grabe, John Lounge, Bryan O'Connor, Richard Richards, Jerry Ross and Robert Springer). The 13th candidate, John Casper, was eventually selected in the Group 11 cadre in May 1984.

Another unsuccessful candidate was Millie Hughes-Wiley (later Hughes-Fulford), who would never join the astronaut ranks, but flew aboard Shuttle *Columbia* as a non-NASA Payload Specialist on the STS-40/SLS-1 mission, launched on June 5, 1991.

The TFNG faced many obstacles along the way during their training and evaluation period prior to the completion of the Shuttle's test program, but everyone saw it through. Then, with the Space Shuttle becoming fully operational, the initial crews were selected. The first of the Group 8 astronauts to fly were listed as crewmembers on the seventh shuttle mission, STS-7. In fact, the only member of that crew who was not one of the TFNG was mission Commander Bob Crippen, flying his second Shuttle mission after successfully serving as Pilot alongside Commander John Young on the first Shuttle orbital mission, STS-1, launched on April 12, 1981. The other four members of Crippen's STS-7 crew were Pilot Fred Hauck and Mission Specialists John Fabian, Sally Ride (making her the first female U.S. astronaut to fly into space) and Norm Thagard.

The crew of the STS-7 *Challenger* mission are: [first row l-r] MS Sally Ride, CDR
Robert Crippen, PLT Frederick H. Hauck [rear row l-r] MS John Fabian and MS Norman
Thagard. STS-7 launched the first five-member crew and the first American female astronaut into space on June 18, 1983. All but Crippen were members of Group 8, the TFNG
astronauts.

As a whole, the Group 8 astronauts performed magnificently, although several
of their number were sadly lost as the years went by. In his memoir *Riding Rockets*,
TFNG Mission Specialist Mike Mullane explained the statistics relating to the
group.

"TFNGs logged nearly a thousand man-days in space and sixteen spacewalks,"
he wrote. "Five became veterans of five space missions (Gibson, Hawley, Hoffman,
Lucid and Thagard). The first TFNGs entered space in 1983 aboard STS-7. Steve
Hawley became the last TFNG in space sixteen years later, when he launched on
his fifth mission, STS-93, in 1999. TFNGs were ultimately represented on the
crews of fifty different Shuttle missions and commanded twenty-eight of those. It
is not an exaggeration to say TFNGs were the astronauts most responsible for taking NASA out of its post-Apollo hiatus and to the threshold of the International
Space Station (ISS)[4]."

This, then, is their remarkable story.

---

[4] Mullane, Mike, *Riding Rockets: The Outrageous Tales of a Space Shuttle Astronaut*, Scribner
Books, New York, NY, 2006

1

Expanding on *'The Right Stuff'*

'The Right Stuff: *the qualities needed
to do or be something, especially
something that most people
would find difficult. [Demonstrating]
the right stuff to be a leader.'*
Cambridge English Dictionary

In July 1975, just six years after Apollo 11 had journeyed to the Moon, the last three NASA astronauts to fly an Apollo spacecraft splashed down in the Pacific Ocean at the end of the joint Apollo-Soyuz docking mission with the Soviets (ASTP). Though it had been known there would be a delay in sending the next American astronauts into space on the Space Shuttle, few would have foreseen the six-year void between ASTP and the first Shuttle Orbital Test Flight.

Despite this, there were already developments at the Johnson Space Center (JSC) near Houston, Texas, to expand the astronaut corps and choose the first new astronauts since 1967[1]. For over 15 years, the Astronaut Office had at its heart a defined military pilot hierarchy, with only a slight amendment to include suitably qualified former military civilian test pilots or, as in 1965 and again in 1967, scientifically educated candidates. Even these academic candidates had to qualify from a United States Air Force (USAF) jet pilot course before going on to train for *possible* flight assignments in Apollo spacecraft.

---

[1] The men who formed NASA's seventh astronaut group in August 1969 had been selected under the criteria for the classified USAF Manned Orbiting Laboratory. Chosen in three groups (November 1965, June 1966 and June 1967), seven of them transferred to NASA two months after the cancellation of the military space station program.

© Springer Nature Switzerland AG 2020
D. J. Shayler, C. Burgess, *NASA's First Space Shuttle Astronaut Selection*,
Springer Praxis Books, https://doi.org/10.1007/978-3-030-45742-6_1

PIONEERING THE SELECTION OF SHUTTLE ASTRONAUTS

It was the daring deeds of the Astronaut Office, emblazoned across popular U.S. media of the time and rooted in the test pilot community of the 1950s, which had given rise to the legend of "*The Right Stuff.*" These were cool-headed, hardened test pilots who could push any plane to its limit and just beyond and, even after flirting with the brink of disaster, would bring the aircraft home safely before repeating the evaluation the next day. Work hard, fly hard and play hard was the culture that found its way into the new Astronaut Office at NASA, first occupied by seven Mercury astronauts in 1959.

This ethos continued for nearly 20 years, so much so that when the new Scientist-Astronauts joined the fraternity in 1965, they were judged to rank below test pilots, jet pilots, astronaut wives and even astro-chimpanzees in the flight hierarchy. By 1975 and the end of the Apollo era, many from the original selections had retired, while most of the Scientist-Astronauts still awaited their first flight (as did some of the Pilot astronauts from later selections). Now, with the Shuttle in the pipeline, there were rumors of a new selection; a need for more Pilot astronauts who did not have to be test pilot trained, and of a new category called the Mission Specialist (MS) who could be chosen from academia. Even more concerning to some who had waited years to fly were the Payload Specialists (PS) – not even full-time astronauts – who could come along with their experiments, train for a few months, take a prized flight seat to space, and then return to their previous occupation after just one mission, perhaps even before some of the still-rookie veterans had the chance to strap in on the launch pad. On top of this, NASA was promoting the idea of expanding the cadre to include minority candidates and the first women in the NASA astronaut program. At the Astronaut Office at JSC, the mid- to late-1970s were a time of great change and challenges.

In 1976, NASA issued its first call for new astronauts for a decade, one with a significant change to past selections. These candidates would be chosen and trained specifically for flights on the Space Shuttle, then still under development and some years from its first test flight. This was the dawn of a new era at NASA's Astronaut Office and the successful group would follow a very different training program and career to those chosen during the 1960s. Though not evident at the time, this very different and diverse eighth group of astronauts would play a major part in evolving the roles of a Space Shuttle crew, and in the twilight of their astronaut careers many would move on to senior managerial roles within NASA or the space industry. But that was very much in the future. In July 1976, the search for these new and very different types of American astronauts was set in motion (see sidebar: *If at first…*).

The call goes out

Issued simultaneously by NASA Headquarters, Washington D.C., and JSC in Houston, Texas, the space agency's News Release No. 76-044, dated Thursday, July 8, 1976, announced a recruiting call for pilots and Mission Specialists to live and work on the space agency's forthcoming fleet of Space Shuttle Orbiters. Applications, it said, would only be accepted if postmarked prior to midnight on June 30, 1977 – a year hence – and all those who applied would be advised of the success or otherwise of their application as the process moved through to a conclusion, likely to be sometime in December 1977.

"At least 15 Pilot candidates and 15 Mission Specialist candidates will be selected," the news bulletin stated, "to report to the Johnson Space Center on July 1, 1978, for two years of training and evaluation. Final selection as an astronaut will depend on satisfactory completion of the evaluation period." [1]

IF AT FIRST…

With each NASA astronaut selection, there are those who make the cut, the successful candidates who become NASA's new astronauts. Then there are those who reach the final shortlist but fall at the final hurdle for one reason or another. Some try again and are more successful next time around. For others it would take more than one application over many years before they made it. Hundreds never did, but perseverance is one of the natural traits of an astronaut candidate – and a dose of good fortune never hurts either – as this side-story of the 1978 astronaut selection, by fellow spaceflight researcher Bert Vis, reveals [2]:

NASA Class of 1978 Astronaut Applicant Mario Runco Jr.

Persistence would eventually pay off for one candidate, who first applied to NASA for the Group 8 selection. Mario Runco Jr., was born in The Bronx, New York, in 1952 and lived there until his family moved to the neighboring town of Yonkers in 1963. He graduated from the City College of New York, where he played collegiate ice hockey, and earned a Bachelor of

(continued)

Science (BSc) degree in 1974. He then went on to earn his master's degree at Rutgers University and spent a year working as a research hydrologist with the U.S. Geological Survey, prior to joining the New Jersey State Police in 1977. He would serve as a New Jersey State Trooper until he joined the U.S. Navy in June 1978. He served as a Geophysics Officer in the Navy, working in the fields of meteorology, oceanography, and geodesy, earning his ship Surface Warfare Officer designation while serving aboard the amphibious assault ship USS *Nassau* (LHA-4). He also served at sea aboard the hydrographic survey vessel USNS (United States Naval Ship, the designation given to non-commissioned ships which remain the property of the USN) *Chauvenet* (T-AGS-29) as Commanding Officer of Oceanographic Unit 4.

By the time he joined the Navy, NASA had already been advertising for prospective space explorers to apply for Pilot and Mission Specialist astronaut training for the Space Shuttle program. Even though Runco had barely a few months of Naval service, and the Navy required a minimum of five years of active duty for its applicants, he applied anyway. "That way," he said, "the Navy would know that I wanted to do this." He was quite philosophical about his rejection when the news reached him, knowing it had been a long shot and that there was much else in life to pursue, but it would not prevent him from trying again… and again, and again.

Runco first met his future wife, Susan Kay Friess, from Sylvania, Ohio, in 1978 while they were both stationed at their respective first duty stations in Monterey, California. Susan was also a fairly new Naval Geophysics Officer, at the Fleet Numerical Weather Center, and Mario was at the Naval Environmental Prediction Research Facility. It turned out that both had an interest in the space program, but it was Susan who first learned that the Navy and NASA were accepting applications for the astronaut corps and told Mario about it. They both applied that year but neither was selected. Undaunted, Mario applied again as a Mission Specialist for the astronaut selection of 1980 but again was not selected. By the time of the 1984 group selection, however, the two were married and had finally achieved the requisite five years of active duty, so they both submitted applications while they were assigned back in Monterey, California, after tours of duty on the East Coast in Norfolk.

Mario was serving as an instructor with the Navy Geophysics Readiness Laboratory at the Naval Post Graduate School and Susan was a student there pursuing her master's degree. The Navy selected only Mario and sent his

(continued)

name and application to NASA along with the names of the other Navy selectees. He was not invited to interview by NASA that year, but true to his Calabrian heritage *(used with permission)* he stubbornly applied once more in 1985 and was again selected by the Navy. Once again, NASA never called him for an interview. His fifth attempt was successful, however, and Mario was both selected by the Navy and finally called to be interviewed by NASA in 1987, which he did in March of that year while both he and Susan were stationed in Hawaii. Susan did not submit any more applications after the one in 1984. Mario eventually reported for duty to NASA in August of 1987, while Susan would continue her tour of duty at Barbers Point for another year before joining her husband in Houston. By that time they had already had their first child, who was barely a month old when she made the trip from Hawaii to Texas.

By coincidence, the Navy was keenly interested in performing oceanographic observations from manned spacecraft and there was already a billet at the JSC for a Naval Geophysics Officer. The Navy also had a policy of trying to keep married couples together, so after Mario was selected, the Navy assigned Susan to the billet at JSC, where she would eventually train astronauts in Crew Earth Observations and in meteorology, oceanography, and geography.

During a conversational interview in 2018, while both were attending the XXXI Congress of the Association of Space Explorers in Minsk, Belarus, Mario told Bert Vis that after his selection, he later found out that in joining the astronaut corps as a member of Group 12 (George Abbey's Final Fifteen, aka "The GAFFers") in June 1987, he may have inadvertently upset plans within his Navy community to have one of several other Geophysics officers selected. Up to that point, Mario's career in the Navy had been a meteoric rise, especially after his tour aboard USS *Nassau* (LHA-4). He was even selected to the rank of Lieutenant Commander a year early, which is known in the Navy as being "deep selected", and was immediately assigned as a Commanding Officer to an at-sea unit, Oceanographic Unit 4, operating in Indonesian and adjacent international waters. Shortly after his selection, however, when it came time for Mario to be seconded to NASA by the Navy, the officer in charge of assignments very much tried to dissuade and discourage Mario from the secondment. Mario took the assignment, of course, and the assignment officer, known as a "Detailer", was not pleased. When the time came for Mario to be promoted to Commander, he learnt that this same officer was on his selection board, so he knew that his promotion was unlikely

(continued)

and he was right. As a result, he would eventually retire from the Navy in 1994, fortunately at time when the U.S. military was downsizing and offering incentives for "early-outs". He would continue working as a civilian NASA astronaut and would eventually fly on three Shuttle missions: STS-44, STS-54 and STS-77.

Mario told Bert Vis that he felt fortunate with those three missions to have flown on two different Shuttles, taken off from both launch pads, landed on both coasts, flown ascent and entry from both the flight deck and middeck of the Space Shuttle, operated the Shuttle's robot arm, deployed four satellites, operated many onboard experiments, and performed an EVA.

By1993, though, the Naval oceanography community had lost interest in the manned space program and was intending to close the billet at JSC where Susan was assigned. This would have meant reassignment elsewhere, with the closest place being at the Naval Oceanography Command located, ironically enough, at the John C. Stennis Space Center in Bay St. Louis, Mississippi. This was some 595 km (370 miles) and six hours by car away and there was certainly no guarantee of her being assigned there. By then, the couple had had their second child, and their daughter and son were only five and three years old respectively. Being separated was not an option for them, so Susan also decided to leave the Navy during the period of "early-out" incentives. Before doing so, however, she had already tentatively lined up an oceanography position with Lockheed (LH), also in Houston at JSC. However, having left active duty to became a reserve officer and about to take on the LH position, a NASA position became available. Susan was hired and she continued to train astronauts in Crew Earth Observations as before. Ironically, she was promoted to Commander in the U.S. Naval Reserve (Individual Ready Reserve) shortly after leaving active duty. Mario related to Bert that, in the end, "I think George (George Washington Sherman Abbey) was very good to us!"

MEETING THE ASTRONAUT CRITERIA

Pilot applicants were required to possess a BSc degree from an accredited institution in engineering, physical science or mathematics, or to have completed all the requirements for a degree by December 31, 1977. An advanced degree or equivalent experience was preferable. They had to have accumulated at least 1,000 hours first pilot time, with 2,000 hours or more increasing their chances. Experience in high performance jet aircraft and flight testing was also considered desirable. Additionally, they had to be able to pass a NASA Class 1 space flight physical, and range in height between 64 and 76 inches (163 to 193 cm).

Applicants for Mission Specialist candidate positions were not required to have any prior piloting experience. Educational qualifications were much the same as for Pilot applicants, except that biological science degrees were included. These applicants had to be able to pass a slightly less rigid NASA Class 2 flight physical. A height of between 60 and 76 inches (152 to 193 cm) was one specific requirement. There had also been a little relaxation in some of the physical requirements from previous astronaut groups; even those wearing glasses were now acceptable candidates, if the glasses corrected the user's vision to 20/20.

The space agency was not only keen to emphasize that the Space Shuttle would usher in a whole new era of qualified people working and testing new procedures in space – rather than training pilots to do these things – but also in rebuilding lagging public support for the space program by involving a far broader slice of America's population. The news release thus declared that "NASA is committed to an affirmative action program with a goal of having qualified minorities and women among the newly selected astronaut candidates. Therefore, minority and women candidates are encouraged to apply." [1]

Pathway to a whole new career
Previously, during the 1960s, NASA's Astronaut Office had taken on board two groups of astronaut candidates who were fully qualified in different disciplines of science and medicine, and were given the designation of Scientist-Astronaut. Only a select few, however, would ever achieve a flight into space in the pre-Shuttle era, with many opting out of the astronaut corps in frustration – mostly due to the cancellation of expanded Apollo and Skylab missions as well as ongoing delays in the Space Shuttle program – and returning to their academic fields. That would not be the case here. Apart from recruiting some additional pilots, the Shuttle program was very much a viable, ongoing and exciting alternative career option for any young scientist or medical practitioner, with the first flights due within a year or two of the conclusion of their astronaut training.

As Carolyn L. Huntoon, then NASA's JSC Deputy Chief for Personnel Development, put it: "This time, the agency wanted applicants who understood what would be expected of them and were willing to devote most of their careers to the program. They had to be very good in what they were doing and yet they had to be willing to give it up to do more general things. They had to understand that they would be rookie newcomers when they got here. That's a bit difficult to accept when people have excelled in their fields." [3]

Fig. 1.1: Dr. Carolyn L. Huntoon, Deputy Chief for Personnel Development, NASA JSC

Dr. Huntoon certainly knew what she was talking about. JSC Director Chris Kraft had personally asked her to consider applying for the candidate group, but after chewing over the prospect for a while she declined. Instead, she became the only woman on the selection board for the Group 8 astronauts, although the lead-up to this phase of the process had proven difficult.

"One of the things we had trouble with at that time was getting people to apply; getting women to apply particularly," Dr. Huntoon recalled in a 2008 interview. "I would say women and minorities, but particularly women, because we had never had any women in the Astronaut Corps. They did not expect that they'd have a chance to be selected. Even though we were saying it, it was hard. So we went out on quite a few recruiting trips to various universities and did a TV show in Chicago and around the country, talking about the selection and what they'd be expected to do; the jobs when they came here, and flying on the Shuttle, and that NASA was serious about considering women and minorities.

"This was a time in our country when also a lot of things changed for women, and you could no longer discriminate against women for their jobs," Dr. Huntoon continued. "I recall thinking how difficult it was to convince women [that] we were serious about selecting women." As part of her recruitment strategy, Dr. Huntoon would visit high schools, colleges, Lions and other community clubs and even retirement centers, trying to impress upon her audiences that NASA had a broader recruitment focus, which now included women and minorities. "They would say, 'Oh, sure,' kind of thing. A lot of guys applied, but not that many women at first." **[4]**

Fig. 1.2: (main) Nichelle Nichols discusses the Space Shuttle program with students in NASA's Mission Control Center. (inset) Nichols as Lt. Uhura in the *Star Trek* series.

One of the people NASA employed to tour and promote the diversity of the astronaut recruitment process was African-American actress Nichelle Nichols, well known for her role as Lt. Uhura in the *Star Trek* TV series. Nichols revealed that NASA had specifically asked her to recruit women and minorities for the Space Shuttle program. She relayed her response to NASA with a mischievous twinkle in her eye, saying, "I am going to bring you so many qualified women and minority astronaut applicants for this position that if you don't choose one… everybody in the newspapers across the country will know about it." Nichols credited *Star Trek* with the success of her recruiting efforts. "Suddenly, the people who were responding were the biggest Trekkers you ever saw. They truly believed what I said… it was a very successful endeavor. It changed the face of the astronaut corps forever." **[5]**

Kathryn Sullivan had been awarded a BSc degree in Earth Sciences from the University of California in 1973, and a PhD in Geology from Dalhousie University (Nova Scotia, Canada) in 1978. While at Dalhousie, she participated in a variety of oceanographic expeditions under the auspices of the U.S. Geological Survey (USGS). Though fascinated with human space activities in her youth, she had never dreamt that she might one day tread a similar path into space, and working in Canada meant she was unaware of NASA's call for a new cadre of astronauts. "In fact, I didn't know about it at all until I went home to California to visit family at Christmas in 1976.

"I saw that little ad that finally helped me draw the parallel with expedition operations, clipped it out or whatever one did, dutifully sent off this little postcard that said, 'Hey, yes, I'd like to look into this.' I recall that if you sent the first inquiry in, they sent you back a postcard or short communiqué that basically read, 'This is just to make really sure that you're really sure,' and listed a few more of the medical requirements and other things. It clearly was a 'Really, please don't bother us if you don't fit this description. Think about it again.'

"If you thought about it again and either were going to ignore the conditions or knew that you passed them, then you sent another card in, and wrote, 'No, no, really, I would like an application package.' Then you got the big old giant application package. The bulk of the application package is a Mark-1 standard U.S. government civil service form; thorough, comprehensive. At some other levels, especially being a grad student in Canada at this point in time, I remember just being bemused and bewildered by parts of it." **[6]**

Another of the ultimately successful female applicants was Sally Ride, who had once given serious consideration to becoming a professional tennis player. Fortunately – even though she excelled at the sport – she found a far more satisfying career through her graduate studies in physics at California's Stanford University, where she became deeply involved in researching X-ray astronomy and free-electron lasers. On January 12, 1977, Ride was reading the campus newspaper, *The Stanford Daily*, which contained an interview with Margaret Collins, director of Stanford's Center for Research on Women. Under the headline, "NASA to recruit women," it stated that NASA was seeking applicants from qualified people – including women and minorities – interested in becoming Mission Specialist astronauts. Ride was not to know it at the time, but NASA's future plans involved transmitting energy from space stations back to Earth, which was her chosen field of study.

Enticed by the exciting prospect of working in space, Ride decided to apply, despite knowing little about what it might involve or how it might affect her career path, but with the determination that "I can do that." As she later revealed, her future direction was suddenly not quite as clear-cut as she had planned. "I thought

I was going to get a job… doing research in free-electron laser physics, and then work at a university doing research and teaching. That's what physicists do." There was also a lot of mystique involved in the prospect of training to be an astronaut. "I didn't know whether they were going to throw us into a centrifuge or hang us from the ceiling by our toes." **[4]**

Fig. 1.3: (left) Sally Ride in her teenage years, excelling at tennis and with a deepening interest in science and astronomy (Image courtesy of Afflictor.com). (right) Rhea Seddon taking part in the residency program at the University of Tennessee medical school (Image courtesy of the University of Tennessee).

A question of astronaut pay

Despite the lengthy lead-in time of 12 months, many enquirers waited until the last minute to apply. Altogether, a total of 24,618 enquiries were received by NASA in that period, and of these, 20,440 requested and were sent application packages. **[7]** Eventually, 8,079 applications were received by the due date.

The question of wages for the civilian candidates was based on the Federal Government's General Schedule pay scale from grades GS-7 through to GS-15, with approximate salaries from $11,000 to $34,000 per year. The candidates were informed that they would be compensated based on individual scientific achievements and experience. Other benefits included vacation and sick leave and participation in the Federal Government retirement, group health and life insurance plans.

Rhea Seddon, MD (also one of the successful candidates), was prepared to give up a potentially highly-paid career as a surgeon in order to take on an entirely new role as a Mission Specialist astronaut, which only offered a starting salary of $24,700 per annum. "Some of my doctor friends will never get over it," she said at the time. "They think I'm crazy." [4]

Any military personnel were asked to apply through their respective departments using procedures that would be disseminated later in the year by the Department of Defense (DOD). The astronaut recruits would be assigned to JSC, but would essentially remain on active military status with regard to pay, benefits, leave and other military matters.

At the time the call for applicants was issued, NASA still had 31 astronauts available as crewmembers for future Space Shuttle missions, including nine scientists. Of these, 28 were assigned to JSC, while three others held government positions in Washington D.C. [8]

PREPARING TO TEST THE APPLICANTS

On April 1, 1977, some four weeks prior to the deadline for applications, JSC medical personnel completed a week of practice in applying standards and conducting tests that would be used in the medical and psychological screening of the astronaut hopefuls. With the assistance of 20 volunteer subjects, the medical team worked out the logistics for the comprehensive evaluation during the Astronaut Medical Selection Exercise, directed by flight surgeon Dr. Sam L. Pool. Dr. Pool was serving at that time as NASA's Division Chief for the Medical Science Division at JSC.

Assisting in the evaluation tests were Drs. George Behaine, Michael Berry and Jerry Hordinsky of NASA; Joseph Harasimowicz of the USAF; and Charles Pickett of the U.S. Navy (USN). NASA records indicate that the 20 volunteer subjects – NASA personnel and outside consultants – were John Arnold, Frances Barbee, Linda K. Bromley, Gil Chisholm, James L. Cioni, James B. Costello, Roger W. Ellsworth, Paul O. Ferguson, Virginia Gibson, Thomas J. Graves, Kathleen Hosea, Samuel E. Jones, Gary Kane, Ronald R. Lanier, Stella Luna, Julie Mattheaus, Mary Lee Meider, Judy Olson, Dale Sauers and David K. Stoughton.[9]

Fig. 1.4: Joseph D.C. Degioanni, PhD, MD, who not only worked in the Aerospace Medicine Division at JSC, but was also a candidate himself. (Image courtesy of Ed Hengeveld).

Testing the candidates and a delay

Dr. Joseph Degioanni, NASA's physical administrator for the recruitment exercise (and himself one of 208 candidate finalists), pointed out that the same medical evaluation would be applied to both Pilot and Mission Specialist candidates. The medical testing involved 24 different procedures, including a general examination by a NASA flight surgeon.

The evaluation included an examination of a candidate's medical history – illness, injury, surgery, and so forth – and carried through to psychological, psychiatric, ophthalmological, neurological, dental, musculoskeletal, body chemistry, and ear, nose and throat examinations, Also included was a battery of tests administered using the Lower Body Negative Pressure device, a rotating chair (the cupulogram, used to measure vestibular function and susceptibility to motion sickness), a treadmill, a Holter monitor (to measure heart rate over an extended period), and the 34-inch (86-cm) Shuttle Personal Rescue Sphere, to gauge a subject's susceptibility to claustrophobia.

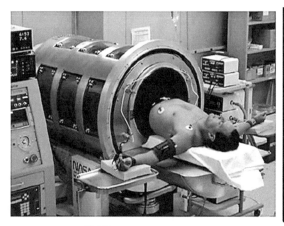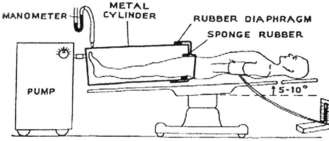

Fig 1.5: (main) A subject is tested in the Lower Body Negative Pressure device (Image courtesy of Science Direct). (inset top right) Demonstrating the Shuttle Personal Rescue Sphere.

Specific studies of body functions such as vision, hearing, exercise capacity and muscular strength were all regarded as important factors in the overall health evaluation. "We want to make this the best of all health screenings because it's for astronauts," Dr. Degioanni said at the time. **[9]**

Many of the applications received by NASA were obviously mailed more in hope than expectation, with a selection board swiftly eliminating over 2,000 enquiries that did not meet the rigid qualifications. With the number now reduced to 5,680, the panel's work began in earnest, weighing up each applicant using a rating system based on such factors as their background, education, abilities, degrees and workplace experience, while in the case of the pilot applicants, their flight hours and/or participation in combat were also taken into consideration. The applicants has also been asked to supply references, and contact was made with these referees in order to gain an insight into such areas as the applicant's personality, work ethic, talent and enthusiasm. This eliminated several more names.

It was a long and involved process, but eventually the list was whittled down to a workable 208. A schedule was drawn up, in which these applicants would be contacted and asked to travel to Houston for a week devoted to full-on medical examinations, psychological tests, and face-to-face interviews with members of the selection board.

The 208 candidates were then split into 10 interview groups and each applicant was provided with information as to when they needed to be in Houston. Their week began once they had checked into the assigned hotel, the Kings Inn Ramada on El Camino Real in Clear Lake, quite close to JSC. There would be an evening reception and pre-brief at the inn for the assembled candidates. Many of them already knew each other from science communities or military circles, but most were still getting used to the unexpected novelty and understandable bewilderment associated with making it through to this phase of the selection process. As Kathryn Sullivan recalled: "I remember thinking, 'Kath, enjoy this week a lot, because these people kind of really seem to have some sense of what the hell is going on here, and you really don't. So have a really good week, because this may be the end of the road. Enjoy it a lot'." [6]

Following the tests and interviews, 149 of the 208 applicants were found to be medically qualified for the astronaut program and indicated that they were still interested in pressing forward in their bid to make the final cut (see Table 1.1).

Table 1.1: GROUP 8 CANDIDATE BREAKDOWN

PILOTS

QUALIFIED APPLICANTS	INTERVIEWED	MEDICALLY QUALIFIED	QUALIFIED/ INTERESTED	SELECTED
147 MILITARY (4 MINORITY)	76 MILITARY (3 MINORITY)	71 MILITARY (2 MINORITY)	70 MILITARY (2 MINORITY)	14 MILITARY (1 MINORITY)
512 CIVILIAN (6 MINORITY, 8 FEMALE)	4 CIVILIAN	4 CIVILIAN	4 CIVILIAN	1 CIVILIAN

MISSION SPECIALISTS

QUALIFIED APPLICANTS	INTERVIEWED	MEDICALLY QUALIFIED	QUALIFIED/ INTERESTED	SELECTED
161 MILITARY (6 MINORITY, 3 FEMALE)	45 MILITARY (4 MINORITY, 2 FEMALE)	34 MILITARY (3 MINORITY, 2 FEMALE)	34 MILITARY (3 MINORITY, 2 FEMALE)	7 MILITARY (2 MINORITY)
5,519 CIVILIAN (332 MINORITY, 1,248 FEMALE)	83 CIVILIAN (8 MINORITY, 19 FEMALE)	43 CIVILIAN (4 MINORITY, 12 FEMALE)	41 CIVILIAN (4 MINORITY, 12 FEMALE)	13 CIVILIAN (1 MINORITY, 6 FEMALE)

NASA Administrator Robert A. Frosch had intended to complete a review of the data presented to him on the selection of the Group 8 astronauts in December 1977. However, at a meeting held with officials responsible for the selection program on December 12, including JSC Director Chris Kraft, Frosch said that he was still involved in current budget activities and would not be able to complete his review until the following month. The announcement of the selection of the new astronauts would therefore be delayed until January 1978. [10]

There would be one small but significant difference this time, for those who were chosen. In the past, the selectees had immediately been called 'astronauts' as soon as they were chosen. For the Shuttle group, however, who would have to undergo a mandatory training and evaluation period, they would initially be known in NASA parlance as 'astronaut candidates,' or 'Ascans.' They would retain this title for the entirety of their envisaged two-year training program. Once each Ascan had successfully passed their training and evaluation period, they would officially earn the rank of 'astronaut' – although some pedants still maintain that a person cannot claim to be a true astronaut until they have flown into space.

AN APPLICANT REFLECTS

William D. (Bill) Heacox is a Professor Emeritus of astronomy at the University of Hawaii at Hilo, where he founded the University's undergraduate astronomy program and the Hōkū Ke'a observatory on Mauna Kea. He has published original research in both physics and astronomy, and has held research appointments at several institutions. At the time of the astronaut selections for the Shuttle program, he was a National Academy of Sciences/National Research Council (NAS/NRC) post-doctoral fellow at NASA's Goddard Space Flight Center.

"I was one of the sixth group of astronaut finalists in the 1977 Space Shuttle recruitment, sometimes referred to as the Space Science group," Dr. Heacox recalled. "Most people in the group, including me, had advanced degrees in astronomy or related sciences. I was somewhat unusual in also having a background in military aviation, having flown Navy jets for several years prior to entering into astronomy graduate education. On the face of it, I suppose I was an attractive candidate. The purpose of bringing applicants to JSC for interviews and tests was probably to weed out the chaff from the wheat among such promising looking candidates. Our Space Science group was interviewed and tested at JSC the second week of October.

"As it happened, these visits of applicants to JSC served a dual purpose: while NASA people were evaluating us, we were evaluating them and the astronaut program. There was a lot of contact between serving astronauts and applicants, some of it planned and some informal. Especially with the Space Science group, there was more than a little mutual mistrust. The astronauts viewed with some suspicion these pointy-headed, apparently unqualified applicants who had probably made it this far only because of political correctness in the NASA front office, and the applicants were often put off by the perceived chauvinism of gung-ho flyers, all of them male and white.

Fig 1.6: (left) Catherine and Bill Heacox on their wedding day in 1966. (right) Heacox at 14,000 feet altitude on Mauna Kea, on the Big Island of Hawaii, circa 1981 (Images courtesy of W. Heacox).

"On more than one occasion, I listened to an astronaut trying to impress what he took to be a mere academic with frightening tales of flying jets in death-defying situations, before telling him that I was quite experienced in such things – which probably didn't help my case for selection. But my favorite such story had to do with Sally Ride. As one of the few women in the group, she was subject to more than a little abuse from male astronauts, until one of them challenged her to a racquetball match. The semi-pro tennis player cleaned his clock, much to the amusement of the other male astronauts who – to their credit – thought the matter hilarious and treated Sally with some respect thereafter.

"But there were also astronauts who took seriously their responsibility to inform applicants as to the nature of the job they were applying for. I especially appreciated the many discussions I had with Bob Parker, an astronomer selected as an astronaut near the end of the Apollo era and who at that time had been serving for more than a decade without flying into space – a consequence of the shift from Apollo to the Shuttle. From him and others I was able to get some appreciation for the nature of an astronaut's career, which was mostly devoted to helping with vehicle and program development, and only partially with spaceflight itself. I was especially interested in the computer-controlled systems in the Shuttle since

I had some experience in computer control of telescopes, a field in which my alma mater (University of Hawaii) had been something of a pioneer during my graduate student days there. Had I been selected as an astronaut, I probably would have concentrated on this area of Shuttle development.

"As the week at JSC wore on, I became increasingly disenchanted with the prospect of a career as an astronaut. By this point my prospects as a career scientist were becoming promising – I was then employed as a post-doctoral scientist by the NAS, and working at the Goddard Space Flight Center – and it was becoming clear to me that I probably could not continue that career as an astronaut. Of equal concern was what I could see as the astronaut culture in NASA. On more than one occasion, our NASA guides made it clear that astronauts were expected to be cultural representatives to the public, and must conduct themselves appropriately. In my imagination, I could see stretching before me an endless array of Rotary lunches to attend in a blue flight suit, and publicity pictures on all occasions. As a rather private person, that prospect appalled. It was not going to be easy to balance that aspect with the enthralling prospect of flying in space – if I could ever get there. The example of Bob Parker's flightless years served as a cautionary tale (Bob did eventually fly in the Shuttle, several years later).

"Throughout this week of mutual examination, there was a palpable tension among the applicants: only a few of us (there were 20 in our class) would be selected as astronauts, if that. There was no real venue for outright competition; we just all tried our individual bests to impress the many people evaluating us. Toward the end of our week at JSC – filled with physical and mental examinations and interviews – the Astronaut Office sponsored a formal dinner for us applicants, at which Chris Kraft and John Young were invited speakers.

"I don't remember what Chris Kraft said, but John Young's speech greatly impressed me and has stuck with me all these years. He said that his talks with us individually had sensitized him to our collective anxiety about our professional futures. At that time, astronomy jobs were thin on the ground, and a large percentage of recent astronomy PhD graduates (as were many of the applicants in the audience) could not find positions in their field, let alone jobs as astronauts. He wanted to assure us that we had nothing really to worry about: he was personally impressed with every one of us, and anyone who could make it this far in the astronaut selection game had what it took to be a success in whatever they cared to take on. So stop worrying and get on with your lives – professionally, you're going to be fine. I was struck with what a gracious thing this was for the world's leading astronaut to say, and have been a John Young fan ever since. I was saddened by his recent death.

"On the last day of our week at JSC, George (Pinky) Nelson and I shared a dinner at a local restaurant. Part way through it, Sally Ride came in and joined us. She was depressed because her latest blood draw had been ruined somehow (the JSC medical people were big on blood analyses, and all of us applicants were down a few quarts), so she not only had to stick around another day for a new blood draw and test but couldn't join Pinky and me in the bottle of Mateus we were sharing. We cheered her up as best we could, and finished the dinner promising to stay in touch no matter what the selection outcome was. But we never did; that was my last contact with either of them.

"Two weeks later, back in my office at the Goddard Space Flight Center, I wrote to the Astronaut Office to withdraw my application. It wasn't an easy decision to make, and I have had second thoughts on many occasions since. But there is no doubt in my mind that it was the right and prudent thing to do: I don't know if I would have been successful as an astronaut, but I'm pretty sure I would have been unhappy in that career. It was a poor match between job and applicant.

"Toward the end of November – nearly two months after the visit to JSC – I was called by the Astronaut Office and asked if I was still interested in the job. It was as if they had not received my withdrawal letter, or chosen not to acknowledge it; or perhaps it was just a functionary checking off items on a list. Whatever the case, I confirmed my withdrawal and have heard nothing further from JSC.

"About a month later, NASA announced its selections for the new astronaut group, including three from my finalist group: Jeff Hoffman, Pinky Nelson, and Sally Ride. I had been quite impressed with all three and was not surprised at their selection. Rumor had it within NASA – in which I was still installed at Goddard – that the initial selection decision had been rejected by NASA headquarters (or someone higher) as being insufficiently diverse, both racially and in terms of gender, and sent back to JSC for reconsideration. Could be. In my opinion, nearly all the finalists in my group were very impressive and would probably have done well as astronauts. It was a real privilege to be one of them, however briefly." [11]

OTHER CANDIDATE VOICES

Lawrence (Larry) Pinsky, PhD, MA, BSc, is still active in the space program. Currently a Professor of Physics at the University of Houston, he also works for NASA and CERN (the Geneva-based European Organization for Nuclear Research). In addition to his full-time teaching duties, he is also involved in research in the areas of space radiation simulation, relativistic and intermediate energy heavy ion physics, charged particle detector development for space radiation dosimetry, Grid Computing, and cosmic ray astrophysics.

Professor Pinsky was also in the sixth group of astronaut applicants and, like Bill Heacox from his group, he remembers the amazing prowess of fellow applicant Sally Ride in a session of racquetball.

Fig 1.7: (left) Professor Lawrence Pinsky in 2011 (Image courtesy of Gallerie Utef). (right) Dr. Wilton Sanders III (Image courtesy of Wilton Sanders III).

"One anecdote I recall is that we were given access to the astronauts' gym facilities, among other freedoms to roam around to get the feel of the Space Center. Since I had worked there previously, as had another of the candidates, we were not interested in exploring the ins and outs of the place, so we agreed to play some racquetball on one of the astronauts' courts. While we were playing, Sally Ride knocked and came in and asked if she could play a game of 'cut-throat' with the two of us. Not realizing her athletic past had included women's tennis in college, and I believe even some professional experience, we agreed. She proceeded to destroy both of us, leaving us as defeated sweaty lumps in our respective corners before bowing out as calm and composed as when she had entered, thanking us and saying she couldn't spare any more time because she had to go running! Needless to say, we were both very impressed, and her selection was clearly a very correct choice in the long run.

"Regarding the selection process, my application was ultimately rejected by the medical board because I have a congenitally high blood sugar baseline, which statistically predicted that there was a significant probability that I would develop diabetes later in life, but still within the time projected to have to remain qualified to fly. I still do not have diabetes, but no regrets. I believe the selection process was fair." **[12]**

When the time came for his participation in the interview process at JSC, Capt. Loren Shriver was on active duty with the USAF, based at the Flight Test Center at Edwards Air Force Base (AFB), California, and fully involved in test work with the McDonnell Douglas F-15 *Eagle*. He reported to JSC along with the fourth group of 20 candidates on September 15, 1977.

"The week-long evaluation consisted of a number activities," Shriver recalled, "including a very thorough physical exam (much more included in it than our annual Air Force flight physicals), psychiatric evaluations, evaluations of new equipment NASA was developing for possible use, and the grand finale of an hour and a half or longer personal interview with a whole team of NASA personnel, any of whom could ask any question that came to mind. Oh, and interspersed through-out the week were 'social' events, or gatherings of other NASA personnel, mostly current astronauts, their spouses, and some other accomplished NASA Space Operations personnel.

"I have to say that it was a most interesting, exciting, stressful, and yes, kind of 'fun' week. For example, I had never really been evaluated [by], or talked with, a psychiatrist before, but during this week there were sessions with two different psychiatrists (Drs. Terry McGuire and Edward Harris), and it turned out to be pretty much one was the 'good guy,' (McGuire) and one was sort of a 'bad guy,' with the first asking just general questions and having a nice discussion, and the other (Harris) being more demanding and 'technical' in his approach, to see if I was easily flustered by his demands and questions. The 'social events' could have been easily dismissed by young fighter pilots, but I had decided that everything that happened during the week would be part of my 'interview,' so while being 'social' I was not going to overdo it or make a complete fool of myself, and I believe that was a good approach.

"I have to admit that the prospect of the interview with all the NASA folks and me was on my mind a good deal from the moment I knew I was going to Houston. Who would be there? What were they going to ask? How technical would it be? My actual interview turned out to be a series of discussions with the NASA participants about things they were interested in as we went along. It varied from quite technical questions and answers from me about my experiences, my flight test programs, the F-15, my specific favorite areas of flight testing, and all the way to how farming in Iowa had changed since my early days as a kid, and even the fact that I played fast pitch softball on a team at Edwards AFB! It actually was a very civilized, straight-forward discussion of many topics of interest to everyone in the room." **[13]**

Dr. Wilton Sanders III is a senior scientist in the Astrophysics/Science Mission Division at NASA Headquarters in Washington, D.C. In the summer of 1977, he was one of the 208 astronaut finalists flown to JSC in Houston for the week of physical, mental, and other tests designed to determine if he was suitable to con-tinue in that program. He was in the fifth group of 20 candidates to be tested and evaluated, beginning on September 26. In that group, 17 were Pilot candidates.

"Even though I was a Mission Specialist candidate, the bulk of my group were candidates not for Mission Specialist, but rather for Pilots... Consequently, most of the guys in my group were military pilots who were quite impressive physical specimens, although my roommate was Steve Hawley who eventually was selected as a Mission Specialist. I have since spoken with Hawley to see if he remembers me, but he does not.

"In any event, it was a busy week, with every day booked with various tests and interviews. The physical tests ranged from chin-ups and sit-ups, at which the military pilots excelled, to cardio treadmill runs and lung capacity measurements. In my case, they also included eye tests. I am far-sighted and slightly cross-eyed and have little depth perception because my brain does not fuse the individual images from each eye. This condition seemed to fascinate the eye doctors and led to several follow-up examinations. There were also psych exams, interviews with two different psychiatrists whom we called Good Shrink and Bad Shrink, and claustrophobia tests where we were put inside a small inflated sphere for a while and told to evaluate it as an 'escape pod.' Towards the end of the week, there was a formal interview with a panel of five or six high-ranking somebodies, the only one of whom I remember was Chris Kraft. They seemed to focus on my lack of the *'Right Stuff.'* My father and brother were Navy pilots, but I had never flown a plane, or piloted a glider, or parachuted for fun, or climbed a mountain, or been scuba-diving, or apparently done anything dangerous or adventurous, so I guess they were concerned that I would chicken out at some inopportune moment. In the end, my rejection letter identified my eyesight as the reason for not being selected, and that is probably valid, but I have carried the suspicion that the lack of the *'Right Stuff'* was just as important.

"Overall it was a busy week, but not overly stressful unless you allowed it to be. There were some candidates who were so invested in being selected that they made it quite stressful for themselves. Others of us saw it as an adventure, a once in a lifetime experience, and just tried to get into it, enjoy it even." **[14]**

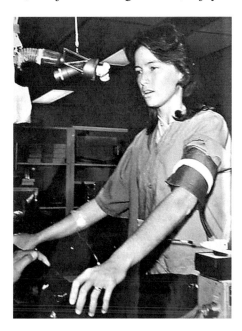

Fig 1.8: Anna Sims (later Fisher), undergoing trials on a treadmill

Twenty-eight-year-old Anna Sims (later Fisher) had recently undertaken her internship in the MD/PhD program at the University of California, Los Angeles (UCLA) when she first heard about the NASA call for astronauts through a colleague, Dr. Mark Mecikalski.

"He used to follow the space program avidly. He got all the NASA newsletters and all those things. I remember he had lunch with my then fiancé, later husband, Bill, and said, 'Hey, NASA is looking for people. You and Anna have always talked about how you're interested'. I remember Bill paging me over the loudspeaker system, getting his call, and saying, 'We have three weeks to apply before the deadline'. So we got our applications ... In those days it wasn't an electronic application, it was just a regular civil service application that you filled out by hand. I remember where you put in the title, what you were applying for, and it was astronaut. It felt kind of weird. I think I got mine in the day before the deadline, because you had to get transcripts and all that kind of stuff. It was a pretty arduous process. So I got mine in maybe the day before the deadline. I think Bill got his application postmarked the day of the deadline.

"For me it was a real struggle in that time period, because we had both accepted surgical internship positions. Bill was a year ahead of me. At the last minute we sat down – well, not the last minute, but fairly early in the process of when you get your internship – and we talked about it. I said, 'Here I am. I'm going to be a surgical resident. Is that the kind of life I really want to lead where I'm going to be on call at all hours of the night?' We started having second thoughts even before the NASA application came along.

"So anyway, we decided to spend a year practicing ER [Emergency Room] medicine while waiting for a decision from NASA. We wound up practicing in emergency medicine for the year in Los Angeles while we were waiting to find out what happened, which was a really hard year. Just practicing emergency medicine in Los Angeles was a very challenging thing."

During their time in ER, Anna and Bill began making wedding plans. "We were just sitting there making plans and flight plans and all that sort of stuff when the phone rang. It was [someone] – I don't remember who called anymore, probably Duane Ross – asking if I wanted to come [for an] interview. This would have been on a Friday I guess, Thursday or Friday. Would I want to leave not that Sunday but the following Sunday and come for an interview? That was the week Bill and I were targeting to plan our wedding ... We were sitting there and I said 'It's NASA. They want me to come [for an] interview'. He said, 'Say yes, we'll figure it out'. So I said yes.

"This was a Friday ... We got married on Tuesday. We went to San Francisco for that night and came back the next day. I think I worked a shift on that weekend and another shift. It was supposed to be my tenth high school reunion that Saturday, and I really wanted to go. That was it. I just ran out of steam that Saturday night having to be ready to go the next day. I said, 'If I'm going to go for this interview, I've got to give it everything I have'. It's my only reunion that I missed for high school, but they were very understanding as well. Then I got on a plane that Sunday morning. I was exhausted. Because I was so busy, I didn't really have time to think about the interview process, study, or research … All of a sudden, I realized, 'Jeez, I'm on my way to do this, and I really haven't researched it, other than I know I want to do this with all my heart'. So I'm going, 'Jeez, what have I gotten myself into?'

"I had my interview, and again my attitude on all of that was: 'NASA is this big entity that knows everything about you'. They already had done background investigations. I knew from some of my friends that people had come and talked to them. I just figured, 'Well, I'd better be honest'. I even remember saying, 'I want to have children, so if that's a factor in your selection, I definitely do want to have children'. I even said that at my interview. Then you leave. After you've been here for a week, then you really want it. Before, I was able to keep it at a distance, but those months from August till when they announced [the final 35] in January were probably some of the hardest months in my life." **[15]**

Thirty-five-year-old Major (later Colonel) Gary Matthes was a highly-trained test pilot and former combat pilot with the USAF, stationed at Edwards AFB, California. Hailing from St. Louis, Missouri, he was selected to attend the USAF Test Pilot School (formerly the USAF Aerospace Research Pilot School) at Edwards in July 1972, entering as a member of Class 72B. Following his graduation, he remained at Edwards in their Flight Test Center. During this time he was twice selected by the Air Force to try out for NASA's astronaut corps, a goal he had hoped to achieve since graduating from the Air Force Academy. The first such occasion was in 1977. He dared to dream of being selected, and the extensive medical test phase held few fears for him.

"Since I had taken the astronaut physical to get into Test Pilot School in 1972, I wasn't intimidated by it, as it was just what I expected," he recalled. "There were some additional 'measurements' taken that the doctor told me were more to expand the database for the future than to influence the selection process."

Ultimately, though, he did not make the final selection, but would try again three years later. "Frankly, my memories are very sparse and I don't think it is just because of my age. I have to admit that since I was interviewed twice, once in 1977 and then again in 1980, I may be combining the two experiences. One part of the process that did stay in my memory all these years was the interview with the psychologist. He was good at finding out just about everything about me, and upon our second meeting in 1980, he picked up where he had left off three years before. It was like a meeting of two old friends. I also recall that on the Friday night after the end of the process, George Abbey and some of the astronauts invited my applicant group to a pizza parlor for beer and pizzas.

"I remember when I got the call telling me that I had not been selected. I was told that I was in the top 20 and so would be very competitive for the next selection. Obviously, the results of that second interview were the same. I did apply one

Fig 1.9:   (inset left) USAF Academy Class of 1964 graduate Gary Matthes. (main) Circa 1983, Lt. Col. Gary Matthes is congratulated on having racked up more than 1,000 hours in the F-16 when he was director of the LANTIRN (Low Altitude Navigation and Targeting Infrared for Night) test force. LANTIRN was a combined navigation and targeting pod system for use on the USAF's premier fighter aircraft. Initial operational tests and evaluations of the LANTIRN navigation pod were successfully completed in December 1984 (Images courtesy of Gary Matthes).

more time and my name was put forward by the Air Force, but NASA declined to interview me that time.

"Although applying to be an astronaut was a big thing back then, the process really didn't leave many lasting memories for me," Matthes said, summing up. "I have come to realize that not being selected was the best thing that could have happened to me. I love flying (still do) and I especially enjoyed test flying. Had I been selected as an astronaut, I would have missed out on some of the most exciting and challenging flying of my career.

"After the *Challenger* disaster, many of the military astronauts were returned to their military services. Roy Bridges returned to Edwards AFB as Commander of the 6510th Test Wing. I was his Vice Commander. He spent a great deal of time talking to me about the astronaut experience, especially about the politics within the group headed by George Abbey. If I had had any doubts about my great fortune in not being selected, Roy's insights blew them away. I could not have survived in that type of organization."

Gary Matthes retired as a colonel from the USAF in August 1992. Six months later, he joined the Lockheed Fort Worth Company (which became the Lockheed Martin Aeronautics Company) as a project manager on the Greek F-16 Program. He subsequently became involved in managing F-16 programs for the USAF. He retired from Lockheed Martin at the end of 2006, and still enjoys the recreational pleasures of flying light aircraft.

TABLE 1.2:   GROUP 8 SELECTION BOARD MEMBERS

CHAIRMAN:
George W. S. Abbey – Director of Flight Operations, JSC

RECORDER:
Jay F. Honeycutt – Assistant to Director, Flight Operations, JSC

PILOT PANEL:
John W. Young – Chief, Astronaut Office, JSC
Vance D. Brand – Astronaut
Martin L. Raines – Director of Safety, Reliability, and Quality Assurance
Joseph D. Atkinson – Chief, Equal Opportunity Programs Office, JSC
Jack R. Lister – Personnel Officer, JSC
Donald K. Slayton – Manager for Approach and Landing Test, JSC

MISSION SPECIALIST PANEL:
Dr. Joseph P. Kerwin, MD – Chief, Mission Specialist Group, Astronaut Office, JSC
Dr. Robert A. Parker – Astronaut, Mission Specialist Group
Dr. Edward G. Gibson – Astronaut, Mission Specialist Group
Dr. Carolyn L. Huntoon – Chief, Metabolism and Biochemistry Branch, Space and Life
    Sciences Directorate, JSC
Joseph D. Atkinson – Chief, Equal Opportunity Programs Office, JSC
Jack R. Lister – Personnel Officer, JSC
Dr. James H. Trainor – Associate Chief, High Energy Astrophysics Laboratory, GSFC
Robert O. Piland – Assistant Director for program development, Engineering and
    Development Directorate, JSC

Candidates as 'Guinea Pigs'

At the time of his candidacy as a Mission Specialist, Frank Harnden, Jr. was a research astrophysicist for the Harvard-Smithsonian Center for Astrophysics (CfA) in Cambridge, Massachusetts. Forty years on and recently retired from his work as a Principal Investigator at CfA (and NASA), he still has many recollections of that time.

"One particularly fond memory is that of meeting Sally Ride. Like mine, her graduate work was in X-ray astronomy. They put all the candidates up (20 per week) in the same hotel, and Sally was billeted in the room right next to me. We walked back and forth together quite a bit."

Dr. Harnden, who was in the sixth group of candidates, also recalls the experience of two psychological tests. "The 'shrink' had a typical psychologist's office but with a somewhat disturbing difference. Although the upholstered armchair we were invited to sit in appeared normal, it wasn't; the whole chair was unusually large. It was so wide that one could not rest both arms on the chair simultaneously. And it was so high that one's feet did not reach the floor but instead dangled off the front. Presumably, we guinea pigs were expected to strike a 'passable' response to this awkward situation. The other psych test was more fun: we got to choose a favorite song and favorite animal. My favorite song was Gordon Lightfoot's 'Canadian Railroad Trilogy' and favorite animal, a seagull – I'd recently read Richard Bach's account of Jonathan Livingston Seagull's flying." **[16]**

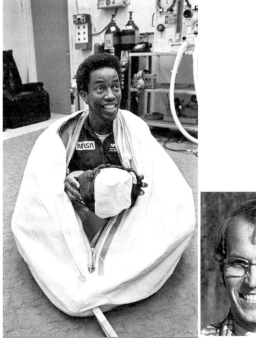

Fig 1.10: (main) Guion Bluford demonstrates how he squeezed into the rescue ball during his psychological evaluation tests. (right) Dr. Frank Harnden Jr., circa 1980 (Image courtesy of Frank Harnden Jr.).

One potentially daunting test the candidates had to endure – as earlier related by Dr. Sanders – was NASA's Personal Rescue Sphere. A spherical enclosure known simply as the "*rescue ball*," the device had been developed for any possible emergency transfer of astronauts between Shuttle vehicles[2]. Once described as a

---

[2] Formally known as the Personal Rescue Enclosure (PRE), although it was used in astronaut selection processes for many years, none ever flew on a Shuttle mission. **[17]**

claustrophobic's nightmare, the 36-inch (86-cm) diameter sphere had one tiny window to prevent total sensory deprivation. The endangered astronaut would clamber into the ball, squeeze up into a fetal position and don an oxygen mask, ready for crewmembers to zip it up and place the sphere into the depressurized airlock, from where they would be transported across to the rescue Shuttle by a space-suited astronaut. **[18]**

As part of their psychological testing, the astronaut candidates were required to be zipped into a rescue bag for an undisclosed period to see how they would cope, although they were given the impression that it was an exercise in which they were to evaluate the sphere as an escape device and submit a report on it.

"At barely 5' 9", I typically faced handicaps, e.g., in high school football (though not in college gymnastics)," Harnden reflected. "When it came to this test, however, I had a definite advantage. I had no trouble curling myself into the three-foot, utterly dark (and quiet) beach ball, whereas my colleague Jeff Hoffman was about six feet tall and must have been extremely cramped. Those running the test refused to tell us how long it would run. Instead they gave us a headset and said we should request termination if we 'wanted out.' I don't know how long it lasted, but it was actually quite a pleasant experience."

Unfortunately for Dr. Harnden and fellow science candidate Wilton T. Sanders, the Astronaut Office eventually rejected both men, citing their lack of binocular vision. As Harnden explained, this was "amblyopia (sometimes called 'lazy eye') that wasn't corrected early enough to enable our brains to fuse the images from our two eyes."

It proved to be a difficult time for Frank Harnden. "I received my rejection phone call the day after Super Bowl Sunday, 1978. Even though my team had won (I grew up in Dallas), the NASA news sent me into a severe bout of depression. It took me several weeks to recover from it, but that one episode has fortunately remained my only experience of such darkness. The successful launch the following November of NASA's Einstein X-ray Observatory (for which I served as one of its Instrument Scientists) also helped me put my astronaut disappointment behind me." **[16]**

Dr. Joseph K.E. (Ken) Ortega was in the final group of 24 candidates called to Houston. Ken is a Professor Emeritus of the Department of Mechanical Engineering at the University of Colorado, Denver. He holds a PhD, MSc and BSc degrees, all from the University of Colorado, Boulder, and all in Aerospace Engineering.

"I remember how excited I was when I received written notification inviting me to fly to Johnson Space Center for interviews and tests," he recalled. "This was my dream coming true. I had dreamed about being an astronaut since I was in middle school during the Mercury program. Growing up on a small ranch in the small coal mining community of Sarcillo in southern Colorado, it did not appear to me that I could have a chance to become a NASA astronaut. But I put in motion a plan to try. The plan would address educational and physical qualifications. I began to run three miles a day (rain or shine), did 90 pull-ups a day (three sets of 30), 150 push-ups a day (three sets of 50), 150 sit-ups a day (three sets of 50), and began a weight-lifting program.

Fig 1.11: (inset) Dr Joseph (Ken) Ortega. (main) Wearing Deke Slayton's Apollo space suit (Images courtesy of Ken Ortega).

"I applied for and was accepted into an Aerospace Engineering program at the University of Colorado, Boulder. I graduated with a BSc degree and during that time I conducted research that I published later with my advisor in *Science* (*Science* and *Nature* are the top scientific journals in the world). Afterwards, I became a Missile Launch Officer in the USAF and won first place in the four basic (4 BX) exercise program, beating out at least 500 officer training candidates. The Air Force sent me back to complete my MSc degree in Aerospace Engineering at CU Boulder. After an honorable discharge from the USAF, I obtained my PhD in Aerospace Engineering, CU Boulder, doing my thesis in Bioengineering. Subsequently, I did a three-month post-doctorate in Biochemistry and then a one-year post-doctorate in Turbulence Fluid Mechanics before I went to work for Martin Marietta Aerospace in the Thermo Physics Department. It was my idea to get research experience in diverse fields to prepare me for a Mission Specialist position.

"I was a Senior Engineer at Martin Marietta when I was invited to fly to JSC. Boy, I was flying high [very excited], and my wife, Alice, was happy for me. At JSC, we were organized into small groups of about ten people. We met with Astronaut John Young, whom we thought 'walked on water.' We were introduced to different aspects of the Space Shuttle program and had an opportunity to observe the Space Shuttle simulator in action. Embedded in these activities were trips to various medical doctors and their staff. We did pull-ups, push-ups, and other physical exercises. They measured various body functions and gauged our psychological aptitude and perspective. I remember a psychologist asking me, 'If you were an animal, what animal would you like to be?' My response was, 'a dolphin.' He then asked why, and I responded that they were very intelligent and I thought they might perceive and understand their surrounding better than other animals. I later discovered that most of my colleagues had said that they would want to be an eagle, so they could fly high and fast.

"The best part of my visit was getting into Mercury astronaut Deke Slayton's Apollo space suit. I was beginning to believe that my dream was becoming a reality. The worst part was the return visits to the doctor for blood pressure tests. In the back of my mind, I was becoming suspicious that something was not going as planned. However, after I returned home, I was still very optimistic. So later, when I received notification that I did not make the final cut, I was devastated. I was very depressed for months subsequent to the notification. I was informed that my blood pressure was high and I should have it checked by my doctor. I did and I have been on blood pressure medication ever since. I was sabotaged by my body.

"Subsequently, I worked as a Group Manager and Principal Scientist at the Solar Energy Research Institute (which later became the National Renewable Energy Laboratory) and then I became a faculty member at the University of Colorado, Denver, in the Department of Mechanical Engineering. During the first part of the 1980s, even though my career was progressing well, it felt as if a dark cloud was over my head – the sun was not shining. As a young Assistant Professor, I met with Ellison Onizuka at CU Boulder. Ellison was now an astronaut with one Space Shuttle mission under his belt. Ellison and I were undergraduates in Aerospace Engineering at the same time, and we double-dated with our respective soon-to-be wives. Alice and I attended Ellison and Lorna's wedding. I remember the mixed emotions I felt as I spoke with Ellison. I was happy for him that he made it and was now an astronaut, but my dark cloud, which was becoming less dark, got darker again.

"On one unforgettably awful day, I was at a Gordon Research Meeting in Santa Barbara, California. Before the morning meetings on January 28, 1986, I had informed a few of my colleagues that my friend Ellison Onizuka was scheduled to lift off that day. During one of the talks, one of my colleagues told me that I should go watch the TV right away. On the broadcast, they were rerunning the tape of the Space Shuttle *Challenger*'s lift-off. The terrible *Challenger* accident added another ingredient into the mixed emotions I was having. I remember thinking about the family Ellison left behind, and I thought about my wife and two daughters and what it would mean to them if I were on that flight. In the subsequent years, these sobering thoughts helped me to shed the dark cloud over me. Recently, I retired after 35 years as a faculty member. Now the sun shines for me and I believe that a career of research and teaching as a professor was my true calling in life." [19]

THE THIRTY-FIVE NEW GUYS (TFNG)

By Monday morning, January 16, 1978, the applicants who were still hopeful of being selected had received the long-anticipated phone call, which in most cases came very early in the morning. For those who had been selected, the call came from George Abbey, then chief of the Flight Crew Operations Directorate (FCOD), the division of NASA at JSC that included the Astronaut Office. Phoning from his office on the eighth floor of Building 1, he simply asked if they were still interested in coming to JSC as an astronaut. Following their acceptance, they were asked to keep the news strictly secret until at least noon. While most of the successful candidates were scattered across the United States, two of the calls were to overseas numbers: Kathy Sullivan was finishing up her doctorate work in Halifax, Nova Scotia, while Steve Hawley was engaged in post-doctoral research studies in Chile. A direct phone call could not be made to Dave Walker, who was serving as an F-4 *Phantom* pilot aboard the aircraft carrier USS *America* (CV-66) in the Mediterranean. Contact was eventually achieved and he accepted. At the same time, the unsuccessful candidates received a phone call from a member of the selection panel.

According to later, widely-spread reports, there were some last-minute changes to the make-up of those initially selected when it was revealed that only one woman had made the cut. The story stated that five pilots were subsequently dropped from the list and replaced with five women Mission Specialists. This would have been a rather curious anomaly on the part of the selection panel and is not borne out by the facts. Initially, NASA JSC was authorized to choose 40 new astronaut candidates for the 1978 selection, consisting of 20 Pilots and 20 Mission Specialists. By around November 1977, the interviews and selection had progressed to some sort of tentative list. It was then that NASA Administrator Robert Frosch and the Associate Administrator for Manned Space Flight, John Yardley, looked at the Shuttle schedule and the number of unflown pilots still waiting for assignments, including those transferred to NASA from the Manned Orbiting Laboratory (MOL) group. They decided that the agency did not need quite as many new Pilot candidates. Consequently, Kraft and Abbey were instructed to cut the number of pilots down to 15. This administrative decision had no bearing at all on the number of women selected.

Not surprisingly, the five pilots dropped from the list made it into the next draft of astronaut candidates in 1980. As there were six unsuccessful pilot applicants in the (overall) 208 interviewed at JSC who made it in to the 1980 group selection, it is uncertain which one was not among the five taken off the list of successful candidates in 1978. The six pilots were John Blaha, Roy Bridges, Guy Gardner, Ron Grabe, Bryan O'Connor and Richard Richards. Bryan O'Connor later said: "[As] I look back, I don't think I would have flown the Shuttle any sooner had I been picked up in that first class… They were going to hire 40 and they cut five pilots out, because they didn't need them, right at the last minute. I found out much later that I was one of those five. The reason they didn't need them is because in that '78 time-frame, the *Columbia* [Orbiter] was having problems. All the tiles had fallen off on a flight across country, one of the main engines had blown up on a test stand, and the whole Shuttle program was slipping.

"So if they had picked me up in 1978 and I had joined the Astronaut Office, I probably would have eventually flown the same time as I did anyway… I figured [it would give me] a couple more years of flight test experience, and if I scored fairly high among this [1978] group… they got hired by NASA, so I don't have all those people in front of me anymore, so maybe I'll be [more] competitive next time. And, sure enough, I applied in 1980 and I was picked up that time." [20]

Dr. Rhea Seddon was one who received the call directly from George Abbey, while working as a surgical resident at the VA Hospital in Memphis, Tennessee. Despite being initially stunned by the news, she was quick to accept. As she later recalled, "I would soon meet the other 34 astonished selectees who received the same call that day. They would become my close friends, my crewmates, my team; one would turn out to be my life partner and father of my children. That day, January 16, 1978, would become for all of us '*Who, me? day*', the date our lives changed forevermore." [21]

Capt. Jerry Ross of the USAF was a Flight Test Engineer with the B-1 bomber program at Edwards AFB, and one of many military personnel based there waiting on that all-important phone call. When it came, it was not the good news he had been hoping for, as he recalled in his 2013 book, *Spacewalker*: "I finally got my phone call. It was from Ed Gibson, who had flown on the final Skylab mission. I don't remember what he said, but it hurt… I was so close. I thought I had achieved my goal, but then it was snatched away.

"Six flyers at Edwards got calls from Mr. Abbey. Test Pilot School classmate Brewster Shaw was selected. Ellison Onizuka was also on the list. He was an instructor at Test Pilot School, and we played on the same softball team. El and I had gone to Houston together for our interview week. Another softball teammate, and a Purdue graduate, Loren Shriver got the call. The other three were Air Force test pilots Dick Scobee and Steve Nagel and Army test pilot Bob Stewart… When I saw my friends leave for Houston, it was hard." [22]

Although Jerry Ross had missed out, George Abbey encouraged him to apply again, and he was subsequently selected as a Mission Specialist in the Group 9 astronaut intake. When he eventually retired from NASA in 2012, Col. Ross had been launched into space seven times. On his final mission, STS-110 in April 2002, he became the first person to complete seven space flights – a record that stood for just two months before it was equaled by Franklin Chang Díaz on the following STS-111 mission. Over his seven flights, Jerry Ross had accumulated 1,393 hours in space, of which 58 hours and 18 minutes were invested in an impressive total of nine spacewalks. [23]

Fig. 1.12: Photo montage of the 35 successful candidates of NASA's Astronaut Class of 1978.

On that Monday afternoon, January 16, with all the acceptances now registered well before the 13:00 deadline, NASA Administrator Robert Frosch was finally able to announce the names of those chosen for the space agency's eighth group of astronauts. There were 35 names in all – made up of 15 Pilot and 20 Mission Specialist astronaut candidates. The group comprised 14 civilians and 21 military officers. Six of those chosen were women, and four were minorities. They were named as:

Pilot Ascans

Daniel S. ("Dan") Brandenstein, Lieutenant Commander, USN, age 34
Michael L. ("Mike") Coats, Lieutenant Commander, USN, age 32
Richard O. ("Dick") Covey, Major, USAF, age 31
John O. ("JO") Creighton, Lieutenant Commander, USN, age 34
Robert L. ("Hoot") Gibson, Lieutenant, USN, age 31
Frederick D. ("Fred") Gregory, Major, USAF, age 37
Stanley David ("Dave") Griggs, Civilian, age 38
Frederick H. ("Rick") Hauck, Commander, USN, age 36
Jon A. ("Jon") McBride, Lieutenant Commander, USN, age 34
Steven R. ("Steve") Nagel, Captain, USAF, age 31
Francis R. ("Dick") Scobee, Major, USAF, age 38
Brewster H. Shaw, Jr., Captain, USAF, age 32
Loren J. Shriver, Captain, USAF, age 33
David M. ("Dave") Walker, Lieutenant Commander, USN, age 33
Donald E. ("Don") Williams, Lieutenant Commander, USN, age 35

Military Mission Specialist Ascans

Guion S. ("Guy") Bluford, Jr., Major, USAF, age 35
James P. ("Jim") Buchli, Captain, USMC, age 32
John M. Fabian, PhD, Major, USAF, age 38
Dale A. Gardner, Lieutenant, USN, age 29
Richard Michael ("Mike") Mullane, Captain, USAF, age 32
Ellison S. ("El") Onizuka, Captain, USAF, age 31
Robert L. ("Bob") Stewart, Major, U.S. Army, age 35

Civilian Mission Specialist Ascans

Anna L. Fisher, age 28
Terry J. ("TJ") Hart, age 31
Steven A. ("Steve") Hawley, PhD, age 26
Jeffrey A. ("Jeff") Hoffman, age 33
Shannon W. Lucid, PhD, age 35
Ronald E. ("Ron") McNair, PhD, age 27
George D. ("Pinky") Nelson, age 27

Judith A. ("JR") Resnik, PhD, age 28
Sally K. Ride, age 26
Margaret ("Rhea") Seddon, MD, age 30
Kathryn D. ("Kathy") Sullivan, age 26
Norman E. ("Norm") Thagard, MD, age 34
James D. ("Ox") van Hoften, PhD, age 33

The Stats

Candidates in previous selections had mostly been aged in their 30s, which had allowed them time to acquire sufficient educational attainments and work experience, and normally to get married and have a young family. This tended to yield well-balanced, educated and settled individuals, and at the same time portrayed the image of the clean-cut, dedicated and patriotic American military pilot or academic professional, which helped establish the "*Right Stuff*" ethos in the Astronaut Office from the late 1950s and throughout the 1960s.

By the mid-1970s the iconic image of "The Astronaut" had wavered a little, with several departing the program having been unable to adjust to the lifestyle, disenchanted with the reduction in available missions and the lack of clear direction, or having endured the breakup of their families under the strain of living the role. By the time the 1978 group joined the Astronaut Program, both the formula and the agency had changed, with the former era of *All-American Boys* now a thing of the past as a new generation of *All-American Astronauts* began to come to the fore.

The 15 pilots had been selected from 659 applicants, of which 147 were military and 512 were civilian (eight were women and ten were minorities). The 20 Mission Specialists were chosen from 5,680 applicants, of which 161 were from the military and 5,519 were civilians (1,251 were women and 338 minorities).

While the majority of the new group was once again in their 30s, there were eight who were in their 20s. They were a talented mixture of individuals with credible flying and academic credentials. The selection net had been spread wider, with the youngest candidates aged 26 (Steven Hawley, Sally Ride and Kathryn Sullivan), and the oldest aged 38 (Dave Griggs, John Fabian and Richard Scobee).

At the time of their selection to the astronaut program, 29 of the 35 were married, with all but three of these (Ron McNair, Anna Fisher and Steve Nagel) having children, though in time those three would also become parents. Over four decades later, many are still married to the same person, but inevitably some in the group divorced, some re-married and a few also experienced tragedies within the family. The six single members at selection were John Creighton, Steven Hawley, Sally Ride, Judy Resnik (divorced), Rhea Seddon and Kathryn Sullivan, with all but Resnik and Sullivan marrying in later years.

Fig 1.13: The next generation of America's astronauts. Seven years after Apollo 15 carried an Air Force crew to the Moon in 1971, Guion Bluford in USAF uniform stands next to *Endeavour*, the Apollo 15 Command Module (Image courtesy USAF).

A new era dawns in the Astronaut Office

By 1978, NASA was 20 years old. In the agency's first ten years, 66 astronauts (49 with a prominent piloting background and 17 being primarily scientifically trained) had been chosen in six groups. As the new group of 35 were to begin their basic training program, so the agency was preparing to celebrate the tenth anniversary of the Apollo 8 mission around the Moon during Christmas time in 1968.

It seems hard to imagine today, but barely ten years separated the achievement of Apollo 8 and the selection of the first astronaut group to fly the Space Shuttle. So much had happened, and indeed changed in those ten years. The agency had been committed, by Presidential decree in 1961, to land a man on the Moon within a decade, achieving it at the first attempt and with a year to spare. But in its second decade, NASA was forced to distance itself from reaching for the Moon, which it had prioritized during 1960s.

Now, in the later 1970s during austere times for human space flight, the desire was to look back at Earth instead of striving for missions deeper into space. Looking beyond Skylab, America's first national space station, and pioneering international cooperation with the Soviets on Apollo-Soyuz, it was time to develop better ways to access and utilize space, discarding one-shot rockets and single-mission spacecraft. That was both the beauty and the downfall of the Space Shuttle; in the promise of being the all-encompassing national space launch system, but in the end falling far short of that goal. Once in orbit, the Shuttle Orbiter performed beautifully with relatively few failings, though maintaining the vehicles and getting them into orbit and back again on time and safely was always a challenge, one which proved fatal for two crews. But early in 1978, all this was in the future. The Shuttle had only flown off the back of a jumbo jet and had yet to prove it could fly in space, let alone repeatedly. However, after the lack of an astronaut in orbit since Apollo, there remained optimism and the hope of a new beginning for American human space flight, and while still untrained and unproven, at the forefront of that optimism were the 35 members of the astronaut Class of 1978.

Americans had not ventured into space for three years. Previously, only two calendar years had passed without an American in space since 1961: between the Mercury and Gemini programs in 1964; and in the aftermath of the loss of the Apollo 1 astronauts in 1967. This third lull in American spaceflight was the longest since the era of human space flight had begun in 1961. It had also been over a decade since the last NASA (and the second scientist) astronaut selection of 1967.

At the time of the announcement of 35 new astronaut candidates, the agency had an active roster of 17 Pilot and 10 Scientist astronauts from the earlier groups, many of whom had yet to fly their first mission and with some on leave of absence in other temporary roles while the Shuttle was being developed. The number selected in January 1978 had been worked out with a view to supplement the veteran astronauts, forming early Shuttle crews following the Orbital Flight Tests.

But how did the definition of a 'Shuttle crew' emerge? What impact did these new, very different and diverse individuals make on the established traditions and hallowed halls of the Astronaut Office in Building 4 at JSC? These questions are explored in Chapter 2 and Chapter 5.

References

1. NASA News Release 76-044, NASA HQ/JSC, *"NASA to Recruit Space Shuttle Astronauts,"* issued July 8, 1976.
2. Compiled by Bert Vis, with thanks to Mario and Sue Runco
3. The Staff of the Washington Post newspaper, *Challengers: The Inspiring Life Stories of the Seven Brave Astronauts of Shuttle Mission 51-L,* Pocket Books, New York, NY, 1986.
4. Carolyn L. Huntoon interview for NASA JSC Oral History Program, Houston, TX, April 21, 2008.
5. Kelly Knox online article for Wired.com, *"Star Trek Week: How Nichelle Nichols Changed the Face of NASA,"* September 26, 2012. Online at: https://www.wired.com/2012/09/nichelle-nichols.
6. Kathryn D. Sullivan interview for NASA JSC Oral History Program, Columbus, Ohio, May 10, 2007.
7. NASA *Roundup* magazine, Johnson Space Center, Houston, TX, *"More than 8,000 applicants vie for 30-40 astronaut slots,"* issue Vol. 16, No. 15, July 22, 1977.

8. NASA *Roundup* magazine, Johnson Space Center, Houston, TX, "*NASA Issues Recruiting Call for Shuttle Pilots, Mission Specialists*," issue Vol. 15, No. 14, July 16, 1976.

9. NASA *Roundup* magazine, Johnson Space Center, Houston, TX, "*JSC's medical team practices astronaut applicant screening*," issue Vol. 16, No. 8, April 15, 1977.

10. NASA *Roundup* magazine, Johnson Space Center, Houston, TX, "*Astronaut selection is delayed*," issue Vol. 16, No. 26, December 23, 1977.

11. William Heacox, email correspondence with Colin Burgess, January 11, 2018.

12. Lawrence Pinsky, email correspondence with Colin Burgess, January 27, 2018.

13. Loren Shriver, article for U.S. Air Force Academy, "*The NASA Astronaut Selection Process*," available online at: http://usafa67.org/pdf/429/33.pdf.

14. Dr. Wilton Sanders III, email correspondence with Colin Burgess, January 17, 2018.

15. Dr. Anna Fisher, interview for NASA JSC Oral History Program, Houston, TX, February 17, 2009.

16. Frank R. Harnden, Jr., email correspondence with Colin Burgess, January 11, 2018.

17. *The Space Shuttle Rescue Ball*, in **Space Rescue, Ensuring the Safety of Manned Spaceflight**, David J. Shayler, Springer-Praxis, 2009, pp. 10-14.

18. Mark Wade, Encyclopedia Astronautica, online at http://www.astronautix.com/r/rescue-ball.html.

19. Joseph K.E. Ortega, email correspondence with Colin Burgess, January 18, 2018.

20. Bryan O'Connor, interview for NASA JSC Oral History Program, Washington, D.C., March 17, 2004.

21. Rhea Seddon, from Astronaut Rhea Seddon website. Online at: http://www.astronautrheaseddon.com/who-me.

22. Jerry Ross with John Norberg, **Spacewalker: My Journey in Space and Faith as NASA's Record-Setting Frequent Flyer**, Purdue University Press, West Lafayette, Indiana, 2013.

23. Jerry Ross, email correspondence with Colin Burgess, January 30–February 3, 2018.

2

Who should fly?

When NASA announced the call for a new group of astronauts in July 1976, it was made clear right from the start that the agency was seeking up to 20 (later reduced to 15) pilots to train to fly the Space Shuttle Orbiter, and an equal number of Mission Specialists (MS) to meet the objective of developing new working practices in orbit rather than simply training pilots to fulfill those roles.

Unlike the two earlier Scientist-Astronaut selections in 1965 and 1967, intended to provide scientific crewmembers for future (and mostly cancelled) Apollo and Apollo Applications objectives, these new MS were not required to 'fly' the Orbiter and therefore would not have to undergo a rigorous military jet pilot course, as the majority of the Scientist-Astronauts had endured before progressing in their astronaut careers[1]. In another change, the new astronauts would not immediately be called 'astronauts'. Previously, all members of the earlier selections had taken that title on day one at NASA, without having to do any training or receive an assignment. However, starting with the Class of 1978, all future selections, both pilots and MS, would be known as 'candidates' until they passed a new two-year Astronaut Candidate (or 'Ascan') basic training program. They would then take up more technical assignments and, hopefully, progress to advanced training prior to be being considered for selection to a flight crew. In addition to the

---

[1] Some of the Class of 1978 had gained civilian pilot licenses earlier in their careers

© Springer Nature Switzerland AG 2020   39
D. J. Shayler, C. Burgess, *NASA's First Space Shuttle Astronaut Selection*,
Springer Praxis Books, https://doi.org/10.1007/978-3-030-45742-6_2

significant changes in selection criteria, it was clear that the path from candidate to orbit was also not going to be easy, straightforward, or indeed guaranteed to culminate in a flight into space.

Despite the difficulties, this system, pioneered by the 35 members of Group 8, proved very effective for NASA. Between 1978 and 2004, a total of 246 candidates (89 pilots and 156 MS) were chosen in seven groups to train specifically as crewmembers on the Space Shuttle. From those groups, crewmembers for 129 Shuttle missions were chosen between STS-7 in 1983 and the retirement of the Shuttle fleet with STS-135 in 2011.

Fig. 2.1: 1970s artist's impression of a Shuttle launch

THE EVOLUTION OF A SHUTTLE CREW

Across the 30-year program, the designations of Shuttle *Pilot* or *Mission Specialist* became familiar to those who followed the program. But how did those roles evolve from the era of the Apollo Command or Lunar Module Pilots? We need only to go back to the late 1970s, when that first Shuttle Ascan group was announced to the world. Just a few years earlier, NASA was emerging from the Apollo era and commencing detailed preparations for flying the Shuttle Orbiter in a series of test flights. Delays in preparing the major components of the new Space Transportation System for flight meant that it would be over five years after their selection before the first four members of the 1978 group (one pilot; three MS) would fly, on the seventh flight of the Shuttle program. So how did the positions, designations and roles of a Shuttle crew evolve in the decade prior to that first selection in 1978?

NASA's Astronauts of the First Generation

In the 1960s, the primary role of the crew of America's first generation, single-mission-designed orbital spacecraft, was to have astronauts with superior piloting skills 'fly' or 'control' the vehicle. It was only in the latter stages of the Apollo era that skills other than flying were considered for the crewing process.

Mercury: For the one-man Mercury series, flown by astronauts with background in testing military aircraft, the designation was simple: 'Pilot'[2].

Gemini: The introduction of a two-man crew required a lead astronaut – hence the designations of Command Pilot and Pilot, with the Command Pilot responsible for rendezvous and docking with an unmanned target, among other things, and the Pilot for assisting him and performing any Extra-Vehicular Activity (EVA, or spacewalking) objectives.

Apollo: Originally, the designations in 1966 for the Command and Service Module (CSM)-only Apollo Block I Earth orbital missions were Command Pilot, Senior Pilot (ideally previously flown) and Pilot (normally a rookie). Then, in a memo from Deke Slayton dated November 29, 1966, the designations for the more advanced Block II (lunar) missions were amended to Commander (CDR), Command Module Pilot (CMP) and Lunar Module Pilot (LMP) respectively, reflecting the more specialist requirements for the increased challenge of the lunar missions ahead. Though capable of flying the Lunar Module, the LMP was more of a systems operator, officially leaving the control and handling of the ungainly vehicle to the CDR.

---

[2] The Mercury astronauts argued strongly against calling their space vehicles "capsules," insisting on using the word "spacecraft" over one they equated to something associated with medication.

Skylab: Under the Apollo Applications Program (AAP), the Commander (CDR) and CSM Pilot (CMP) designations were retained. For the third, scientific member of the crew, the designation was initially to be known as the Mission Module Pilot (MMP), but by the time AAP became Skylab this had changed. When the crews were announced in January 1972, the third member was identified as the Science Pilot (ScPLT).

Apollo-Soyuz: The final Apollo flight in 1975 still featured a Commander and CMP, but instead of flying a Lunar Module there was a smaller Docking Module unit that allowed the CSM to link up with, and the crew transfer to, the Soviet Soyuz. Without the LM there was no need for the LM Pilot designation, so the third position for this final flight of Apollo was changed to the not-very-original Docking Module Pilot (DMP), even though there was actually nothing to 'pilot' in the DM.

These designations worked fine for Mercury, Gemini and Apollo, crewed by a small number of pilots and pilot-trained scientists in the Astronaut Office, but for Space Shuttle, new criteria needed to be devised for its multiple crewmembers.

Early developments
This summary of the evolution of the Shuttle 'crew' takes up the story from early 1968, a decade before the first of those chosen to fulfill the majority of the flight crew roles were named.

On February 29, 1968, NASA Administrator James E. Webb appointed the Director of Langley Research Center, Dr. Floyd L. Thompson, as his Special Assistant and head of the NASA Interim Working Group based at NASA Headquarters in Washington D.C. Thompson was given the remit to examine future manned space programs beyond Apollo, including studies looking at the development of a Space Shuttle and, by default, its crew. [1]

On August 10 of that year, Dr. George E. Mueller, NASA Associate Administrator for Manned Space Flight, speaking at the British Interplanetary Society in London on the occasion of his award of Honorary Fellowship of the Society, told the audience that he believed the future expansion of space exploration was severely limited by the cost of placing an object in orbit and the inaccessibility of payloads once they have been launched. Mueller forecast "that the next major thrust in space will be the development of an economical launch vehicle for shuttling between Earth and the installations such as orbiting space stations which will be operating in space." [2]

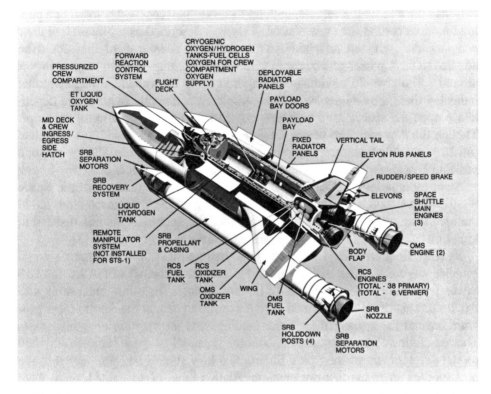

Fig. 2.2: The components of the Shuttle 'stack', consisting of an Orbiter (OV) with its three Main Engines (SSME), the External Tank (ET) and twin Solid Rocket Boosters (SRB).

In his speech, held at a special meeting of the members of the BIS at Imperial College, Dr. Mueller outlined the concept of a reusable shuttle vehicle designed to perform a range of missions in Earth orbit. These projected 'missions' included recycling; repair; inspection and transport; supporting the creation of larger space stations beyond those planned for Apollo Applications (which were to utilize the third stage of the Saturn V as an orbital workshop); and rotating station crews. Mueller explained that the aim of the shuttle vehicle was to reduce the cost of deploying a payload in orbit significantly, by using reusable hardware and there-fore lowering the launch costs. At this point there were no details of crewing, but hints that the system would operate "in a mode similar to that of large commercial air transports," capable of operating in all weathers, and that the cockpit "would be similar to that of large intercontinental jet aircraft." An accompanying illustra-tion shown during the presentation suggested that NASA was thinking of a two-man crew to fly the vehicle, with up to seven 'passengers' enclosed in a pay-load module behind the crew station. This was described as a *pressurized crew compartment*, with an attached *mission payload compartment* that could be adopted to ferry passengers or cargo with "comfort comparable to large transport aircraft." Pilots would obviously be required to fly the vehicle, though Mueller indicated that "the landing would be completely automatic, with prime depen-dence upon the spacecraft's guidance system but with ground control back-up." Though the nomenclature of the crewmembers were not revealed, the definitive 'mission' module would clearly require specialists trained to handle the cargo, a possible genesis for what became known as the 'Mission Specialists.'

Eight months later, on April 24, 1969, during the first meeting of the Space Shuttle Task Group held at NASA HQ in Washington D.C., the discussions led to the identification of a number of future directions, requirements and discussion topics on "a nominal 12-person capability for crew and passengers." [3]

By May 30, just six days after Apollo 10 had returned from the Moon, George Mueller was back at the BIS in London to accept, on behalf of NASA, the BIS Silver Trophy for the achievement of the first manned flight into lunar orbit with Apollo 8. This time, he explained NASA's desire to extend the Apollo lunar pro-gram beyond the envisaged ten lunar landings and the creation of a lunar base. This was followed by a discussion about the potential of space stations, and then on the development of the Space Shuttle and how studies had progressed in the nine months since his last presentation. At this time, he explained, there were more than ten different designs proposed for the shuttle system, but despite this Mueller estimated that the rapid development on the program to date indicated that "the Space Shuttle can be operational in seven years," i.e., by 1976. [4]

On July 28, 1969, four days after the return of the Apollo 11 astronauts, a pre-liminary statement of work was circulated in advance of the Phase B Space Transportation System (STS) contracts, which noted that the ultimate goal of STS was to "introduce a new, more mature and routine mode of space transportation than past programs. Men [*authors' note*: no mention of females or minorities was made at this stage] will travel into space and return in a shirtsleeve environment similar to the present-day commercial aircraft." In the accompanying design refer-ence mission, a crew of two astronauts would fly the 'shuttle', together with a crew of up to *ten* to occupy the envisaged 12-man space station. [5]

In September that year, the Space Task Group issued its *Post-Apollo Space Program: Directions for the Future* report to President Richard M. Nixon. This blueprint for the next decade or so highlighted the need to develop a new space transportation system, with emphasis upon vital factors of *commonality, reusabil-ity, and economy.* In relation to the Shuttle, this meant incorporating fewer major systems across a wide variety of missions (commonality); the repeated use of the same systems over many missions (reusability) and a reduction in 'throwaway' elements (economy). This, the report suggested, would improve the overall cost and operational capability, and encourage the ability to *carry passengers* and items of hardware to and from orbit in airline-like routine operations.

As the design of the vehicle was being debated, the media at the time picked up on the variety of missions to be flown, the reusability aspect and the projected reduction in launch-to-orbit costs. Largely overlooked in all this was the crew configuration, which was not surprising as the actual decision on crewing was still some way off. This omission did not stop the media putting forward suggestions that access to space would soon be opened up for all. One typical contemporary account, published in 1970 and aimed at teenage readers (who may have dreamed of flying on the Shuttle themselves one day), suggested the Shuttle would "open the way for ordinary men and women to go into space." However, the idea of scores of commercial passengers being ferried to and from huge space stations, or even hotels in orbit, in the near future remained more in the minds of those involved with reporting on the Shuttle, rather than those who were actually developing it.

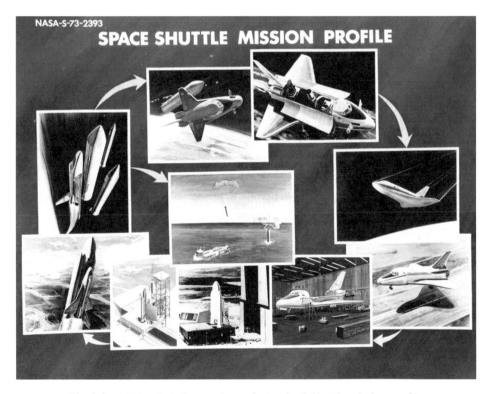

Fig. 2.3: 1973 artistic impressions of a 'typical' Shuttle mission profile.

The 'big sell' to politicians, the press (and through them to the public) and potential customers, was in the projected ability to fly a mission after a turn-around of just two weeks, thus reducing the launch costs, and by operating a fleet of between four and six shuttles which would complete up to 60 missions a year "between 1977 and 1990." It was reported that the older Orbiters would be replaced by new vehicles every five or six years. This suggested an Orbiter production line of 18 to 24 vehicles, which clearly was not accurate. In these accounts, each Orbiter would feature a large, 60ft by 15ft (18.28 by 4.57 meters) cargo bay that could not only be adopted to carry satellites or packed with freight ferried to a space station, but also transport up to 50 passengers in a pressurized compartment. For a younger generation enthused by the success of Apollo this looked like a bright future, but unfortunately, as the decade progressed, the realities of trying to develop and deliver on those broad claims became apparent within in the program. [6]

Schedule Analytics Chart circa early 1970s

To demonstrate a more accurate, but still ambitious, mode of planning being developed for Shuttle crewing, even in the early 1970s, an analytics chart (undated and shown in Figure 2.4) included these assumptions:

- There would be FIVE operational Orbiters, with the first flight under the orbital test program completed during 1978–1980 (by OV-102). OV-101 would then join the fleet, followed by OV-103, OV-104 and finally OV-105.
- The payload program was to be operational by October 1, 1979
- Each Orbiter would fly on five-month cycles
- There would be one flight each month from KSC.

This planning forecast would also see a Pilot and Co-Pilot for each of the Orbiters (five pairs), with a 'mission team' of 2–4 (only men were included in this projection) made up of either cargo handlers, Flight Engineers, Payload or Mission Specialists, Principle Investigators or passengers.

In this plan, the Orbiter 1 Pilot/Co-Pilot pairing, for example, would fly with mission Team A (and Payloads 1, then 16, then 31 etc.); Team B (payloads 6, 21, 36); and Team C (Payloads 11, 26, 41) etc. The pairings for the other four operational Orbiters would follow a similar profile. Under this plan, therefore, 15 teams of astronauts would be required (three teams for each of the five Orbiters), each dedicated to a specific payload (presumably with a built-in attrition rate within the team) and with projections of 45 payload missions between the Fourth Quarter of 1979 and the Second Quarter of 1983.

As analytically useful as this may have appeared on paper, it was not long before such paper studies were amended, scrapped or revised, and then changed again many times before any astronaut's name appeared on a manifest.

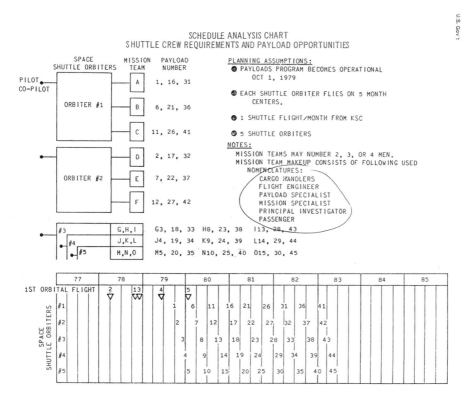

Fig. 2.4: Schedule Analysis Chart for Shuttle Crew Requirements and Payload Opportunities circa early 1970s. Note the early mention of Cargo Handlers, Mission and Payload Specialists, Principle Investigators and Passengers.

An early status report

On January 5, 1972, the year of the final Apollo missions to the Moon, President Richard M. Nixon announced the nation's decision to develop the Space Shuttle. That same month, the U.S. House of Representatives Committee on Science and Astronautics published a timely 1,000-page status report entitled *Space Shuttle–Skylab, Manned Space Flight in the 1970s*, compiled using information attained from NASA field centers, key industrial contractors and from briefing transcripts, describing the current status of both the Skylab Orbital Workshop and the Space Shuttle programs (then undergoing Phase B preliminary design studies). As the letter of submittal by Thomas N. Downing, Chairman of the Sub-Committee on NASA Oversight, stated in the report, the hope was "that this status report will contribute to those decisions necessary to determine our nation's role and participation in our future space-related activities." Of note was a report on the development of Shuttle payloads by Mr. William T. Carey of Marshall Space Flight Center (MSFC), Huntsville, Alabama, dated October 26, 1971, which focused upon the analysis of the first ten Shuttle missions.

One of these featured missions was the Large (later Hubble) Space Telescope, for which Carey suggested that it would be more cost effective to modify the telescope on orbit than to return, refurbish and then re-launch it, adding "you can also put up an astronomer once in a while with the Shuttle." He also suggested, rather strangely, that perhaps even a *graduate student could be flown* [author italics] as part of the crew on their educational career path towards becoming a professional astronomer! Carey added that one or two MS "would be ideal" on many of these first flights. **[7]**

Fig. 2.5: 1970s cutaway image of the Orbiter crew compartment showing the upper flight deck and lower middeck living area.

A second presentation, this time by North American Rockwell Corporation Space Division at Downey, California, dated November 19, 1971, focused on the status of the program at Space Division and gave an early indication of some of the various designations being assigned to the crew. In this presentation, the basic Shuttle crew configuration was given as a two-man flight crew, supported by one or two astronauts trained in a particular mission and identified as "Mission Specialists (MS)," or by one MS and one crewmember who would have an in-depth knowledge of a particular payload, identified here as a "Payload Specialist (PS)." As Mr. Franklin stated in the presentation: "In the early days of the Shuttle …we would like to look at what is the most logical way to introduce man into the pertinent operations of the Shuttle and to make that investment in such a way that it leads us in that direction." Part of that work was being conducted at Langley Research Center, Hampton, Virginia, where the types of payload that should be included on the Shuttle were being evaluated in a program called the Shuttle Orbital Applications and Requirements (SOAR). The SOAR program looked at various payloads which involved the crew in their operation, or those which did not require manned intervention. A second, parallel program to define potential research on the Shuttle was being conducted at MSFC. Called the Research Application Module (RAM), this was a pressurized module located in the Shuttle payload bay (which would eventually evolve into the European-built Sortie Module that became Spacelab), within which an "experiment management crew" would work to conduct a fairly wide variety of missions in what was described as "an inexpensive way." [8]

Fig. 2.6: The four-seat flight deck showing the pilots and Mission Specialists in their launch and (as shown here) entry positions. In this 1970s image, a multi-racial crew is evident, as is the lack of pressure suits.

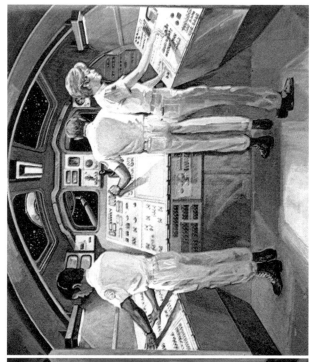

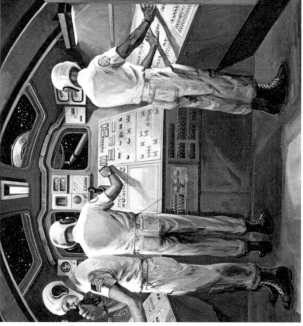

Fig. 2.7: These two interesting, almost mirror images (note the RMS and payload positions out of the windows) show on-orbit activities at the aft flight deck. Compare these 1970s artistic impressions of what Shuttle activities were expected to look like to actual views inside the Orbiter on real missions.

Getting rid of irrelevant pilot training?

On March 2, 1972, NASA Administrator James C. Fletcher was at the Kennedy Space Center (KSC, Florida), addressing an audience of 200 at the Equal Employment Opportunities Conference. He stated: "We are working on plans to get members of minority groups into space. The Space Shuttle, which is the keystone in all our future space programs, will be an important factor in accomplishing this goal." [9] This was all well and good, but for some time any appearance or speech given by an astronaut had prompted questions about when NASA would send female or black astronauts into space. With the authorization of the Shuttle in April 1972, concerns were being raised over this same question, especially following recent 1972 amendments to the 1964 Civil Rights Act, which subjected the U.S. Government (including NASA as a government agency) to equal opportunity legislation. In his response to reporter questions at the KSC event, Fletcher commented that where Equal Employment Opportunities were concerned, he was attempting to cut out all the red tape and remove the symbolic blocks that had hindered progress in this area. Later, in the *New York Times* dated March 28, 1972, University of Michigan astronomer James A. Louden wrote that one of the Shuttle's most important aspects was its ability to carry passengers: "For the first time, scientists will be able to perform experiments in space without spending years in irrelevant pilot training first." With the authorization of the Shuttle program and the reduced launch and landing loads that its design projected, the case for recruiting non-piloting astronauts in the space program – including females and minorities – gathered pace.

Early ideas for crewing

For several years, the Crew Training and Simulation Division (Code CE) at MSC/JSC conducted numerous reviews into exactly what the crew duties would be on a 'typical' Shuttle mission[3] and who would be responsible for what. Just *one* example of this early planning can be found in a March 1972 memo, in which the established four-person crew baseline was adopted and was intended to be "quite general so as to accommodate subsequent discussions pertaining to the crew skills, mix and more specific duties for particular missions." This suggested that the thinking was to assign crewmembers with a medical background to a medical mission, astronomers to astronomy missions etc., training certain crews to fly specific mission profiles, such as deployment of a series of payloads (TDRS, comsats), or focusing on particular skills, such as EVA servicing, Remote Manipulator System (RMS) operations and so on. [10]

---

[3] In reality, there was no such thing as a 'typical mission' for the Shuttle. While each mission flown may have looked similar, they were all highly individual.

Part of this long-term planning, according to the memo, included the early application of a definitive nomenclature and formal description of duties, to create training plans and subsequent flight operations and procedures. The suggestions put forward in this 1972 memo, a decade prior to the first Shuttle crews flew missions, are detailed in Table 2.1.

Table 2.1: SPACE SHUTTLE FLIGHT PERSONNEL NOMENCLATURE AND DUTIES [FEBRUARY 1972] ALL PAYLOAD CLASSES

Crew Nomenclature	Flight Dynamics - Aero	Flight Dynamics – Space	Flight Procedures	Flight Plan	Booster Subsystems	Orbiter Subsystems	Emergency Procedures	Payload Subsystems	Payload Instruments	Payload Operations	Rendezvous	Docking	Orbital Mechanics	EVA	Network
Commander	1	1	1	1	1	1	1	2	3	2	1	1	1	2	1
Pilot	1	1	1	1	1	1	1	1	3	2	1	1	1	2	1
Mission specialist	2	2	2	1	-	2	1	1	2	1	2	2	2	1	2
Payload specialist	-	-	3	2	-	-	2	2	1	2	-	-	2	3	3
Passenger/observer	NOT CONSIDERED AN ACTIVE CREW MEMBER														

Adapted from attachment to March 7, 1972 memo from Crew Training and Simulation Division [from the NASA JSC History Collection; copy held in AIS files] Key:
1. Primary training and responsibility
2. Direct support responsibility, specialized training
3. Operational support

In an April 1972 attachment to the memo, the preliminary duties identified for each position on the Shuttle crew were assigned as follows: [11]

The *COMMANDER* would be in command of the flight and responsible for the overall space vehicle, personnel, payload flight operations and vehicle safety. They would be proficient in all phases of the vehicle flight, payload manipulation, docking, and subsystem command, control and monitor operations. A commander would also require knowledge of payload and payload systems as they related to flight operations, communication requirements, data handling and vehicle safety.

The *PILOT* would serve as second in command and hold duties essentially equivalent to those of the Commander.

The *MISSION SPECIALIST(S)* would be responsible for the interface of payload and Orbiter operations and management of payload operations. The MS (and there may be more than one on each flight) would also need to be proficient in vehicle and payload subsystems, flight operations and payload communications data management.

The *PAYLOAD SPECIALIST(S)* would be responsible for the application, technology and science payload/instrument operations. They would have a detailed knowledge of payload instruments, operations, requirement, objectives, and supporting equipment.

The *PASSENGERS/OBSERVERS* which, according to the document, meant "anyone who has no direct part in Shuttle operations." Though not detailed in this document, they would (presumably) have to undergo Shuttle briefings, safety, and crew equipment training sessions to become part of 'a crew' and thus be safe to fly with.

Even at this early stage, MS and PS assignments were being applied to some of the long-term planning charts and documents.

On July 25, 1972, North American Rockwell (later Rockwell International) was named as lead contractor for the design, development and construction of the Orbiter element of the Space Shuttle stack. It was stated at that time that the first manned orbital flight of the Shuttle was planned for 1978 and that 445 flights were projected to be completed by 1990[4].

Meanwhile, NASA was already evaluating who might be suitable for its next astronaut intake, and were in fact planning to go much further than just eliminating the jet-pilot training qualification that the Scientist-Astronauts had needed to acquire for its next class of astronauts. On July 29, just four months after Louden's letter in the *New York Times*, and four days after Rockwell was named leading contractor, NASA Administrator James C. Fletcher spoke in Baltimore, Maryland, at the annual meeting of the National Technical Association (NTA)[5]. In his speech, Fletcher informed the audience that NASA planned to expand opportunities for black and minority groups from within the space agency, including possible flights on the Shuttle: "We are planning now for Shuttle flights," he stated, adding that "it will carry minority and women scientists into orbit." [12]

By 1973, several astronauts had been working in development and support roles for the Space Shuttle for some time. This was part of a restructuring of all departments in the Astronaut Office to take into account expected retirements once the Skylab and Apollo-Soyuz Test Project (ASTP) phases had been completed, while at the same time utilizing the talents of those who remained, in particular the Scientist-Astronauts. On December 7, 1973, in support of the pending retirement of the Apollo-Saturn spacecraft system and the introduction of the

---

[4] Less than a year later, on April 2, 1973, NASA HQ had refined the Shuttle flight manifest to show six orbital flights in 1978, 15 in 1979, 24 in 1980 and 40 in 1982, increasing to 60 each year from 1983 to 1987, and then dropping to 28 in 1988. This was still clearly an over-optimistic goal, as over the next decade funding remained a challenge, and qualifying the systems proved much more time consuming than first envisaged.

[5] The NTA was a non-profit organization aimed at aiding members of the black and other minority groups to prepare for technical or scientific careers.

Space Shuttle, the eleven remaining Scientist-Astronauts in the office were reassigned either to the Science or the Application of Life Science directorates. Though more changes were forthcoming, many of the Scientist-Astronauts assumed support roles in the emerging development of Space Shuttle payloads and experiments, most notably in Spacelab, the European contribution consisting of a pressurized science module and unpressurized support pallets to be flown in the payload bay of the Shuttle. Spacelab was designed to be operated by a 'science' crew made up of NASA astronauts and 'guest' foreign astronauts from member countries of (from 1975) the European Space Agency (ESA). To aid this development, a series of ground and airborne simulations of Spacelab missions were conducted during the mid- to late-1970s that provided valuable information on how the orbital Spacelab missions might be planned, prepared and flown, as well as the roles and responsibilities the science crewmember could hold on those missions. [13]

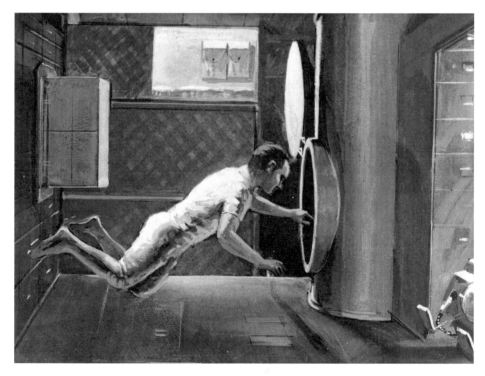

Fig. 2.8: One of the promoted advantages of the Space Shuttle over previous American crewed spacecraft was the extra room available. This image shows an almost empty middeck and airlock. After the four OFT missions, the middeck on real missions was rarely so devoid of equipment.

Fig. 2.9: Meal times 1970s style, illustrating the multi-cultural and mixed gender crews often portrayed in these concept images as the search for a new class of astronauts was pursued.

Defining the roles

The flying capability of the Orbiter indicated the two main designations almost immediately. With two pilots seated on the flight deck, the left seat would be occupied by the lead pilot, assisted from the right-hand seat by his 'co-pilot', as is traditional in large commercial, military or transport aircraft. For the Shuttle, that meant the Commander (CDR) on the left and the Pilot (PLT) on the right. With an expected crew compliment of between 4–6 additional crew members, depending on the mission, the designations of the other astronauts were defined by the specialty skills required, hence MS or PS. This evolution took countless memos, matrixes and meetings over a period of time and is highlighted here in another example from the JSC Space Shuttle Archive. The on-going topic of Shuttle crewing was discussed in a meeting held at JSC on October 24, 1972, between members of the Engineering Directorate (Mail Code E), Flight Operations (Mail Code C), and personnel from simulator contractor Ling Temco Vought, who created a baseline for further discussion. This baseline was summarized as:

- A *basic* crew would consist of two NASA astronauts: CDR and PLT.
- A maximum of four additional crew members, including NASA astronauts as MS, and non-career PS, as required.

- Total maximum in-flight personnel (crew (CDR, PLT, MS, PS) and 'passengers') would be no more than 10.
- Two crewmembers (MS, and CDR or PLT) would conduct EVAs.
- Responsibility for the crew assignments were listed as:

  - CDR or PLT were responsible for all activities affecting the Orbiter, including operation of the Remote Manipulator System.
  - MS would be responsible for Orbiter/payload interface and payload/instrument operation
  - PS would be responsible for application, technology and science payload or instrument operation.
  - Passengers [undefined in the document at that time] would have no responsibility for Space Shuttle or payload operations, raising the question of their purpose (for example, artist, politician, foreign dignitary, celebrity, etc.).

The document generated from that meeting also defined an early study of the initial STS flight crew assignments. For the six orbital test flights prior to commencing the operational phase, only a two-person *basic* crew (CDR/PLT) would be flown, after which the crew complement would expand as required. When the full operational period began, a flight rate of 60 missions per year was projected. There was also a suggestion for *two* crews to be assigned to each mission (prime and back up, and by flying only a *basic* crew, both astronauts in each crew would be trained and equipped to complete any EVA required).

The prediction for subsequent operational flights was for a *basic* crew annual flight rate of between 4–6 missions *each year*, with the MS flying 3–4 missions per year. If the low rate for the *basic* crew and high rate for the MS astronauts (minimum of two) were taken into account (i.e., a team of four astronauts flying four missions each year as a unit) then at the least *15 crews would be in training each year, totaling 60 astronauts* [author italics]. This was without taking into account natural attrition (retirement from active status), down time (proficiency training, vacation or illness), injury, or loss of life.

Using this 1972 format to surmise a 'typical' Shuttle mission, planned for an average of seven days (some less, others more), meant that a crew could be back in the training cycle every two or three months. Having the same crew flying with similar payloads could be both advantageous and disadvantageous to training, and raises the question of the desire and motivation of the crew to fly such repetitive missions so frequently. In reality, it was found that such a rate was highly impractical, financially prohibitive and operationally impossible, as training for an 'average' Shuttle flight was found to take many months. The option of rapid turnaround was investigated early in the program, however.

A decade later, in 1982, astronaut Bob Crippen was interviewed for JSC *Roundup* and was asked this very question. The interviewer suggested that such a rapid turnaround would represent a couple of weeks of jubilation in flying the mission, a week of debriefing and one day of rest, followed almost immediately by months of training for the next flight. Crippen admitted that the crew rotation system remained a work in progress even in 1982, but as flight rates increased it was expected that the crew turnaround time would decrease, though it was still far from decided if this was a path NASA would take. There had indeed been some planning done in the Astronaut Office about what the ideal gap between flights should be, what additional training, if any, would be required to fly a similar mission, or whether generic crews would be implemented in the future, but this was still being hotly debated. By the early 1980s, it was found that it took about 12 months to train a Shuttle crew, though it was hoped this could be reduced to a turnaround of between 3–6 months. This was still being worked on two years later when Crippen was assigned to his fourth mission, his second of 1984. Around this time, a DOD stand-by crew had been created, reflecting the delays in the program and changes in launching classified payloads. This 'launch-ready' crew had to be ready to go at short notice, but in the end was not utilized in that role. The 1986 *Challenger* accident, program changes and hardware delays finally put paid to this idea of a defined crew flying multiple missions based on short turnaround training. On some missions, however (e.g. Space Radar Laboratory; Spacelab Life Science; Atlas; Hubble Servicing), at least one crew-member from an earlier crew was assigned to the next flight in the series to help with the payload and training preparations. [14]

In a second document, a rough draft (undated but filed with the above 1972 memo) of 'job descriptions' for Shuttle crewmembers suggested as early as 1972 that "all crewmembers that will fly will be highly trained individually as capable of contribution to the success of the operational or scientific missions being flown." The document then went on to suggest that the Shuttle had the capacity to fly two or more crewmembers that were "not directly responsible for the safe return of the vehicle," thereby utilizing their expertise instead on training in Orbiter-to-experiment operations. This would vary mission by mission, between payload and availability, but three levels of 'Shuttle crew training' were being defined for the Shuttle vehicle and payloads/experiments as early as 1972.

For example:

Shuttle Pilots: Both the CDR and PLT (the flight deck crew; collectively referred to here as 'the pilots') would have intensive knowledge of the Shuttle vehicle and spaceflight operation in general.

- The CDR retained primary responsibility for safety of the vehicle and entire crew.

- The Orbiter could be operated by the CDR and PLT without additional assistance (the workload experienced during ascent and entry across the four Orbital Flight Tests (OFT) contributed to the decision to assign an MS as a Flight Engineer (FE), to assist the CDR and PLT from the first operational mission, STS-5).
- The pilots were to be responsible for system operation, fault isolation and necessary maintenance on the vehicle and its systems.
- In support of the payload, the pilots would ensure operation of the fuel cells and management of consumables, as well as distribution of power to support the payload and experiments.
- The pilots would be responsible for opening the payload bay doors, deploying the payload, grappling satellites and payload stowage.
- The pilots may also support payload operations in a limited manner without requiring additional training, such as gathering of data in multi-shift operations.
- The PLT may act as 'buddy' during EVA operations, performing as IV (Intra-Vehicular) assistant and safety observer to the MS performing the EVA while the CDR monitors the EVA from inside the Orbiter. (In the early days of the program, participation in EVAs had been suggested as part of the role of a Shuttle Pilot, but their increasing responsibility and workload dictated that they should instead fulfill a support role for both planned and contingency spacewalks.)

Creating the criteria

Plans for selecting the next group of astronauts were therefore challenging, not only because the program they would be chosen for was unlike any that had preceded it, but also because the established method of selecting top test pilots as candidates, or selecting suitably qualified scientists to be trained as jet pilots before assigning them to a mission, needed to be changed.

Flying experience would always be a required skill for those assigned to land the Orbiter on a runway at the end of the mission, but for the majority of the Shuttle mission profile, the emphasis would be on scientific and operational activities in orbit, and this would require a new breed of astronaut. NASA was also under pressure to be seen to include suitably qualified female and minority groups, and doing so meant that the selection process would take longer than the earlier intakes. Therefore, the program to select what would become NASA's next astronaut group actually began five years before they were announced, during a two-day meeting in September 1972. This was not even at a NASA field center, but away from the glare of publicity and the media.

At the Peaks of Otter Lodge, located in the Blue Ridge Mountains north of Roanoke, Virginia, the directors of the NASA field centers associated with the *manned* space flight program gathered to discuss developing a plan to define the requirements for a new group of astronauts to train specifically for the Space Shuttle, as well as the timeline for such a plan. They were guided by Jim Fletcher, who indicated right from the start that full consideration should be given to including minority groups and female candidates. Leading this effort would be the Manned Spacecraft Center in Houston, Texas (MSC; renamed the Lyndon B. Johnson Space Flight Center – JSC – in 1973, in memory of the 36th U.S. President who died early that year). Over the next four months, MSC would create a range of staffing plans for Space Shuttle astronauts, to be presented to Dale Myers, the Associate Administrator for Manned Spaceflight, by February 1, 1973. The creation of these plans would be based upon the most accurate and up-to-date projections for the number and frequency of projected Shuttle flights, as well as an estimate of the need for pilots to fly the vehicle over the first few years during the test and early operational phases.

The Shuttle program had only just been authorized, so it was still many years away from the first flight, and as NASA already had a cadre of veteran and rookie astronauts from earlier groups, it was expected that at least some of these men would still be available to crew the first few test flights to qualify the system for operational missions. In addition, past experience allowed officials at Houston to realize that there was no immediate need to employ a new group of astronauts at this point, as a long wait for a flight opportunity would diminish their enthusiasm. The precedent was the departure of several of the Scientist-Astronauts from the 1965 and 1967 selections before they had flown a mission. There was also the limitation of training equipment to be considered and indeed the lack of a defined training program, which was still under development. Added to this was concern that over-training and maintaining proficiency would be detrimental if a new group was selected too soon and was forced to wait for years before achieving their first space flight.

Therefore, it was announced in June 1973 that the new selection would not take place for five years (1978), when the opportunity for them to fly the Shuttle was expected to be much closer. In the meantime, the current astronauts would be employed filling roles in support of the design, development and testing of the Shuttle hardware and procedures (see below). Dr. Fletcher, commenting on these discussions, said that NASA had no problem in recruiting, but now was not the time, and that serious consideration would be given to recruiting female and minority astronauts in the future. [15]

A few months later, in October 1973, timely results from a five-week experiment held at Ames Research Center, Moffett Field, California, designed to determine female qualifications for space flight, unsurprisingly found that women were as physically fit for weightless flights as men. Eight Air Force nurses had endured two weeks of total bed rest and their post-test results were comparable to those of male volunteers. It represented a step in the right direction for the future. [16]

The following month, the latest ongoing efforts to define the activities that a Shuttle crew might be expected to be involved with during the orbital phases was uncovered in another memo (later researched by the authors). This was the result of ongoing in-house studies at JSC and revealed how complex the topic had become when compared to the memo of March 1972, just 20 months previously. **[17]** In this memo, the crew complement was stated to be directly influenced by the complexity of the Shuttle missions, the types and numbers of payloads flown, the skills of the crewmembers, and the proposed duty cycle. In reviewing the progress to date, the memo stated that "crew cross-training will most likely be mission dependent." The attached matrix (see Table 2.2) of crew functions circa 1973 showed a level of cross-training. The document was recommended to be used for planning purposes, but with the caution that "as the Shuttle and payload designs mature, and the subsystems and operational criteria/constraints evolve, it is expected that this matrix will be updated," as it inevitably was… several times.

Over the following months, the staff at JSC continued to review and revise plans concerning a new group of astronauts. The emphasis at this time was, logically, for Pilot astronaut candidates, as mastering flying and landing the vehicle was prioritized in the objectives of the Approach and Landing Test (ALT) and OFT programs. As with Apollo, it seemed that science would have to wait until the

Table 2.2: SPACE SHUTTLE OPERATIONS CREW FUNCTION MATRIX

Flight personnel	Shuttle Flight Operations											Manipulator Operations				Rendezvous/ docking operations				
	Flight dynamics - aero	Flight dynamics - space	Rendezvous/station keeping	Guidance, navigation, control	Shuttle system management	Safety of flight monitor/control	Emergency procedures	Flight procedures	Mission flight plan	Shuttle consumables management	Network communications	Payload bay doors	Payload restraints	Manipulator controls	CCTV/direct viewing	Orbiter attitude/translation control	Docking target/COAS	docking systems controls	Direct viewing	Monitor/control rendezvous sensors
Commander	P	P	P	P	P	P	P	P	P	P	P	B	B	B	B	P	P	P	P	P
Pilot	B	B	B	P	P	P	P	P	P	P	P	P	P	P	P	B	B	B	B	B
Mission specialist	S	S	S	S	S	S	P	F	S	F	S	S	B	S	S	F	-	-	S	S
Payload specialist	-	-	-	-	-	S	P	F	S	F	S	-	F	-	F	-	-	-	F	-

(continued)

Table 2.2: (CONTINUED)

Flight personnel	Orbiter/payload interface				Payload operations								Habitability			
	Interface support & consumables management	Safety of flight c&w/control	Communications management	Subsystem maintenance/repair	Payload operations & monitoring	Experiment annunciation, monitoring & control	Experiment senor maintenance/repair	Payload data management	Activity scheduling	Autonomous payload consumables management	EVA/IVA	Vehicle attitude control	Food preparation and cleanup	Equipment stowage/unstowage	Waste management	Sleep and exercise
Commander	S	P	S	F	F	S	F	F	S	F	S	P	P	P	P	P
Pilot	B	P	B	F	S	S	F	F	S	S	P	P	P	P	P	P
Mission specialist	P	P	P	P	B	B	B	B	P	P	P	B	P	P	P	P
Payload specialist	S	S	S	B	P	P	P	P	B	P	F	F	P	P	P	P

Key:
PARTICPATION
P Prime,
B Back Up,
S Support,
F Familiarization

ORBITER UPPER FLIGHT DECK STATIONS
• Forward
- Flight operations
• Aft Area of Upper Flight Deck
- Orbiter/Payload Operations
- Payload Operations
- Rendezvous/Docking/Manipulator

REF: NASA-S-73-3208; -3209 & -3210 dated June 11, 1973

test program was completed and the orbital vehicle was certified to fly. As test pilots had the most appropriate skills for that task, the priority was to define new Pilot astronaut selection requirements, focusing more on operational experience rather than test piloting skills and on devising a suitable training program to address that change and requirements.

By 1974, the Shuttle design had been decided upon. Gone were the huge fly-back booster and the liquid-fueled rocket boosters that replaced it. Also gone were hopes of using the Shuttle to construct and supply a large, 50-crew space station. Instead, more modest plans were being designed for the Shuttle; to deploy and repair satellites, dispatch space probes across the solar system, conduct short science missions in a pressurized laboratory in the cavernous payload bay, support the interests of national security in a series of secret missions for the U.S. Department of Defense (DOD), and in using the transportation system, its components and resources, to promote the benefits of developing new commercial ventures in space. To attain these bold, far-reaching plans, the 'Shuttle' vehicle now consisted of the reusable delta-winged Orbiter, launched by means of a (one-flight) External Tank (ET) fueling three exchangeable main engines on the Orbiter, and assisted by twin recoverable Solid Rocket Boosters (SRB). This design was a far cry from the originally envisaged grand plan of the Space Task Group just five years previously, but would still require a new class of astronauts to fly and operate it.

A status report update
In the spring of 1974, the Committee on Science and Astronautics released a new status report, featuring briefings from various NASA field centers and primary Shuttle contractors on the state of Apollo-Soyuz, the planned (and subsequently abandoned) Space Tug and the Space Shuttle. With the final Skylab mission recently recovered after a record 84 days in space, progress toward the joint flight with the Soviets underlined that this would be the final time Americans would venture into orbit until the advent of the Shuttle. Progress in the Shuttle program was presented in the report and a lot had changed in two years, most notably the reduction in the overall weight of the Orbiter. The former 170,000 lbs. (77,112 kg) dry weight was now revised to 150,000 lbs. (68,040 kg) thanks to a 40-point weight reduction program. This included taking 24 inches (60.9 cm) out of the crew compartment, and moving the EVA airlock into the crew compartment as the central payload bay area was reduced by 14 inches (36 cm). Just one RMS (the Shuttle's Canadian-built Remote Manipulator System robotic arm) was planned instead of the original two, now moved to the port longeron (left side) but with provision to add the second on the opposite longeron if required (which it never was). These seemingly small adjustments affected the internal volume inside the crew module and the facilities provided for the crew.

Aaron Cohen, NASA's Orbiter Project Office Manager, explained to the Committee that "the Orbiter cabin arrangement is well defined. In fact… we've got a group of people out at Rockwell to freeze the design on the Orbiter display and flight control for the Pilot station and the Payload Specialist station." In the accompanying presentation diagram (dated 1973), the four seats on the flight deck illustrated the aft starboard MS workstation, the central docking station and the payload monitoring station. There were provisions for up to six seats on the mid-deck, to accommodate the maximum of ten crewmembers (only for emergency/space rescue modes). Interestingly, in a later briefing, the labeling of a similar diagram identified the CDR seat in the front left (port) position, the PLT at front starboard (right), an MS seat at aft starboard (behind the PLT) and a 'Payload Monitor' in the central seat behind and between the CDR and PLT (which, from STS-5 in 1982 through to the end of the program in 2011, would be assigned to the MS-2/FE).

The overall mass of a payload is always a major consideration in trying to launch anything into space, or indeed return it to the ground, with weight reduction programs featuring in the development of every spacecraft over the years and continuing to this day. Payload mass therefore has a direct influence on the crew complement flown. A standard Shuttle crew of four, together with their equipment and provisions for a flight of seven days, on average, was a significant contribution to this mass. The addition of more crewmembers affected this calculation and had knock-on effects across other areas, most notably environmental control and life support. Consumables were baselined at 28 man-days total (4 crew x 7 days) but the storage areas were sized for 48 man-days, with allowance for a contingency mission extension capability built in while still remaining within the total launch mass capability and the vehicle's Center of Gravity (CofG) calculations for ascent and landing. As early as 1974, there were studies into longer missions. If a 30-day mission was flown, this would add a significant amount to the overall weight of consumables and the volume required in the payload bay for extra cryo-tank kits (which eventually became the Extended Duration Orbiter – EDO – package for missions of up to 17 days, flown for the first time in 1992). The desire was to fly an average of seven persons for seven days on every mission to maximize the return, but the balance between achieving this and the limits of available power, consumables and capabilities was a factor in assessing payload manifests and assigning the suitable crew complement to support them. In addition to the difficulties of launching, flying and returning a mission into space, it was also expensive in terms of budget and manpower, so no mission would be flown for longer than necessary. [18]

On June 5, 1974, Dr. John E, Naugle, NASA's Deputy Associate Administrator, confirmed in a press briefing on the status of the transportation system in Washington D.C., that the construction of the first Orbiter was underway, the ALT remained on schedule for 1977 and that the first manned OFT was being planned for 1979. At the same briefing, Elwood W. Land Jr., of the Space Shuttle

Systems Office, stated that a core of about 25 active astronauts could be maintained (in training) to fly a proposed rate of *one orbital mission each week* [author italics].

The view from JSC
In September 1974, MSC Director Christopher C. Kraft wrote a letter to John F. Yardley, NASA Associate Administrator for Manned Spaceflight, stating that the feeling at JSC was that there were an adequate number of pilots available already in the Astronaut Office, based on current estimates (such as those mentioned by Naugle), who could cover prime and back-up crew assignments in the ALT and OFT programs, and probably for the first few operational missions. He added that the Center would *not* be actively seeking replacement pilots before 1982.

Experience from past selections in the 1960s had shown to the management at JSC that it took a *minimum* of 20 months between the first advert appearing in the press calling for new astronauts and one of the successful applicants actually flying in space. This suggested that the call for new pilot astronauts would not go out much before 1980, when there would be an estimated 20 positions open for new candidates. This figure was derived from estimated natural attrition rates of about two pilots each year and the (rather optimistic) expectation that by then *each* pilot would be flying no more than *six missions a year.* It was also assumed that back-up assignments would no longer be required following the OFT program, as there would be a sufficient pool of pilots available to step into a vacant crew position should the need arise. There was also the fact, largely overlooked in published accounts at the time, that the training system and resources would not be able to cope with duplicating the number of crew expected to be in training. The primary crew would therefore be priority, with a pool of suitability prepared crewmembers following generic, support and proficiency training until called upon to join a flight crew.

For the first Shuttle missions, Kraft believed that the unique stability and control demands on a Shuttle flight deck crew during launch, entry and the various abort profiles would require pilots with considerable flight test experience. As the program evolved and flight experience and maturity in the systems was demonstrated, then the strict piloting requirements could be relaxed for new astronaut selections, he thought, without impacting safety or heightening the risks. Kraft also expressed his desire to include the current Scientist-Astronauts (chosen in 1965 and 1967) as crewmembers on the early flights of the Shuttle, which was logical given their decade of involvement in the program and the range of developmental work many of them had accomplished. The four remaining members of the 1965 selection had flown a single mission each, while all those still active from the 1967 class were still awaiting their first mission, though all had passed a USAF jet pilot course prior to acceptance into the team and had maintained their proficiencies in the years since. Despite this, Kraft informed Yardley that studies at JSC assumed that the Scientist-Astronauts would be assigned as MS on the Shuttle, rather than as pilots.

This is a contentious issue. As qualified jet pilots with (at the time) nearly a decade of experience in flying jets with NASA, it could be argued that their assignment as a Shuttle PLT would be an option, perhaps on a dedicated science mission, but it was not to be. Despite the fact that Scientist-Astronaut Story Musgrave would eventually log more flight time in the T-38 than any of his pilot colleagues, not one MS (Scientist-Astronaut or otherwise) flew as PLT or CDR on a Shuttle mission. By contrast, two of the 1978 pilot astronauts (Griggs as MS-1 on STS-51D and Nagel as MS-2 on STS-51G) flew as MS on their first space flights. Unfortunately, Griggs was killed before he could make a second space flight, but Nagel progressed to the PLT and CDR seats on his next two missions. Whatever the reasons – and astro-politics certainly played a part – pilots retained the command of the Shuttle flight deck until the end of the Space Shuttle program in 2011. Some MS were eventually assigned as commanders from the 1990s, but for the payload not the vehicle. Subsequently on the International Space Station (ISS), non-pilot astronauts would take command of long expeditions on that facility, though not of the vehicles which carried them to or from the station.

In 1974, Chris Kraft at JSC was very clear about the future direction that NASA should take in selecting new pilot astronauts. They needed to be highly motivated and personally dedicated individuals, while the missions offered should be challenging but safe. It was important to keep the time gap between selection and flight relatively short, and if NASA did this then retaining pilots would not be a problem. Kraft informed Yardley that he would review and update this assessment annually. In the meantime, a definite plan for future crewing would be created by March 1975 that included a detailed review of the currently-proposed role of both the PLT and the MS on a Shuttle crew, as well as the criteria to be used in selecting these personnel, the training methods and requirements to prepare them for Space Shuttle flights, and the staff levels required to support the first three years of (orbital) flight operations.

This was the first detailed planning to include the MS. In reply, Deputy Administrator George Low expressed his concern that the emphasis had been mainly on the pilots, stating "I think an even more important subject is the acquisition of scientists and others who will fly the Shuttle, *but not as pilots* [author italics]."

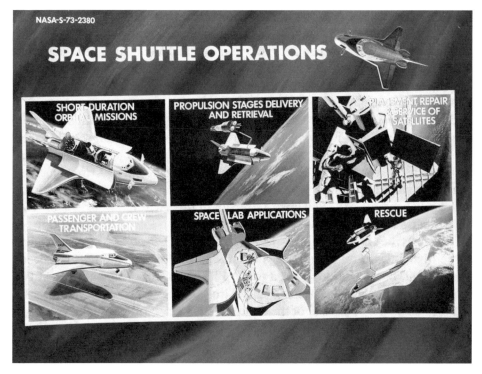

Fig. 2.10: A collection of artist's impressions highlighting the variety and scope of missions the Shuttle program was planning to fulfill.

Scientist-Astronaut's role

On November 18, 1974, NASA asked the Space Program Advisory Council (SPAC) to study the role that Scientist-Astronauts could fulfill within the Shuttle program. As a number of them were already heavily involved in the development of Spacelab-type missions, experiments and procedures, the advice from the Council only underlined the importance of their early involvement. It was highly recommended that the current Scientist-Astronauts could be a valuable asset in the development and execution of experiment and payload integration issues, for Spacelab-type missions, between the Principle Investigators (PI) and the Astronaut Office. It was also suggested that they could serve as the primary experiment operator for dedicated Spacelab missions. [19] At this time, the on-going debate remained about what to term the non-pilot crewmembers on a Shuttle mission. From the ground-based Spacelab simulations, the term "Experiment Operator" was coined, which evolved over the next few years into the more defined roles fulfilled by the MS and PS.

In a response to the SPAC findings, Louis C. Haughney, Geophysics Program Manager of the Airborne Science Office at NASA Ames, suggested drawing upon the broad experience of the Scientist-Astronauts, feeling that it would be worthwhile considering them as managers for the Spacelab flights and assigning them early in the planning process[6]. Then, when the flight was manifested, the Scientist-Astronaut could become a full time manager of that mission, working in close cooperation with the other MS assigned to the flight crew. During the mission, the 'senior' Scientist-Astronaut could serve as an on-orbit flight director responsible for all Spacelab systems relating to the experiment package, working with a back-up Scientist-Astronaut assigned as Assistant Mission Manager who would coordinate ground activities. Though this was not exactly the way things turned out, it did create the concept of a science crew working the payload and scientific objectives and an Orbiter crew handling the Space Shuttle systems. During the 1970s, the designation of Senior Scientist-Astronaut was commonly used until they merged into the collective MS designation and role. A decade and a half later, in January 1990, these ideas evolved into the role of Payload Commander (PC), with members of the Thirty-Five New Guys (TFNG) becoming the first assigned to this role. In 2002, NASA introduced the designation of NASA Science Officer onboard the ISS, to coordinate research primarily on the U.S. segment. **[20]**

The future crewing plan
By March 1975 this plan was in place, but with recommendations to expand the next selection to encompass applications from a wider field of experience than those first considered. This included JSC employees with initiative and ambition, even if they had no flying skills or had not scientifically qualified through the National Academy of Science (NAS). The plan supported the idea that there was no immediate urgency in selecting a new group, but reasoned that within the next ten years many of the current astronauts in the office would have turned 50 and would probably be looking to retire.

It is clear that the process to define the criteria needed for a new group was painfully slow, but it had to be. From the start, deciding to broaden the criteria to enable a much wider segment of the U.S. population to apply meant that those selected would require an extensive – and thereby longer – training

---

[6]An in-depth background to the participation during the 1970s of the Scientist-Astronauts in the simulated ground and airborne Spacelab development 'missions' is covered by the authors in their 2007 title **NASA's Scientist-Astronauts** [pp. 283–331]. While connected to the development of Shuttle MS, this falls outside the scope of this current work.

program based at JSC Houston in order to attain the suitable qualifications required to fly. It was hoped that this new, positive, flexible approach to the selection criteria would encourage minority and female candidates to apply, though there was no guarantee of this. As authors Atkinson and Shafritz noted in their 1985 book: "This concept was compatible with the idea of permitting current NASA personnel who were not astronauts to enter the program," and that, "some of those selected might never reach flight readiness, but the training, evaluation and counseling could minimize this risk, unless a candidate simply lost interest or decided that an alternative career would be more personally beneficial." **[21]**

One of the most debated items in establishing the selection criteria was the minimum educational qualifications for the MS; that of a doctorate or equivalent in engineering, life or physical sciences or mathematics. This was seen not as a barrier to candidates, but as a way of refining the number of applicants to make the whole system more manageable, as well as the fact that such applicants would have several years of academic and practical (teaching) experience behind them. Chris Kraft thought that not requesting an advanced degree would be a mistake. JSC desired a doctorate, while Headquarters suggested a minimum of a BSc degree. Kraft wrote on March 12, 1976 that if an advanced degree was not stipulated, then "In my opinion… the number of people that will apply will be almost impossible to deal with."

The original selection plan
It was quickly realized that the selection would be an intensive program on several levels over a considerable period of time, but if necessary could begin as early as June 1975. After allowing three months to file applications, the screening of those applications could begin in the fall of 1975. After a series of medicals, further interviews, and security and background checks, the announcement of a new group of astronauts could be made by the first or second quarter of 1976. Expecting an intake of 20–25 pilot and MS candidates, it was envisaged at the time that the candidates would arrive at JSC in the summer of 1976 and then begin a program designed to enhance their qualifications over a period of *three to six years*. In the period after training, the Ascans could either be reassigned to different NASA field centers to gain an insight into other branches of NASA, be allowed study for higher academic qualifications, or work in different branches at JSC, such as Mission Control or Aircraft Operations. Then, depending on attrition rates among the veteran astronaut and the individual progress of each Ascan, the candidates would be phased back into the program as fully qualified (but unflown) astronauts. NASA also revealed at this time that the intention was to continue subsequent astronaut selection programs at regular intervals.

Fig. 2.11: A cooperative venture with the Europeans resulted in Spacelab, and the opportunity to fly invited Payload Specialists and passenger/observers into orbit alongside the NASA career astronauts.

Reality hits home

This was all fine on paper and in planning documents, but difficulties in developing the Shuttle pushed the first flight back several times. As a result, JSC was not that enthusiastic about bringing in another group of scientists to train as astronauts while many of those selected a decade earlier were still waiting for their first flights. In addition to the very evident technical and hardware issues facing the program, NASA also had to answer public feedback from minority groups and Congress. But it took time. The plan to select the pilots took two years and the plan for the MS took another year, to the end of 1975. On top of this, the specific provisions for female and minority applicants were not finalized until early 1976.

Finally, in March of that year, a definitive plan emerged to recruit a new group of astronauts and the Group 8 Astronaut Selection Board was created. It consisted of senior management, scientific and pilot personnel from JSC, together with Dr. James H. Trainor from Goddard Space Flight Center as a scientist representing the scientific community. The board held their first meeting on March 24, 1976.

In the spring of 1976, it was felt that the Astronaut Office and the Science and Applications Branch should participate in the forthcoming ASSESS II airborne mission to define the function and role of the MS. **[22]**. Maintaining that the MS should be "professionally knowledgeable in the prime discipline of the mission to which he was assigned," the Science and Application Branch saw the role as "in-flight integration of experimental objectives and overall for the successful completion of the experiment mission objectives and to reduce training loads."

On March 3, 1976, the *Chicago Tribune* reported a *Baltimore Sun* article that NASA was formally about to announce the acceptance of women into the astronaut corps that July. Though the 'exact' number of new astronauts was not disclosed, the prediction was fairly accurate in stating that women would "form part of a group of 15 Mission Specialists." A program of screening, evaluation and physical examination would be completed within the next year, and the candidates chosen by the end of 1977, to start training in July 1978 and expected to be operational by 1980. **[23]**

Following a three-day meeting between NASA and ESA officials two months prior to issuing the call for Space Shuttle astronaut candidates, John F. Yardley, Assistant Administrator for Manned Spaceflight, informed the press at a Paris news conference on May 15, 1976, that current Shuttle planning envisaged 200 missions in the project, some of which would take three NASA and four European crewmembers on space science missions lasting between a week and a month. He reported that in addition to the 30 astronauts available for the program, the agency would soon select 30 more, including women. Strangely, he also suggested that participants might be selected from the USSR, although no such offer had been indicated or planned for. The schedule at the time envisaged 600 Shuttle missions over a 12-year period from 1980, with a fleet of five Orbiters flying 60 missions each year. A three-person NASA crew would consist of CDR, PLT and an MS, together with up to four PS. The PS could expect to fly one or two missions, but NASA career astronauts could be expected to fly up to 20 or 30 times in their career. **[24]** Two months later, on July 8, NASA issued its call for 15 pilots and 15 MS, to be selected by the end of 1977.

On May 20, 1976, Eugene Kranz, the Deputy Director of Flight Crew Operations, wrote a memo addressed to a number of JSC directorates concerning a recent ASSESS and Shuttle planning meeting held at NASA HQ in Washington D.C. Kranz voiced his frustration at the deep-rooted disputes between field centers regarding science and engineering objectives on the Shuttle program, which had carried over from the Apollo program. "Again, history is repeating itself," he wrote, especially referring to the question of science objectives on the engineering test flights under the OFT program, intended to establish the data that proved the Shuttle system worked as designed *before* committing major scientific payloads or programs to the flights. Kranz was at the receiving end of some flak for this approach in stating that science working groups had "never satisfied the need to be involved" in the preparation of what he termed "a real mission." It was becoming clear, at least for the Astronaut Office, that crew assignments for the first missions under the OFT program would be filled by veteran pilots from the earlier groups. Noting that the exact role of the MS was still in debate, Kranz stated that "the Mission Specialist job description and functions are not recognized by some branches of NASA, and even some of the personnel at the Office of Space Flight are unclear as to the intent in this area." He also suggested that more work was required in this area across the agency, which was interesting as NASA began advertising for its first group of Shuttle astronauts to train as MS just a couple of months later.

Fig. 2.12: Satellite servicing and maintenance was a major factor in promoting the benefits of the Space Shuttle system. Together with system reusability, a balance of national defense missions, exclusive scientific research and developing commercial opportunities were all highlighted as program priorities.

But how many?

Funding was one issue that required consideration before requesting new astronauts, both in terms of training costs and whether the funds and facilities were in place to support the process, despite the fact that the Shuttle was still some years from launch. Balancing the administrative plans and budgets against the reality was a fine line and a delicate issue. There were many questions to be answered: How many astronauts would be required? What would be the most productive balance between current astronauts and new candidates? When would the systems and facilities be in place to support the training? And for how long could veteran and new astronauts be expected to remain active? This last point would be an important factor when timing the new Ascan selection and training programs to replace departing astronauts without affecting the crewing roster or flight schedule.

One question which had to be addressed before deciding on the number of new astronauts to select was to determine how many of the veteran astronauts had chosen to remain to support the early flights. In addition, there were four astronauts not directly working in the Astronaut Office but on special assignment elsewhere, who had been assured that they could return to active flight training should they desire it and advised that training would commence for the OFT in 1977. These four, all detailed to Washington D.C., were Joseph P. Allen (unflown Group 6 Scientist-Astronaut), who was Assistant Administrator for Legislative Affairs at NASA Headquarters; William A. Anders (Group 3 pilot astronaut; LMP Apollo 8), currently serving as Chairman, Nuclear Regulatory Commission; Russell L. Schweickart (Group 3 pilot astronaut; LMP Apollo 9), serving as Director, User Affairs, Office of Applications, NASA HQ; and John L. Swigert (Group 5 pilot astronaut; CMP Apollo 13), who was on the Committee on Science and Astronauts, U.S. House of Representatives. In June 1976, James Fletcher wrote to each of them advising that a new recruitment drive would be announced the following month, and that he needed to know if they planned to return to NASA and the Astronaut Office. Each was eligible for assignment on OFT, but needed to advise the administration by July 1, 1977 as to whether they would be returning to the Astronaut Office in January 1978. These four did not have a significant impact on the selection of the 1978 group, with only Joe Allen electing to return and the other three formally withdrawing from the astronaut program in the interim, as did former Skylab 4 Commander Jerry Carr and Apollo 17 CMP Ron Evans.

Following a paper trail
There are scores of documents, memos, charts and plans filed in the STS archive boxes at the JSC History Office, and we have sampled just a few to illustrate how crew definitions, roles and requirements evolved many times during the years leading up to the first Shuttle flights. Even some of those produced as late as 1980 forecast that *when* shuttle flights exceeded 20 missions per year (which it would never come close to in its 30-year history), the core 'three-person crew' teams (CDR, PLT, MS) would be recycled intact within a year to make the most of the training process. If this plan had come to fruition, the manifests of 22 missions annually would have required at least 11 four-person crews, or 44 individual astronauts, in training *each year*. Clearly, had this occurred, it would have been a massive undertaking that would have been challenging to maintain and would have

placed enormous strain on resources and capabilities. Reading these documents with the benefit of hindsight, their necessity for minimum and maximum long-term planning is obvious, but with the real-time difficulties encountered during the 30-year program, even overlooking the two tragic accidents with *Challenger* and *Columbia*, it is difficult to see how the program could ever have sustained this rate, even in the best-case scenarios without any setbacks, failures or delays. These plans are useful for reference, in understanding how things changed as new documents superseded the previous versions, and how different the actual program became during its lifetime.

The 1980 planning forecasts would later be used to justify the selection of a further (ninth) astronaut group later in 1980, and annually after that. This was well before any of the TFNG had been assigned to their first mission or ventured anywhere near a Shuttle on the launch pad.

PIONEERING SPACE SHUTTLE ASTRONAUT ASSIGNMENTS (1969–1980s)

Before we review the backgrounds of the 1978 selection and discuss their early assignments in more detail, we summarize here the support assignments held by 'veteran' astronauts of previous selections who, over more than a decade, pioneered and refined the training and crew duties assigned to the new astronauts once their Ascan program had been completed. The TFNG may have been the pioneers of the Ascan training program and fulfilled early ground support roles for the first Shuttle mission, but their path was made much easier by the dedicated work of some of NASA's 'original' astronauts, many of whom would never get the opportunity to fly on the vehicle they had worked so hard to make a reality.

To remain or to depart?
In the spring of 1972, the Space Shuttle had become a realistic proposition. Apollo had landed men on the Moon five times and was about to end the program with its sixth landing mission. Many of the 'veteran' astronauts who had flown the earlier missions on Mercury, Gemini and Apollo to attain this success had long since retired, or were planning to do so very soon. The trio of three-man Skylab crews had been announced in January of that year and were in training, and with just a single, joint docking mission with the Soviets in 1975 on the manifest, but not yet crewed, the short-term flight opportunities were zero.

As the search for new astronauts began a year after ASTP, the hope was still to orbit the Shuttle by 1978 or 1979 at the latest, but difficulties in developing the new technology and hardware, especially the three main engines, pushed this into the early 1980s. By then, most of the astronauts who had been selected during the first decade of NASA had long gone, frustrated by such a long wait for a flight seat, unwilling to endure months of training and simulations, or expressing a desire to pursue new challenges. The size of the NASA astronaut corps was significantly diminished. For some, after flying to the Moon, a seat on the Shuttle could never replace the magic of that lunar trip. For others, any space flight was a *good* one as long as things went well, and so they decided to stick it out and wait for a seat on the Shuttle. However, the limited availability of astronauts to crew the expected increase in missions after the test program, and the ever-aging original population of astronauts, meant that NASA needed to instigate a wider search to bring in a new generation of astronauts.

Over a decade had passed since the last selection of astronauts by NASA in 1967. It had been nearly two decades since the first group of seven had been chosen, and a decade since seven transferred from the cancelled USAF Manned Orbiting Laboratory (MOL). When the Group 8 candidates arrived at JSC in January 1978, there were very few of the original 73 astronauts left from the first seven groups chosen between April 1959 and August 1969. It was also a very different NASA. The work may not have made the headlines, but since 1969 a cadre of astronauts had fulfilled a variety of important, often mundane roles in the early development of the Shuttle vehicle, systems and procedures, helping to define the role of 'the astronaut' in the program. These included: evaluating the early designs of the controls, displays and equipment they intended to use during a mission; the roles they would employ in flying the vehicle; deploying and retrieving satellites; operating scientific equipment; performing EVA; and servicing satellites. Their work was significant in defining the criteria that would prepare Shuttle pilots or MS, or what should be expected from a PS or Observer. Often overlooked, these assignments were far reaching and are an integral part of the story of Group 8 and beyond, thanks to their input and participation in Shuttle support roles years before the Shuttle became a reality. It is a similar story today, as current astronauts participate in simulations of lunar and Martian missions and crew activities, train in extreme environments and offer their experience to mission development and for future flights they will never participate in.

Some of the veterans from the early selections, who remained at JSC long enough to see the creation of a Shuttle branch of the Astronaut Office and the start of vehicle development and testing, occupied managerial rather than flight positions. From the original Group 1 selection (1959 Mercury astronauts), **Donald K. 'Deke' Slayton** flew on the 1975 ASTP docking mission after having been medically grounded for 16 years, then went on to serve as manager of the ALT and OFT programs between 1975 and 1982. Only one member of the second selection in 1962 remained in the Office. Gemini and Apollo veteran **John W. Young** had assumed the role of Chief Astronaut from Alan Shepard after America's first astronaut had retired in 1974. Young would remain in that position for the next 13 years, taking breaks only to train for his two Shuttle missions, before assuming an administrative role until leaving NASA in 2004. Two astronauts from the third selection in 1963 remained at NASA during the late 1970s. Apollo and Skylab commander **Alan L. Bean** had focused on Shuttle training issues since 1975 and subsequently served as Chief Astronaut in Young's absence while he flew STS-1. Bean was the Group 8 and 9 training supervisor prior to retiring from the program in 1981 to pursue a career in painting. Gemini and Apollo veteran **David R. Scott** had left the Astronaut Office in 1972 but was subsequently appointed Director of Dryden Flight Research Center in California during the ALT program, finally resigning from NASA shortly after that series of flights had been completed in October 1977.

From the late 1960s, many of the active astronauts who were between crew assignments, or at the end of their astronaut careers, were assigned to the Space Shuttle Program Office. On October 17, 1969, *MSC Roundup* reported that Mercury and Gemini astronaut **L. Gordon Cooper** had been named as Assistant for the Shuttle Program Flight Crew Operations Directorate (FCOD) at MSC. In this position, he was responsible for the flight crew training program, astronaut input into the design and engineering, and was the directorate representative in hardware development and tests of the Space Shuttle, though it is doubtful if his heart was really in his new assignment. Officially still active, he lost the command of an Apollo lunar mission to fellow Mercury astronaut Alan Shepard and elected to leave the agency in July 1970. Gemini and Apollo astronaut **Edwin E. 'Buzz' Aldrin** worked on defining the early designs of the Shuttle as part of "theoretical committees for a manned booster," an idea he rejected. Another Gemini and Apollo veteran, **Richard F. Gordon**, who had lost his own Apollo lunar mission due to budget cuts, became Chief of Advanced Programs in August 1971, working on the design and development of the Shuttle until he left the agency in January the following year. In November 1976, Apollo 9 astronaut **Russell L. Schweickart** worked as Assistant for Payload Operations in the Office of Planning and Program Integration, Space Shuttle Office at JSC, where he helped define polices for Shuttle payload operations until he decided to leave the agency in the summer of 1977.

Several astronauts of the fifth (1966) pilot selection worked on a variety of early Shuttle development issues in one form or another. Between November 1975 and March 1978, former ASTP crewmember **Vance D. Brand** worked on mission planning and studied the entry phase of the Orbiter for OFT, for which he was assigned in 1978. Skylab 4 commander **Gerald P. Carr** was Head of the Shuttle Design Support Group at JSC from May 1974 until leaving NASA in 1977. He worked on payload support (mission phase), crew station hardware, RMS development and evaluation of emergency egress procedures. In 1973, Apollo 16 LMP **Charles M. Duke** was assigned as a Technical Assistant to the Manager for Shuttle Orbiter Integration, and later as Deputy Manager for Advance Planning STS. As he explained in his 1990 autobiography *Moonwalker,* these were mostly paperwork assignments combined with a lot of meetings. Former X-15 pilot **Joe H. Engle** was assigned to the Shuttle Program Office in 1971, shortly after losing the Apollo 17 LMP position to geologist Jack Schmitt. He became involved with mission phase control and display hardware, and served as a liaison for the Shuttle Training Aircraft (STA). He also worked hard on the preparations for test flying the Orbiter in ALT, for which he was selected in 1976. Apollo 17 astronaut **Ronald E. Evans** joined the Astronaut Office Shuttle Development Branch in July 1975 and became responsible for operational aspects of the OFT Ascent Phase. Apollo 13 LMP **Fred W. Haise** joined the Shuttle Development Branch in April 1973 as Technical Assistant to the Manager of the Shuttle Orbiter Project, working towards a test flight assignment on ALT in 1976. **Don L. Lind** was busy with Shuttle development assignments during 1974 and 1975. After a one-year sabbatical, he returned to JSC in the fall of 1976 and was assigned to the Operational Mission Development Group responsible for the development of payloads for OFT and early operational missions.

Skylab 3 astronaut **Jack R. Lousma** worked on Shuttle Development from July 1975 until 1978, assigned to cockpit layout issues in the design support team prior to being named to the OFT training group. Apollo 16 CMP **Thomas Kenneth 'Ken' Mattingly II** joined the Shuttle project in 1973 and was assigned a variety of roles, including Head of the Operation and Engineering Support Group until 1978. He then became the Technical Assistant to the Manager of OFT (Deke Slayton) and also worked on Extravehicular Mobility Unit (EMU) issues for several years. **Bruce McCandless** was instrumental in the development of the Shuttle Manned Maneuvering Unit (MMU), following on from his work on the Skylab demonstration astronaut maneuvering unit in 1973. He also worked on developing EVA equipment and procedures, the prelaunch and ascent phases, in developing crew input for the Inertial Upper Stage (IUS), and in evaluating servicing methods and procedures for the Large (later Hubble) Space Telescope. Skylab 4 pilot **William R. Pogue** worked on Shuttle development from June 1974, on the launch and abort mission phase and the pilots' handbook, until he left the agency in 1975, returning for short time in 1977. One of his lasting commitments, before leaving again in 1978, was to argue in favor of keeping the new Ascans in the Astronaut Office rather than, as he wrote in 2011, having them "farmed out" to different NASA field centers to gain experience while awaiting their first crew assignments. Apollo 14 CMP **Stuart A. Roosa** worked on crew training issues from January 1973 until leaving the agency in 1976. Apollo 13 CMP **Jack L. Swigert** had an early Shuttle assignment in 1971, using the Lockheed simulator for "flying and landing" a delta-shaped lifting-body Shuttle configuration, along with Group 7 astronauts Gordon Fullerton, Hank Hartsfield, and Don Peterson, before leaving the Astronaut Office in 1973 and NASA altogether four years later. Apollo 15 CMP **Alfred M. Worden** was also briefly assigned to Shuttle development issues at JSC from May 1972 until he transferred to NASA Ames in California in August that year, where he put his test pilot and astronaut experiences to good use including on various Space Shuttle vehicle simulations. Finally from this group, Skylab 2 pilot **Paul J. Weitz** worked on crew station design hardware and Earth resources issues, as well as completing underwater zero-g simulations of contingency EVAs related to forthcoming Spacelab missions. He also became technical coordinator, Shuttle Support Office and remained active long enough to command an early Shuttle mission.

Of the remaining active members of the 1965 Scientist-Astronaut selection, Skylab 4 science pilot **Edward G. Gibson** had originally left NASA after his Skylab mission in 1974 and worked as consultant for ERNO West Germany on Spacelab development between 1976 and 1977. He returned to NASA as Chief of MS Selection and Trainer for the 1978 Group, and served as a member of the 1978 selection board. He was assigned as the ascent Capcom for STS-1 and in line for an early flight as an MS, but resigned for a second time in 1980. Skylab 2 science pilot **Dr. Joseph P. Kerwin** worked on Shuttle crew station design, controls and medical monitoring from 1974. He took part in the selection of the new MS in 1977 and as their first training supervisor in 1978. He then worked on planning operational missions, including the role of the astronaut in rendezvous, satellite deployment and retrieval, and RMS operations.

Several members of the second Scientist-Astronaut selection chosen in 1967 were still at JSC a decade after their selection, all unflown. Many fulfilled a number of key roles in Shuttle development while awaiting their first space flight on the Shuttle. **Joseph P. Allen** was nominated as a candidate for airborne simulation of Spacelab missions (ASSESS) and worked on Spacelab development, and payload support crew station controls and displays for the physical sciences. **Anthony W. England** returned to the Astronaut Office in 1979 after a seven-year absence and was assigned to the Operations Mission Development Group, working in hardware and experiments for the Spacelab 2 mission. He also worked on developing the Shuttle computers and software. **Karl G. Henize** was first assigned to Shuttle development issues in 1974. His assignments included participation in ASSESS simulations, Shuttle-borne astronomy payloads, payload support and handling, and assisting in the development of crew station controls and displays for physical sciences. His work on the Spacelab 2 payload had begun in 1977. **William B. Lenoir** worked in Shuttle development from 1974 on payload crew station controls and displays for the physical sciences, payload deployment and retrieval systems and procedures, development of the Extravehicular Mobility Unit (EMU) and Portable Life Support System (PLSS), and supporting the OFT series. He also worked on Powersat studies, man's role in the remote sensing of the Earth, and trained on using the RMS. **F. Story Musgrave** also began Shuttle development work in 1974, in payload support crew stations controls and displays for the life sciences. He worked on the design and development of all Shuttle EVA-related equipment, including the PLSS airlock suits, toolkit and MMU. He participated in ground simulations of Spacelab missions, and early design and development work on the Shuttle Avionics Integration Laboratory (SAIL). **Robert A. Parker** became involved in early designs of Spacelab from 1973 and worked in the Payload

Operations Working Group, which involved meetings and negotiations between representatives of NASA and ESA. He also worked on payload support and crew sciences, served as chief Scientist-Astronaut, participated in an ASSESS flight and was a member of the MS selection board during 1977. **William E. Thornton** worked on Shuttle development from 1974, mainly in payload support crew stations controls and displays for life sciences. He completed Spacelab ground simulations and, realizing that his first space flight might be several years away, applied unsuccessfully for a seat as PS for Spacelab 1. He also worked on deployable payloads, but mainly became deeply involved in monitoring crew health on orbit, developing techniques and procedures to investigate those phenomena.

Of the seven transfers from MOL in 1969, **Karol J. Bobko** was assigned to development issues on the program almost as soon as he joined NASA. From 1975, he was assigned as support crew and alternate Capcom/chase pilot for ALT. He then worked on OFT preparation and mission definition issues and as lead astronaut, test and checkout at KSC, for the first launch of *Columbia*, defining the roles which became known as the Cape Crusaders. **Robert L. Crippen** was assigned to the Shuttle branch in 1975, where he worked on OFT mission phase hardware and software integration and became a lead in the Shuttle General Purpose Computer and OFT training. From 1973, **C. Gordon Fullerton** worked on development issues concerning recovery, mission phase and control and displays, prior to his assignment to ALT in 1976. **Henry W. Hartsfield** worked on flight control and simulation hardware in the Shuttle Development Office from February 1974 to April 1981. As a member of the OFT astronaut support group, he was involved in the development of entry flight control systems and associated interfaces. **Robert Overmyer** joined the ALT team after completing his support work on ASTP in the summer of 1975. He flew the first flight of the (converted Gulfstream) STA in October 1976 and alternated as Capcom and chase pilot during the 1977 ALT program. During 1979/1980, Overmyer served as the Deputy Orbiter Manager for OV-102 *Columbia* during its preparations and processing for the STS-1 flight. **Donald H. Peterson** was assigned to Orbiter systems, including mission phase, navigation, and communications and tracking hardware, as well as for Orbital Maneuvering System (OMS) or Reaction Control System (RCS) issues, from May 1972 until December 1981. In addition, he served as a member of the OFT Group (missions), responsible for engineering support, man/machine interface and safety assessment for the OFT program. Finally, **Richard H. Truly** was assigned to ALT duties immediately after completing his support work on ASTP in 1975.

Senior astronauts who also flew the Shuttle
A total of 23 astronauts from the earlier selections flew as MS, PLT or CDR in the first decade of Shuttle missions between April 1981 and December 1990, a decade in which all of the 1978 members flew at least their first Shuttle mission and many their second or third, bridging the gap between those eras.

No astronaut from the first selection (1959) remained active to participate in the early orbital missions, although John Glenn did return in 1998, aged 77, to fly one mission (STS-98) as a PS. From the second selection (1962), John Young would command the first mission in 1981 and the first Spacelab mission in 1983. No one from the 1963 selection flew on the Shuttle, and only one (Scientist-Astronaut Owen Garriott) from the fourth selection completed one mission, the first Spacelab flight in 1983.

From the fifth selection in 1966, seven astronauts flew on the Space Shuttle: Vance Brand (3 missions); Joe Engle (2 missions); Don Lind (1 mission); Jack Lousma (1 mission); Ken Mattingly (2 missions); Bruce McCandless (2 missions); and Paul Weitz (1 mission). Seven members of the second Scientist-Astronaut selection in 1967 (Group 6) also flew Shuttle missions: Joe Allen (2 missions); Tony England (1 mission); Karl Henize (1 mission); Bill Lenoir (1 mission); Story Musgrave (6 missions); Bob Parker (2 missions); and Bill Thornton (2 missions).

After losing the opportunity to fly on the USAF MOL program, all seven who transferred to NASA in 1969 went on to fly Shuttle missions: Karol Bobko (3 missions); Bob Crippen (4 missions); Gordon Fullerton (2 missions) Hank Hartsfield (3 missions); Bob Overmyer (2 missions); Don Peterson (1 mission); and Dick Truly (2 missions).

Passing the baton

These veteran astronauts lent their experience to very early Shuttle development issues, mainly from the early 1970s, as can be seen from the brief explanations of the varied roles they fulfilled over many years prior to the vehicle even flying in the atmosphere, let alone in orbit. Though the 1978 group were the first chosen specifically to train for and fly the Space Shuttle, almost a decade of work had already been completed in establishing the roles the crew would perform, input into the design and development of crew-related systems and procedures, and in identifying where further input was required.

When the 1978 candidates participated in the pristine syllabus of the newly-created Ascan training program over just 12 months instead of the expected two years, their individual capabilities were evident and reinforced the decision to select them in the first place. Following the completion of the Ascan program, and in part to further challenge and prepare them for space flight, the whole group were given various technical roles in support of the first orbital flight of the Shuttle. But this was by no means virgin territory. While many of the astronauts from the 1959–1969 selections who had forged the path from the design boards to the launch pad were now retired, their input, achievements and dedication made the transition from Ascan to astronaut much smoother for the TFNG. They would now participate in support roles in a rapidly developing program, just months prior to both the Shuttle's first flights into orbit, and also their own.

References

1. **Astronautics and Aeronautics 1968**, NASA SP-4010, 1969, p. 51.
2. *Manned Space Flight: The Future*, George E. Mueller, *Spaceflight*, Vol. 10 No. 12, December 1968, pp. 406–413, British Interplanetary Society, London.
3. Memo [Report] from Maxime A. Faget MSC to Robert R Gilruth MSC, April 28, 1969, pp. 1–35; **A Shuttle Chronology 1964–1973**, Abstract concepts to letter of contract Volume 1, NASA JSDC 23308, December 1988.
4. *Apollo and Beyond*, George Mueller, *Spaceflight* Vol. 11 No. 9, September 1969, pp. 298–303.
5. MSC Statement of Work, Space Transportation System Program Definition (Phase B), July 28, 1969, minutes pp. III–20; **A Shuttle Chronology 1964–1973**, Abstract concepts to letter of contract Volume 1, NASA JSDC 23308, December 1988.
6. **Peter Fairley's Space Annual**, TV Times Publication, 1970, pp. 54 & 58.
7. *Space Shuttle–Skylab, Manned Space Flight in the 1970s*, Status Report for the Subcommittee on NASA Oversight of the Committee on Science and Astronautics, U.S. House of Representatives, 92nd Congress, 2nd Sessions Serial N, January 1972, pp. 111–137.
8. Reference 7, pp. 561–652.
9. **The Real Stuff, A History of NASA's Astronaut Recruitment Program**, Joseph D. Atkinson Jr & Jay M. Shafritz, Praeger, 1985, p. 134.
10. Memo from Carroll H. "Pete" Woodling, Chief, Crew Training and Simulation Division (Code CE), MSC March 7, 1972, to Warren J. North, Alan B. Shepard, Joseph P. Algranti, Dean F. Grimm and James W. Bilodeau. Located in the JSC History Office Files, Shuttle series, Section 40, Box 5, Crew Station/Payload documents; copy on file AIS Archives.
11. Adopted from a single-page document dated April 13–14, 1972, filed with the March 7, 1972 memo & February 11 matrix. Located in the JSC History Office Files, Shuttle series, Section 40, Box 5 Crew station/payload documents; copy on file AIS Archive.
12. NASA Activities, August 25, 1972, Vol. 3 No. 8, p. 162.
13. **NASA's Scientist-Astronauts,** David J. Shayler & Colin Burgess, Springer-Praxis, 2007, pp. 286–288.
14. **The Last of NASA's Original Pilot Astronauts**, David J. Shayler & Colin Burgess, Springer-Praxis, 2017, pp. 331–332.
15. Reference 9, pp. 138–139.
16. **Aeronautics and Astronautics 1973**, NASA SP 4018, 1975, p. 296; & *"Physiological Response of Women to Simulated Weightlessness,"* NASA SP-403 (1979).
17. Memo, dated November 7, 1973, from Warren J. North, Assistant Director for Space Shuttle [undisclosed] distribution list. Located in the JSC History Office Files, Shuttle series, Section 40, Box 5 Crew station/Payload Documents, Copy on file AIS Archives.
18. *Space Shuttle Space Tug, Apollo-Soyuz Test Project – 1974*, Status Report for the Committee on Science and Astronautics, U.S. House of Representatives, 93rd Congress, 2nd Session, Serial K, February 1974, pp. 385–386, 405, 563, & 626.
19. Memo from Homer E. Newell, Chairman SPAC Ad Hoc Subcommittee on Scientist-Astronauts, Washington D.C., dated November 18, 1974, Robert A. Parker Files, NASA History Collection, JSC, ASSESS Box 70, copy on file AIS archives.
20. Reference 13, pp. 296–298.
21. Reference 9, pp. 141–142.
22. Reference 13, pp. 299–303.
23. **Aeronautics and Astronautics 1976**, NASA SP 4021, 1984, p. 46.
24. Reference 23, p. 95.

3

The new Pilot astronauts

*"My most compelling reasons for wanting
to become an astronaut are a desire to
extend and use the engineering and test pilot
experience I have gained, to hopefully aid
in the success of the space program"*
Francis Richard "Dick" Scobee, comments
written in his NASA application to become an astronaut.
From *Silver Linings,* by June Scobee Rodgers.

By late 1977, out of a total of 8,079 submissions, NASA had received applications from 659 pilot candidate hopefuls, made up of 147 military applicants and 512 civilians. By December, the total number of candidates had been reduced to 208, of whom 128 were Mission Specialist (MS) applicants and 80 were pilots – 76 military and four civilian.

On January 16, 1978, the names of the 35 successful pilot and MS applicants were announced, with 15 of that number selected as future Space Shuttle pilots. Nine were graduates of the U.S. Naval Test Pilot School at Patuxent River, Maryland, although one (S. David Griggs) had then resigned from the U.S. Navy (USN) and was subsequently selected as a civilian pilot astronaut by NASA. A further six had graduated from the U.S. Air Force (USAF) Test Pilot School at Edwards Air Force Base (AFB), California.

The 15 pilot astronauts selected in NASA's Group 8 were:

DANIEL C. BRANDENSTEIN

In October 1971, Dan Brandenstein graduated from the U.S. Naval Test Pilot School at Patuxent River, Maryland. Another graduating member of Class 59 was Fred H. Hauck, who would also go on to become one of the Group 8 astronauts.

© Springer Nature Switzerland AG 2020

D. J. Shayler, C. Burgess, *NASA's First Space Shuttle Astronaut Selection*,
Springer Praxis Books, https://doi.org/10.1007/978-3-030-45742-6_3

Hauck would fly as Pilot (PLT) on STS-7 and Brandenstein as PLT on the following mission, STS-8.

Lieutenant Commander (LCdr.) Daniel Charles Brandenstein, USN, was born on January 17, 1943. He graduated from Watertown High School, Watertown, Wisconsin, in 1961, and four years later received a Bachelor of Science (BSc) degree in Mathematics and Physics from the University of Wisconsin. He subsequently entered active duty with the USN in September 1965 and was attached to the Naval Air Training Command for flight training. He was designated a naval aviator at Naval Air Station (NAS) Beeville, Texas, in May 1967, and was then assigned to Attack Squadron 128 (VA-128) for A-6 fleet replacement training. From 1968 to 1970, while attached to VA-196 flying A-6 *Intruders*, he participated in two combat deployments to Southeast Asia on board the USS *Constellation* (CV-64) and the USS *Ranger* (CVA-61), flying 192 combat missions. In subsequent assignments, he was attached to VX-5 to conduct operational tests of A-6 weapons systems and tactics, and to the Naval Air Test Center where he conducted tests of electronic warfare systems in various Navy aircraft following graduation from the U.S. Naval Test Pilot School in 1971. Brandenstein completed a nine-month deployment to the Western Pacific and Indian Ocean on board the USS *Ranger* while attached to VA-145, flying A-6 *Intruders* during the period March 1975 to September 1977. Prior to reporting for his selection as an astronaut candidate, he was attached to VA-128 as an A-6 flight instructor. He had logged 6,400 hours flying time in 24 different types of aircraft and had made around 400 carrier landings.

MICHAEL L. COATS

Mike Coats was another aviator who passed through the U.S. Naval Test Pilot School, graduating from the 25-strong Class 66 in November 1974. A fellow graduate from the class was another future astronaut, LCdr. Michael J. Smith (selected by NASA in Group 9, 1980), who was the PLT of the ill-fated flight of Shuttle *Challenger* in January 1986.

LCdr. Michael Lloyd Coats, USN, was born on January 16, 1946, in Sacramento, California, but has always considered Riverside, California, to be his hometown. Coats graduated from Ramona High School in Riverside in 1964. He received a BSc degree in Naval Science from the U.S. Naval Academy in 1968, a Master of Science (MSc) degree in Administration of Science and Technology from George Washington University in 1977, and an MSc degree in Aeronautical Engineering from the U.S. Naval Postgraduate School in 1979. He graduated from Annapolis in 1968 and was designated a naval aviator in September 1969. After training as an A-7E pilot, he was assigned to VA-192 aboard the USS *Kitty Hawk* (CV-63) from August 1970 to September 1972, with whom he flew 315 combat missions in the Vietnam War. He served as a Flight Instructor with the A-7E Readiness

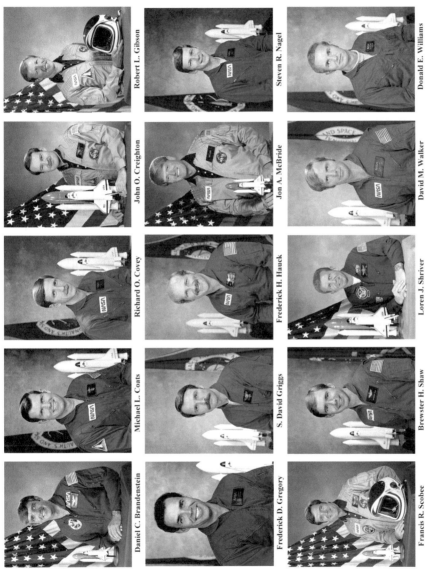

Fig. 3.1: The 15 new pilot candidates.

Training Squadron (VA-122) at NAS Lemoore, California, from September 1972 to December 1973, and was then selected to attend the U.S. Naval Test Pilot School in Maryland. Following test pilot training in 1974, he was Project Officer and test pilot for the A-7 and A-4 aircraft at the Strike Aircraft Test Directorate, then served as a Flight Instructor at the U.S. Naval Test Pilot School from April 1976 until May 1977. He then attended the U.S. Naval Postgraduate School at Monterey, California, from June 1977 until his selection for the astronaut candidate program on January 16, 1978, his 32nd birthday. At the time of his selection, he had logged more than 5,000 hours flying time in 28 different types of aircraft and had completed more than 400 carrier landings.

RICHARD O. COVEY

As a member of the USAF Test Pilot School Class 74B, Richard ("Dick") Covey was a classmate to two very prominent people. The first was Captain (Capt.) Jane Holley, the first female pilot to enter (and graduate from) the Test Pilot School, while the second, Capt. Ellison Onizuka, would become a fellow NASA Group 8 astronaut, later to perish in the loss of Shuttle *Challenger*.

Capt. Richard Oswalt Covey, USAF, was born on August 1, 1946, in Fayetteville, Arkansas, but calls Fort Walton Beach, Florida, his hometown. "My father was a World War II and Korea fighter pilot and then did flight tests for some period of time," he recalled for his NASA Oral History interview, "so I grew up in an environment with military aviators, and it was a logical thing for me to want to go do." He was 14 when Alan Shepard made his first flight. "I was enamored with the idea that men were riding atop rockets, and I consumed everything I could find about the early astronauts. They all turned out to be military test pilots, so I connected with that, because my father worked with the military test pilots and was a military pilot. So at that time I started thinking that would be something I would like to do." [1]

He graduated from Choctawhatchee High School, Shalimar, Florida, in 1964. He then received a BSc degree in Engineering Sciences with a Major in Astronautical Engineering from the USAF Academy in 1968, followed by an MSc degree in Aeronautics and Astronautics from Purdue University in 1969.

Between 1970 and 1974, Covey was an operational fighter pilot, flying the F-100, A-37, and A-7D. He flew 339 combat missions during two tours in Southeast Asia. At Eglin AFB, Florida, between 1975 and 1978, he was an F-4 and A-7D weapons systems test pilot and Joint Test Force Director for electronic warfare testing of the F-15 *Eagle*. At the time of his selection as a NASA pilot astronaut, he had flown over 5,700 hours in more than 30 different types of aircraft.

JOHN O. CREIGHTON

Class 57 of the U.S. Naval Test Pilot School, Patuxent River, graduated 22 aviators in February 1971, with the average number of aircraft assigned to them during their almost seven months of intensive training being 35. Only one of their number, John Creighton, would go on to become a NASA astronaut.

LCdr. John Oliver Creighton, USN, known as "John O," was born on April 28, 1943, in Orange, Texas, but considers Seattle in Washington State to be his hometown, graduating from Ballard High School there in 1961. He subsequently received a BSc degree from the U.S. Naval Academy in 1966 and an MSc in Administration of Science and Technology from George Washington University in 1978. He began flight training following his graduation from Annapolis and received his gold wings as a naval aviator in October 1967. He was then assigned to VF-154 from July 1968 to May 1970, flying F-4Js. He flew two combat deployments to Vietnam aboard the USS *Ranger* (CVA-61). From June 1970 to February 1971, he attended the U.S. Naval Test Pilot School at Patuxent River, Maryland, and upon graduation was assigned as a project test pilot with the Service Test Division at NAS Patuxent River. During this two-year tour of duty, he served as the F-14 engine development Project Officer.

In July 1973, Creighton began a four-year assignment with VF-2 and became a member of the first F-14 operational squadrons, completing two deployments aboard the USS *Enterprise* (CVN-65) to the Western Pacific. He returned to the U.S. in July 1977 and was assigned to the Naval Air Test Center's Strike Directorate as Operations Officer and F-14 Program Manager. Prior to his selection by NASA, he had logged over 6,000 hours flying time, the majority in jet fighters, and had completed 500 carrier landings and 175 combat missions.

ROBERT L. GIBSON

Class 71 at the U.S. Naval Test Pilot School knew they were in for a tough few months, so their class motto was *Illegitimi non Carborundum*, or "*Don't Let the Bastards Grind You Down.*" Nevertheless, all 27 trainees graduated in June 1977, including future Shuttle astronaut Robert Gibson, who goes by the nickname "Hoot," after the famed cowboy movie star. "I'm very fond of saying I got it because of the expression 'not worth a hoot', but it was from the cowboy star of the 1930s." When Gibson arrived at his first Navy fighter squadron, the nickname ended up on his airplane, his name tag, and even his coffee mug, and has stuck ever since. [2]

Lieutenant (Lt.) Robert Lee Gibson, USN, was born on October 30, 1946, in Cooperstown, New York (home of the National Baseball Hall of Fame), but regards Lakewood, California, as his hometown. Gibson graduated from Huntington High

School, New York, in 1964. He then attained an Associate degree in Engineering Science from Suffolk County Community College in 1966 and a BSc degree in Aeronautical Engineering from California Polytechnic State University in 1969. He entered active duty with the Navy in 1969, and in July that year was in Officer Candidate School watching the first Moon landing, with three other candidates, on a black-and-white TV in a motel room they had rented for the night just to watch the moonwalk. He was fascinated by the event, but as an aviator, he had no more than a casual interest in the space program – since Apollo had no wings on it – until a few years later when he saw a picture in *Aviation Week* of the future Shuttle, coming in to land like an aircraft. This rekindled childhood memories of science fiction images of future space planes. "I was interested in a space ship that had wings on it and flew a gliding re-entry and landing on a runway," and he recalled saying at the time, "Oh man, I have got to get me one of those."

Hoot Gibson received primary and basic flight training at NAS Saufley Field, Pensacola, Florida, and NAS Meridian, Mississippi, before completing advanced flight training at NAS Kingsville, Texas.

From April 1972 to September 1975, he was assigned to Fighter Squadrons 111 and 1, and saw duty aboard the USS *Coral Sea* (CVA-43) and USS *Enterprise* (CVN-65), flying combat missions over Southeast Asia. He is a graduate of the Naval Fighter Weapons School, known as "Top Gun." On May 10, 1972, Gibson was engaged in a large combat mission at the same time as two of his future NASA classmates: "We had a number of Space Shuttle astronauts that were involved in the Vietnam War, and even specifically May 10 [1972]." Gibson was part of a big strike group from the *Coral Sea*, while Mike Coats was flying in a strike group from USS *Constellation* and Brewster Shaw was involved in the same action as an Air Force F-4 pilot. During Gibson's tour on the USS *Enterprise*, another Group 8 astronaut, Rick Hauck, was serving as Air Wing Operations Officer. On his return to the United States, Gibson received an assignment as an F-14A instructor pilot with Fighter Squadron 124. He graduated from the U.S. Naval Test Pilot School in June 1977, and later became involved in the test and evaluation of F-14A aircraft while assigned to the Naval Air Test Center's Strike Aircraft Test Directorate. His flight experience includes in excess of 6,000 hours in over 50 types of civil and military aircraft, with more than 300 carrier landings.

FREDERICK D. GREGORY

A number of different services were represented in the U.S. Naval Test Pilot School Class 58 at Patuxent River. As well as two pilots from the U.S. Army, there were two from the civil service, two from the U.S. Marine Corps (USMC), and one each from the air forces of Australia, Italy and Japan. A solitary pilot from the USAF was also undergoing the naval aviation training: Fred Gregory.

Fig. 3.2: Fred Gregory stands next to an F-4 *Phantom.*

Major (Maj.) Frederick Drew Gregory, USAF, was born on January 7, 1941, in Washington, D.C., and graduated from Anacostia High School in 1958. He then attained a BSc degree from the USAF Academy in 1964, and an MSc degree in Information Systems from George Washington University in 1977. After graduating from the Air Force Academy, he entered pilot training and attended undergraduate helicopter training at Stead AFB, Nevada. He received his wings in 1965 and was subsequently assigned as an H-43 helicopter rescue pilot at Vance AFB, Oklahoma, from October 1965 until May 1966. In June 1966, he was assigned as an H-43 combat rescue pilot at Danang Air Base, Vietnam. When he returned to the United States in July 1967, he was assigned as a missile support helicopter pilot, flying the UH-1F at Whiteman AFB, Missouri. In January 1968, Gregory was retrained as a fixed-wing pilot, flying the T-38 at Randolph AFB, Texas. He was then assigned to the F-4 *Phantom* Combat Crew Training Wing at Davis-Monthan AFB, Arizona. He attended the U.S. Naval Test Pilot School at Patuxent River, Maryland, from September 1970 to June 1971. Following completion of this training, he was sent to the 4950th Test Wing, Wright-Patterson AFB, Ohio, as an operational test pilot flying fighters and helicopters. In June 1974, he was detailed to the NASA Langley Research Center in Hampton, Virginia. There, he served as a research test pilot until selected for the astronaut program in January 1978. By this time, Gregory had logged more than 6,976 hours flying time in over 50 types of aircraft, including 550 combat missions in Vietnam.

S. DAVID GRIGGS

On his first space flight, STS-51D in April 1985, Dave Griggs carried out the first unscheduled and unrehearsed spacewalk in the American space program. He had subsequently been assigned to a second mission for the Department of Defense, STS-33, but was killed in an off-duty airplane crash, his vintage T-6 crashing into a field in Arkansas while he was rehearsing aerobatics for an airshow.

Stanley David ("Dave") Griggs was born on September 7, 1939, in Portland, Oregon. He graduated from Lincoln High School, Portland, in 1957 and later from Annapolis in 1962 with a BSc degree. He entered naval pilot training shortly thereafter. In 1964, he received his wings as a naval aviator and was attached to VA-72, flying A-4 *Skyhawks*. He completed one Mediterranean cruise and two Southeast Asia combat cruises aboard the aircraft carriers USS *Independence* (CV-62) and USS *Franklin Roosevelt* (CV-42). In 1967, he entered the U.S. Naval Test Pilot School at Patuxent River, Maryland and, upon completion of test pilot training, was assigned to the Flying Qualities and Performance Branch, Flight Test Division, where he flew various test projects on fighter and attack-type aircraft. In 1970, he received an MSc in Administration from George Washington University, and resigned his regular U.S. Navy commission before affiliating with the Naval Air Reserve, in which he achieved the rank of rear admiral.

As a naval reservist, Rear Admiral Griggs was assigned to several fighter and attack squadrons, flying A-4 *Skyhawk*, A-7 *Corsair II* and F-8 *Crusader* aircraft while based at NAS in New Orleans, Louisiana and Miramar, California. He logged 9,500 hours flying time, 7,800 hours in jet aircraft, and flew over 45 different types of aircraft, including single- and multi-engine prop, turboprop and jet aircraft, helicopters, gliders, and hot air balloons. He made over 300 aircraft carrier landings and was a certified flight instructor.

In July 1970, Griggs was employed at the Johnson Space Center (JSC) as a research pilot, working on various flights test and research projects in support of NASA programs. In 1974, he was assigned duties as the Project Pilot for the Space Shuttle trainer aircraft, and participated in the design, development and testing of those aircraft pending their operational deployment in 1976. He was appointed Chief of the Shuttle Training Aircraft Operations Office in January 1976, with responsibility for the operational use of the Shuttle trainer. He held that position until his selection as an astronaut candidate by NASA in January 1978.

Fig. 3.3: TFNG Mission Specialist Ascan Ron McNair receives instruction in T-38 back seat procedures from instructor and fellow TFNG Dave Griggs.

FREDERICK H. HAUCK

Rick Hauck was one of two future astronauts, together with Dan Brandenstein, who graduated from the 20-strong Class 59 that attended the U.S. Naval Test Pilot School from February to October 1971.

LCdr. Frederick ("Rick") Hamilton Hauck, USN, was born on April 11, 1941, in Long Beach, California, and graduated from St. Albans School in Washington, D.C. in 1958. He received a BSc in Physics from Tufts University in 1962 and an MSc in Nuclear Engineering from the Massachusetts Institute of Technology (MIT) in 1966. A Navy Reserve Officer Training Corps (ROTC) student at Tufts University, Hauck was commissioned upon graduation in 1962 and reported to the USS *Warrington* (DD-843) where he served for 20 months, qualifying as Underway Officer-of-the-Deck. In 1964, he attended the U.S. Naval Postgraduate School in Monterey, California, to study mathematics and physics, and for a brief time in 1965 he studied Russian at the Defense Language Institute, also in Monterey. It was while at Monterey during 1965 that Hauck learned that NASA was recruiting scientists to become astronauts, and as he had avidly followed the Mercury and Gemini flights, he wrote off to NASA, saying "I am in graduate school. You could tailor my education however you saw fit to optimize my benefit to the program, and I'd be very interested in becoming an astronaut." The reply he received back from NASA thanked him for his letter, but was essentially worded as "Don't call us; we'll call you." Twelve years later, he was accepted for astronaut training. [3]

Selected for the Navy's Advanced Science Program, Hauck received his MSc in Nuclear Engineering from MIT the next year. He began flight training at NAS Pensacola, Florida, in 1966, and received his wings as a naval aviator in 1968. As a pilot with VA-35, he was deployed to the Western Pacific with Air Wing 15 aboard the USS *Coral Sea* (CVA-43), where he flew 114 combat and combat support missions.

In August 1970, Hauck joined VA-42 as a visual weapons delivery instructor in the A-6. Selected for test pilot training, he reported to the U.S. Naval Test Pilot School at Patuxent River in 1971. Following his graduation, he began a three-year tour in the Naval Air Test Center's Carrier Suitability Branch of the Flight Test Division. During this period, Hauck served as a project test pilot for automatic carrier landing systems in the A-6 *Intruder*, A-7 *Corsair II*, F-4 *Phantom* and F-14 *Tomcat* aircraft, and was Team Leader for the Navy Board of Inspection and Survey aircraft carrier trials of the F-14. During a test flight in 1973, he was forced to eject from an RA-5C *Vigilante* at low altitude when its fuel tank exploded. In 1974, he reported as Operations Officer to Commander Carrier Air Wing 14 aboard USS *Enterprise* (CVN-65). On two cruises, he flew the A-6, A-7, and F-14 during both day and night carrier operations. He learned of the call from NASA for new astronauts to fly the Space Shuttle during his second cruise on *Enterprise*, and on that ship with him were three others who were also selected in the 1978 class: Hoot Gibson, Dale Gardner and John Creighton. Three of the 15 pilots selected were from Air Wing 14 and Dale Gardner was chosen as an MS, Hauck recalled in his NASA Oral History. He found it interesting that three of 15 pilots chosen in the group had served in the same Air Wing. "Three of 15... What's that? Twenty percent came from that ship," Hauck realized. In February 1977 he reported to Attack Squadron 145 as Executive Officer, and in January 1978, NASA selected Hauck as a pilot astronaut candidate.

JON A. MCBRIDE

Unlike the other Group 8 Navy pilots, Jon McBride did not attend the Naval Test Pilot School, instead joining Class 75A at the USAF Test Pilot School at Edwards AFB, California. He graduated in 1975, along with fellow future astronauts Guy Gardner (selected by NASA for Group 9, 1980), Loren Shriver and Steve Nagel – the latter two joining him as Group 8 astronauts in 1978.

LCdr. Jon Andrew ("Big Jon") McBride, USN, was born on August 14, 1943, in Charleston, West Virginia, but considers Beckley, West Virginia, to be his home-town. He graduated from Woodrow High School in Beckley in 1960, before undertaking further studies at West Virginia University from 1960 to 1964. He received his BSc in Aeronautical Engineering from the U.S. Naval Postgraduate School in 1971. McBride's naval service began in 1965 with flight training at Pensacola, Florida. A fan of the space program since Sputnik in 1957, he and his friends would design, build and launch model rockets, and he later became fascinated with the careers and achievements of the USN astronauts chosen in the 1960s. [4] After winning his wings as a naval aviator, McBride was assigned to Fighter Squadron 101 based at NAS Oceana, Virginia, for training in the F-4 *Phantom II* aircraft. He was subsequently assigned to Fighter Squadron 41, where he served for three years as a fighter pilot and Division Officer. He has also served tours with Fighter Squadrons 11 and 103. While deployed to Southeast Asia, McBride flew 64 combat missions. In 1975, he attended the USAF Test Pilot School at Edwards AFB, prior to reporting to Air Test and Development Squadron Four at Point Mugu, California, where he served as Maintenance Officer and Sidewinder Project Officer. He has flown over 40 different types of military and civilian aircraft and piloted the Navy's bicentennial-painted "Spirit of '76" F-4J *Phantom* at various air shows during 1976, 1977, and 1978. He has logged more than 8,800 hours flying time, including 4,700 hours in jet aircraft.

STEVEN R. NAGEL

Although he would have preferred to fly in the capacity he was selected for, as a pilot astronaut, Steve Nagel accepted the role of MS when he flew into space for the first time on STS-51G in June 1985. This flight carried an interesting Payload Specialist (PS) in the form of Sultan Salman Al Saud from Saudi Arabia, the first Arab, the first Muslim and the first member of a royal family to fly into space. However, Nagel did achieve his goal on his second mission just five months later, when he flew as PLT on STS-61A in November 1985, the last successful mission of Shuttle *Challenger*, which carried the NASA/European Space Agency (ESA) Spacelab module into orbit with more than 75 experiments. On his final two missions (STS-37, April 1991, and STS-55, April 1993) he flew as mission Commander (CDR).

Steven Ray Nagel was born on October 27, 1946, in Canton, Illinois. He graduated from Canton Senior High School in 1964, and received a BSc in Aerospace Engineering (high honors) from the University of Illinois in 1969. Nagel received his commission in 1969 through the Air Force Reserve Officer Training Corps (AFROTC) program at the University of Illinois. He completed undergraduate pilot training at Laredo AFB, Texas, in February 1970, and subsequently reported to Luke AFB, Arizona, for F-100 training.

From October 1970 to July 1971, Nagel was an F-100 pilot with the 68th Tactical Fighter Squadron at England AFB, Louisiana. He served a one-year tour of duty as a T-28 instructor for the Laotian Air Force at Udorn Royal Thai Air Force Base (RTAFB) in Thailand, prior to returning to the United States in October

1972 to assume A-7D instructor pilot and flight examiner duties at England AFB. He then attended the USAF Test Pilot School at Edwards AFB, California, from February to December 1975 as a member of Class 75A. In January 1976, he was assigned to the 6512th Test Squadron located at Edwards. As a test pilot, he worked on various projects, including flying the F-4 and A-7D. He would later receive his MSc in Mechanical Engineering from California State University in 1978. When selected by NASA, he had logged 12,600 hours flying time, 9,640 hours of which was in jet aircraft.

With ambitions to become a Shuttle Pilot, Nagel was pleased that the qualification requirements for applicants had been liberalized. "In the old days, I wouldn't be here," he commented following his selection. At 6 feet 2 inches (1.88 meters), he would have been rejected as being too tall in previous astronaut applications. "I'm here now because the Shuttle cockpit is bigger. The flight deck's the same size as an airliner." [5]

FRANCIS R. SCOBEE

If Air Force Maj. Dick Scobee ("Scobe") was looking for someone to inspire him as a future NASA astronaut while attending Class 71B of the Aerospace Research Pilot School (ARPS, previously the USAF Test Pilot School, TPS) at Edwards AFB in 1971, he needed to look no further than the school's commandant at that time, Col. Edwin E. ("Buzz") Aldrin, who had walked on the Moon two years earlier. In fact, Scobee would be the only member of Class 71B to become a NASA astronaut[1].

Maj. Francis Richard ("Dick") Scobee, USAF, was born on May 19, 1939, in Cle Elum, Washington, and graduated from Auburn Senior High School, Washington, in 1957. He subsequently enlisted in the USAF in 1957, and trained as a reciprocating engine mechanic. He was later stationed at Kelly AFB, Texas, where he attended night school and acquired two years of college credit which led to his selection for the Airman's Education and Commissioning Program. He graduated from the University of Arizona with a BSc in Aerospace Engineering in 1965 and was commissioned that same year. After receiving his wings in 1966, Scobee completed a number of assignments, including a combat tour in Vietnam. He returned to the United States and attended the Air Force ARPS at Edwards AFB, California. After graduating in 1972, he participated in various test

---

[1] In 1961, the USAF Test Pilot School at Edwards AFB in California expanded its operations to include the Aerospace Research Pilot Course at which military test pilots received preparatory training for operational or managerial assignments in the nation's space programs. Between 1961 and 1972, when the course was eliminated, 37 TPS graduates were selected for the U.S. space program under the civilian NASA or USAF Manned Orbiting Laboratory (MOL) programs. Of these, 26 would earn their Astronaut Wings participating in missions flown under the X-15, Gemini, Apollo and Space Shuttle programs.

programs, flying aircraft such as the Boeing 747, the X-24B (see sidebar: *The X-24*), the transonic aircraft technology (TACT) F-111, and the C-5. At the time of his selection by NASA in January 1978, he had logged more than 6,500 hours flying time in 45 types of aircraft.

Fig. 3.4: X-24B lifting body pilots (l-r): Einar Enevoldson, John Manke, Dick Scobee, Tom McMurtry, Bill Dana and Mike Love.

The X-24
The X-24 was an experimental aircraft developed from a joint USAF/NASA program called PILOT (PIloted LOw speed Test), part of the Lifting Body research program which ran from 1963 until 1975. The objective was to design and build lifting-body concepts for unpowered re-entry and landing, a technique later used by the Space Shuttle. The test flight began with an air drop from a carrier aircraft at high altitudes before igniting its rocket motor. After expending the on-board fuel, the pilot would glide the X-24 to an unpowered landing. The original X-24 was rebuilt as the X-24B. In 1975, towards the end of the program. Captain Dick Scobee, the only member of any post-1970 selection to have flown a lifting body aircraft, logged two unpowered familiarization glide flights in X-24B #1:

1. October 21, 1975; 32nd flight of X-24B; 0.700 Mach/463 mph (745 kph); 45,000 ft. (13,716 m); glide flight.
2. November 19, 1975: 35th & penultimate X-24B flight; 0.714 Mach/471 mph (757.8 kph); 45,000 ft. (13,716 m); glide flight. [6]

Fig. 3.5: During a 1982 refueling stopover at Ellington AFB, the crew of the Shuttle Carrier Aircraft (SCA 747) NASA 905 are seen at their stations inside the aircraft. (l-r) Joe Algranti, pilot; Dick Scobee, co-pilot; Louis E. Gidry, flight engineer.

BREWSTER H. SHAW

Two of the pilots assigned to Class 75B at the USAF Test Pilot School at Edwards AFB in July 1975 were destined for selection in NASA's Group 8. But while Capt. Mike Mullane entered the astronaut corps as an MS, Brewster Shaw was named as a pilot astronaut. Yet another classmate, Capt. Jerry Ross, would be selected as an MS astronaut in Group 9 in 1980.

Capt. Brewster Hopkins Shaw Jr., USAF, was born on May 16, 1945, and grew up in Cass City, Michigan. He graduated from Cass City High School in 1963 and received a BSc in Engineering Mechanics from the University of Wisconsin-Madison in 1968. The following year, he completed his MSc in Engineering Mechanics, also at UW-Madison. Shaw entered the USAF in 1969 after completing Officer Training School, and attended undergraduate pilot training at Craig AFB, Alabama. He received his pilot wings in 1970 and was then assigned to the F-100 Replacement Training Unit at Luke AFB, Arizona. In April 1973, Shaw reported to George AFB, California, for F-4 *Phantom* instructor duties with the 20th Tactical Fighter Squadron. He then attended the USAF Test Pilot School at Edwards AFB, beginning in July 1975. On completion of this intensive training period, he remained at Edwards as an operational test pilot with the 6512th Test Squadron. He subsequently served as an instructor at the Test Pilot School from August 1977 until his selection by NASA. By this time, Shaw had accumulated more than 5,000 flying hours in more than 30 different aircraft types, including 644 hours of combat flying in F-100 and F-4 aircraft.

LOREN J. SHRIVER

Class 75A at the USAF Test Pilot School, Edwards AFB, produced three graduates who would become NASA pilot astronaut candidates in January 1978: Jon McBride, Steven Nagel, and Loren Shriver. As mentioned earlier, a fourth graduate, Guy Gardner, became a pilot astronaut in the Group 9 selection in 1980.

Capt. Loren James Shriver, USAF, was born on September 23, 1944, in Jefferson, Iowa. He received a BSc in Aeronautical Engineering from the USAF Academy in Colorado Springs, Colorado in 1967, and an MSc in Astronautical Engineering from Purdue University, Indiana, in 1968. He was commissioned in 1967 upon graduation from the Air Force Academy, and from 1969 to 1973 served as a T-38 academic instructor pilot at Vance AFB, Oklahoma. Shriver completed F-4 *Phantom* combat crew training at Homestead AFB, Florida, in 1973, and was then assigned to an overseas tour in Thailand until October 1974. In 1975, he attended the USAF Test Pilot School at Edwards AFB, and upon completion of this training program was assigned to the 6512th Test Squadron at Edwards, participating in the Air Force development test and evaluation of the T-38 lead-in fighter. In 1976, he began serving as a test pilot for the F-15 Joint Test Force at Edwards.

Loren Shriver has flown in 30 different types of single- and multi-engine civilian and military fixed-wing aircraft and helicopters, and has logged over 6,200 hours in jet aircraft.

DAVID M. WALKER

It was not unusual to find officer pilots from the other services involved in the USAF ARPS based at Edwards AFB. This was the case in Class 71A, where Lt. David Walker was one of two naval aviators undertaking an exhaustive six-month program. The school placed heavy demands on the technical competence, ingenuity and managerial skills of the students, while they learned about performance and flying qualities of a number of aircraft, in order to conduct an effective overall evaluation on them. Under the guidance of school commandant Edwin "Buzz" Aldrin, David Walker would successfully graduate with the rest of Class 71A, and was assigned as a test pilot to the Naval Air Test Center in Patuxent River, Maryland.

LCdr. David Mathieson Walker, USN, was born on May 20, 1944, in Columbus, Georgia. He graduated from Eustis High School, Florida, in 1962. He then attended the U.S. Naval Academy, receiving his BSc on June 8, 1966. Following his graduation from Annapolis, he received flight training from the Naval Aviation Training Command at bases in Florida, Mississippi, and Texas, and was designated a naval aviator in December 1967. He then proceeded to NAS Miramar, California, for assignment to F-4 *Phantoms* aboard the carriers USS *Enterprise* (CVN-65) and USS *America* (CV-66). From December 1970 to 1971, he attended the USAF ARPS at Edwards AFB, California, and was subsequently assigned in January 1972 as an experimental and engineering test pilot in the Flight Test Division at the Naval Air Test Center, Patuxent River, Maryland. While there, he participated in the Navy's preliminary evaluation and Board of Inspection and Survey trials of the F-14 *Tomcat* and tested a leading edge slat modification to the F-4 *Phantom*. He then attended the U.S. Navy Safety Officer School at Monterey, California, and completed replacement pilot training in the F-14 *Tomcat* at NAS Miramar.

In 1975, Walker was assigned as a fighter pilot to Fighter Squadron 142, stationed at NAS Oceana, Virginia, and was deployed to the Mediterranean Sea twice aboard the USS *America*. At the time of his selection by NASA, he had logged in excess of 5,500 flying hours, including 5,000 hours in jet aircraft, primarily the F-4, F-14 and T-38.

DONALD E. WILLIAMS

On average, each student in Class 65 at the U.S. Naval Test Pilot School was assigned to 36 different aircraft types between July 1973 and May 1974. One of those pilots – and the only one who would join the ranks of NASA astronauts – was Don Williams.

LCdr. Donald Edward Williams, USN, was born on February 13, 1942, in Lafayette, Indiana, and raised in the nearby town of Green Hill. He graduated from Otterbein High School in Otterbein, Indiana in 1960. Williams then went on to earn a BSc in Mechanical Engineering from Purdue University in 1964, and also received his commission through the university's Naval ROTC program. He completed flight training at Pensacola, Florida, Meridian, Mississippi and Kingsville, Texas, receiving his wings as a naval aviator in May 1966. After A-4 *Skyhawk* training, he completed two Vietnam War deployments aboard the aircraft carrier USS *Enterprise* (CVN-65) with VA-113. He subsequently served as a flight instructor with VA-125 at NAS Lemoore, California, for two years and transitioned to the A-7 *Corsair II* aircraft. He made two additional Vietnam deployments aboard the USS *Enterprise* with Carrier Air Wing 14 staff and VA-97. He completed a total of 330 combat missions.

Williams attended the Armed Forces Staff College in 1973, and graduated from the U.S. Naval Test Pilot School in Maryland in June 1974. He was then assigned to the Naval Air Test Center's Carrier Suitability Branch of the Flight Test Division. From August 1976 to June 1977, following reorganization of the Naval Air Test Center, he was head of the Carrier Systems Branch Strike Aircraft Test Directorate. He reported next for A-7 refresher training, and was assigned to VA-94 at the time of his selection as an astronaut candidate by NASA. By then, he had accumulated over 5,700 hours of flying time, including 5,400 hours in jet aircraft, and had completed 745 carrier landings.

Fig. 3.6: Class of 1978 pilot astronaut hopefuls (clockwise from top left) Capt. Jane L. Holley, USAF; Maj. Gen Claude M. Bolton Jr., USAF; Col. Richard S. Couch, USAF: Capt. William V. Cross II, USN.

SOME NOTABLE BUT UNSUCCESSFUL PILOT APPLICANTS

The undeniable quality associated with the backgrounds and suitability of the 208 finalists probably made the task of reducing this number down to the final 35 a source of intense debate for the selection board. Many candidates held (or would hold) prodigious qualifications and were unfortunate to have missed the cut. Here are just a select few of the military pilots who failed to make the final number.

Capt. Jane Leslie Holley, USAF

On March 17, 1971, Second Lieutenant Jane Holley from Shreveport, Louisiana, became the first woman to be commissioned through the Air Force ROTC scheme, graduating through Auburn University, Alabama, where she also received a BSc degree in Aerospace Engineering. Twenty-two months after entering the Air Force, Capt. Holley applied for admission to the USAF TPS and was accepted early in 1974. As a Flight Test Engineering student in Class 74B at Edwards AFB, she gained the distinction of being the first woman to graduate from the TPS. By the end of the 44-week course, Capt. Holley had logged more than 600 hours in the classroom and more than 100 hours in the air.

At the heart of the TPS course was the aim of establishing cooperation and understanding between pilot and engineer. "In the past," she said at the time, "when a test pilot returned from a mission, he sometimes had difficulty conveying to the engineer the problems he encountered during certain tests. They really had no basis for a common language. As a flight test engineer, I am given the opportunity to ride in the back seat during a mission. I know what tasks the pilot is required to perform and through observation I'll know exactly what he means when he says he experienced a 'slight' vibration: we will have a common point of reference."

At the time of her astronaut application (the only Air Force female nominated), Capt. Holley was working with the USAF Tactical Fighter weapons center at Nellis AFB, Nevada.

Capt. Claude Milburn Bolton Jr., USAF

Capt. Bolton was born in 1945 in Sioux City, Iowa, joining the USAF in June 1969. During the Vietnam War, he flew 232 combat missions in the F-4 *Phantom*, 40 of them over North Vietnam, for the 497th Tactical Fighter Squadron ("*Nite Owls*"). After the Vietnam War, he transitioned to the F-111 at Cannon AFB, New Mexico, and RAF Upper Heyford in Oxfordshire, England. In 1977, he attended the USAF TPS. While undertaking testing from Eglin AFB, Florida, he set a low altitude, high speed record.

After attending the Defense Systems Management College in 1982, Bolton transitioned to program management. He became the first program manager for the Advanced Tactical Fighter Technologies Program, which evolved into the F-22 *Raptor* System Program Office. Following his distinguished military career, Maj. Gen. Bolton was nominated by President George W. Bush for the position of Assistant Secretary of the Army for Acquisition, Logistics and Technology, and served in that eminent position from 2002–2008. A veteran of more than 32 years of active military service, Maj. Gen. Bolton died unexpectedly at home on July 28, 2015. At the time of his death, following a highly decorated career as a test pilot and an acquisition officer, he was serving as the Executive-in-Residence for the

Defense Acquisition University (DAU), assisting in supporting the Congressional mandate to recruit, train and educate the Department of Defense acquisition workforce.

Capt. Richard S. (Rick) Couch, USAF
Rick Couch entered the Air Force in 1968 and flew 199 combat missions in Southeast Asia as a forward air controller. After graduating from the USAF TPS in Class 75A, he was assigned to the 4950th Test Wing at Wright-Patterson AFB, where he flew test missions developing the all-weather landing system for the C-141 and other research and development programs. In 1978, following his unsuccessful bid to become an astronaut, Couch returned to Edwards AFB as a TPS Instructor.

In 1985, he became the first commander of the B-2 Combined Test Force at Edwards, and would later be the pilot on the maiden flight of the B-2 stealth bomber, which was first revealed to the public on November 22, 1988 when AV-1 (82-1066) was unveiled at Palmdale, California. Taxi tests began on July 10, 1989 and the first flight of the B-2 took place on July 17 from Palmdale, crewed by chief test pilot Bruce J. Hinds and Col. Couch. The flight lasted 112 minutes and ended with a landing at Edwards AFB. Couch also participated in early development activities of the B-2.

In 1990, he was presented with the Iven C. Kincheloe Award, which recognizes outstanding professional accomplishment in the conduct of flight testing. He later became Test Wing Vice Commander at Edwards, then served as Deputy Director of the Tri-service Standoff Attack Missile System Program Office at Wright-Patterson. After retiring from the USAF in 1992, Col. Couch joined Martin Marietta (now Lockheed Martin).

LCdr William V. (Bill) Cross II, USN
Bill Cross, from Omaha, Nebraska, enjoyed an active duty U.S. Naval career for 33 years, retiring as a two-star Rear Admiral on February 1, 2000. His military experience included combat tours in Vietnam and Operation Desert Storm, and four operational commands including an F-14 fighter squadron, an amphibious assault ship, a nuclear aircraft carrier (USS *Dwight D. Eisenhower*) and an aircraft carrier strike group. When he applied for astronaut training in 1977, he was 31 years old and serving on the USS *Nimitz*.

Cross became the Navy's first Program Executive Officer for Aircraft Carriers, leading the initial design of the advanced command and control system for the Navy's newest class of aircraft carriers. His other military positions included Navy test pilot, TPS flight instructor, engineering manager for the Navy's F-14 programs, Director of Plans and Policy for the U.S. Transportation Command, and

Director of Operations for the U.S. European Command. As Commander of Carrier Group Six in 1995/96, Cross directed numerous carrier air strikes against targets in Serbia and Bosnia from the flagship USS *America*, flying from the Adriatic Sea.

References

1. Richard O. Covey, NASA Oral History, November 1, 2006.
2. Robert L. Gibson, NASA Oral History, November 1, 2013.
3. Frederick H. Hauck, NASA Oral History, November 20, 2003.
4. Jon A. McBride, NASA Oral History, April 17, 2012.
5. "Astronauts of the '80s," <u>Washington Post</u>, Thomas O'Toole, issue 20, July 1980.
6. **The X-Planes X-1 to X-45**, Jay Miller, Midland Publishing, Hinckley, England, 3rd Edition, 2001, p. 268.

The authors would like to acknowledge, with appreciation, the use of biographical information contained within NASA's official astronaut biographies in the compilation of this chapter.

4

The first Mission Specialists

"Step past our places of comfort,
to walk over to the edge of our abilities
and then move beyond that edge.
The unknown is mysterious.
The unknown is frightening.
But you can only become a winner
if you are willing to walk over to the edge
and dangle over it just a little bit."
Ronald E. McNair, STS-41B post-flight lecture at MIT, 1984.
From *Ronald McNair, Astronaut*, by Corinne Naden

Between August 2 and November 21, 1977, the astronaut applicants were subjected to tests, interviews, evaluations and careful scrutiny at the Johnson Space Center (JSC) in order to assess their eligibility as astronaut candidates. Of the 128 applying for the newly-created role of Mission Specialist (MS), 21 were women. In all, 208 applicants came to Houston, in ten groups, for their week of medical and psychological testing, as well as facing a selection panel armed with probing, analytical questions both personal and technical.

On January 16, 1978, NASA announced the selection of 20 MS, which included six pioneering women. The 20 successful candidates assigned for training as NASA Group 8 MS were:

GUION S. BLUFORD, JR.

When selected as a member of NASA's Group 8 astronauts, Guion ("Guy") Bluford was a veteran of 144 combat missions during the Vietnam War. In his youth, he had spent many happy hours building aircraft, with a determination that

© Springer Nature Switzerland AG 2020   101
D. J. Shayler, C. Burgess, *NASA's First Space Shuttle Astronaut Selection*,
Springer Praxis Books, https://doi.org/10.1007/978-3-030-45742-6_4

his future was in aircraft design or even mechanical engineering, like his father. "It wasn't until I got in the Air Force ROTC [Reserve Officer Training Corps] program at Penn State that I started to develop a strong interest in flying," he once told *Time* magazine. "My thinking was that if I were a pilot, I would be a better engineer." [1]

Guion Stewart Bluford Jr., PhD, was born in Philadelphia, Pennsylvania, on November 22, 1942. He graduated from Overbrook Senior High School in 1960, and in 1964 he received a Bachelor of Science (BSc) degree in Aerospace Engineering from Pennsylvania State University, also graduating as a distinguished Air Force Reserve Officer Training Corps (AFROTC) graduate. He then attended pilot training at Williams Air Force Base (AFB), Arizona, and received his pilot wings in January 1966. He went on to F4C combat crew training in Arizona and Florida, and was assigned to the 557th Tactical Fighter Squadron, based in Cam Ranh Bay, Vietnam. Of his 144 combat missions, 65 were over North Vietnam. In July 1967, he was assigned to the 3630th Flying Training Wing, Sheppard AFB, Texas, as a T-38A instructor pilot. He served as a standardization/evaluation officer and as an assistant flight commander. In early 1971, he attended Squadron Officers School and returned as an executive support officer to the Deputy Commander of Operations and as School Secretary for the Wing.

In August 1972, Bluford entered the Air Force Institute of Technology (AFIT) residency school at Wright-Patterson AFB, Ohio. On graduating in 1974, he was awarded his Master of Science (MSc) degree with distinction in Aerospace Engineering, and was assigned to the Air Force Flight Dynamics Laboratory at Wright-Patterson AFB, Ohio, as a staff development engineer. He served as Deputy for Advanced Concepts for the Aeromechanics Division and as Branch Chief of the Aerodynamics and Airframe Branch in the Laboratory. In 1978, again through AFIT, he would also receive his doctorate (PhD) in Aerospace Engineering with a Minor in Laser Physics. While completing his PhD, he responded to NASA's call for astronauts, believing that – if selected – it would give him the chance to combine his interests in flying and engineering.

JAMES F. BUCHLI

As it turned out, James Buchli was the sole representative of the U.S. Marine Corps (USMC) selected for NASA's Group 8 astronaut cadre. Previously, only five Marines had become NASA astronauts – John Glenn, C.C. Williams, Jack Lousma, Jerry Carr and Bob Overmyer.

Captain (Capt.) James Frederick Buchli, USMC, was born on June 20, 1945, in New Rockford, North Dakota. He also considers Fargo, North Dakota, as his hometown. Buchli graduated from Fargo Central High School in 1963, and received a BSc in Aeronautical Engineering from the U.S. Naval Academy in

Fig. 4.1: The first 20 Mission Specialist candidates

1967, during which time he also completed a stint aboard a submarine[1]. Buchli received a commission in the USMC following his graduation from the Naval Academy in 1967. He subsequently graduated from the USMC Basic Infantry Course, after which he was sent to the Republic of Vietnam for a one-year tour of duty. Once there, he served as Platoon Commander, 9th Marine Regiment, and then as Company Commander and Executive Officer, "B" Company, 3rd Company, 3rd Reconnaissance Battalion. He returned to the United States in 1969 for naval flight officer training at Pensacola, Florida, and spent the next two years assigned to Marine Fighter/Attack Squadron 122 (VMFA-122), at Kaneohe Bay, Hawaii, and Iwakuni, Japan. In 1973, he was assigned to VMFA-115 at Namphong, Thailand, and Iwakuni. After completing this tour of duty, he returned to the United States and participated in the Marine Advanced Degree Program at the University of West Florida, where he received his MSc in Aeronautical Engineering Systems in 1975. He was then assigned to VMFA-312 at the Marine Corps Air Station, Beaufort, South Carolina, and, in 1977, to the U.S. Test Pilot School at Patuxent River, Maryland.

At the time of his selection, he had logged over 4,200 hours flying time, 4,000 of them in jet aircraft.

JOHN M. FABIAN

A highly experienced aeronautical engineer, John Fabian had always thought about becoming an astronaut, but knew that his height would preclude him from achieving his ambition. Then, in NASA's call for astronauts in June 1977, he noticed that the height limit had been raised from 5 feet 11 inches (1.80 meters) to 6 feet 4 inches (1.93 meters). He subsequently requested an application form for the category of MS, filled it out and sent it off to NASA.

John McCreary Fabian, PhD, was born on January 28, 1939, in Goosecreek, Texas, but considers Pullman, Washington, to be his hometown. He graduated from Pullman High School in 1957, and received a BSc in Mechanical Engineering from Washington State University in 1962. While studying at the university, he became an AFROTC student, and was commissioned upon graduation. He would then receive an MSc in Aerospace Engineering from the AFIT at Wright-Patterson AFB, Ohio, in 1964. Following further duties at AFIT, he was assigned as an aeronautics engineer in the Service Engineering Division, San Antonio Air Materiel

---

[1] When he flew STS-51C in 1985, Buchli became, by definition, the first submariner in space due his short-term service on a submarine while at the U.S. Naval Academy. However, Michael J. McCulley (NASA Class of 1984, Group 10) is recognized everywhere as the first *qualified* submariner in space when he flew on STS-34 in 1989, with Stephen Bowen (NASA Class of 2000, Group 18) becoming the second by flying STS-126 in 2008.

Area, at Kelly AFB, Texas. He then attended flight training at Williams AFB, Chandler, Arizona, and subsequently spent five years as a KC-135 pilot at Wurtsmith AFB, Michigan. Fabian lost his original pilot training slot in July 1965 to newly-selected Scientist-Astronaut Owen Garriott and had to wait six months to join the next class. He saw action in Southeast Asia, flying 90 combat missions before returning home. Fabian was airborne above Laos in Southeast Asia, listening to *Voice of America* over the onboard long-range radio they had on the aircraft, when Apollo 11 launched from the Cape on July 16, 1969, and again four days later as Armstrong and Aldrin landed *Eagle* on the surface of the Moon. "That was an interesting place to be," he recalled in 2006, "over Laos and listening to something which is as world shaking as the first lunar landing." **[2]** Following additional graduate work at the University of Washington, he earned a PhD in Aeronautics and Astronautics in 1974. Prior to his selection by NASA, he served on the faculty of the Aeronautics Department at the United States Air Force (USAF) Academy in Colorado. As a pilot, he had logged 4,000 hours flying time, including 3,400 hours in jet aircraft.

ANNA L. FISHER

When she mailed her application to NASA in 1977, Dr. Anna Tingle was working as an emergency room physician at Harbor General Hospital in Torrance, California. In her youth, she had decided there would be a need for physicians on future space stations, and this was too good an opportunity to miss. On her only Shuttle mission in 1984, Anna Fisher made the headlines as the first mother to fly into space.

Anna Lee Fisher, MD, was born in New York City on August 24, 1949. Dr. Fisher considers San Pedro, California, to be her hometown. On May 5, 1961, she was in seventh grade listening to a small transistor radio her teacher had brought into the class to hear the flight of Alan Shepard, America's first man in space. "Wow, I would love to go and do something like that," she remembered thinking at the time, as she recalled in her 2009 Oral History, "but of course all the astronauts at the time were male. They were all test pilots." That did not interest her, but she did think that someday there would be a space station and, always gravitating towards math and science, she looked towards a career in science or medicine. **[3]** After graduating from San Pedro High School in 1967, she attended the University of California Los Angeles (UCLA), and received her BSc in Chemistry in 1971. After graduating from UCLA, Fisher spent a year in graduate school working in the field of X-ray crystallographic studies of metallocarboranes. She co-authored three publications relating to these studies for the Journal of Inorganic Chemistry. She began medical school at UCLA in 1972 and achieved her Medical Doctorate (MD) in 1976. Following her graduation that year, she began a one-year internship at Harbor General Hospital in Torrance, California. After completing her internship in 1977, she specialized in emergency medicine and worked at several hospitals in the Los Angeles area.

Fig. 4.2: Anna and Bill Fisher, the first married couple to be chosen to train as astronauts. Bill Fisher was chosen in the Class of 1980 (Group 9).

DALE A. GARDNER

In 1984, Dale Gardner made space flight history when he participated in the world's first space salvage operation during his second Shuttle mission. He also became the last of six astronauts to use the Manned Maneuvering Unit (MMU) jetpack, as he and spacewalking companion Joe Allen worked to retrieve two malfunctioning satellites for their return to Earth. In a humorous moment, Gardner acknowledged the successful retrieval of the satellites by holding up a large "For Sale" sign, for what has become one of the iconic images of the Shuttle program.

Dale Allan Gardner was born on November 8, 1948, in Fairmont, Minnesota, and grew up in Sherburn, Minnesota and Savanna, Illinois, although he considered his hometown to be Clinton, Iowa. He graduated as Valedictorian of his class from Savanna Community High School, Illinois, in 1966, and received his BSc in

Engineering Physics from the University of Illinois (Urbana-Champaign) in 1970. After graduating, he entered into active service with the U.S. Navy and was assigned to the Aviation Officer Candidate School at Pensacola, Florida. He was commissioned an ensign and selected as the most promising naval officer from his class. In October 1970, he began Basic Naval Flight Officer training with the VT-10 squadron at Pensacola, graduating with the highest academic average ever achieved in the history of the squadron. He proceeded to the Naval Technical Training Center at Glynco, Georgia, for Advanced Flight Officer training and was selected as a Distinguished Naval Graduate and awarded his Naval Flight Officer wings on May 5, 1971. He then attended the Naval Air Test Center at Patuxent River, Maryland, from May 1971 to July 1973, where he was assigned to the Weapons Systems Test Division and involved in initial F-14 *Tomcat* developmental test and evaluation as Project Officer for Inertial Navigation and Avionics Systems.

Gardner's next assignment was with the first operational F-14 squadron (VF-1) at Naval Air Station (NAS) Miramar, San Diego, California, where he participated in two Western Pacific and Indian Ocean cruises while deployed aboard the aircraft carrier USS *Enterprise* (CVN-65). From December 1976 until July 1978, he was assigned to Air Test and Evaluation Squadron 4 (VX-4) at NAS Point Mugu, California, involved in the operational test and evaluation of Navy fighter aircraft.

TERRY J. HART

The first time Terry Hart experienced the mysteries of space was the night when Sputnik, the first Soviet satellite, flew over his childhood home in Pittsburgh, Pennsylvania, in October 1957. He recalls stepping outside to catch a glimpse of the beach ball-sized satellite traversing the evening sky and was immediately captivated. But not everyone was quite as enthralled. "My neighborhood was panicked," he reflected. Americans were fearful that similar Soviet satellites could be armed with nuclear weapons, ready to rain down on them, but Sputnik fascinated Hart. "The space race was on," he said. [4]

Terry Johnathan Hart was born on October 27, 1946. He graduated from Mt. Lebanon High School, Pittsburgh, Pennsylvania, in 1964 and then graduated with a BSc in Mechanical Engineering from Lehigh University in 1968.

From 1968 to 1978, Hart was employed as a member of the Technical Staff of Bell Telephone Laboratories (now known as AT&T) in Whippany, New Jersey, where he received two patents. During those years, he also earned an MSc in Mechanical Engineering from the Massachusetts Institute of Technology (MIT) in 1969, and entered active duty with the USAF Reserve in June 1969. He completed undergraduate pilot training at Moody AFB, Georgia, in December 1970, and from then until 1973 he flew F-106 interceptors for the Air Defense Command at Tyndall AFB, Florida, at Loring AFB, Maine, and at Dover AFB, Delaware.

In 1973, he joined the New Jersey Air National Guard. During this time, Hart – who by now had logged 3,000 hours flying time, 2,400 of them in jets – saw an advertisement calling for new NASA astronauts in a magazine for the National Guard, and decided to apply.

STEVEN A. HAWLEY

Steven Hawley applied to be an astronaut after seeing a flier on a bulletin board in graduate school at the University of California Santa Cruz. Although he had followed America's space program in his youth – creating his own cardboard box spacecraft – and had given serious thought to becoming a pilot, he decided to choose a career in astronomy and took on an undergraduate degree at the University of Kansas, where he received a BSc in Physics and Astronomy in 1973 with the highest distinction. Four years later, NASA began recruiting scientists to become astronauts at just the right time for Hawley. In his astronaut career, he launched the Hubble Space Telescope (HST) and flew another mission to work on it. He later said that, as an astronomer, it gave him a great thrill to be involved directly with the science that the telescope helped to uncover.

Steven Alan Hawley, PhD, was born on December 12, 1951, in Ottawa, Kansas, but considers Salina, Kansas, to be his hometown. He graduated from Salina (Central) High School in 1969. He then attended the University of Kansas, and received Bachelor of Arts (BA) degrees in Physics and Astronomy (graduating with highest distinction) in 1973. He spent three summers employed as a research assistant: in 1972 at the U.S. Naval Observatory in Washington, D.C.; and in 1973 and 1974 at the National Radio Astronomy Observatory in Green Bank, West Virginia. Hawley attended graduate school at Lick Observatory, University of California Santa Cruz, where he was awarded his PhD in Astronomy and Astrophysics on August 26, 1977. Five months later, he was named as one of America's newest astronauts, and at the tender age of just 26, one of the youngest.

Prior to his selection by NASA in 1978, Hawley was a postdoctoral research associate at Cerro Tololo Inter-American Observatory in La Serena, Chile.

JEFFREY A. HOFFMAN

Jeffrey Alan Hoffman, PhD, was born in Brooklyn, New York, on November 2, 1944, but considers Scarsdale, New York, to be his hometown. Growing up with a love of science fiction, he can recall one clear night watching the first Soviet Sputnik fly over with his dad and a few friends, in the football field of the Scarsdale High School from which he subsequently graduated in 1962. He recalled that the older brother of his best friend was a ham radio operator, and was able to tune in to Sputnik's 'beep beep' signal.

Hoffman received a BA in Astronomy (graduated *summa cum laude*) from Amherst College in 1966, and in 1971 he was awarded his PhD in Astrophysics from Harvard University. Dr. Hoffman's original research interests were in high-energy astrophysics, specifically cosmic gamma rays and X-ray astronomy. His doctoral work at Harvard was for the design, construction, testing, and flight of a balloon-borne, low-energy, gamma ray telescope.

From 1972 to 1975, during postdoctoral work at Leicester University in England, he worked on several X-ray astronomy rocket payloads. "That was very interesting because I got some insight into the European space program. I have had a long-standing interest in European affairs," he stated in 2009, which culminated in him becoming the NASA European representative for four years towards the end of his astronaut career. [5] From 1975 to 1978, Hoffman worked in the Center for Space Research at MIT, as project scientist in charge of the orbiting HEAO-1 A4 hard X-ray and gamma ray experiment launched in August 1977. His involvement included pre-launch design of the data analysis system, supervising its operation post-launch, and directing the MIT team undertaking the scientific analysis of the flight data being returned. He was also involved extensively in analysis of X-ray data from the SAS-3 satellite being operated by MIT. His principal research was the study of X-ray bursts, about which he authored or co-authored more than 20 papers.

An avid mountaineer and sky diver, he once found himself in dire trouble when his parachute failed to open fully and he began to spin wildly as he hurtled to the ground. Remaining calm despite the circumstances, he cut the half-open parachute from his back, knowing from his altimeter that he only had seconds to spare. Still spinning, he then deployed his reserve chute which, to his immense relief, popped open. He was just 244 meters from the ground.

SHANNON M. W. LUCID

The oldest of the Group 8 female candidates, Shannon Matilda Wells Lucid, PhD, was born in Shanghai, China, on January 14, 1943, the daughter of Baptist missionaries J. Oscar and Myrtle Wells. She came into the world at the height of World War II, and was just six weeks old when her parents were captured at gunpoint by advance troops of the Japanese army. She was held in Shanghai's Chapei Civil Assembly Center prison camp over the next year and a half. Eventually, late in 1944, the family would be transferred back to the United States aboard the Swedish ship *Gripsholm* as part of a peaceful exchange of noncombatant citizens. Following the Japanese surrender and subsequent withdrawal from China, the family returned to Shanghai where her parents wanted to continue their missionary work. Unfortunately, they were captured and held once again, this time by the advancing Communist Chinese, and thrust into yet another prison camp. Shannon was just six years old. Eventually, the family was expelled from China and returned home. After living for a time in Lubbock, Texas, they moved back to Bethany, Oklahoma, the family's original hometown.

Fig. 4.3: A young Shannon Lucid (left) during her upbringing in China

Shannon Wells graduated from Bethany High School in 1960, and was accepted at Wheaton College in Illinois, where she majored in chemistry before transferring to the University of Oklahoma. She attained her BSc in Chemistry from there in 1963. While studying at university, her passion for airplanes led her to take flying lessons, paid for by babysitting and cleaning college dormitory rooms. Following her graduation, Shannon took on work at the Oklahoma Medical Research Foundation while continuing her flying lessons, eventually accumulating sufficient flying hours to be awarded a commercial pilot's license with instrument and multi-engine ratings. In 1970, she decided to resume her studies and obtain her doctorate in biochemistry. She would be awarded her MSc and PhD in Biochemistry from the University of Oklahoma in 1970 and 1973, respectively, before returning to her work at the Foundation.

Dr. Lucid's experience includes a variety of academic assignments, such as teaching assistant at the University of Oklahoma's Department of Chemistry from 1963 to 1964; senior laboratory technician at the Oklahoma Medical Research Foundation from 1964 to 1966; chemist at Kerr-McGee in Oklahoma City from 1966 to 1968; graduate assistant at the University of Oklahoma Health Science Center's Department of Biochemistry and Molecular Biology from 1969 to 1973, and research associate with the Oklahoma Medical Research Foundation from 1974 until her selection to the astronaut candidate training program.

JSC Director George Abbey later stated that Dr. Lucid's motherhood (of three children) had not been taken into consideration when judging her application to become an astronaut. Interestingly, Dr. Lucid had been so keen to join NASA's astronaut corps that her application was the first to reach JSC.

RONALD E. McNAIR

A softly-spoken African-American physicist, Ron McNair grew up in rural Lake City, South Carolina, the son of an auto body repairman and an elementary school teacher. It was here that he developed an interest in and a drive for everything, but his early education was very much up to himself. "It was catch-up all the way for me," he said during his astronaut training. "I learned most of my high school science on my own. It was truly a disadvantaged situation. Even though I graduated first in my class, I had lots of catching up to do and I knew it. I was a hungry kid, academically." His specialty was research on lasers and molecular spectroscopy, but he was prepared to set that aside for the chance to fly into space. "I had a game plan at one time," he added, "but this has kind of interrupted it. After this, it will be hard to go back to the lab and turn knobs." [6]

Ronald Erwin McNair, PhD, was born on October 21, 1950, and as a first-grader who was always talking about the Soviet Sputnik satellite, he gained the nickname "Gizmo." [7] He graduated from Lake City's Carver High School in 1967 and subsequently received a BSc in Physics from North Carolina A&T State University in 1971 and a PhD in Physics from MIT in 1976. An accomplished saxophone player and a sixth-degree black belt in taekwondo, he co-authored a study, published in the *Scientific American,* which analyzed the physical aspects of a karate strike and its interaction with a target. One of the images in the paper showed McNair breaking three concrete patio slab blocks (each measuring 40 cm long x 19 cm wide and 4 cm thick) using the heel of his right palm. [8]

While at MIT, Dr. McNair produced some of the earliest developments in chemical HF/DF and high-pressure CO_2 lasers. His later experiments and theoretical analysis on the interaction of intense CO_2 laser radiation with molecular gases provided new understandings and applications for highly excited polyatomic molecules. Earlier, in 1975, he had studied laser physics with many authorities in the field at École d'Été de Physique Theoretique in Les Houches, France. Following his graduation from MIT in 1976, he became a staff physicist with Hughes Research Laboratories in Malibu, California. His assignments included the development of lasers for isotope separation and photochemistry, utilizing non-linear interactions in low-temperature liquids, and optical pumping techniques. McNair also conducted research on electro-optic laser modulation for satellite-to-satellite space communications, the construction of ultra-fast infrared detectors, ultraviolet atmospheric remote sensing, and the scientific foundations of the martial arts. He published several papers in the areas of lasers and molecular spectroscopy and gave many presentations in the United States and abroad.

In 1978, Dr. McNair was presented with an honorary Doctor of Law degree from North Carolina A&T State University.

RICHARD M. MULLANE

Growing up in New Mexico with a passion for space and rocketry, Richard ("Mike") Mullane would pour his dreams of satellites and space travel into a number of model rockets that he built as a hobby during his high school days in Albuquerque. After building them, Mullane would take them out into the desert and fire them off. He even won first prize for his rockets in a high school science fair. Although he knew his poor eyesight would preclude him from becoming an astronaut – or so he thought at the time – he had always been keen on joining the USAF, but his eyesight also prevented him from qualifying as a pilot. Instead, he turned his attention to aeronautical engineering, and following his graduation from the U.S. Military Academy, he trained as a weapons systems operator and served in Vietnam, where he completed 150 combat missions. He was subsequently awarded six Air Medals and the Air Force Distinguished Flying Cross, among many other honors.

Richard Michael Mullane was born on September 10, 1945, in Wichita Falls, Texas, but considers Albuquerque, New Mexico, to be his hometown. He graduated from St. Pius X Catholic High School, Albuquerque, in 1963. Setting his sights on a career in the USAF, he achieved a BSc in Military Engineering from the U.S. Military Academy at West Point in 1967. Following graduation, he elected to join the Air Force, was commissioned a second lieutenant and was then assigned to Mather AFB, California. He then completed 150 combat missions as an RF-4C weapon system operator while stationed at Tan Son Nhut Air Base, Vietnam, from January to November 1969. He subsequently served a tour of duty from 1970 to 1973 with the 32nd Tactical Reconnaissance Squadron of the Royal Air Force in Alconbury, England.

On his return home, Mullane was awarded an MSc in Aeronautical Engineering from AFIT in 1975. Then, in July 1976, upon completing the USAF Flight Test Engineering Course (Class 75B) at Edwards AFB, California, he was assigned to the 3246th Test Wing at Eglin AFB, Florida, as a flight test weapon system operator. He was serving there when he applied for a position as an MS with NASA.

GEORGE D. NELSON

George ("Pinky") Nelson flew on the crews of Shuttle missions STS-61C and STS-26, the flights which immediately preceded and followed the loss of Shuttle *Challenger* in January 1986. His first space flight was on STS-41C in April 1984, when he and fellow Group 8 astronaut James 'Ox' van Hoften walked in space twice to repair a malfunctioning satellite named Solar Max, earlier deployed to make observations of solar phenomena. In doing so, Dr. Nelson became one of only six astronauts (four from the 1978 selection) to fly untethered in space using NASA's MMU[2], and the first astronaut to repair a satellite in orbit. On STS-61C, Nelson attempted to capture light-intensified images of Halley's Comet. Though he saw the comet, the light intensifier failed to work, frustrating the astronomer who hoped for the best as he took regular 30-second exposures.

With his youthful good looks, bright blond hair and blue eyes, Nelson looked more like a champion surfer than a dedicated scientist who once did research in solar physics in a New Mexico observatory. He also came into the astronaut cadre with the nickname "Pinky," which he said he received shortly after his birth due to his complexion, and it just stuck (his middle name, Driver, was his mother's maiden name).

George Driver Nelson, PhD, was born in Charles City, Iowa, on July 13, 1950, but considers Willmar, Minnesota, to be his hometown. In his NASA Oral History, Nelson admitted that "I can't remember the time when I didn't want to be an astronomer, from the time I was [four at least]." **[9]** He graduated from Willmar Senior High School in 1968 (after attracting the interest of baseball scouts), and achieved a BSc in Physics (with distinction and departmental honors) from Harvey Mudd College in Claremont, California, on June 4, 1972. He would later attain his MSc in Astronomy from the University of Washington in 1974. The following year, he became a researcher in solar physics at the Sacramento Peak Solar Observatory in Sunspot, New Mexico.

---

[2] The six, in order of first MMU flight were: Bruce McCandless, *Bob Stewart*, *Pinky Nelson*, *Ox van Hoften*, Joe Allen and *Dale Gardner* (Group 8 astronauts in *italics*).

In 1976, Dr. Nelson travelled to Europe, where he spent a further year performing astronomical research at the Astronomical Institute in Utrecht, the Netherlands, and the University of Göttingen Observatory in West Germany. On March 17, 1978 he received his PhD in Astronomy from the University of Washington with a dissertation entitled, "Convection in the Surface Layers of the Sun and Stars."

At the time he was selected by NASA, Dr. Nelson was a postdoctoral research associate at the University of Colorado's Joint Institute for Laboratory Astrophysics in Boulder, Colorado.

ELLISON S. ONIZUKA

Ellison Onizuka began reaching for the stars early in his life. As a young boy, he used to cut out pictures of airplanes from magazines, telling his grandmother he wanted to be a pilot when he grew up. He once said that he first thought of becoming an astronaut when he was just 13 years old. His grandfather patiently told him not to waste his time, saying he should study something useful, like medicine or dentistry. Back then, flying into space was still the stuff of dreams for any youth, but he loved nothing better than looking up at the stars and planets through a telescope at Honolulu's Bishop Museum. His fascination with space travel really took root in October 1962, when Mercury astronaut Wally Schirra splashed down near Hawaii after completing his epic six-orbit space flight and spent a few hours in Honolulu before flying back to the mainland.

Ellison Shoji Onizuka was born on June 24, 1946, in Kealakekua, Kona, Hawaii. He graduated from Konawaena High School, Kealakekua in 1964. An Eagle Scout, he considered his experiences in the Boy Scouts of America program a positive influence in his life: "The leadership, self-confidence and development of personal attributes ingrained [at] youth by the scouting program is tremendous," he wrote to his former Boy Scout Advisor Norman Sakata in November 1982. [10] He attained his BSc and MSc in Aerospace Engineering in June and December 1969, respectively, from the University of Colorado. He then entered active duty with the USAF in January 1970, after receiving his commission at the University of Colorado through the four-year ROTC program as a distinguished military graduate. As an aerospace flight test engineer with the Sacramento Air Logistics Center at McClellan AFB, California, he participated in flight test programs and systems safety engineering for the F-84, F-100, F-105, F-111, EC-121T, T-33, T-39, T-28, and A-1 aircraft. He attended the USAF Test Pilot School from August 1974 to July 1975, receiving formal academic and flying instruction in performance, stability and control, and systems flight testing of aircraft as a member of Class 74B. Also in his class was future fellow Group 8 astronaut Dick Covey, as well as Capt. Jane L. Holley, the first female pilot to attend the elite pilot school.

Fig. 4.4: Ellison and Lorna Onizuka on their wedding day, June 7, 1969 (Image courtesy of the Onizuka family).

In July 1975, Capt. Onizuka was assigned to the Air Force Flight Test Center at Edwards AFB, initially serving on the USAF Test Pilot School staff as Squadron Flight Test Engineer and then later as Chief of the Engineering Support Section in the Training Resources Branch. His duties involved instruction of USAF Test Pilot School curriculum courses and management of all flight test modifications to general support fleet aircraft (A-7, A-37, T-38, F-4, T-33, and the NKC-135 *Stratotanker* military aerial refueling aircraft) used by the Test Pilot School and the Flight Test Center. Prior to his assignment as a NASA astronaut, he had logged more than 1,700 hours flying time.

JUDITH A. RESNIK

Dr. Judy Resnik, who attained her PhD in Electrical Engineering from the University of Maryland in 1977, was an obvious choice for NASA. She came to the space agency not only as a gourmet cook and classical pianist, but also as an outstanding student. While attending Firestone High School in Akron, Ohio, she had excelled in solid geometry and calculus in an age when girls rarely studied math, earned a perfect score of 1600 in her SATs, and graduated as Valedictorian of her class. She attended Carnegie-Mellon University in Pittsburgh, initially as a math major but later switching to engineering.

Judith Arlene Resnik, PhD, was born in Akron, Ohio, on April 5, 1949. As a child, she attended Hebrew School at Beth El Synagogue, where she celebrated becoming Bat Mitzvah. She then graduated from Firestone High School in Akron in 1966, and attained her BSc in Electrical Engineering from Carnegie-Mellon University in 1970. Upon graduating from university, she was employed by RCA in Moorestown, New Jersey, and in 1971, transferred to the RCA operation in Springfield, Virginia. Her projects while with RCA as a design engineer included circuit design and the development of custom integrated circuitry for phased-array radar control systems; specification, project management, and performance evaluation of control system equipment; and engineering support for NASA sounding rocket and telemetry systems programs. She authored a paper concerning design procedures for special-purpose integrated circuitry.

From 1974 to 1977, Dr. Resnik was a biomedical engineer and staff fellow in the Laboratory of Neurophysiology at the National Institute of Health in Bethesda, Maryland, where she performed biological research experiments concerning the physiology of visual systems. Also in 1977, she received her PhD in Electrical Engineering from the University of Maryland. Immediately prior to her selection by NASA in 1978, she was a senior systems engineer in product development with the Xerox Corporation at El Segundo, California.

SALLY K. RIDE

At the age of 11, with the encouragement of her mother, Sally Ride took up the sport of tennis. She learned quickly under the instruction of Alice Marble, a four-time U.S. Open women's singles champion, and within a few years she was a regular on the junior tennis circuit, ranked 18th nationally. Her confidence and prowess in the sport won her partial scholarship to Westlake, an exclusive prep school for girls in Beverley Hills, California. While studying there, she found the subject of physics intriguing, and was able to combine the two interests until her graduation in 1968, when she enrolled at Swarthmore College in Pennsylvania as a physics major. After three semesters, however, she dropped out to concentrate on her tennis. Still maintaining an interest in physics, she later signed on for a physics course at the University of California. At one stage, tennis champion Billie Jean King tried to steer her into the professional tennis circuit, but Ride had decided to return to her studies and took on a double major in Physics and English Literature at Stanford University, graduating with two BSc degrees in 1973. From there, she moved into the field of astrophysics, and tennis took a back seat to her exciting new career. [11] Years later, when a child asked her what made her decide to become a scientist rather than a tennis player, she laughed and said, "A bad forehand."

Sally Kristen Ride, PhD, was born in Encino, Los Angeles, California, on May 26, 1951. She attended Portola Junior High (now Portola Middle School) and then Birmingham High School before graduating from the private Westlake School for Girls in Los Angeles on a scholarship. She then attended Swarthmore College for three semesters, took physics courses at UCLA, and then entered Stanford University as a junior, graduating with BSc degrees in English and Physics. She would later earn an MSc in Physics in 1975 and a PhD in Astrophysics in 1978, also from Stanford. Her graduate work involved research on the interaction of X-rays with the interstellar medium, with astrophysics and free electron lasers as her specific areas of study.

Dr. Ride was finishing studies at Stanford University and on the lookout for a job when she happened to see an advertisement in the *Stanford Daily* that said NASA was calling for a new group of astronauts, including women. "A light flashed," she said of that momentous day. "As soon as I saw the ad, I knew that's what I wanted to do." She read the list of qualifications, said to herself, "I'm one of those people," and wrote off for an application form.

M. RHEA SEDDON

When she was selected by NASA in 1978, Rhea Seddon, who stands at just 5 feet 2 inches (1.57 meters), was the smallest person to become an astronaut. As a result, she struggled to cope with ladders and spacesuits that were designed for more traditionally-sized pilot types. A medical doctor and trained surgeon, Rhea Seddon flew on three Space Shuttle missions, logging more than 722 hours in space and orbiting the Earth 480 times. In her role as an MS, she performed numerous scientific and medical experiments, which included recording the first ultrasound of a human heart in space.

Dr. Seddon's medical background, she believes, was what helped to raise her profile over many others interviewed by the NASA selection panel. That, and the fact she held a private pilot's license.

Margaret Rhea Seddon, MD, was born in Murfreesboro, Tennessee, on November 8, 1947, and graduated from Central High School in Murfreesboro in 1965. As a young girl, like so many others in that era, Seddon was interested in the space program and would have liked to have been a part of it, but she also knew that NASA only hired male pilot astronauts, and that women were still excluded. Instead, she decided to go to medical school and pursue surgery, because she knew she would find the work interesting and challenging. Nevertheless, she maintained the hope that one day things would change at NASA, and felt that a background in surgery would make her an attractive candidate. With this determination in mind, she also undertook flying lessons at a local flight school, at the end of which she received her private pilot's license. During July 1969, Seddon's father took her and some friends water skiing near to their home. After an exhausting day out in the sun and water skiing, they stayed up all night to watch the Apollo 11 moon-walk on TV.

In 1970, Seddon received a BA in Physiology from the University of California, Berkeley, and her Medical Doctorate from the University of Tennessee College of Medicine in Memphis in 1973. After medical school, Dr. Seddon completed a surgical internship and three years of a general surgery residency in Memphis, with a particular interest in nutrition in surgery patients. Between her internship and residency, she served as an Emergency Department physician at a number of hospitals in Mississippi and Tennessee, also serving in this capacity in the Houston area in her spare time. She also performed clinical research into the effects of radiation therapy on nutrition in cancer patients.

ROBERT L. STEWART

When selected by NASA in January 1978, Robert ("Bob") Stewart became the first astronaut who was attached to the U.S. Army. On his first flight into space in February 1984, he became one of the first two American astronauts to make an untethered spacewalk using the MMU.

Robert Lee Stewart was born on August 13, 1942, in Washington, D.C. In his youth he absolutely loved anything to do with aviation, checking out as many books from the local public library as he could. His mother was not happy about finding aviation books everywhere she looked, "…under a pile of dirty clothes on the floor, I stuck them in the refrigerator and everywhere around [the house]," he said in 1984. She eventually gave up harassing him and told him to "Go fly if you want to." [12]

Stewart graduated from Hattiesburg High School, Mississippi in 1960, the same year he received his private pilot's license, and taking part-time civilian flying jobs gave him extra money while studying for his math degree. He attained his BSc in Mathematics from the University of Southern Mississippi in 1964. He then entered active duty with the United States Army in May 1964, completing the Air Defense Officer Basic Course later that year and following that with an assignment as an air defense artillery director at the 32nd NORAD Region Headquarters (Semi-Automatic Ground Environment – SAGE), Gunter AFB, Alabama. In July 1966, after completing rotary wing training at Fort Wolters, Texas, and Fort Rucker, Alabama, he was designated an army aviator. He flew 1,035 combat hours from August 1966 to 1967, primarily as a fire team leader in the armed helicopter platoon of "A" Company, 101st Aviation Battalion (designated 336th Assault Helicopter Company).

In November 1968, Stewart became an instructor pilot at the U.S. Army Primary Helicopter School (USAPHS), Fort Wolters, Texas, serving one year in the pre-solo/primary-1 phase of instruction and about six months as Commander of Methods of Instruction Flight III, training rated aviators to become instructor pilots. He is a 1969 graduate of the U.S. Army's Air Defense Artillery School Air Defense Officers Advanced Course, and Guided Missile Systems Officers Course. He was subsequently reassigned from the USAPHS in March 1972, the same year that he attained his MSc in Aerospace Engineering from the University of Texas at Arlington.

From March 1972 through April 1974, Stewart served in Seoul, Korea with the 309th Aviation Battalion (Combat) as Battalion Operations Officer and Battalion Executive Officer. He next attended the U.S. Naval Test Pilot School at NAS Patuxent River, Maryland, completing the Rotary Wing Test Pilot Course with Class 65 in June 1974. He was then assigned as an experimental test pilot to the U.S. Army Aviation Engineering Flight Activity at Edwards AFB, California. His duties there included Chief of the Integrated Systems Test Division, as well as participating in engineering flight tests of UH-1 and AH-1 helicopters and U-21 and OV-1 fixed-wing aircraft. He also served as Project Officer and Senior Test Pilot on the Hughes YAH-64 advanced attack helicopter during government competitive testing, and participated with Sikorsky Aircraft Corporation test pilots in developing an electronic automatic flight control system for the new army transport helicopter, the UH-60A *Black Hawk*. He came to NASA with military and civilian experience in 38 types of airplanes and helicopters and had logged approximately 6,000 hours total flight time. In 1980, while still a serving U.S. Army officer on secondment to NASA and after completing Ascan training, Stewart attended the U.S. Army Command & General Staff College. **[13]**

KATHRYN D. SULLIVAN

Among her many outstanding accomplishments with NASA, MS Kathy Sullivan entered the history books as the first American woman to complete a spacewalk. History also records that she was beaten to the honor of becoming the world's first woman to complete that feat by just under three months, by cosmonaut Svetlana Savitskaya who left the Salyut-7 space station on a 3-hour 35-minute EVA on July 25, 1984. During the STS-41G mission, Dr. Sullivan spent an equivalent amount of time on her spacewalk, together with fellow astronaut Dave Leestma, successfully completing a satellite refueling test outside Shuttle *Challenger* on October 11, 1984.

Kathryn Dwyer Sullivan, PhD, was born on October 3, 1951, in Paterson, New Jersey, but considers Woodland Hills, California to be her hometown. For a short time in elementary school, she and fellow Group 8 candidate Sally Ride were classmates. Sullivan graduated from Taft High School, Woodland Hills, California, in 1969. With a deep interest in maps and oceanography, she went on to become an Earth Sciences major at the University of California Santa Cruz, then spent 12 months from 1971–1972 as an exchange student at the University of Bergen in Norway. Her BSc (with honors) in Earth Sciences was awarded in 1973 from the University of California. Five years later, in May 1978, Sullivan would receive her PhD in Geology from Dalhousie University, in Halifax, Nova Scotia.

Her doctoral studies at Dalhousie University had included her participation in a variety of expeditions, under the auspices of the U.S. Geological Survey, Woods Hole Oceanographic Institute, and the Bedford Institute. This research included the Mid-Atlantic Ridge, the Newfoundland Basin, and fault zones off the coast of southern California. Further adding to her impressive résumé, she taught second-year and first-year labs and tutorials at Dalhousie from 1973 to 1975 and worked for the Geological Survey of Canada as a research student in the summer of 1975. She also became a lieutenant commander in the U.S. Naval Reserve.

In 1977, Dr. Sullivan learned of NASA's call for several new astronauts from her younger brother who was a jet pilot, and decided to fill out an application. Her brother did likewise, but was unsuccessful.

NORMAN E. THAGARD

Norman Earl Thagard, MD, was born in Marianna, Florida, on July 3, 1943, but considers Jacksonville, Florida, to be his hometown. He once told his high school classmates at Paxon Senior High School in Jacksonville that he wanted to be a medical doctor, a fighter pilot, an engineer, and an astronaut. He became all four. Thagard graduated from high school in 1961, after which he attended Florida State University where he received BSc and MSc degrees in Engineering Science in 1965 and 1966, respectively, and subsequently performed pre-med course work.

Thagard joined the USMC Reserve in September 1966, achieved the rank of captain the following year, was designated a naval aviator in 1968, and was subsequently assigned to duty flying F-4s with VMFA-333 at Marine Corps Air Station Beaufort, South Carolina. From January 1969 to 1970, he flew 163 combat missions while serving with VMFA-115 in Vietnam. For his service, he was awarded 11 Air Medals, the Navy Commendation Medal with Combat "V", the Marine Corps "E" Award, the Vietnam Service Medal, and the Vietnamese Cross of Gallantry with Palm.

After returning to the United States, Thagard received an assignment as Aviation Weapons Division Officer with VMFA-251 at Marine Corps Air Station Beaufort, South Carolina. He resumed his academic studies in 1971, pursuing additional studies in electrical engineering, and a degree in medicine. He went to the University of Texas Southwestern Medical School, earning his Medical Doctorate in 1977. Prior to coming to NASA, he was interning in the Department of Internal Medicine at the Medical University of South Carolina.

As a pilot, Dr. Thagard had logged over 2,200 hours flying time, the majority in jet aircraft.

JAMES D. A. VAN HOFTEN

On the STS 51-I mission in August 1985, James van Hoften (nicknamed "Ox") performed the first manual grapple and manual deployment of a satellite in orbit. During their seven-day mission, the crew aboard Shuttle *Discovery* was tasked with salvaging, repairing and redeploying the ailing Navy Syncom IV-3 satellite. This would all be achieved during an EVA by MS van Hoften and Bill Fisher. Attached to the spindly robot arm, or Remote Manipulator System (RMS), van Hoften was able to grasp the satellite and coax it into *Discovery*'s payload bay, where Dr. Fisher carried out the necessary repairs. Then, still perched on the end of the RMS, van Hoften managed to heave the hot-wired, seven-ton satellite back into orbit, concluding a breathtaking service operation in orbit. "'There that bad boy goes!" van Hoften reported after re-launching the now fully-functional communications satellite.

James Dougal Adrianus van Hoften, PhD, was born on June 11, 1944, in Fresno, California, but considers Burlingame, California to be his hometown. He graduated from Mills High School in Millbrae, California, and then entered the University of California Berkeley, where he earned his BSc in Civil Engineering in 1966. He then did his graduate work at Colorado State University, and earned his MSc in Hydraulic Engineering in 1968. Several years later, he resumed his graduate studies, completing his PhD in Hydraulic Engineering in 1976. The eight-year gap between his master's and doctorate degrees was due to his participation in the Vietnam War as a pilot with the USN.

"Ox" van Hoften enlisted in the USN in 1969, and received his initial pilot training in Pensacola, Florida, before completing his jet pilot training in Beeville, Texas, in November 1970. He was then assigned to NAS Miramar, California, where he flew the F-4 *Phantom*. In 1972, he was assigned to the VF-154 Air Group on the aircraft carrier USS *Ranger* (CV-61), and participated in two cruises to Southeast Asia where he flew approximately 60 combat missions during the Vietnam War. He remained on active duty status with the USN until 1974, when he was able to resume his academic studies.

Following the completion of his PhD at Colorado State University in August 1976, van Hoften accepted a position as Assistant Professor in Civil Engineering at the University of Houston, where he taught fluid mechanics and performed research into artificial heart valves. To earn extra money, he also sold swimming pools [14].

While he was a graduate student during his doctoral studies, he was able to remain in the USAF/USN Reserve, and so was able to maintain his flight status. Initially, he flew F4Ns with the Navy's Reserve Fighter Squadron 201 at NAS Dallas for three years from 1974 to 1977, before becoming a member of the Texas Air National Guard with the 147th Fighter Interceptor Group, and serving as a pilot in the F4C from 1977. He was at the University of Houston when selected by NASA, by which time he had logged more than 3,300 flying hours, most of it in jet aircraft.

Fig. 4.5: Class of 1978 Mission Specialist astronaut hopefuls. (clockwise from top left): Millie Hughes-Wiley; Byron Lichtenberg; Dr. Richard Terrile (image courtesy of SpaceFacts.de); Dr. Carolyn Griner; Dr. Mary Helen Johnston

SOME WHO NARROWLY MISSED SELECTION

The NASA selection panel had to deal with selecting 20 MS candidates out of a pool of over 5,000 qualified hopefuls from an all-encompassing range of scientific and medical fields. Inevitably, some superbly proficient and disciplined people had to miss out at the last hurdle. These are just a few of those talented candidates.

Millie Elizabeth Hughes-Wiley, BSc, PhD

At the time of her JSC interviews and tests in August 1977, 31-year-old Millie Hughes-Wiley from Mineral Wells, Texas, was working at the Veterans Administration Hospital in San Francisco. She was later divorced and married once again, this time to George Fulford. Although never selected as a NASA astronaut, Dr. Hughes-Fulford flew into space in June 1991 aboard Shuttle *Columbia* as a Payload Specialist (PS) on the STS-40 Spacelab Life Sciences (SLS-1) crew. This was the first Spacelab mission dedicated to biomedical studies. The SLS-1 crew completed more than 18 experiments during a nine-day period, bringing back more medical data than any previous NASA flight. STS-40/SLS-1 would prove to be her only space flight. Today, Dr. Hughes-Fulford is a professor at the University of California Medical Center in San Francisco, where she continues her research.

Byron Kurt Lichtenberg, Sc.B, Sc.D, S.M

Originally from Stroudsburg, Pennsylvania, Byron Lichtenberg is an American engineer and former fighter pilot who, like Millie Hughes-Fulford, was never selected as a NASA astronaut but flew instead as a PS, in his case on two NASA Space Shuttle missions (STS-9 and STS-45). In 1983, he and German Ulf Merbold became the first PS to be launched into space when they flew on the STS-9/Spacelab-1 mission. Over the next ten days in orbit, Lichtenberg conducted multiple experiments in life sciences, materials sciences, Earth observations, astronomy and solar physics, and upper atmosphere and plasma physics. His second flight was on the ATLAS-1 (STS-45) Spacelab mission aboard *Atlantis* in 1992, conducting 13 experiments in atmospheric sciences and astronomy.

Richard John Terrile, PhD

Today, Richard Terrile is an astronomer and the director of the Center for Evolutionary Computation and Automated Design at NASA's Jet Propulsion Laboratory (JPL). He has a Doctorate in Planetary Science from the California Institute of Technology (CalTech) and has developed simulated missions to Mars and the outer solar system. He would also make unsuccessful applications for NASA astronaut groups 9 and 10 and was a semi-finalist for a Spacelab mission, but failed to be selected. Dr. Terrile is the discoverer of four moons around Saturn, Uranus and Neptune, and took the first pictures of another solar system around the nearby star Beta Pictoris. His other interests include planetary rings, planetary geology, evolutionary computation, and the development of medical instrumentation for tissue identification during neurosurgery.

Carolyn Spencer Griner, BSc

In 1964, Carolyn Griner, from Granite City, Illinois, joined NASA as a co-op student, becoming one of only three women employed in technical positions at the Marshall Space Flight Center (MSFC) in Huntsville, Alabama. They were outnumbered by their male engineering counterparts at a ratio of more than 1,000 to one. She received her BSc degree in Astronautical Engineering from Florida State University in 1967 and completed graduate work in industrial and systems engineering at the University of Alabama. Dr. Griner then progressed to positions of increasing responsibility within several key program areas at NASA, and served a term as Acting Director of MSFC in 1998. Before and after this, she was Deputy Director at MSFC before retiring from NASA in 2000. She has received numerous awards throughout her career, including the NASA Exceptional Service Medal in 1986, the Presidential Rank of Meritorious Executive in 1992, and the Presidential Rank of Distinguished Executive in 1995. In 1999, Carolyn Griner also received the space agency's highest honor, the Distinguished Service Medal, for her exemplary contributions to NASA's missions.

Fig. 4.6: Carolyn Griner (at front), Mary-Helen Johnston (left rear) and Ann Whitaker wearing scuba gear at the MSFC Neutral Buoyancy Simulator.

Mary Helen Johnston, BSc, PhD

Mary Helen Johnston (now McCay) was born in September 1945 in West Palm Beach, Florida. She gained her BSc degree in Engineering Science from Florida State University in 1969 and her Doctorate in Metallurgical Engineering from the University of Florida, Gainsville in 1973. Dr. Johnston began working at MSFC in Alabama and, following unsuccessful bids to join the astronaut corps in 1977 and 1980, was one of those chosen for Scientific Payload Specialist Selection 3 in 1984. She was then named as Alternate PS (together with Eugene Trinh), to the prime PS Taylor Wang and Lodewijk van den Berg, for Shuttle mission STS-51B/Spacelab 3. The mission eventually flew with Wang and van den Berg on board, and then the loss of the *Challenger* Orbiter precluded her from flying any later Spacelab missions during the Shuttle program.

In 1974, Johnston was joined by Carolyn Griner, Ann Whitaker and fellow researcher Doris Chandler for a series of NASA/MSFC experiments. Their participation in a five-day simulated Spacelab experiment that year proved the benefit of having a highly-trained crew who were ready for anything, from identifying minor malfunctions to salvaging experiments.

In December 1975, Johnston again joined Carolyn Griner and Ann Whitaker, this time in the MSFC Neutral Buoyancy Simulator as part of their efforts to design and develop new experiments for astronauts to conduct in microgravity. Each specialized in a different field: Dr. Griner in material science; Dr. Whitaker in lubrication and surface physics; and Dr. Johnston in metallurgical engineering. Their job was to identify what tasks could be successfully completed in microgravity, and which would require additional foot- or hand-holds or specialized tools. The experiments they designed were eventually flown and operated within Spacelab, the European-built science module carried in the spacious cargo bay of the Space Shuttle.

Ann Whitaker was also a PS candidate for a time, along with Craig Fischer, Michael Lampton, Robert Menzies, Byron Lichtenberg and Rick Terrile. Although she never flew on the Space Shuttle, her materials experiments did, and these pioneering experiments provided data used to design the International Space Station (ISS). Today, Dr. Whitaker is head of the Science Directorate at NASA's MSFC and leads a team of 700 researchers.

References

1. "NASA Readies a Nighttime Dazzler," Anastasia Toufexis, *Time* magazine, issue August 29, 1983, Vol. 122, No. 9.
2. John M. Fabian, NASA Oral History, February 10, 2006.
3. Anna L. Fisher, NASA Oral History, February 17, 2009.
4. Sarah K. Satullo, online article, "*Lehigh University professor says his path to becoming a NASA astronaut 'all serendipity'*," 13 March 2014, online at: http://www.lehighvalleylive.com/bethlehem/index.ssf/2014/03/retired_nasa_astronaut_and_leh.html].
5. Jeffrey A. Hoffman, NASA Oral History, April 2, 2009.
6. "An Astronaut's Hunger," unaccredited article, *San Francisco Chronicle*, issue 24 August 1979, p. 17.
7. *Challenger daughter's journey to know her hero dad*, Tom Patterson, CNN, June 16, 2012.
8. *The Physics of Karate*, Michael S. Feld, Ronald E. McNair and Stephen R. Wilk, *Scientific American*, Vol. 240, No. 4, April 1979, pp. 150–161.
9. George D. Nelson, NASA Oral History, May 6, 2004.
10. "Scouting helped him to 'be prepared'," In Memoriam issue, *Hawaii Tribune-Herald*, Sunday, February 9, 1986, p. 111.

11. **Men and Women of Space**, Douglas B. Hawthorne, Univelt Publishers, San Diego, California, 1992.
12. "Robert Stewart: First Army pilot to fly in space," Olive Talley, UPI Archives, February 2, 1984.
13. Additional biographical data from U.S. Army Career Brief, undated but circa 1984, on File, AIS Archives.
14. Reference 11, p. 754.

The authors would like to acknowledge, with appreciation, the use of biographical information contained within NASA's official astronaut biographies in the compilation of this chapter.

5

All change in the Astronaut Office

"There is nothing wrong with change,
if it is in the right direction."
Winston L. Spencer-Churchill (1874−1965)
Two-term Prime Minister of the United Kingdom
(1940−1945 and 1951−1955)

During the two weeks after their names were announced to the world, life became a bit of a whirlwind for the 35 individuals publicly identified as NASA's latest group of astronauts. As the reality dawned that they had succeeded in making the selection, being figuratively 'rocketed' from relative obscurity to making national and international news reports and creating the headline in local hometown papers was, for most of them, well outside their comfort zone.

In her 2015 biography *Go for Orbit,* Rhea Seddon revealed that the day before the official announcement of her selection, ABC News correspondent Jules Bergman told her that he had been reliably informed by his "sources in Houston" that she was going to be one of the names revealed. In the middle of what could sometimes be a 100-hour working week in the Memphis VA (Veterans Affairs) Hospital, and naturally cautious about such unconfirmed news, Seddon graciously deferred an out-of-the-blue offer by ABC to fly her to New York to appear on the TV show *Good Morning America* the day after the formal announcement. Such a conversation was so out of the ordinary that she asked herself "Is my life about to change so radically?" Wondering whether she had indeed made the cut for selection was already enough to deal with, so she put any crazy thoughts of an interview on a big U.S. TV morning show to the back of her mind and focused on the reality of her work in the hospital.

Of course Bergman was right, and the next day Seddon was informed of her selection by George Abbey, who also warned that after the official announcement

© Springer Nature Switzerland AG 2020
D. J. Shayler, C. Burgess, *NASA's First Space Shuttle Astronaut Selection*,
Springer Praxis Books, https://doi.org/10.1007/978-3-030-45742-6_5

there might be some press interest. That was something of understatement, as she later recalled "After noon, the press began to call and arrive at the hospital in droves. The VA Public Affairs staff set up a press conference. What a circus it was! I learned about the click-wheeze of the camera motor every time I looked up and smiled." Seddon was careful not to say anything about topics she knew nothing about. Fortunately, all went well, as did the promised appearance on the *Good Morning America* show the following day. "Having my face on the front of newspapers and on television meant my week was filled with calls day and night." Many of these were from family or friends expressing their congratulations and best wishes, but there were also strangers who called at all hours, introducing Seddon to the downside of fame and requiring her to seek an unlisted number. Her life, and those of her 34 colleagues, was indeed about to change forever. [1]

INTRODUCING THE THIRTY-FIVE NEW GUYS

On February 1, 1978, just two weeks after the announcement of their names, the 35 new Shuttle-era astronauts found themselves standing self-consciously on stage in front of an audience of NASA employees and reporters, gathered inside the Olin E. ('Tiger') Teague Auditorium of NASA's Building 2 on site at the Johnson Space Center (JSC)[1]. They were gathered for their official welcome as NASA's newest astronauts, and to take in a three-day period of briefings and orientation. As flashbulbs popped, the 35 were warmly welcomed by Dr. Chris [Christopher C.] Kraft. Following this, their names were read out one after another by a NASA Public Relations officer, with each announcement greeted by a round of applause – which was noticeably louder for the six women candidates. The women were undoubtedly the principal focus of media attention that day as, to a lesser extent, were the minority candidates. The others were almost ignored and mostly reduced to spectators as the media's fascination with the six women continued. "I could have mooned the press corps and I would not have been noticed," reflected Mike Mullane. "The white males were invisible." [2]

Kathy Sullivan recalled how Carolyn L. Huntoon and Ivy F. Hooks were excellent counselors to the six females. Though all six were very accomplished career women who were clearly very smart, capable and confident, none of them had ever experienced anything like what they were going through when being introduced as America's first female astronauts. "I don't remember the particulars, like each being called out on the stage," Sullivan recalled in 2007. "I think that's what happened, but I don't have any real vivid memory of standing backstage, hearing my name, and walking out on stage like actors at the Oscars. That didn't register. There we are out

---

[1] Olin Earl 'Tiger' Teague (1910–1981) was a WWII veteran and was congressional representative for the sixth congressional district of Texas for 32 years. A member of the House Committee on Science and Astronautics, he also chaired the Manned Space Flight Subcommittee.

on stage, and there's this 'click, click, quote, quote', and about two minutes later they say, 'Thank you. This is the end of the formal event, and any of the astronaut candidates are now available for the rest of the day for media interviews'.

"We'd been told that was the deal. You walk out, you do appear as a group, and then there are media interviews. Well, you know, the six of us gals plus Fred [Frederick D. Gregory], Ron [Ronald E. McNair], and Guy [Guion S. Bluford] plus Ellison [S.] Onizuka were odd people. There had never been critters that looked like us admitted into the astronaut corps. I remember the conversations before we all went out there, probably some of them with Carolyn telling us gals in particular, 'There's a lot of media interest. There might be some for you guys. It's going to be a long day'. So I braced for that. We eventually came to refer to our class as 'ten interesting people and 25 standard white guys'. This event was one of the things that started that, because the 25 standard white guys were done about four minutes after the formal event, left, and had the whole rest of the day free. They could go run. They could do whatever they wanted. The other ten of us, we were there till Lord only remembers. I don't remember what time – way late." **[3]**

Fig. 5.1: Part of the orientation tour in early 1978 was to get badged and measured for formal NASA flight suits. Here, during a fitting session at Ellington AFB, Judy Resnik tries on the jacket part of the in-flight coverall, while behind her John Creighton is sized for headgear.

Doubling the size of the Astronaut Office

On July 10, 1978, the new astronauts were officially welcomed to the first Astronaut Candidate (Ascan) training program at NASA's JSC in Houston, Texas. Since their selection, the group had had to find accommodation and arrange to move to Houston (see sidebar: *Hunting for Houses in the Heat of Houston*). As the group was so large, not everyone arrived in Houston at the same time, but on that day in July the formal training program began. Originally envisaged as a two-year program, NASA stated that it would determine which of the candidates would be 'promoted' at the end of that period to full astronaut status, while adding that the agency expected all 35 to pass the course. However, until the Ascan program was completed, they were not officially classed as astronauts. Any who did not make the cut would initially be offered alternative employment within NASA should they wish it. Fortunately for this pioneering group every candidate performed well, indeed better than originally expected or hoped for, and passed the course with flying colors.

HUNTING FOR HOUSES IN THE HEAT OF HOUSTON

Moving to the humid, dry heat of Houston was one of the challenges the new astronauts candidates faced in the early summer of 1978, but the reward was seen to be far greater than the pitfalls of the Texas climate, even for those who had been born there. "I wasn't looking forward to coming to Houston," recalled John Fabian in 2006. "I don't like the city and don't like living here. I was born here, so I can say that without being accused of anything adverse. It's not a very pleasant place to live, but it was a great job, and I really loved it." [4] For Fred Gregory and his family, "the heat and humidity was something that we had not been used to. Houston is a lot different than San Antonio [even though both are in Texas], and we'd never been in a coastal city. So I will be honest, it took a couple of years, really, to adapt to the temperature and humidity and lack of true seasons. But once we adapted – and we were there for 14 or 15 years – then it was just absolutely normal." [5] In one of her house hunting trips, Anna Fisher discovered there were very few windows that opened in the house she was looking at. "There was maybe a bathroom window that opened. The real estate agent said, 'This is Houston, you're not going to want to open your windows'. Funny the things you remember." [6]

The task of finding a property that was both suitable and in the right price bracket was another issue to address, as the impression that all astronauts made a lot of money simply by having the title alone or by flying in space is a myth. Besides, the group had yet to begin their training let alone earn their

(continued)

astronaut wings. "We went out with this one realtor," Dick Covey recalled in 2006, "and she was trying to figure out where we might want to live and what we were going to do. And we're saying, 'Well, God, I don't know really'. And she says, 'Well, tell me how much you make, because that will help me'. Now, she's thinking, 'These people are going to be astronauts. They're making lots of money'. I remember the Mullanes were with us in the car, and we say, 'Well, we're Majors in the Air Force. We make x', – whatever it was – 'a year'. She slammed on the brakes, and she turned around, and said, 'Welders make more than that'. I thought, 'Well, they must be well paid, I guess'. It was kind of an eye-opener in that she didn't think much about how much we got paid. And it didn't get any different." **[7]**

"So we drove down there, and they were waiting for us," explained Fred Gregory. "I mean, we went to a bank to get a loan on a house, and they had a very special rate for us. Everybody was extremely helpful in Clear Lake City [Texas]. We rented a house, initially, as we looked for a house to buy [and] a couple of months after we got there, we did buy a house." **[5]**

"The first thing that I remember about the move to Houston is that the housing was so incredibly expensive," explained Mike Mullane, "I had a house that I think we ended up selling for $55,000 in Florida, and when I got to Houston, I mean, to live in a $55,000 house in Houston, you'd better be looking at a doublewide down in some trailer park. The housing seemed to be incredibly expensive. That's the first sensation, the first thing that shocked me when I got down there. But like everybody else, you finally work a way around it and end up getting a house. It wouldn't have bothered me to live in a doublewide if I'd gotten a chance to fly into space." **[8]**

When the new group arrived it was the largest influx of astronauts for a decade, but despite this, there was very quickly a sense of community within the group and where they chose to live near to the space center, especially for some with a new or young family to settle in as well. "It was a little bit of an ordeal," Terry Hart remembered 25 years later. "It was the middle of the summer; I guess we showed up toward the end of June. We had come earlier in the spring, of course, and found a house, and that was a nice experience. We'd met a real nice realtor and got settled in El Lago [Texas] with a nice little home for us to move into. Of course, Jim Buchli was right around the corner from us and Steve Nagel was a block away, so we were all kind of in the same community, and that was all very nice for the families." At this point Hart's wife was six months pregnant. "I remember driving down with the dog, and then she flew down a little bit later with our one-year-old. And I think her mother and her sister came out and helped us get settled. But it was so nice, because it was such a wonderful community. When I'd been in

(continued)

the military before, as a single person, I understood how military families work so well to support each other. It's just a wonderful environment. And I was pleased to see that NASA had that and more, and it was a very supportive environment to bring a family in, under these rather exciting conditions. But [it was] also stressful, for people moving, especially when you're six months pregnant." [9]

Some of the new arrivals were able to purchase houses previously occupied by some of the veteran astronauts who had left the program, while others were now their neighbors. "I bought a house in Nassau Bay, on Back Bay Court," recalled Hoot Gibson in 2013. "It was on a cul-de-sac, and there were three houses in the cul-de-sac, and right next to me was Owen [K.] Garriott, and right next to him, on the other side, was Joe Engle. So this was Astronaut Cul-de-Sac. That was '78. Then Rhea [Seddon] and I married in 1981. She really didn't like that house, and so immediately, she set about looking for another house. That had been Al [Alfred M.] Worden's house originally, the house that I was living in. Rhea set about looking for another house, and she found [one] that was two and a half times as expensive and that had been Rusty [Russell L.] Schweickart's house, originally. We sold my house and bought that house. At the time, it was owned by Craig [L.] Fischer, who was one of our NASA flight surgeon docs here. He was selling it because he had a teenage daughter who was having a lot of trouble with asthma here in Houston. The doctors said, 'You're going to have to get her out of here', so he went and moved the whole family to Palm Springs [California]. He hated to leave. We bought his house. That was a waterfront house over on, I guess it's called Cow Bayou, which is just off of Clear Creek. We had a friend that kept his boat behind our house – we had 100 feet of bulkhead out behind the house – kept his 30-foot cabin cruiser back there. We could sail down Clear Creek and through Clear Lake and out into Galveston Bay from our house. That was the house that Rhea wanted, and so we lived in that one for 14 years."

It was also handy having colleagues in the office to help move house, as Gibson explained, "We actually moved into the house while Rhea was in the hospital giving birth, which was July of '82. She left for the hospital from the house on Back Bay Court and by the time she came home, we had moved into the other house, on Barbuda Lane. [Other] astronauts moved us. The Ace Moving Company is what we were known as. When somebody would be moving, the word would go out, "Hey, so-and-so is going to move on Saturday; everybody show up that can help." You'd have 25 astronauts show up to help move. Usually, it happened in about three or four hours, you'd move a whole entire house. They helped us move." [10]

(continued)

Once the move had been made, the group quickly discovered how focused the local area was to the space program and JSC, as Fred Gregory explained in 2004. "The school system was terrific in that independent school district down there. The majority of the people who lived in Clear Lake City were people who were associated with the Space Center. All the services supported the Space Center. All the major aerospace industries were located down there. Many of the neighbors were. So it was like moving into utopia, and so the transition was great. Now, once you got into the neighborhood, then you were just part of the neighborhood. Our friends were friends because they were neighbor kid friends, and their kids went to the same school, or our kids went to the same school that they went to. So it was a wonderful place. [5]

Standing out from the crowd

In order to distinguish themselves from the earlier astronaut groups, the new cadre had decided to give themselves a nickname, the 'Thirty-Five New Guys,' or TFNG, designing patches and T-shirts in order to foster closer camaraderie (see sidebar: *It's All in the Name*). Judy Resnik designed one with 35 astronauts crammed, sardine-like, into every available corner of a Shuttle, and proudly displaying the Shuttle's "We Deliver" motto that would later become famous around the world.

As Mike Mullane observed in his 2006 memoir, *Riding Rockets*, concerning the group's nickname, military pilots also knew of a commonly-used phrase with a similar acronym. "A sign of our closeness, we now had our class name," he wrote. "There was no official requirement that a new class of astronauts name themselves. It just happened. For us, TFNG stuck. In polite company, it translated to 'Thirty-Five New Guys.' Not very creative, it would seem. However, it was actually a twist on an obscene military term. In every military unit, a new person was an FNG, a F***ing New Guy. You remained an FNG until someone newer showed up, and then they became the FNG. While the public knew us as the Thirty-Five New Guys, we knew ourselves as The F***ing New Guys." [2]

IT'S ALL IN THE NAME

From 1962, the name of a new astronaut group was normally, though not always, assigned by members of the previous class. Up to 1969, it usually reflected the number of candidates rather than a specific nickname (see Table 5.A)

(continued)

TABLE 5.A THE ORIGINAL ERA OF NASA ASTRONAUT SELECTIONS 1959–1969

Class of	NASA group	Month selected	Class members	Selection category	Class name
1959	1	April	7	Test pilots	The Original (or Mercury) Seven
1962	2	September	9	Test pilots	The Next Nine
1963	3	October	14	Jet pilots	The Fourteen
1965	4	June	6	Scientists	The Scientific Six
1966	5	April	19	Jet pilots	The Original Nineteen
1967	6	August	11	Scientists	The Excess Eleven (XS-11)
1969	7	August	7	Ex-MOL pilots	The MOL Guys
Total selected in 7 groups			**73**	**Comprising 56 pilots and 17 scientists**	

The naming of the Group 8 candidates started a new tradition of choosing a group nickname, which has continued in the Astronaut Office ever since. There has never been an official requirement for a class to receive a name, but the only group not to have adopted one was the class selected in 1985 (Group 11). Between 1978 and 2004, twelve groups were chosen specifically to train as crewmembers on the Space Shuttle (see Table 5.B)

TABLE 5.B THE SPACE SHUTTLE ERA OF NASA ASTRONAUT SELECTIONS 1978–2004

Class of	NASA group	Month selected	Class members	Pilots	Mission Specialists	Class Name
1978	8	January	35	15	20	Thirty-Five New Guys
1980	9[1]	May	19	8	11	19+80
1984	10	May	17	7	10	The Maggots
1985	11	June	13	6	7	No Name Chosen
1987	12	June	15	7	8	The GAFFers
1990	13	January	23	7	16	The Hairballs
1992	14[1]	March	19	4	15	The Hogs
1994	15[1]	December	19	10	9	The Flying Escargots
1996	16[1]	May	35	10	25	The Sardines
1998	17[1]	June	25	8	17	The Penguins
2000	18	July	17	7	10	The Bugs
2004	19	May	11	2	9[2]	The Peacocks
A total of 12 selections spread across 26 years			**248**	**91**	**157**	

[1]Additional International Mission Specialist Candidates chosen in group (2 in 1980; 5 in 1992; 4 in 1994; 9 in 1996; 7 in 1998.)

[2]In the 2004 selection, three of the Mission Specialists were former teachers termed Educator Mission Specialists. This group was the final team chosen to train on and fly the Space Shuttle.

When the 20th selection was made in 2009, they became the first group not to train on Space Shuttle systems or procedures as they would not fly until

(continued)

after the Shuttle fleet was retired in 2011. The 2009 group was also the first not to include any pilots in the selection, as training focused on crewing the International Space Station, the Russian Soyuz, and the new Crew Transport Vehicles from Boeing and SpaceX, as well as assisting in the development of future exploration vehicles, procedures and systems for a return to the Moon, Mars, the asteroids and deep space exploration (see Table 5.C).

From 2009, the designation of active astronauts in the Astronaut Office changed. No longer would the division of astronauts between Pilot and Mission Specialist – begun in 1978 with the TFNG – be used with new Ascans. Those qualified to fly in the T-38 front seat would now be called "professional pilots." When the Shuttle program ended in 2011 – and until its replacement flies with a crew – there was no longer a requirement to define astronauts as Pilot or Mission Specialist, only as ISS crew members.

TABLE 5.C: THE SPACE STATION AND EXPLORATION ERA OF NASA ASTRONAUT SELECTIONS FROM 2009

Class of	NASA group	Month selected	Class members	Class Name
2009	20[3]	June	9	The Chumps
2013	21	June	8	The Eight-Balls
2017	22[3]	June	12	The Turtles
			29	

[3]Additional international astronauts chosen (2009, 5 selected; 2017, 2 selected)

In his 2013 Oral History interview, Robert 'Hoot' Gibson was asked if he thought the women in the group felt that they were part of the guys, or if the men thought the women were just one of the guys and part of the 'gang', or whether there was some separation between the two. Gibson did not agree with the latter, stating, "I think I called them guys, too, and I considered them one of the guys. I believe they considered themselves one of the guys. I never heard any of them say, 'We don't want to be called guys; we don't want to be called The Thirty-Five New Guys'. Never did hear that. I suppose it's possible, but it would surprise me. Actually, it would surprise me to hear of any of them saying, 'Well, I object to being called guys'. Never have heard that." [10]

The 1978 group created another new tradition in the Astronaut Office by creating a Group Emblem, something which had not been adopted for the earlier selections. The idea of a group patch was led by Guion Bluford. "During the first year of training, I worked with Bob McCall, the artist, to develop a patch that represented our class," Bluford explained in 2004. "Bob had designed the flight patch for STS-1 and I asked him to do the same for our class. He came up with a design which highlighted the Space Shuttle, the 35 members of our class and 1978, the year that we arrived in Houston."

(continued)

According to Bluford, there was no consideration given to special symbolism or messages in the design of the group emblem. [11]

Fig. 5.2: Designed by Robert McCall, the 1978 Group 8 emblem was the first to be officially adopted by a NASA astronaut class. It started a tradition that has continued ever since.

"As Ascans, we called ourselves the 'TFNGs' or 'Thirty-Five New Guys' and Judy Resnik came up with a T-shirt design that illustrated that identity," Bluford recalled. The TFNG T-shirt was a unique element to this astronaut group. Anna Fisher explained that "Judy Resnik was involved with the T-Shirt with Jim [Buchli]. Jim was actually a little bit of an artist. They designed a Shuttle. There's 35 people all over the Shuttle: on the outside, EVA, hanging from the arm. We have the blue version and the red version of this shirt [the colors the group were split into]. So they were the ones who did all of that. [6] John Fabian thought the T-shirts were wonderful. "It was cleverly done, and I know that Judy was involved in it. I would guess probably Hoot Gibson was involved in it, too. But it was kind of presented to me as a done deal; they said, 'Look. Look what we've done', and it was well done." [4]

In an age before the merchandising bubble burst, the TFNG saw applications for their emblem and T-shirt artwork, to help establish their presence in the Office. "We had our T-shirts, we had our mugs, and we had everything,"

(continued)

recalled Rhea Seddon. "We were the TFNGs. We referred to ourselves as the TFNGs. We referred to those other people [astronauts selected before 1970)] as the old guys. They didn't like that, the old guys or the TFNGs." **[12]**

Fig. 5.3: (top) TFNG T-shirt design. (bottom right) TFNG 'Red' team leader Rick Hauck displays the T-shirt during parabolic familiarization flights on the KC-135, February 1979. (bottom left) 'Blue' team leader John Fabian is seen on the aft flight deck of *Challenger* during the STS-7 mission in June 1983. He is wearing the other version of the TFNG T-shirt.

In which we serve

As with earlier NASA astronauts selections, there was significant representation from the U.S. armed forces within the Class of 1978, including the first member of the U.S. Army to be selected for astronaut training. Flying experience was obviously an important element in choosing the four pilot selections of 1959, 1962, 1963 and 1966. It was less important for the seven former Manned Orbiting Laboratory (MOL) astronauts who transferred across to NASA in August 1969, two months after the United States Air Force (USAF) program had been cancelled. In their case, with around 30 astronauts from the 1965, 1966 and 1967 selections still awaiting their first flights, the qualifying factor for the "MOL-Guys" was the upper age limit of 36, set in expectation of a long wait for a flight.

Of the two groups of scientists selected in 1965 and 1967, those members who had not previously attended and graduated from a military flying course were dispatched to do so shortly after their selection. Despite having qualified from this arduous course and then maintained their flying quota with NASA, none of the Scientist-Astronauts who decided to remain with the agency and eventually flew in space ever commanded a mission. For the Group 8 selection, flying credentials were important for selection as a potential Shuttle Pilot astronaut (PLT), but test flying was not as highly prioritized as in earlier selections. For Mission Specialist (MS) candidates, a prior flying career was not necessary, and since they would not be required to 'fly' the Space Shuttle they would not be subjected to a jet-pilot course during their Ascan training program. Instead, their qualifying criteria depended more upon an accredited academic or engineering background.

Among the 35 candidates, 20 were representatives from the main four branches of the U.S. armed forces, while four of the remaining 15 had previously served in the military at one time (see Table 5.1). Of the 15 pilots candidates selected, 14 were from the military (eight from the U.S. Navy (USN) and six from the USAF) while the other was a pilot with NASA at the time of his selection. One interesting fact is that 20 of the TFNG had previous combat experience.

Among the MS candidates, seven were from the military (four from the USAF, and one each from the USN, the U.S. Marine Corps (USMC) and the first representative from the U.S. Army). Ten of the MS had never served in the military (Hawley, Hoffman, Fisher, Lucid, McNair, Nelson, Resnik, Ride, Seddon and Sullivan, although Sullivan later served as an oceanographer in the USN Reserve from 1988 until 2006).

TABLE 5.1: SERVICE AFFILIATION OF ASTRONAUT CANDIDATES (CLASS of 1978) AT TIME OF SELECTION

Serving Military Personal Seconded to NASA			
Air Force	**Navy**	**Marine Corps**	**Army**
Bluford	Brandenstein	Buchli	Stewart
Covey	Coats		
Fabian	Creighton		
Gregory	Gardner		
Mullane	Gibson		
Nagel	Hauck		
Onizuka	McBride		
Scobee	Walker		
Shaw	Williams		
Shriver			
Civilian Ascans Service Affiliation Prior to Selection			
Air Force	**Navy**	**Marine Corps**	
Hart	Griggs	Thagard	
	Van Hoften		

A bright bunch.
The TFNG as a group held an impressive collection of educational attainments at selection, with no less than 35 degrees (see Table 5.2). There were ten bachelor's degrees, 13 master's degrees and 12 doctorates. Of these degrees, 13 were in engineering, three in medicine, three in aeronautics and/or astronautics, two in physics, two in astronomy, two general Bachelor of Science degrees from the Naval Academy and one each in math/physics, engineering physics, Earth science, administration of science and technology, information systems, management engineering, engineering mechanics, fluid mechanics, biochemistry and astrophysics.

TABLE 5.2: EDUCATIONAL ATTAINMENT OF ASTRONAUT ANDIDATES (CLASS OF 1978) AT TIME OF SELECTION

Bachelor's degrees		Master's degrees		Doctorates	
Name	Degree	Name	Degree	Name	Degree
Brandenstein	Math/Physics	Buchli	A/A[1]	Bluford	Engineering
Creighton	Bachelor of Science	Coats	Admin, Science & Technology	Fabian	A/A[1]
Gardner	Engineering Physics	Covey	Engineering	Fisher	Doctor of Medicine
Gibson	Engineering	Gregory	Information Systems	Hawley	Astronomy
McBride	Engineering	Griggs	Management, Engineering	Hoffman	Astrophysics
Nagel	Engineering	Hart	Engineering	Lucid	Biochemistry
Scobee	Engineering	Hauck	Engineering	McNair	Physics
Sullivan	Earth Sciences	Mullane	Engineering	Nelson	Astronomy
Walker	Bachelor of Science	Onizuka	Engineering	Resnik	Engineering
Williams	Engineering	Ride	Physics	Seddon	Doctor of Medicine
		Shaw	Engineering Mechanics	Thagard	Doctor of Medicine
		Shriver	A/A[1]	Van Hoften	Fluid Mechanics
		Stewart	Engineering		

[1]A/A refers to a degree attained in aerospace systems, astronautics, or a combination of aeronautics and astronautics.

Changing Times
So what did all this mean? For the first time, an astronaut group included both pilots and non-pilots, career military officers and civilians, professional engineers, scientists, doctors and educators, the first female and minority candidates and the first representative from the U.S. Army. The eighth class represented a new era of American astronauts, setting the precedent for subsequent selections over the next

25 years as NASA chose further groups to train for and fly the Space Shuttle. As for the pilots, there was a clear shift towards a broader range of flying credentials and away from the focus on flight test experience that had dominated the earlier selections. The 35 also held a wide range of educational qualifications and experiences, and though none were classed as professional 'teachers', many had given instruction to classes and held 'professorships'[2].

When the new astronauts arrived at JSC, their numbers swelled the Astronaut Office significantly for the first time in years. Not only that, but the makeup of the office immediately changed. Many of the veteran, 'original' NASA astronauts had retired by this point but some of the 'older hands' were still in senior positions, with the Astronaut Office headed by a naval pilot (at that time, John Young). With far broader academic and engineering backgrounds, as well as a range of experience from a new generation of military personnel, and the presence of females in the previously male bastion of the Astronaut Office (other than a few administrative assistants and secretaries), a new approach had to be found, and quickly.

WALKING IN HALLOWED HALLS

For the TFNG, their first entry into the upper floors of Building 4 in July 1978 must have been as exciting and daunting as appearing in front of the press five months previously. Some of them related their experiences of those first few days in their Oral History interviews:

"We were the first group in a long time, nine or ten years, I guess," reflected Loren Shriver in 2002. "I think the folks who were still in the Astronaut Office, which, of course, had been between programs for several years of that period, were glad to see us there on the one hand, because they were all really busy doing the technical things that astronauts do while they're waiting to go fly – various inputs to boards and panels, and safety inputs and crew displays, and all that kind of thing. They were all really busy, and I think they were happy to see us show up so that we would be able to help them and take some of the load.

---

[2]The first professional, qualified teachers chosen for space flight training were Christa McAuliffe and Barbara Morgan, who were selected under the Teacher in Space (TIS) Program in 1985. After the loss of McAuliffe on STS-51L *Challenger*, Morgan continued her association with NASA while resuming her teaching career. In June 1998, Morgan was chosen as an MS with Group 17 and later flew on STS-118 in 2007. In the 1990s, the Educator Astronaut Project was created as a successor to the TIS program. Three teachers, Joseph M. Acaba, Richard R. Arnold and Dorothy Metcalf-Lindenburger, were chosen as MS candidates in 2004 (Group 19) and all three later flew on Space Shuttle missions. Acaba and Arnold subsequently completed long duration missions on the ISS.

"At the same time, I think there was a bit of the 'Oh, no, all these new guys. How are we ever going to get them trained and up to speed? Will they ever be ready to go fly in space?' Well that's kind of a natural reaction from the group of people who have been there and done that a lot. That's a bit of a different aspect of 'We're happy to have them here, but I don't know, it's maybe just a little more work for a while until we get them all checked out'.

"The other thing was, since we were the first in a long time, and since the Shuttle, of course, was a new program, basically they were inventing our training as we went along. Now, that sounds worse than it really was. They didn't really know… A good part of Johnson Space Center is devoted to training and training astronauts, but the Shuttle was such a new vehicle, with a lot of totally new systems, that they were learning, first of all, and then trying to decide what a new group of astronauts needed to know that was important. And, again, they were kind of putting this all together as we went through the training." **[13]**

As Michael Coats recalled, "It was really a special time for us, because this class had a real mixture for the first time. Most of the previous classes had either been pilot astronauts or Scientist-Astronauts, including medical doctors. In our class, we had 15 pilot astronauts, and we had six women, we had engineers, scientists, medical doctors – a real mixture of folks. And the age difference was pretty large too. It was fascinating to see the interaction of that class of 35 people, because nobody was really senior to anybody else, and we outnumbered all the astronauts that were already here. There were only 29 astronauts when we got here, and we were 35 more. Suddenly we dominated, but they were glad to see us because they had a lot of work they needed us to do. It was really fun for us because there were enough Apollo astronauts still left over that they were able to mentor us, and we really enjoyed that. Of course, we were still developing the Space Shuttle and learning how the Orbiter was going to operate. We got involved right away in developing operational procedures and flight rules for this amazing new vehicle; it was really a special time." **[14]**

When asked about bonding with the remaining 'original' members of the Astronaut Office during these first few weeks, and the reception that the group were given by all these astronauts, Rick Hauck recalled, "First of all, all of the 'real astronauts', I'll call them, they couldn't have been better to us. They needed us. Through attrition, there were fewer and fewer astronauts there. I don't know how many – I think there were less than 30 astronauts in the office when we got there. Of course, they'd already started gearing up for Shuttle and they needed help. So we were there to be helpful in any way we can. They wanted to get us as smart about the systems as soon as we could." **[15]**

Anna Fisher, one of the first women to become a member of that elite office, later recalled what the reception was like from these veteran male astronauts as this new very diverse class arrived. "It was really interesting. It was fun. They were all very nice. I'm sure privately they had their thoughts, but they never in any way expressed that. I know that with the flight rate the previous ten years they certainly didn't think they needed a bunch of new astronauts; they could quite adequately take care of it themselves, I'm sure. I'm sure that many of them thought that. In fairness to them I would have probably felt the same way, but they were very gracious." At the time of her interview, Fisher reflected on her participation with selecting the current class, which became Group 20 'The Chumps': "I'm involved in how we're developing the Ascan training flow for this new class that we're in the process of selecting. It's interesting, because it's the first class that's not going to have any Shuttle training and [we're] trying to figure out how to do that. For us [the TFNG], I don't think they really thought it out ahead very well. I was expecting to come in and have it be like a military flight school, but really they were inventing how they were going to train us on the fly." **[6]**

Fred Gregory said that the TFNG were like the freshmen, and they (the older astronauts) were the seniors, so "As such, we paid great respect to them. Al [Alan L.] Bean was assigned as our training official, and so Al Bean set up the training schedule and coordinated all of that. John Young was the chief of the office down there. Our offices were all mixed together, and so [Fred W.] Haise [Jr] was in my office, and then I learned about Apollo 13 and what his role had been there. But there were… a lot of folks down there who had not only been Apollo, but they had been Skylab and obviously had come in from the MOL Program, like Richard Truly and [Donald H.] Peterson and Bob [Robert L.] Crippen and people like that. Then the scientists, really, who'd come in in '67. I think Story Musgrave was in that group. But, you know, they were just all there, but us 35, we were in our own world, and they had us eight hours a day in academics or in some kind of orientation that went on for easily six months. Then after that initial orientation of NASA acronyms and locations and things, then we began some initial training in the single-system trainers that they have and into the Shuttle Mission Simulators, the SMS simulators down there. So we did not interface with these guys daily, though they were kind of our super dads down there to make sure that everything was prepared for us." **[5]**

Steve Hawley thought that his group were the first true civilians inducted into the program without the need to be a qualified jet pilots. "NASA did pick civilians in earlier selections [in 1965 and 1967]," he explained, but "one of the first things they did was send them off to military pilot training and teach them to be jet pilots. [Dr. Harrison H.] Jack Schmitt, for example, was a civilian, and [Dr. Edward G.] Ed Gibson and [Dr. William B.] Bill Lenoir and [Dr. Joseph P.] Joe Allen and all those guys. But they all went off and learned to fly jets. [Ours was] the first group where we were accepted as scientists, engineers, doctors, and they didn't try to make us pilots. So in that sense that was unique and different."

Hawley did not recall a lot of interaction with the older astronauts at the start "We pretty much stayed to our group. I actually don't really remember what they thought of us. It may have been that since we hadn't really been flying a lot by '78 – we had flown ASTP [Apollo-Soyuz Test Project], but that was one crew that got to do that, and Skylab and that was three crews, so since we'd landed on the Moon [1969–72], there'd only been four crews to get to fly – and here's this new bunch of guys walking in the door. I could see how some of the guys that had been around for a while waiting to fly might have been a little resentful. If they were, that didn't come across in any way, because our training was separate from what most every-body else was doing. Everybody else was doing mainstream support of Shuttle and development of everything that needed to be done before STS-1. We would cross paths at the Monday morning meeting, or you'd run into them at the gym or some-thing like that, but mostly we did our own thing. But we did have some interaction with them in the context of the training that focused on the history of the program. They thought it was important, and I think it is, that we hear from people that had flown Apollo and had flown Skylab and ASTP. So I remember we got lectures from some of the guys that were still there, some of the guys that had left but came back to talk to us about their flights and what it was like back then. So we had that kind of interaction... These were my heroes, and to actually get to sit in a room and listen to them talk about their flights was pretty awesome." **[16]**

"I remember the first Monday morning meeting that we went into," recalled Rhea Seddon. "I think we started on July 10, on a Monday morning. It was like, 'Report to work Monday morning'. It was the all-astronaut meeting; I can remem-ber the guys that were there. There were 29 that were still there, I think, and 35 of us. Most of them were considerably older than we were, and there were a lot of people in our class that looked considerably younger than they were; Pinky [George D.] Nelson, probably me, Sally. Not only were we young, but we looked young. There were beards. Jeff Hoffman had a beard; Ron McNair had a beard.

"The old guys are sitting watching us come in, and you could feel that they were going, 'Oh, my gosh, what have we gotten ourselves into? Who are these people?' And women for the first time. We were all new and shiny-faced. [For] The military folks, I think it wasn't a difficult transition. It was just a new squad-ron for them. For those of us coming from very diverse academic backgrounds it was [a case of], 'What do we do? What do we say?' Most of us were smart enough to not say anything. We were asked to introduce ourselves, I think. That was our indoctrination into what was going on in the astronaut program. They did the regu-lar meeting and left us to pick up schedules and start going to orientation-type meetings. What do you all do here, what's the plan for getting you trained to be astronauts?" **[12]**

The arrival of the TFNG in the summer of 1978 was indeed a bold step in the story of the American astronaut program that had begun two decades before. In the late 1950s, it was very different circumstances that guided the selection of the nation's first astronauts, notably what type of person to select, how to prepare them, and where to base them.

Fig. 5.4: The first six female astronauts named by NASA to fly on the Space Shuttle stand by a model of the Orbiter. (l-r) Rhea Seddon, Anna Fisher, Judy Resnik, Shannon Lucid, Sally Ride and Kathy Sullivan.

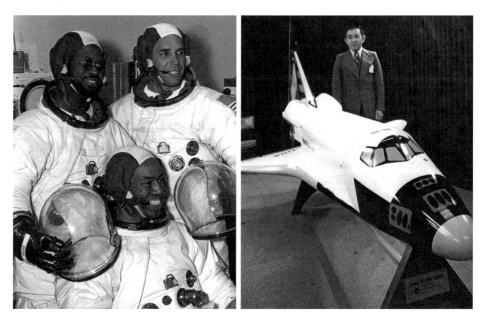

Fig. 5.5: (left) The three African-American candidates, Ronald McNair [left], Guion Bluford [seated] and Fred Gregory [right] pose for photographs wearing Apollo extravehicular mobility units during their orientation visit to JSC [Note: McNair wears the gloves of ASTP astronaut Vance Brand]. (right) The tenth "interesting person", Hawaiian-born candidate Ellison Onizuka

THE ASTRONAUT OFFICE EVOLVES

The original Mercury astronauts were located at the Langley Research Center, Hampton Virginia, from 1959, as part of the Space Task Group (STG). But in 1961 plans were made to support the growing human space flight program at a specific center, and a site in the Clear Lake area of Houston, Texas was chosen. It was constructed between 1962 and 1963 and officially opened in September 1963. Originally called the Manned Spacecraft Center (MSC) it was renamed the Lyndon B. Johnson Space Center (JSC) in 1973 to honor the late Texas-born president.

As the facility was being built, it also took some time to transition the staff to run it, with NASA acquiring a number of sites across Houston. Gradually, however, the various buildings on site at MSC were made ready, and from 1963 the still relatively small astronaut group, comprising the 16 astronauts of the first two selections, moved into their new office in Building 4. Formally part of the Crew Systems Directorate, the 'Astronaut Office' (which used the Mail Code CB) was in fact several offices on the third floor of Building 4. As more astronauts joined the program, so the number of rooms they occupied expanded[3].

After debating the makeup of the first astronaut group – whether military or civilian candidates, what experiences they should have and their level of education – it was decided to select from a cadre of military test pilots who had the required experience and had already passed a high level of security. This set the pattern for the majority of the astronaut selections in the 1960s, although the military test pilot criteria was subsequently relaxed to accept civilians and those with military jet pilot training. Selected in the spring of 1959, the first group of astronauts numbered just seven. All were serving military officers and test pilots, many of whom knew each other from past assignments in their professional careers prior to arriving at NASA. None of them emerged as the formal leader of the group, though strong characteristics were soon evident in several of them, and they agreed to contribute as a single voice representing "the astronauts". Over the next decade, the iconic Mercury Seven were followed by other test and operational jet pilots, as serving military or civilians with prior military service or air industry experience in test or operational assignments, in separate selections in 1962 (9), 1963 (14) and 1966 (19). The pilot groups were completed by the seven transferred from the MOL program in 1969. These selections were all highly professional pilots at the peak of their flying careers, all trained under the military system. Many had actual combat experience, but they all had a wide variety of test and operational flying skills in their military or later civilian roles. The legend of *The Right Stuff* American astronaut was based upon these men.

---

[3] From 1963, the original Building 4 housed the 'astronaut office[s]', its administrative staff and some part-task trainers. The Astronaut Office was relocated to the new Building 4S which opened on November 30, 1992 as part of the Flight Crew Operations Directorate. The new building also included the Mission Operations Support Offices. The old Building 4 became Building 4N and housed the Mission Operations Support Office and the Flight Directors Office.

Creation of an astronaut hierarchy

When a heart condition grounded Mercury astronaut Deke Slayton (USAF) in 1962, preventing him from flying in space for the next decade, he retired his Air Force commission and remained at NASA, assuming the new position as Coordinator of Astronaut Activities from September 1 that year, with responsibility for the day-to-day operation of the Astronaut Office. By default, he also informally held the position of Chief Astronaut. That position was created officially in November 1963 and was filled by the first American in space, Alan Shepard (USN), who had also been medically grounded, in his case with an inner ear ailment, until 1969. When Shepard retired in 1974, he was replaced by John Young (also USN) who held the post until April 15, 1987. Young was succeeded by TFNG and (yet again) naval aviator, Dan Brandenstein.

Despite all the early selections being 'pilots', there were still clear divisions within the Office, with a tendency to favor naval aviators over USAF pilots (rather than Marine pilots). Seniority was seen to be by service, rank, and most notably by selection, so the original seven chosen in 1959 were at the top of the "pecking order", followed by those chosen in 1962, 1963, 1966 and then the 1969 'MOL Guys'. Within this was further division, between test pilots over jet pilots, or those who had more operational or combat experience over more academic or research careers in aviation.

Then there was the introduction of scientists into the office in 1965 and again in 1967, which really did initially create an 'us and them' friction. At the time, all astronauts had to be graduates of a military pilot school, or had to agree to attend and pass a jet pilot course before continuing in their astronaut career. Of the six scientists chosen in 1965 and the eleven selected in 1967, only two (from the 1965 selection) had already earned their pilot wings, so it was off to flight school for the others before returning to continue their astronaut basic training program. This caused some to question their choice in joining the program. Dr. Duane Graveline, from the 1965 selection, left NASA two months after selection due to personal reasons, while from the 1967 selection, Brian O'Leary left in 1968 as did Tony Llewellyn, who had found the jet pilot course too demanding. They were followed by Curt Michel from the 1965 group who left in 1969, disgruntled over such a long wait for a flight.

The Group chosen in 1965 assumed a new nickname as the Scientific Six (until Graveline left, when they became the "Incredible Five") and soon discovered how low down the pecking order they were when it came to crew assignments. The order appeared to be test pilots first, then pilots, and the scientists suspected that even the secretaries and probably the astro-chimps would get to fly before they did.

Two years later, the prospects for a second group of scientists chosen for astronaut training were no better, as they were told on Day One at NASA that they were not really required and it would be a long time before they flew. As a result, they became known as the Excess Eleven (or XS-11). For some, the wait for a flight or assignment to a crew – even a back-up one – was just too long and they left the program. Those who stayed faced years waiting to be assigned. The first of the 1965 group finally flew in space in December 1972 (Jack Schmitt on Apollo 17), with the remaining trio (Joe Kerwin, Owen Garriott and Ed Gibson) each embarking on a long flight aboard Skylab in 1973. Those remaining from the 1967 selection had an even longer wait, with the first representatives of the group (Joe Allen and Bill Lenoir) flying in November 1982 after waiting 15 years. That was just seven months before the first of the TFNG flew, having only been at NASA for five years. The final members of the 1966 and 1967 groups to fly did so in 1985, after all of the MOL group from 1969 and most of the TFNG had flown (some of them twice), as well as one of the few Payload Specialists (PS), McDonnell's Charles Walker. Even the first member of the 1980 selection (Dave Leestma) made it to orbit six months before the last of the 1966 or 1967 astronauts flew[4].

Operational experience, test-flying assignments and even combat experience from tours in Vietnam were common among the pilots of the TFNG (see Table 5.3), but in addition they also held broader education, academic and leadership experience. The aircraft in which they had flown had also changed considerably in the intervening years. Many of the earlier pre-1978 astronauts were veterans of propeller and early jet engine aircraft, and some had flown combat tours during World War II, the Korean War and the early years of the Southeast Asia conflict. Some of the new pilots of the TFNG could also point to their experience of the Cold War, and with a very 'hot war' in Southeast Asia, and like their forebears, some had seen service together in the same units, aboard the same ships or in the same theatre of war prior to moving to NASA. One of the TFNG, Dick Scobee, had also been fortunate to experience flying the experimental X-24 lifting body (unpowered) towards the end of that program. So there were similarities as well as differences between the type of pilots chosen in the TFNG and those of the earlier selections.

However, the biggest changes in selection came with the new MS candidates and what would be required to incorporate them into the previously male pilot-orientated Astronaut Office.

---

[4] The authors have produced four other titles in this series which review the NASA astronaut selection program between 1959 and 1969: *Selecting the Mercury Seven, The Search for America's First Astronauts* (NASA Class of 1959) in 2011; *Moon Bound, Choosing and Preparing NASA's Lunar Astronauts* (NASA Classes of 1962 and 1963) in 2013, both by Colin Burgess; and *NASA's Scientist-Astronauts,* (NASA Class of 1965 and 1967) in 2007, and *The Last of NASA's Original Pilot Astronauts, Expanding the Space Frontier in the Late Sixties* (The USAF MOL Selections and NASA Classes of 1966 and 1969) in 2017, both by David J. Shayler and Colin Burgess.

TABLE 5.3: NASA GROUP 8 - FLYING EXPERIENCE AS OF JULY 1978

Name	Category	Service	Total Hours	Types of aircraft	Hours in jets	Carrier landings	Combat missions	Other
Bluford	MS	USAF	Over 2,200	5	2,200		144	T-38 instructor pilot; FAA Commercial pilot license
Brandenstein	Pilot	USN	2,900	19		400	192	
Buchli	MS	USMC	1,600		1,480			
Coats	Pilot	USN	2,300		2,200		315	
Covey	Pilot	USAF	1,900	8			339	
Creighton	Pilot	USN	2,700			500	175	
Fabian	MS	USAF	2,500		2,000		90	
Fisher, A.	MS	Civilian						Private flying license
Gardner, D.	MS	USN	1,100	20				
Gibson, R.	Pilot	USN	1,700	35	1,500	300	56	Commercial pilot, multi-engine, & instrument ratings; held private rating since age 17
Gregory, F.	Pilot	USAF	3,700	40				Single-, multi-engine fixed & rotary wing military and civilian aircraft including gliders; holds FAA commercial and instrument certificate for single- multi-engine & rotary aircraft
Griggs	Pilot	USN Reserve	5,900	40	4,800	300	300	Single- & multi-engine prop; turboprop and jet aircraft; helicopters, gliders and hot-air balloons; Airline transport pilot license and certified flight instructor; Project Pilot NASA Shuttle Training Aircraft
Hart	MS	USAF Reserve	2,000		1,400			
Hauck	Pilot	USN	3,000		2,500		114+	
Hoffman	MS	Civilian						
Hawley	MS	Civilian						
Lucid	MS	Civilian	1,500					Commercial private flying license; commercial, increment, multi-engine ratings

Name	Position	Background					Remarks
McBride	Pilot	USN	2,800	40	2,300	64	Flown military & civilian aircraft; piloted Navy "Spirit of '76" bicentennial-painted F-4J "Phantom" 1976-1978
McNair	MS	Civilian					
Mullane	MS	USAF				150	RF-4C weapons system operator
Nagel	Pilot	USAF	3,700		1,900		Sports-flying (light aircraft & gliders)
Nelson	MS	Civilian					
Onizuka	MS	USAF					Flight test engineer
Ride	MS	Civilian					
Resnik	MS						Working towards a private flying license
Scobee	Pilot	USAF	5,000	40		Unspecified	Flew the X-24B in two unpowered glide flights
Seddon	MS	Civilian	100				Private license; small single-engine planes
Shaw	Pilot	USAF	3,000	30	644 hrs.		
Shriver	Pilot	USAF	2,600	30			Single- & multi- engine civilian and military fixed wing aircraft and helicopter; holds commercial & private glider ratings
Stewart	MS	USA	4,000	38	1,035 hrs.		Military and civilian aircraft and helicopters
Sullivan	MS	Civilian					Recreational flying; powered and glider aircraft
Thagard	MS	USMCR	800	775		163	
Van Hoften	Pilot	USN	1,500	1,400		100	Private, commercial instrument and multi-engine land pilot certificates.
Walker, D.	Pilot	USN				Unspecified	
Williams, E.	Pilot	USN	3,100	2,900	745	330	

Six months after their selection, NASA issued the first set of detailed biographical data sheets for all 35 candidates. As well as detailing their previous careers, they included facts about their individual flying careers to date. The data did not reveal all the details, but they did show the breadth of the group's experience: from hot-air balloons, gliders and light aircraft, to military jet fighters, heavy transport aircraft, and in one case a lifting body. Naval aviators had completed hundreds of carrier landings, and at least 18 of the 35 had combat experience. Over the next four decades, they would log many more hours in various NASA aircraft, with all of the MS candidates qualified as T-38 crewmembers. To this can be added their flight hours in the fleet of Space Shuttles, supplemented by three of them experiencing the Russian Mir space station, and one flying aboard a Russian Soyuz spacecraft.

Changing the 'old school'

The major difference with the 1978 Group was the arrival women in the office. "I think they didn't quite know what to do about that," recalled Rhea Seddon in her Oral History. "[Some of] the older astronauts had met us when we'd come down for the initial visit. I think it was Wally [Walter M.] Schirra who was quoted in the newspaper, [saying] something stupid about women. I remember one of the astronauts, and I won't mention his name because he would be embarrassed at this point in time, who was asked, 'What do you think about these new women coming in?' and he said, 'Well, some of them are kind of cute'. He didn't mean anything by it. They didn't understand what it was going to be like. There was a little bit of moving around, trying to figure out what was correct, what was offensive and how to work with women as equals. I think there was a broad range of different attitudes in the women about whether we could take kidding or whether we found the things that some people thought were funny were not funny. We had to work that out. I came out of a world where it was all men anyway so to me it wasn't a big deal. I was used to working only with men for the most part. Anyway they worked it out; we worked it out. I think we had to prove that we were serious about what we were doing, that we were willing to do whatever they needed us to do: water survival, parasail training, scuba training. We could carry our own parachutes out to the airplane, and we could handle emergencies. I think they were all watching to see whether or not that was really going to work." **[12]**

In the 1960s, women were simply not allowed to apply for military jet pilot training, hence it was unusual to find any female applicants who had progressed through a flying career which would have given them the required experience to apply to NASA at that time, though several did so. While 13 very capable female pilots unofficially went through the same Mercury medical selection program, and emerged with results which indicated there was no reason why they could not be selected and trained as astronauts, none were. In each of the pilot selections between 1959 and 1966 (and the three MOL selections of 1965, 1966 and 1967), not one female made it to the medicals or shortlists prior to final selection. Interestingly, there was also an absence of female candidates for both of the 1960s Scientist-Astronaut groups.

Aside from the flying criteria, it was also difficult for women to attain the appropriate academic qualifications and experiences, so that even by the early 1970s there were few females who had the right qualifications to be selected into the program, as Rhea Seddon alluded to in her Oral History, "I don't think many of the academics [at that time] had worked with women on their level, because when people applied to the astronaut program, they wanted people with at least a bachelor's degree in science, math, or engineering, preferably advanced degrees, and preferably with experience after that. I remember people saying of all the women that applied, and I think there were about 1,500 of them, many of them didn't have advanced degrees, and their science degrees weren't too strong. There just weren't an awful lot of women in those fields back then. The academics certainly had more experience working with women than say test pilots, but women were still very much in the minority in all those fields, I think. Because if you talk to most of them, they will tell you in their graduate schools it was mostly men. I think all of them had a little bit of difficulty, and I think some of them didn't really want us there. I've heard some stories about people that were upset that they were going to take women. They just didn't feel that needed to be done. That was the old school. I don't think most of them felt that way but some of them, I think, did." [12]

Things clearly had to change at JSC and the selection of the TFNG was instrumental in that change, for both the wider issues and some of the smaller ones as well. "The most obvious was that they needed to add a women's locker room to the astronaut gym," recalled Sally Ride. "That and most other things were accomplished before we arrived. I think JSC worked hard to prepare for the arrival of women astronauts and female technical professionals. The technical staff at JSC – around four thousand engineers and scientists – was almost entirely male. There were just a very small handful of female scientists and engineers, I think only five or six out of the four thousand. The arrival of the female astronauts suddenly doubled the number of technical women at JSC!" [17]

Sally Ride also acknowledged the role played by Carolyn L. Huntoon, who had been on the selection committee, in supporting the six new female astronauts "There was one person who was very important to me at JSC, and it was Carolyn Huntoon. When I was going through the application process, Carolyn was on the selection committee. At the time, she was a PhD biochemist in charge of a small group, but was the highest-ranking technical woman at JSC and therefore deemed to be the expert on everything related to women at JSC. She was the only woman on the selection committee, and it was a large selection committee. Once we arrived, she became almost the de facto liaison to all of the women astronauts; she became a very good friend to all of us and a very important person – especially in helping us steer our way through the first couple of years that we were there.

"If we ever had any problems, we all knew that we could call Carolyn, and we did! This was even as she was rising up through the ranks at JSC and becoming a more and more important person. She was always the person that we could call, and she would always help us solve any problem, no matter how small. So she's one of the very few people that I think I owe my career to. I would hate to think what it would have been like for the six women in our class if she hadn't been there before us, been part of the selection committee, and then been there for us once we arrived. She made our lives much, much easier." [18]

Now that females were to be included in flight crews, one of the questions raised was what would be suitable to put in the Personal Hygiene Kit (PHK) for them on board the Space Shuttle when all the previous crewmembers since 1961 had been male? Answering this question was one of Anna Fisher's early technical assignments, as she explained in 2009. "That was a real interesting assignment, because I had long ago decided when I was in medical school that since I was going to be on call, and having to get up in the middle of the night, that I wasn't going to wear makeup. I didn't want to be one kind of person in the daytime and sleep and have makeup running all over my face; so I went all through medical school with no makeup. I was one of these no-makeup kinds of people. Then there were other people who felt differently about it. It was just really interesting coming up with this kit. My mom is European [and] in Europe they use Nivea® cream a lot. So I had grown up on Nivea® cream, which probably wasn't that big in the United States. I got Nivea® cream on board in the kit. That was one of my actual first fun assignments was to come up with what we wanted in our kit. I remember years later somebody complaining about the Nivea® cream. I said, 'Gee, I don't know how that happened'. It was too greasy or something." Fisher also explained what typical items were included in the kits, such as toothpaste, toothbrush, and creams. "All that stuff then has to go out to White Sands [Test Facility, Las Cruces, New Mexico] and be tested for outgassing and flammability. You can't just take whatever it is you want. So I just let everybody else decide what they wanted, but I picked the cream. That was the thing I cared most about [in that kit]." **[6]**

Sally Ride, the first American female to fly on the Shuttle, added her comments about makeup in the PHK. "It's actually kind of funny, because there was a reasonable amount of discussion about it. The engineers at NASA, in their infinite wisdom, decided that women astronauts would want makeup, so they designed a makeup kit. A makeup kit brought to you by NASA engineers! You can just imagine the discussions amongst the predominantly male engineers about what should go in a makeup kit. So they came to me, figuring that I could give them advice. It was about the last thing in the world that I wanted to be spending my time in training on, so I didn't spend much time on it at all. But there were a couple of other female astronauts who were given the job of determining what should go in the makeup kit, and how many tampons should fly as part of a flight kit. I remember the [male] engineers trying to decide how many tampons should fly on a one-week flight; they asked, 'Is 100 the right number?' I said, 'No. That would not be the right number'. They said, 'Well, we want to be safe'. I said, 'Well, you can cut that in half with no problem at all'. And there were probably some other, similar sorts of issues, just because they had never thought about what kind of personal equipment a female astronaut would take. They knew that a man might want a shaving kit, but they didn't know what a woman would carry. Most of these were male engineers, so this was totally new and different to them." (see sidebar: *When you have to go!*) **[17]**

WHEN YOU HAVE TO GO!

One basic but important function carried out routinely aboard the Space Shuttle was an astronaut's need to visit the toilet occasionally. There was just a single unit aboard, so it had to be unisex, and it had to function perfectly throughout the flight. The crews had been fully trained in its use, in order to avoid any embarrassing and calamitous leakages into the cabin. The one-meter-square, $30 million toilet had no door and was shielded behind a privacy curtain. In order to expel solid waste into the toilet, the astronaut first had to go through the lengthy preliminary procedure of securing themselves using foot restraints and thigh straps to ensure a good seal and the correct alignment on the seat.

Differential air pressure would then act like a vacuum cleaner, dragging the solid waste down and into large containers, where it would be vacuum-dried, chemically treated to remove odors and bacteria, and then compressed. It could not simply be jettisoned from the Shuttle, as it would then be a rather nasty piece of orbiting space junk that could even become a dangerous projectile. Paper was not allowed into the toilet, so this had to be placed into airtight bags for later disposal.

For liquid waste, the male astronauts would urinate into their own personal funnel, placed at the end pf a plastic tube attached to the toilet. The liquid waste would then be vented out into space, where it soon broke down into a harmless gas. They had to ensure, however, that they did not make contact with the funnel with their genitalia while urinating, as this could create a vacuum-like blockage and cause considerable pain to the astronaut. The female astronauts had it slightly easier, as they just had to attach their own self-adhesive cup-shaped funnel before urinating.

When the first females came into the Astronaut Office, the crew systems people had to devise a way for women to use the bathroom during launch, re-entry and on EVA. What they came up with was a diaper system called the DA[C]T, or Disposable Absorption Containment Trunk. This was generally comfortable, as Rhea Seddon observed during her NASA Oral History: "That worked and was acceptable, and they gave us a chance to try it out. I think we flew in the KC-135. I remember the idea was to try to get two voids during the flight so you didn't just have one opportunity. We could try to go with a full bladder and void. They had a little bathroom area or curtained area where we could be. Then [you had to] try to drink enough water that you could pee again before the thing landed. I don't remember an awful lot about it. It was a fairly simple solution that was acceptable. At the time I don't think they had diapers for adults so this was a very expensive thing. They measured you and form-fitted you. It was like a panty girdle with an absorbent middle section. I think there was comment about how much each they cost. We didn't have

(continued)

anything to do with it. We needed something. Obviously, we couldn't use the male variety. We didn't take them on cross-country trips with us or anything like that. We had a few to try out and then once we were happy with it they just put it on board for us. They're fairly comfortable. I think the only thing is [that] because of the padding they make you look heavy. You put your flight suit on, and you looked like a sausage. I never did any of the [EVA] suit training, so I can't tell you whether people were comfortable in the EVA suits. I thought of all of the ideas they came up with that one was the simplest and easiest and most reliable."

Seddon was also involved in evaluating the Shuttle Orbiter Waste Collection System. "I got called over one day to test the seat belt on the toilet. They had several different designs [including] a seat belt-type seat belt. But you couldn't get stable that way. They finally came up with the thigh bars. That was one of the things that I tried out. I went over to the water tank on scuba and just sat on a simulated toilet seat and tried seat belt this way, seat belt that way, thigh bars this way. It was pretty obvious to everybody who tried it that the thigh bars were the things you needed and that they were very simple, easy to use, and kept you very stable [and you] didn't have to hold on to anything. That was my big contribution to the space program.

"I remember testing out the funnel on the KC-135. It wasn't entirely satisfactory. Number one, you had 30 seconds or less. The KC, it's sort of weightless, but because it's a little bit up and down, you can't make it entirely smooth. I'm not sure that we were really comfortable that it would work well for women, the urine funnel, until Sally [Ride] flew and said, 'Works fine'. So that was one that in theory could work; it worked sort of okay on the KC, but you really didn't get a good chance to use it until you got to space. Everybody knew that, and had extra toilet paper, so that if urine got out, you got to go chase it around." According to Seddon, it apparently worked well during STS-7 for Sally Ride. "We talked after her flight and asked a few questions about the stuff that was female-related. She said 'Yes, make sure you do this, make sure you do that'. I don't even remember what it was. She said 'It works fine'.

"A lot of people thought, 'Oh, it's going to be really different'. It just wasn't. When Sally flew, when Judy [Resnik] flew, when Kathy [Sullivan] flew, it was just like you're one of the guys. You get to decide, I think, how modest you want to be or how private you want to be. There was accommodation for everything. If I was going to change clothes, I either went in the airlock or went into the bathroom, pulled the curtain closed. If the guys wanted to change clothes they said, 'I need to change clothes. Turn around'. You just worked out whatever was comfortable for you and your crew and the guys. It wasn't a big deal." [12]

Reflecting on her 30-year career at NASA, Anna Fisher also commented on how the test pilot-based Astronaut Office had changed, to the point where the first female (Peggy A. Whitson) became Chief Astronaut in 2009. "It has changed dramatically in some respects. A lot more rules. When I first came here, [in 1978] you were expected to use good judgment. There weren't rules for every little thing you did, particularly T-38 flying, [but] just across the board. [Maybe] people didn't use good judgment, which resulted in the rules, but there's just a lot more regulations now [2011] and things are spelled out a lot more clearly. The people that I first came into the office with were Vietnam pilots who had also been to test pilot school. They also weren't used to working with women as colleagues. They were used to women in a support role, either as their wives or maybe as secretaries. But the new class of men has grown up with women and are used to women being their colleagues. So that's a difference.

"Definitely with the Shuttle flying, it would be more difficult to have a chief of the office who wasn't a pilot, just because so many of the decisions that the chief of the office has to make required you to be a pilot, and a lot of the things that they do, like being the weather pilot at KSC [Kennedy Space Center, Florida]. They usually tried very hard to have a Mission Specialist be the deputy chief [the first being Steve Hawley] so that both views got represented. I think that's pretty consistent over the entire program. When I first got here it was pretty much all pilots for both the chief and the deputy, but starting probably after *Challenger*, that changed. That's been pretty consistent that there'd be a pilot who was the chief and a Mission Specialist who was a deputy. Of course, now with Station, it's really the Station experience. Particularly those who've launched on a Soyuz, since that is pretty much all that we're launching on now. So somebody with that kind of experience is in a better position to make the critical decisions that need to be made at that level. The needs in that position changed, which allowed a Mission Specialist to actually be the head of the office, but it does create a different feel to the office, particularly with the director of FCOD [Flight Crew Operations Directorate] being Janet [L.] Kavandi, a woman. I remember Judy [Judith A.] Resnik came up with these bright pink bumper stickers that we had made that said, 'A Woman's Place is in the Cockpit', but we wouldn't need those anymore, I don't think." **[19]**

As a way of highlighting how things had changed since 1978, Fisher also recounted an incident in 2011 when attending the launch of STS-133, just as the *Blue Angels* (USN Flight Demonstration Squadron) were flying, and how she shared an elevator with one of the pilots of that team, a lady. "It was just so interesting, because nobody thought anything different of her being a pilot. I was just thinking how far we had come in that short period of time, because nobody thought anything about her being a pilot in the *Blue Angels*." **[20]**

iA

ted States ▰▰

Fig. 5.6: There were three physicians chosen in 1978: Rhea Seddon, Norman Thagard and Anna Fisher.

BROADENING THE OPPORTUNITIES

After two decades as a bastion of male dominance, the arrival of the six female candidates into the Astronaut Office in 1978 was certainly a change that took a while to get used to, but it signaled that times were moving on. As well as the adjustment within the halls of NASA, getting through the obligatory public and press attention for all 35 candidates, especially for the six women and four minority candidates, was new and at times even unsettling. As pioneers of the role of female astronauts in the NASA program, it was obvious that they would create headlines, especially after a few years when the first of them were set to fly. As alluded to by several of the TFNG, the females stood out for the fact that they were female in what was traditionally seen as a male dominated profession. Even though they deserved to be there and had all demonstrated that the decision to select them had been correct, there must have been a desire to see a day when selecting a female to train as an astronaut or fly a mission was routine. Over four decades since their selection, history has recorded that women have been Chief Astronaut, Space Shuttle or Space Station Commander, and fulfilled other senior roles. Women have set world records on the space station, there have been single- and two-female spacewalks (Extra-Vehicular Activity, or EVA), and it is hoped that the first female will step onto the Moon in the near future. Back in 1978, the announcement of those first six females to be chosen to train as career astronauts was new, exciting and unusual. In 2019 this is not so uncommon, and rightly so. All astronauts like to be known as crewmembers, no matter their background, beliefs or gender, and so it should be, but in the late 1970s and early 1980s this was the start of the process and the six female TFNG were the vanguard. It was still a time of change at NASA, while some of the press conferences still irked the astronauts when reporters singled out the female crewmembers because of their gender and not their experiences and skills. It would take time for this to change, but that change began with the 'ten interesting' members of the TFNG.

Fig. 5.7: A trio of professional astronomers was also selected in Group 8. (left) Jeff Hoffman explains to the public the differences between the Apollo Saturn and the Space Shuttle. (right) The other two astronomers chosen in 1978 were Steve Hawley and Pinky Nelson, seen here in January 1986, eight years after their selection, flying their second mission on STS-61C.

THE OTHER 'ODD CRITTERS'

As Kathy Sullivan noted in her Oral History, quoted at the start of this chapter, in addition to the six women the others who 'stood out' to the media in the Group 8 introductory press conferences were the four minority candidates: Guion Bluford, Fred Gregory, Ron McNair and Ellison Onizuka. As with their colleagues in the group, male or female, these four would contribute to significant changes in the Astronaut Office and make their own mark in American human space flight in their careers, in particular McNair and Onizuka who, along with Judy Resnik and their Commander Dick Scobee, were assigned to the ill-fated crew of STS-51L.

Good but not exceptional

The desire to include African-American or other ethnic minorities in the astronaut program goes back to the early 1960s and the selection of the third group of astronauts in 1963. A year before, in June 1962, President John F. Kennedy had established an advisory committee to investigate equal opportunities in the military, removing discrimination on the basis of race, color or creed. President Kennedy was also eager to include a black serviceman in the high profile astronaut corps. [21]. A search of USAF records, however, failed to reveal any black officers who had the required amount of flying hours or the level of academic qualifications needed to be considered for astronaut selection in the third group.

Undeterred, a search was made to find a suitable candidate to be enrolled in the next Aerospace Research Pilot School (ARPS) course at Edwards Air Force Base (AFB) in California. That airman, once qualified, would be nominated to NASA for selection to the third group of astronauts, even without suitable flying hours. That search found one pilot who had requested both assignment as a test pilot and consideration for astronaut training – 28-year-old Captain Edward J. Dwight. As a result, he was included in ARPS Class IV, along with classmates Dave Scott (a later Gemini and Apollo astronaut) and Mike Adams (subsequently chosen for the USAF MOL program and later an X-15 pilot). The plan was to have just eight students in the class, but Dwight's qualifications placed him well below his fellow candidates, so a comprise was agreed that included those who were better qualified ahead of him in the class; hence, ARPS IV included twice the intended number at 16 members. Unfortunately, Dwight's career path succumbed to a change of policy following the assassination of President Kennedy in November 1963. A month prior to graduation, he was reassigned to Germany and never formally graduated from the ARPS course. Dwight left the USAF in 1966 and in later years became a successful and noted sculptor. It has been said that he was not a bad pilot, but not an exceptional one, and that was what was required for him to progress further.

A man for MOL

Four years later, on June 30, 1967 a second African-American was in the headlines as a potential astronaut, this time in the third class of USAF MOL Astronauts. Major Robert H. Lawrence Jr. became the first African-American to be selected for astronaut training by either NASA or the USAF. He had earned a BSc in Chemistry and subsequently a PhD in Nuclear Chemistry, as well as serving as a fighter pilot and instructor, and was later assigned to ARPS Class 66B at Edwards. Tragically, he was killed on December 8, 1967 in the crash of an F-104D piloted by ARPS Operating Chief Major Harvey Royer, who survived. Lawrence was 33 and had he survived, it is widely thought that he would have transferred to NASA as part of the Class of 1969 (Group 7) when MOL was canceled. He would almost certainly have flown the Space Shuttle. **[22]**

The next generation
Guion Bluford, Fred Gregory, Ron McNair and Ellison Onizuka did not exactly follow in the footsteps of Dwight and Lawrence, but were able to grasp their chance and, largely unwittingly at first, found themselves in the spotlight and as early role models for others to follow. Arriving at NASA in 1978 had been a difficult journey for each of them in their own ways, but as males, their acceptance as astronauts was somewhat easier than for the women, who were truly breaking new ground. Nevertheless, each of the four were fully aware of the prominent position their selection had placed them in for inspiring a younger generation and other members of minority ethnic groups to put their names forward in the future. This did not just to apply to astronaut training, but to furthering their education, careers and prospects generally. In subsequent selections, other minority candidates, both male and female, would be recognized and chosen to join the still relatively select few who can call themselves NASA astronauts. So these four pioneers also helped to change the perception of NASA's Astronaut Office from its old ways of the 1960s into what it is today, half a century later.

SPACE SHUTTLE TAXI DRIVERS

Prior to 1978, all astronauts had to have graduated from a military pilot school, at least during some point in their earlier careers, or would be sent by NASA to attend one before progressing in their astronaut career. The reasoning was that the skills of a jet pilot, or ideally a test pilot, would be invaluable on space missions, and while actually flying a vehicle in space was significantly different to flying an aircraft, the discipline and mindset inherent in their piloting skills would come to the fore in difficult situations.

   With the advent of the Shuttle in the mid-1970s, its design meant that pilots would still be required to operate the vehicle during launch, on orbit and especially flying back to a runway landing (as a significantly heavy glider) at the first attempt, which would be no mean feat. But the Shuttle design also generated the idea that the flights could be repetitive and that, while never standard or routine, the ascent and decent profiles would become more operational rather than pushing the boundaries of piloting skills. Indeed, once the vehicle was on orbit, the majority of the roles for completing the mission were handed over to the MS while the Commander (CDR) and PLT monitored onboard systems, moved the Orbiter and then brought it home. In the late 1970s, as the new astronauts arrived in the Astronaut Office, this perceived routine flight profile of launch, orbit, entry, landing, ground processing, then repeat, gave rise to the suggestion, for a short time at least, of new missions every two weeks, with a team of CDR, PLT and MS/FE (Flight Engineer) flying regular missions together as a core trio. Almost like operating a space version of a taxi system to Earth orbit and back, as TFNG Brewster Shaw explained in his 2002 Oral History: "[I] showed up in July of '78 with 34 other Ascans in the '78 group, and I liked everybody. I was real impressed with the people that were there and hired. It kind of made me wonder how the hell I ended up being there. I wondered even more when things would happen. I remember we were having an interview one time, and I was in this interview and there was another pilot in the interview. I can't remember who it was. And there was Kathy [Kathryn D.] Sullivan. Kathy, bless her heart, said something that shocked me at the time but turned out to be pretty much the way it was, when she made the comment that the pilots were going to be just like taxi drivers and that it was the Mission Specialists who were going to do all the significant work on the Space Shuttle Program. Turns out, by golly, she was pretty much right. But at the time, being a macho test pilot, I was a little appalled at her statement. But, bless her heart, she was smarter than I was; probably I'm sure she still is." [23]

OFFICE REGENERATION - AGAIN

With the retirement of the Space Shuttle in 2011, the Astronaut Office required a further regeneration to encompsass new plans and goals in space, and not just in near-Earth orbit. The final class of NASA astronauts to go through Shuttle training and fly on the vehicle was chosen in 2004 and by then all of the TFNG had departed the long line of active astronauts preparing for a space flight. It was a new era, and the serving members of the Class of 1978 were no longer the ones pushing the boundaries of American human space flight, a role they had been proud to fulfill for so many years.

References

1. **Go for Orbit**, Rhea Seddon, Your Space Press, 2015, pp. 37-42.
2. **Riding Rockets: The Outrageous Tales of a Space Shuttle Astsronaut**, Mike Mullane, Scribner Books, New York, 2006, p. 31.
3. Kathryn Sullivan, NASA Oral History, May 10, 2007.
4. John Fabian, NASA Oral History, February 10, 2006.
5. Fred Gregory, NASA Oral History, April 29, 2004.
6. Anna Fisher, NASA Oral History, February 17, 2009
7. Dick Covey NASA Oral History November 1, 2006.
8. Mike Mullane, NASA Oral History January 24, 2003.
9. Terry Hart, NASA Oral History, April 10, 2003.

10. Hoot Gibson, NASA Oral History, November 1, 2003.
11. Guion Bluford, NASA Oral History, August 2, 2004.
12. Rhea Seddon NASA Oral History, 20 May 2010.
13. Loren Shriver, NASA Oral History, December 16, 2002.
14. Michael Coats, NASA at 50 Oral History, January 4, 2008.
15. Rick Hauck, NASA Oral History, November 20, 2003.
16. Steve Hawley, NASA Oral History, December 4, 2002.
17. Sally Ride, NASA Oral History, October 22, 2002.
18. Sally Ride NASA Oral History second interview, December 6, 2002.
19. Anna Fisher, NASA Oral History third interview, May 3, 2011.
20. Anna Fisher, NASA Oral History second interview, March 3, 2011.
21. **Moon Bound, Choosing and Preparing NASA's Lunar Astronauts**, Colin Burgess, Springer-Praxis, 2013, pp. 200–203.
22. **The Last of NASA's Original Pilot Astronauts, Expanding the Space Frontier in the Late Sixties**, David J. Shayler and Colin Burgess, Springer-Praxis, 2017, pp. 264–265.
23. Brewster Shaw, NASA Oral History, April 19, 2002.

6

Ascan Pioneers

"Obviously our group was all very excited.
It was kind of interesting
[because] we were the first group that had
Mission Specialists in addition to Pilots."
Dan Brandenstein, NASA Oral History, January 1999

NASA's Space Shuttle was sufficiently complex to require the new astronauts to be divided into two groups from the start. The Pilots (PLT) would fly the vehicle and the Mission Specialists (MS) would handle the cargo and scientific instruments. The qualifying requirements for both groups had been opened up, so that the MS could now wear glasses, and all the astronauts could be as short as 5 feet (1.52 meters) and as tall as 6 feet 4 inches (1.93 meters).

"Almost every one of these new guys has shown before they even got here that they could perform under stress," said veteran Gemini and Apollo astronaut John Young, who would command the first Shuttle flight in April 1981. "They've got a depth of experience that wasn't available ten years ago. They will need all the experience they can muster." In Young's opinion, the winged Space Shuttle was "an order of magnitude" more complex and difficult to fly than the Apollo 16 spacecraft he took to the Moon and back in 1972.

WHY AN ASCAN?

That level of complexity was why this group of people were chosen. During the selection process, they had demonstrated their aptitude and capabilities to impress the selection board. Now it was time to repay the faith that had been put in them

© Springer Nature Switzerland AG 2020
D. J. Shayler, C. Burgess, *NASA's First Space Shuttle Astronaut Selection*,
Springer Praxis Books, https://doi.org/10.1007/978-3-030-45742-6_6

and deliver on their promise, as they embarked on a revised astronaut training program to prepare them for space flight.

Past experience had taught NASA that it was not always wise to consider their new intakes as fully-fledged 'astronauts' from day one. Indeed, the Russians have long considered that their cosmonauts could only truly be considered to have earned that title once they had completed at least one orbit of the Earth; up to then, they were essentially trainees for space flight. All of the pilots NASA had selected between 1959 and 1969 were, or had been, in military service and had flown some of the most advanced planes in America at the time. Their challenge in transitioning to the new environment of NASA was not so much about the technical aspects of space flight, but more about adjusting to the academy-like campus and dealing with the media. For the Scientist-Astronauts of the Class of 1965 and 1967, the first hurdle was to qualify from a military jet pilot course – if they had not already done so – which some found to be a step too far in their expectations of becoming one of America's pioneering astronauts. Then there were the long training hours, time away from their research and families, and the prospect of not flying in space at all. As a result, some budding Scientist-Astronauts decided that being a 'spaceman' was not for them after all and left the program. By the mid-1970s, NASA was adamant this would not happen again, and for the new selection to find future Space Shuttle crewmembers, a different system was devised.

All the new applicants chosen would be designated as Astronaut Candidates (Ascans) for a given duration (up to two years) in which they would undertake basic training and evaluations. If successful at the end of that period, they would no longer be considered as candidates but as astronauts – if still inexperienced – and would receive their astronaut silver pin. For the previous groups, this had been awarded shortly after selection. Should a candidate, or indeed NASA, decide either mutually or separately that their application or selection for astronaut training had not been the right move, then the candidate could agree to part company with the agency or be reassigned to a different role within NASA. For 40 years this system worked well and, apart from a single unrelated accident[1], no candidate failed to progress beyond the Ascan program. However, some found the transition after that trial period to be even more challenging, with a few completing only one mission, or never flying at all, before leaving the astronaut program.

---

[1] The unrelated accident concerned Ascan Stephen Thorne (Class of 1985/Group 11), who was killed in May 1986 in an off-duty flying accident shortly before completing his Ascan course. In August 2018, Robb Kulin (Class of 2017/Group 22) became the first astronaut in five decades to leave the process voluntarily several months before the end of the training program.

In his NASA Oral History, Steve Hawley recalled his fears at selection: "What was unknown to me was how well I would do. I realized that there were a lot of things that I didn't have any experience in that they were going to expect me to be able to do, like fly jets. Frankly, even when I applied and when I was selected, my biggest concern was whether or not I would be able to make it. We still say today, as we did then, this is a probationary period. You're called 'candidate' for a reason [and] I realistically thought there was a chance in a couple of years they might get rid of me. So I was concerned about that. I knew this was going to be an interesting mix of people, and I knew that there were going to be people that knew a lot more about stuff that was important than I knew, and there were going to be these pilots and all that other stuff, so I was a little concerned how we would all get along. But I think primarily I was just concerned about [whether I would] be able to really do the things that would be expected of me. So that was the big unknown."

When each candidate was informed of their success in being selected to the astronaut program, they were also told to be back in Houston before the end of January 1978 for a week of orientation and a press conference to introduce them. On Monday January 30, the 35 received their day of orientation at the Johnson Space Center (JSC), and they were introduced to the press the following day at the Teague Auditorium. Center Director Chris Kraft delivered the formal welcome, followed by a member of the Public Relations Office introducing each of the 35 in turn. **[1]**

Each time a name was announced, the NASA employees in the audience applauded. When it came time to name Mike Mullane, the enormity of the situation dawned on the future astronaut: "My heart was trying to make like an alien and explode out of my chest. I still couldn't believe this was for real," he wrote. "When he got to it, I expected the announcer to pause on my name, look bewildered, consult with Chris Kraft, and then say 'Ladies and Gentlemen, there's a mistake on the list. You can scratch R. Michael Mullane. He's a typo'. Then two burly security guards would grab me by the elbows and escort me to the gate." **[2]** To his surprise, the announcer did not pause, there was no mistake, and Mullane was supposed to be on the list: "It's really official now, I thought. I had to believe it. I was a new astronaut… candidate."

In all, the group would spend a week at JSC undertaking a host of administrative tasks, such as receiving security badges and posing for their first official portrait. They also visited Ellington Air Force Base (AFB), just north of JSC, to be fitted into flight overalls during a clothes fitting session. This included being measured up for the flying helmet they would need as a crewmember on the T-38 aircraft. Then it was back home to make arrangements to move to Houston.

In her 2015 biography, Rhea Seddon summed up this part-indoctrination, part-house hunting trip, which was all still new and a bit of a whirlwind to most of the group. "We would all get a chance to meet one another and start figuring out what kind of group we were." Some had served together in the military, while others had met during the selection process, but Seddon still wondered, as probably did most of the others, "What would happen to all of us? Would we all stay with NASA? Would we all get to fly in space? Who would be my friends? Who would fly first? Would some get killed?" She realized that "this was the beginning of the first chapter in a book of unknown length, the unfolding of a fascinating story." [3]

Arriving at JSC

According to NASA's *JSC Roundup* news magazine of July 7, 1978, the 35 new astronaut candidates were scheduled to arrive at the center by Monday, July 10, ready to begin a two-year training and evaluation period (later revised to just one year). Their formal training would begin the next day, with all 35 participating. As Guion Bluford remembered: "Each of us in our class was assigned an office, which we shared with a fellow astronaut, on the third floor of Building 4A, and a scheduler and training coordinator was assigned to our group to facilitate our training. The role of the coordinator was to ensure that all of us received the training we needed in order to qualify as NASA astronauts." For the Class of 1978, this was to be Apollo 12 moonwalker and Skylab veteran, Alan L. Bean. He would ensure that there was an effective training plan for the group, and that mentors from the active astronaut group were assigned to the class as points of contact to discuss any issues or procedures.

For the rest of July, most of their training focused on familiarization with the T-38 that the astronauts used to fly to different training, mission-related or contractor locations around the country. This would be followed by their first (water) survival training course and then several weeks of lectures delivered by other astronauts, engineers, management and support contractors at JSC. They would then receive their first technical assignments and introductions to the various components of the Space Transportation System (STS), sessions on simulators, and visits to various NASA locations, in support of the effort to fly the first Shuttle missions under the Orbital Flight Test (OFT) program, and payloads destined to fly later on the Shuttle. [4]

Times were certainly changing from the early astronaut days. For one thing, the centrifuge in Building 29 at JSC had been dismantled and a water tank the size of the Shuttle cargo bay was being built in its place to enable the astronauts to simulate the feeling of weightlessness. "It was good experience, but we don't need the centrifuge anymore," observed Deke Slayton. "The Shuttle only pulls 3 g on launch and 1.5 g on entry. Hell, anybody can take that."

The centrifuge was not the only thing to disappear from astronaut training. Jungle and desert survival courses were no longer required thanks to the Shuttle's flight profile around the Earth. However, the water survival course remained because exiting the Shuttle cockpit (and ejecting from the T-38 aircraft) could drop the pilots in the ocean if their lift-off was aborted. Gone, too, were the extensive geology field trips that were synonymous in preparing the Apollo astronauts to explore the lunar features, principally because the Shuttle was not flying to the Moon. These were replaced by Earth-focused geological lectures. There would also be less flying time in the T-38 jet trainer due to soaring jet fuel costs. This meant that the astronauts were limited to 15 hours of flying time per month, so they used that time to practice aerobatics and formation flying, which was a lot more demanding than cross-country flights.

SCHOOLING THE ASCANS

"I do remember one of the first days at work going to Building 1, the ninth floor [and the] big conference room up there, [Room] 966. Dr. Chris Kraft briefed our group," recalled Steve Nagel. "It was kind of a big deal for our group, because we were the first group of Shuttle astronauts. It marked a big change in the program. If I've got my numbers right, there were 26 or 27 astronauts [in the Astronaut Office] prior to our group. That's how small that office was. My group was 35, so we more than doubled the size of the [corps], plus the first women, first minorities, everything. So it was just a huge change. But it was very well accepted, and as far as I could tell there never was any problem." Nagel also remembered that "Dr. Kraft was really nice to us, gave us a good briefing, and I just remember him saying, 'The important thing is you have fun while you're here', which I did. He said, 'Work hard, but have fun at it, too'."

In his NASA Oral History, Hoot Gibson recalled another problem specific to the military candidates when they reported to NASA on Day One. "Part of it was kind of traumatic for us military guys because you had to decide what to wear. Previously, you got up and you got showered, shaved and dressed and you put on your uniform and you went to work. [Mike] Mullane talks about that in his book. He didn't know how to dress or what to dress [in] and [his wife] Donna had to take him to the store and pick out some [civilian] clothes for him. All of a sudden, you had to have clothes, and none of us had any of that. I will never forget, I studied the pictures of the 35 of us before I showed up so that I'd know who [they were], because we had only just met them for that one or two days back in January. Now it's June, and I showed up at Building 4. Of course [I] first had to go through the badge office and get my badge. I parked in the old Building 4 [parking lot] and went up the stairs, opened the door and stepped into the hallway, and there was Dick [Francis R.] Scobee and Judy Resnik. They were the first two that I saw, and walked up to them and I said, 'Oh, my gosh, it's Scobee and Resnik,' and shook their hands. I don't remember a whole lot about the rest of that day. I guess the rest of that was finding out where my office was going to be."

The Red and the Blue
The arrival of 35 new astronaut candidates, which, as Nagel indicated, effectively doubled the size of the Astronaut Office, created an administrative and logistical challenge for the support staff at JSC. To help alleviate any congestion, it was decided to split the candidates into two groups – the Red Team and the Blue Team (with appropriate team T-shirts provided) – and as John Fabian (Blue) and Rick Hauck (Red) were the most 'senior' (oldest) astronauts in the class, they became the team leaders. Dick Scobee was older but lacked the seniority as he was originally an enlisted man. These two would be the ones who would receive a call from George Abbey if there was an issue which needed to be resolved, as John Fabian recalled: "I really don't remember exactly how it happened, but we were the oldest, and in our military ranks we were the senior military people in the class. And the civilians were mostly much younger, so I guess it was just natural that the old fogies would be tasked with this. There were no rewards and very few responsibilities, so it really wasn't a big deal. [But] We got to know the people that were in our group."

Fabian also noted that there was a simple, practical reason to divide the group up, as even the largest classroom could not hold all 35 Ascans and the support staff. The people in each group got to know each other better than the people in the other group, particularly in the first three or four months, but as Fabian explained, "what we found out very quickly was that all of these people, whether they were the youngest in the group or the oldest in the group, they were all extraordinarily bright, extraordinarily capable, and very, very eager to succeed in what it was that they were doing. So we didn't have any dim bulbs in the group. At the same time, I will tell you that in any group, doesn't matter how fine it is, there's always going to be a dim bulb, okay? But it appears that way only because it is a comparison with the others. This was an extraordinarily talented group of people that we were working with." Instead of having a large class of 35, therefore, they would teach a class to the Blue Team and then the same class to the Red Team, although the two were mixed and matched when needed.

During that first year of training, the plan was to provide a baseline of knowledge that they needed to operate in the Astronaut Office. For some, this meant revisiting old courses they had gone though in their previous careers, while others were breaking new ground all the time. As Dan Brandenstein recalled: "We had aerodynamics courses which, for somebody who had been through a test pilot school, was kind of a 'ho-hum, been there, done that [situation],' but for a medical doctor, that was something totally new and different. But then the astronomy courses and the geology courses and the medical-type courses we got, all that was focused on stuff we'd have to know to operate in the office, and at least understand and be reasonably cognizant of some of the importance of the various experiments that we would be doing on the various missions. So I found that real fascinating.

"The way the courses were approached, certainly in the early months, was to go to a classroom for half a day and then you did some of these other things on the other half of the day. So one half of the class was doing something in the morning and then the other half did it in the afternoon, and then they switched, to make the best use of the available assets."

Dick Covey remembered that Ascan training was as much orientation as it was training. "Seems like we were always on road trips and going to NASA centers or to NASA contractors. I just seem to remember more about those events than I do a lot of the classroom stuff. Obviously, the classroom activities were focused on orienting us to the Space Shuttle, to JSC, to mission operations, and all the aspects of human space flight that we would not have had any background in just coming in cold."

Terry Hart agreed. "It was a great experience. I think a lot of people would think of the training program as being somewhat stressful, testing, and everything of that nature, but it was very professional, but not in a stressful way. It was just all mission oriented, and learning what you needed to learn, and the degree of professionalism among the training people and the way they worked you through the syllabus and everything was very reassuring. The task is a little bit daunting, in the sense that the astronauts all have such different backgrounds. I mean, roughly half were pilots and half were scientists or doctors, and yet they had to bring everybody up to some common level of understanding of engineering and science disciplines. Most were familiar with some, but not all. It was just so much fun to go through all that, and of course not having the pressure of exams like you do in a university environment made it that much more pleasant. And the camaraderie and the sense of teamwork that you build up with your mutual experiences, and the social life, and everything… it's just a wonderful experience."

Becoming a T-38 crewmember

Aside from the initial administration chores, the more important task of finding a place to live, and starting the academic program which would continue concurrent to other activities until the spring of 1979, their major activities scheduled for the remainder of July, according to the news releases, would include aircraft life support and ejection seat training for the T-38 jet trainer, aircraft physiological training, T-38 aircraft systems and operations, and T-38 check-out flights (see sidebar: *The NASA T-38*).

The NASA T-38

NASA has operated a small fleet of Northrup T-38 *Talon* two-seat jet trainers since the 1960s. Stationed at Ellington Field, just north of JSC, the T-38s are used by the astronauts to fly to various NASA field sites, training locations and contractor sites across the United States. During the Shuttle program, they were also employed as chase planes, and it became a tradition for crews to arrive at the Kennedy Space Center (KSC) a few days prior to launch in the T-38s.

In the 1960s, four astronauts were killed in separate T-38 accidents: Ted Freeman in 1964; Elliot See and Charles Bassett in 1966; and CC Williams in 1967. There had also been several non-fatal incidents over the years involving astronauts, including one of the Thirty-Five New Guys (TFNG), Brewster Shaw: "I got sent out to Downey, to Rockwell Space Systems Division, to work on safety enhancements for the Orbiter. So I spent a lot of time out there and flying back and forth to California, landing at Los Alamitos Army Air Field out there by Seal Beach, which was the closest place to the Downey facility we could land a T-38.

"I remember one day, Rob [Robert A.] Rivers [of the Aircraft Division at JSC] and I were flying out there. He was just giving me a ride out. He was in the front seat. I was in the back. We were jetting down to Los Alamitos and we got hit by lightning, and the lightning created an arc in the fuel tank on NASA 914 and blew the whole back end spine of the airplane, blew that completely off. All these fuel bladders were sitting there full of JP4 fuel cooking away. We had fire lights on both engines. Rob shut one of them down. We had smoke in the cockpit, so he turned off the generators, and we landed, no flaps, at Los Alamitos. On fire, [we] rolled down to the end of the runway, stopped the airplane, [and] unstrapped. Jeez, there's fire all around. The fuel's underneath the airplane spreading out and the back end is just blazing away there. So we jump over the side and run away, and the guys come and spray cow guts on the airplane. That's what they use; the foam is made up of cow guts or something like that. And [they] put the fire out. So this T-38 was the second time I'd been in a burning jet; same sort of sound indications when a fuel bladder or a flame front propagates in an airplane. I heard that on that T-38 and I thought to myself 'Oh, this is not good. I've heard that sound before, and I know what it means.' And it wasn't good. But fortunately, we got the airplane on the ground, and that airplane is flying again today, because the miraculous guys out here in Ellington put it back together."

(continued)

Fig. 6.1: (top) Steve Hawley receives instruction on T-38 rear-seat procedures. (bottom) Judy Resnik, sitting in the rear seat of a T-38, prepares for a familiarization flight with Dr. Richard Laidley, Ellington AFB, May 1978.

Most of the Navy and Air Force pilots had flown the T-38 at some point as a part of their initial training, but while some of the MS had private pilot's licenses, most of them had never flown in a high-performance airplane before. They needed to be checked out to fly in the back seat as a crewmember and team up with the pilots flying in the front, and together they would learn to operate the aircraft as a crew for the first time in a stressful environment.

"I think the most unique thing was learning to fly the T-38s, because I was totally new," recalled Jeff Hoffman. "I had never flown. I had jumped out of airplanes, but I'd never flown. To fly a high-performance jet is really a pretty incredible thing. It took a lot of getting used to. It was a completely different world. When I first got up there, I couldn't understand what anybody was saying on the radio. I would get fixated on certain things and forget what was going on. That's one of the reasons why they put us in that environment, because space flight is very much an extension of high-performance aviation. For the first time, we didn't all have to become pilots, so they didn't send us off to a military flight school like they had done the previous group of astronaut scientists back in '67, but we had a complete ground school and a flight training program. So there was a syllabus of 15 or 20 different flights that we had to complete with a lot of different skill sets. It was really exciting. It was just a lot to learn. After a while, you can become a little bit blasé about 'Oh jeez, I got to get my flight hours this month, what am I going to do, I got too much other stuff', but at the beginning, any flight I could get to do anything, anywhere, it was just the most exciting thing that we could imagine."

Sally Ride agreed that this was probably one of the more challenging aspects of the training, just because it was so new. But she thought it was still "an awful lot of fun! I really enjoyed both the ground school and then the flight training itself. Flying was completely out of my experience, whereas learning a schematic for an electrical system was something that I knew how to do. And it was just a matter of putting in the time to learn it."

Anna Fisher was expecting to get military flight school training, but found that NASA was inventing how they were going to train the group as they went. "What helped was that Jim Buchli and Dale Gardner were both qualified military 'back-seaters', so together they devised the training plan for the Mission Specialists in the group who had absolutely no aviation experience," Fisher said. "They put together the course and taught the civilian MS how to do the navigating and talk on the radios. Thirty years later, the pilots out at aircraft operations at Ellington handle all the training of both the pilots and the Mission Specialists, but back then it wasn't exactly that way. Each Mission Specialist who had little or no aviation background got assigned to one of the pilots in the astronaut corps. I was assigned to [Thomas K.] Ken Mattingly. It pretty much then became a matter of pride for them how their Mission Specialists did when they talked on the radios and how quickly they progressed. [So] Jim and Dale came up with this ten-flight syllabus that you would go through. On this flight you went out, and you concentrated on visual flight rules. On this one, instrument rules. It was just a ten-card syllabus; something like that is still in existence today."

Homestead AFB

Some of the women were beginning to wonder if they had made the right choice in their desire to become an astronaut. They were scientists and physicians who had achieved instant celebrity status when their names were announced in January, and they were finding the attention unrelenting even as they began their astronaut training.

On July 31, a total of 16 candidates (the six females: Fisher, Lucid, Resnik, Ride, Seddon and Sullivan; and ten male colleagues: Gardner, Hart, Hawley, Hoffman, McBride, McNair, Nelson, Onizuka, Thagard and van Hoften), arrived at

Homestead AFB in south Florida for three days of Air Training Command water survival exercises, under the supervision of the 3613th Combat Crew Training Squadron. In an earlier news release, NASA indicated that the other 19 candidates had already received water survival training prior to entering the astronaut program, so they were excluded. It also stated that the 16 carrying out the exercises would receive classroom lectures on water survival techniques as well as actual water training, including participating in what was known as the "Drop and Drag" exercise. This involved sliding down a wire into Biscayne Bay from a 15-foot-high (4.5 meter) boat platform while wearing a tethered parachute harness. They would then be towed through the water, simulating a parachute dragging them across the surface, until they could stabilize themselves before releasing the parachute riser.

Before they made their way down to Homestead AFB, NASA's press agents had assured the 16 participants that they would not be subjected to media scrutiny during this phase of their training. But when they arrived in Florida the press was already there, unimpeded and eager for photographs and interviews. "My husband and I knew the attention would be intense for a while," Anna Fisher said of the experience, "but we never expected it to go on and on like this." **[5]**

The male candidates in the training group seemingly got off lightly, with most of the media attention firmly focused on the six female astronaut trainees. The women were understandably annoyed, both with the constant barrage of media questions and requests for photos, and with the space agency for allowing this to happen when they had been guaranteed privacy to conduct what was a very serious training session. It was later widely reported that one television reporter had been so eager for a comment from Rhea Seddon that he called out, "Hey, miss!" at which she gave him a withering stare and snapped out, "It's 'Doctor'!"

One exercise required the candidates to be towed aloft under a parasail canopy, which would be released 400 feet above the water. After landing in the water, they would be retrieved by the returning boat. On their final plunge into the choppy water they were dropped wearing their full survival gear, after which they had to inflate their small life raft and fire off a flare to indicate their position before being plucked from the water by helicopter.

Sally Ride remembered thinking to herself one day, "What am I doing here? I'm supposed to be a smart person." **[5]**

According to Anna Fisher [in 2009], parasailing off a board at Homestead was "really neat. I'm not sure if they do it to this level of depth [with more recent Ascan classes]; where you actually parasail and then they cut you loose. The guy gave the signal and you hit these two handles, they fall away, and then you do the last part of a water entry as though you were ejecting from a T-38. Then you had to get into your raft. Then a helicopter actually came with the water swirling around. So you actually got to see what it would be like to be in a water rescue. You got on the little thingy [helicopter retrieval platform] and saw what it would have been like to hoist you up. Then they left you there. Then they went on to the next person. So it was actually really interesting training."

While at Homestead, Fisher remembered an Air Force pilot named Tom [Thomas D.] Jones. "He was doing his water survival training at the time that we, the six women, came through, with all the reporters that were there covering it. It was really interesting that he wound up coming [there] and many years later being selected as an astronaut [Group 13, 1990]. We laughed about that a lot, because he and I wound up working together on Space Station [issues] many years later."

Fig. 6.2: (top) Survival course classroom studies at Vance AFB, Enid, Oklahoma, August 1978. Together with their training officer and two JSC physicians, 11 Ascans are briefed prior to their practical training. Ascan manager Harold E. 'Bud' Ream [left, row one] also participated in the exercises. Sitting next to him on the front row is Seddon. [row 2, l-r] Onizuka, Dr. Michael Barry JSC Medical Science Division, Hoffman, McNair; [row 3, l-r] Fisher (obscured), Sullivan, Nelson and Hawley; [row 4, l-r] Resnik (partially obscured), Ride, Lucid and Dr. Joseph Degioanni JSC Medical Science Division. (bottom) Attending the Water Survival School, Homestead AFB, Florida, in July 1978. [from far left to right] Lucid, Hawley, Gardner, McNair and Seddon. Note the looks of concentration, maybe anticipation, or even trepidation for what they were about to participate in.

Vance AFB

Three weeks later, on Monday, August 28, having barely had time to unpack and dry out, 11 of the 35 candidates attended parasail training at Vance AFB, Enid, Oklahoma. This session was designed to familiarize them with the correct procedures for landing by parachute should they ever have to eject from a T-38 over land. Twenty-four of their colleagues had already completed this type of training in their careers, so the six females were joined this time by just five of their male classmates. The 11 attending this course were: Fisher, Hawley, Hoffman, Lucid, McNair, Nelson, Onizuka, Resnik, Ride, Seddon and Sullivan. **[6]**

Anna Fisher also recalled this trip to Vance AFB years later in her Oral History: "We were sent to Enid, Oklahoma, to do land emergency training, which they don't do anymore, thankfully. They now send people for winter survival, where you learn to navigate and eat bugs and all that kind of stuff. Luckily we didn't do that. I'm not sure I would have passed that one. I think I would have rather died than eat bugs. They took us to Enid, Oklahoma, which is Shannon Lucid's hometown area. In Oklahoma, they [used] a pickup truck [to tow them] instead of a boat. Suddenly, you're going across the land. It was actually pretty scary. I was kind of light. Some other people weren't as light as me. Some of them were really struggling to get up into the air. I remember poor Shannon was getting dragged for a while. So I think after that experience people perhaps looked at the safety aspects of that. I don't think any other class did that; I don't remember. But it was probably one of the more interesting parts of our training. Then you actually did a parachute landing, because they cut you loose and you did the same thing."

The pilot astronaut candidates were checked out flying the NASA T-38s as well as the Shuttle Training Aircraft (STA), the Gulfstream aircraft that were modified to fly like the Space Shuttle during approach and landing. "For those Mission Specialists who had graduated from military flight school, we were trained as pilots in the aircraft," said Guion Bluford, "For me, flying the NASA T-38 as a pilot was a great responsibility and a great privilege. I had been a T-38 instructor pilot at Sheppard Air Force Base in Wichita Falls, Texas, six years earlier, so it was easy for me to transition back into the aircraft. I accumulated over 4,600 hours of flight time in T-38s during my NASA and Air Force career."

Flying in the T-38 would be a regular task during their years at NASA, so for those MS without flying experience the plan was to qualify them as a crewperson. Dan Brandenstein explained the role the MS played as a T-38 crewmember: "They had to be trained on not being a pilot necessarily but being a crew person; what procedures you had to use to fly in the air ways around and, once again, to learn the systems of the T-38 and the various skills and functions of a crew person. And the T-38 had a stick in the back so they obviously got to fly some. They weren't allowed to land because they didn't get trained for that, and didn't want to break the airplanes, but we could take it off and up and away and swap back and forth who was flying. A lot of the flying was business-type travel, but also there was an area just outside of Houston over the Gulf of Mexico, [where] we could go out and do what we called 'turn and burn', which is do aerobatics and just do loops and

Fig. 6.3: The survival training course at Homestead and Vance AFBs gave the Ascans the skills required to cope with ejection from the T-38. This training featured familiarization with the parachute and harness, simulated water and land parachute landings, and parascending towed by a boat with parachute descent and recovery from the water.

rolls and chase around clouds and stuff like that. But all the time that's a way of maintaining your piloting skills.

"Obviously for people who had never flown before, that was – well, it's a kick for people that flew thousands of hours – but for somebody who never had flown before, or had very little experience, or just a little experience in a small light airplane, it was a real kick, because you could go supersonic with those. You'd pull 7 g and it all happened. It was all new and different to them. And they flew with guys like myself. Most of us were military pilots… at least somewhere in our career. We'd go out and do simulated combat and show them what it's [like]. As I said, all the pilots had been test pilots before, so we'd go out and run them through the wringer, showing them the various things you'd do if you're testing a new airplane, like having a dog fight and all those sorts of things. So that was fascinating to them, as [it was for] me sitting down with an astronomer or a doctor and finding out about the types of things they did."

Though actually flying in the T-38 might be daunting for the new MS, it had been made clear during the interviews that this was to be part of the Ascan program. "They'd taken us out to Ellington [Field, Houston, Texas] and let us sit in a T-38 and told us that we would have to do flight training," said Rhea Seddon. "They weren't going to put us through the Air Force pilot training, they were going to just train us on the NASA syllabus to be back-seaters, but that was about it. They began rolling out these different things. Water survival was one of the first things that we did, because if we were going to fly in the T-38s we had to understand how to use the ejection seats and then be capable of parachuting down in the water and staying alive. So they sent us to that.

"For a female that hadn't done a lot of really exciting things, that was pretty exciting and probably not something that I would have ever chosen to do. But it was a rite of passage. I certainly wasn't going to chicken out. It turned out the folks at Homestead Air Force Base were used to training a lot of people coming through, probably 18-year-olds and little people and big people. They made it easy. They said, 'Here's what we're going to do; here's what you need to do. Here's what will happen. If you have any trouble, we'll pick you up and we'll do it again.' It was very graduated. 'First you do this, that's a little hard, and then you do this, that's a little harder. Then you do these things, that look exciting, but here's what you do. Tell me the steps. I just told you; you tell them back to me. We have a safety boat here.' It wasn't as difficult as I imagined it would be. We all just got through it, had a good time, had a chance to get to know each other, and have a beer after work. It was part of the adventure.

"Then we did parasailing in Oklahoma. We didn't actually parachute out of planes, but we did parasail. There was one of the women who had some difficulty with that because you had to be able to land on the ground and do a parachute roll. She ended up spraining an ankle or a knee or something like that. I came home with bruises all over the place, but I wasn't going to show those to anybody or complain about it at all. That was physically demanding."

In fact, the women fitted in faster than anyone had anticipated. They underwent the same T-38 ejection seat training, the same parasailing instruction and the same water survival course as their male counterparts. The only difference was that they flew in the back seat of the T-38s, as did as any of the male MS who were not qualified jet pilots. They were not expected to be qualified jet pilots and therefore were not required to take the training that would have qualified them.

"I've always maintained since Day One there wasn't any reason to have women [in the astronaut corps] just because they were women," said Deke Slayton at the time. "The only reason they weren't here before is that they didn't meet the qualifications we had. Now they do."

Fig. 6.4: Most of the TFNG are visible, listening to an astronomy and astrophysics presentation by Dr. Harlan Smith (standing, left), University of Texas, at JSC December 1978. From the rear of the room, table by table and clockwise from the far side of each, the attendees were: Table 1: Onizuka, Scobee, Thagard, McBride, Gibson, Griggs, Fabian. Table 2: Shaw, Coats, Lucid (obscured), Nelson, Stewart, Hart and Shriver. Table 3: van Hoften, Williams, Nagel, Covey, Sullivan, Ride, Buchli and McNair. Table 4: Seddon, Creighton, Brandenstein, Fisher (hand to face in the corner at right), Gardner, Gregory, Hauck and Resnik (obscured near left foreground). (Image courtesy of Ed Hengeveld.)

Back in class

The extensive classroom training extended over their first eight months at NASA and was given by instructors from universities or other NASA centers. It included lectures on the history of space flight, technical assignment methods, and procedures within the Astronaut Office. There were also presentations on manned space flight engineering, the Space Shuttle program and vehicle systems, ascent and entry aerodynamics, flight operations, orbital mechanics, tracking techniques, materials science, space flight physiology, biology, anatomy, geology, astronomy and so on. Veteran astronauts delivered lectures on a variety of topics such as 'How to be a Capcom', while other NASA personnel took the lead in sessions with such topics as 'Washington Round Up', 'Evolution of a JSC Budget', or 'People and Requirements, it Takes a Bunch to Make Things Work in NASA'.

The 'real' veteran astronauts

Chris Kraft invited 31 former astronauts to participate in technical briefings and updates on the status of NASA programs, primarily that of the Shuttle including the U.S. Air Force (USAF) role in the program, at JSC during August 21−22². The group would tour the crew systems area, the Remote Manipulator System (RMS) training facility, the full-scale Orbiter mockup in Building 9, and the Orbiter aero flight and mission simulators in Building 5. They also had the opportunity to meet the new astronauts. [7] One of the new intake's favorite pastimes was listening to the veteran astronauts tell tales about space flight.

In his role as group coordinator, Alan Bean told the group about his time on the Moon and his time in Skylab, as well as the thrill of each new day and night in the black seas of space: "I can remember one time I was cleaning off a Skylab lens and I looked down and I could see Italy and Egypt, and then Israel, and then, in darkness, the fires off the Arabian coast," Bean told several of them at one get-together. "Can you see yourselves passing over New York City at 17,000 miles an hour and you're outside assembling some solar panels? People are going to look up and see these things each night get bigger and bigger. You know what that's going to be like? It's going to be incredible, that's all." [8]

Eight months into the Ascan program, astronomer Steve Hawley commented: "This morning, I was counting the years I've been in a class like this." [9] Unwittingly forging a new path at NASA, in their first few months the TFNG participation in the classroom program was so successful that it was decided by their training team to compile a videotape library (long before the days of digital capture or on-line streaming) for use by other JSC employees and as a reference for subsequent selections.

While mundane for some, and challenging for others, the hours spent in the classroom this early in their astronaut careers provided a foundation for all the things astronauts are required to do on a mission, or in support of one. "They're more like briefings than classes," said candidate training coordinator Tom Kaiser

---

² The 31 veteran astronauts who participated were: Buzz Aldrin, Bill Anders, Neil Armstrong, Frank Borman, Scott Carpenter, Jerry Carr, Gene Cernan, Mike Collins, Pete Conrad, Gordon Cooper, Walt Cunningham, Charlie Duke, Donn Eisele, Tony England, Ron Evans, John Glenn, Dick Gordon, Jim Irwin, Jim Lovell, Jim McDivitt, Ed Mitchell, Bill Pogue, Stu Roosa, Wally Schirra, Jack Schmitt, Rusty Schweickart, Dave Scott, Al Shepard, Tom Stafford, Jack Swigert and Al Worden.

Fig. 6.5: Some of NASA's 'original astronauts', from Project Mercury to ASTP (1959–1975), visit JSC for orientation on the Space Shuttle program and to meet the new Ascans. Rear (l-r): Gene Cernan, Charlie Duke, Gordon Cooper, Neil Armstrong, Stu Roosa. Middle (l-r): Wally Schirra, Rusty Schweickart, Donn Eisele, Jim Lovell, Dave Scott, Al Worden, Tom Stafford, Bill Pogue. Front (l-r) Al Shepard, Ed Mitchell, Dick Gordon, Pete Conrad, Buzz Aldrin, Ron Evans, Jim Irwin, Walt Cunningham and Mike Collins.

at the time, "[and] this is the first time we've had this thorough a training program." **[9]** There were lots of viewgraph presentations, wiring diagrams, and schematic drawings provided. However, there were initially no Space Shuttle training workbooks available, and to many who had come from an organized military system, the training seemed haphazard at times, despite the effort that had gone into giving the astronauts the required technical knowledge. The training was intense and wide-ranging, designed to give the candidates a "little bit for everybody." There was also the tradition that astronauts never had to take written exams.

"They started us in classes, and they were fun classes," recalled Mike Coats. "They brought in some experts [in] different fields… things that you would expect if you have to look back at the Earth; oceanography and geology, celestial mechanics and astronomy. They brought in all these experts to teach [us], and it was fascinating, because they didn't give us any tests. They said your test will be when you fly. Nowadays, there's a lot more testing, so I think we were spoiled back then.

Nowadays, they come in and they're evaluated on everything they do and ranked. That really wasn't the case back then. The classes were fun because you could sit there and learn and soak it all in and ask dumb questions and not have to worry about a test. So we enjoyed that. Right off the bat, they started teaching us what the systems were on the Shuttle. Of course, they gave us training materials by the piles and piles, and we pored over those. We all wanted to fly as soon as possible."

"We'd spend about three hours in the morning, break for lunch, and come back for two or three hours in the afternoon," said John Creighton. "They brought in a number of speakers to talk to us in general about spacecraft design, and how NASA was organized, and a little bit about all of the different centers. Then we got progressively into more and more detail on how the Space Shuttle was designed at that time."

"We had a lot of professors," added Fabian. "Probably for the first three, four months or so, we had lectures by these professors who were experts in their particular area, whether it was geology, or meteorology, or oceanography, and they were there to educate us on how space affected their particular science; how it affected geology, what we could do from space to help geologists, what we could see and record from space that would help oceanographers, that type of thing. And these professors were also renowned in their areas. So we had some top-drawer instruction there early on. It was kind of a shotgun: a little bit of science, a little bit of past astronaut experiences, a lot of insight into the Shuttle."

Brewster Shaw described it as "being spoon fed a lot of information. [We] learned a tremendous amount about NASA, about the program, about the Johnson Space Center, about all the people and the history of human space flight and how we were just going to be kind of passing through… The people who were here at the Johnson Space Center, who had done Mercury, Gemini, Apollo, Skylab, and now were going to do Space Shuttle, they were the real people that made all this happen, and if we were nice to them, they might let us fly their spacecraft. That was a good perspective. It's pretty accurate, I think."

One aspect which surprised many astronauts was the lack of formal testing during the Ascan year. van Hoften commented: "I don't know if I would have done it this way, but they were careful that we never took any tests. We'd sit through classes endlessly and never had a test to see whether we learned anything. I think they were careful to not try and grade people, or putting anybody with any kind of a structural grading. [But] it lent an air of real mystery, as this went on, about how you get selected for a flight. That's the big *question du jour* in NASA, and no one knew."

To Fabian, the systems appeared unstructured, unlike the military, which surprised him: "In the military, the training was always very structured and was always, when you're talking about flying an airplane, delivered by pilots or back-seaters who flew that airplane. So I had this vision that when I got there the astronauts would be training us. They were the ones that flew the Shuttle or had flown the vehicles before the Shuttle. So we would have these older astronauts that would be training us, and that it would be very structured as it was in the Air Force.

"In the Air Force, before you fly a new airplane, you have this ground school [and] you have workbooks. I mean, you have to study stuff, and you take tests on it. You have to pass these tests before you move on to the next level. You fly with an instructor pilot before you're ever turned loose in the plane by yourself. You have a certain number of minimum hours you have to get. It's very, very structured, and I kind of expected that when I got to NASA, that it would be a classroom. You'd go to a classroom, and you would sit there and follow a workbook or these various documents, and that the astronauts would be teaching us, and I was really surprised to see how unstructured it was at that time. I think it probably is more structured now. But at that time, everybody was coming up to speed on how to train these new astronauts to fly the Shuttle."

"Obviously, most of the folks who put the books together were not military," recalled Fred Gregory, "[or] did not have an orientation in the military, but I think that the folks in the Astronaut Office who did have the military orientation edited [them]. So I felt very comfortable when I got these books, specifically the training books for the simulators. I thought they were very well put together. The academic books were very much like I had experienced in colleges and universities. So I thought that they had done a very, very good job preparing the literature and in preparation for the lectures. I felt very comfortable in that environment. I was not surprised at all. I *was* surprised at how complex the Shuttle was. I had never been in any kind of an airplane that was that involved."

High-flying extra-curricular studies
One of the additional science trips, taken to photograph a solar eclipse, was explained by Steve Hawley: "There were several of us: Sally [Ride] and me, Hoot [Gibson], and Mike Coats, and Pinky [Nelson]. There may have been one other; Jeff [Hoffman], probably. As part of our science training, we concocted this boondoggle trip. We would go chase the solar eclipse and take photographs of it from the airplanes. I guess we did a good enough job selling it that they let us go do that.

"So we actually did go, and we flew up to North Dakota, I think, because the path of totality went through the northwest part of the U.S., through North Dakota and Montana and Wyoming. We went up there and [did] the calculations for where the shadow would be, because… at 30,000 or 40,000 feet… it's not exactly in the same place… as it is on the ground. So you can figure that out and figure out what flight path you have to fly. And actually it was pretty good training, and we did get some pictures. The thing I remember about it, and it was sort of humiliating, is that I think the best pictures were the pictures that Hoot took, and he was flying one of the jets. Sally was in his back seat. I was in Mike Coats' back seat. So we had the cameras, and we had the time, and we could do all this. He was just flying along and taking pictures out the canopy, and he ended up with the best eclipse pictures. I'm not quite sure how that worked. He thought it was skill. I think it was luck. But that was an interesting trip."

Fig. 6.6: (left) Kathy Sullivan shown wearing a full pressure garment in front of a NASA WB-57 high-altitude aircraft on the day she set an unofficial sustained altitude record for a woman. (right) Sullivan in the role of scientific operator from the rear seat of the WB-57.

As well as qualifying as a T-38 crewmember, two MS were certified as scientific operators for the WB-57 [F] high-flying aircraft. This was based upon the Martin/General Dynamics RB-57F *Canberra*, the specially designed strategic reconnaissance aircraft originally operated by the USAF, three of which were transferred to NASA as part of the agency's Airborne Science Program that included a fleet of aircraft to support Earth science research. The three aircraft began operating out of Ellington Field in the early 1960s.

The two astronauts who qualified for this role early in their careers at NASA were Kathy Sullivan and Pinky Nelson. "My reasoning was [that] since I was working on the suit, I wanted to get as much experience in pressure suits, period, as I could," explained Nelson. "Plus, flying in the WB-57 sounded really cool. So Kathy Sullivan and I both got checked out in the WB-57, and my reason for doing it was to get experience working in a pressure suit. It's a different kind of suit, but similar kind of experience. The price to pay for doing that was to learn about how the airplane worked and how the missions that the airplane did worked, and then to actually pull together and run a couple of missions. So I spent three weeks down in Miami at Homestead Air Force Base doing microwave observations over the tops of thunderstorms for some scientist. I actually got a publication in *Science* magazine out of it.

"Then Kathy and I both went to South America to do an air sampling mission with the WB-57, and flew from Houston to Panama, to Lima [Peru], to…

Montevideo [Uruguay], down south over the Falkland Islands. It was neat. That was a fun trip. Long flights. Seven hours in a cockpit in a pressure suit. This is the classic one that people always think that the astronauts in the Shuttle have, eating out of a toothpaste tube and things like that. That's what you really did in those suits. You have your apple sauce out of a tube [and your] water out of tube, sticking in the hole in your helmet.

"The main job was actually organizing the expedition, getting all the people together. The maintenance guys, the suit guys, the airplane, coordinating the ground operation, making sure you had fuel, and all that kind of stuff, just being in charge of the mission. NASA had all these people who knew how to do this stuff; you just had to pull them together. Then flying the mission itself was pretty straightforward. You just had to program what was the equivalent of a GPS [Global Positioning System] back then. You had to program the little navigation device so the airplane would fly where you planned it to fly, and then when you were ready, just operate the instrument; basically, just be a technician and throw switches. The pilot in the front was doing all the real work; he was flying the airplane."

During a four-hour WB-57F flight on July 1, 1979, and wearing a high-altitude pressure suit, Kathy Sullivan set an unofficial sustained altitude record for a woman of 63,300 feet (19,294 meters). The flight was out of Ellington AFB and the aircraft was flown by Jim Korkowski, a pilot from the NASA Airborne Instrumentation Research Program, with Sullivan flying as scientific operator. During the flight, she was responsible for operating color infrared cameras and multispectral scanning equipment for 90 minutes, as the WB-57F flew over the Big Bend area of West Texas. **[10]**

Such flights were supplemental to their Ascan training assignments and afforded the candidates experience in working as part of a team on a mission with defined research objectives, using scientific equipment while wearing restrictive pressure garments, and hands-on knowledge of maintaining and operating equipment similar to what they would encounter on their future Shuttle missions.

Touring NASA field sites and contractors

In their first year, the group conducted several trips across the United States, visiting NASA field centers in order to become familiar with different elements of the agency and its major contractors. Most of it became a little tiring after a while, but it was useful public relations for NASA and invaluable to the candidates to meet and speak with different employees across the NASA family, as well as talking directly to the numerous contractors that were developing and building the Space Shuttle they had been selected to fly. "We saw a good slice of the whole program," recalled van Hoften, "so we got to see almost every main contractor for the Shuttle. We got to crawl over all the hardware. We learned all the systems, in and out."

Fig. 6.7: The TFNG during an orientation tour of the Building 30 Mission Control Room at JSC. (foreground to rear): Onizuka, sitting at the Guidance (GUIDO) console, Resnik, Hawley, Ride, Hart and Fabian.

The group toured the facilities across the JSC site, many of which they would spend hundreds of hours using during their careers in training and supporting numerous Shuttle missions over the next three decades. This would include being briefed on the operations inside Mission Control (Building 30) in January 1979. Arrangements were also made for group visits to the Lunar and Planetary Institute, right next door to JSC, as well as to Congress and NASA Headquarters in Washington D.C.

Other familiarization tours included at least one geology field trip to northern Arizona, while astronomy training was supplemented by several visits to the Houston Planetarium. They toured Ames Research Center, Moffett Field, California; Marshall Space Flight Center, Huntsville, Alabama; Goddard Space Flight Center, Greenbelt, Maryland; and Lewis Research Center, Cleveland, Ohio. On September 27, the group visited the Rockwell International Corporation's Palmdale facility in California, where they were able to examine the Space Shuttle Orbiters under construction close up.

Fig. 6.8: During September 1978, the Ascans visited the Rockwell facility at Palmdale, California. (main) Four of the TFNG seen touring the facilities at Palmdale during OV-102 *Columbia* assembly. [l-r] Sullivan, Thagard (back to camera), McNair and Ride. The Ascans are seen on a work platform above the open payload bay of *Columbia*. (inset, clockwise outer row) Nagel, Onizuka, Walker, Hoffman, Lucid, Mullane, Shaw, Resnik, Brandenstein, Buchli, two unidentified Rockwell personnel, Shriver, Sullivan (partly obscured), Fisher, Gardner and Hauck (obscured). Seated at tables are (clockwise from left) Creighton, Coats, Thagard, Nelson, Gregory, Hawley, Ride, McNair, Scobee, Group 8 training coordinator Al Bean, Seddon and Griggs.

During December 1978, the group went to Cape Canaveral, Florida, to witness the launch of an expendable rocket. It was during this trip that Steve Hawley gained a new title. "As part of your astronaut candidate training, as a group, you go to the different NASA centers, and you're culturally broadened by seeing what other people do within the agency. We were down in Florida at Kennedy [Space Center], and Senator [Adlai Ewing] Stevenson [III] was there, [and] for whatever reason we got all herded into an auditorium where he addressed the group of new astronaut candidates. As I remember, we had to go around the room and all introduce ourselves. It would be 'I'm Sally Ride, and I'm a physicist', and 'I'm Anna

Fisher, and I'm a doctor', and 'I'm Mike Coats, and I'm a pilot'. They got to Hoot Gibson, and Hoot said, 'I'm Hoot Gibson, and I'm a Navy fighter pilot'. Dan Brandenstein was next, as I recall, and he said, 'Well, I'm Dan Brandenstein, and I'm a Navy attack pilot'. So it seemed natural that when it got to me that I would be 'Steven Hawley, attack astronomer', and unfortunately it stuck. That's how that came to be."

To be briefed on NASA's robotic space program, they visited the Jet Propulsion Laboratory (JPL) in Pasadena, California. "That was an interesting visit," recalled Hoot Gibson. "We got to JPL and the Center Director briefed us. He basically told us that he didn't believe that we needed [a] manned space [program] at all. Talk about what I think nowadays would be considered 'an inappropriate subject' when you're speaking to a bunch of manned space people! It'd be like me going there and telling him that we didn't need robots for anything. We did get to see a bunch of fascinating things, but I'll just never forget the director getting up and telling us that, 'We really didn't need you guys; we really don't need manned space'. It's okay to feel that way, but just don't say it!"

"Basically, being an astronaut, it's not real narrow [and] highly specialized," said Dan Brandenstein. "It's very diverse. Your missions carry a variety of experiments. So, as I came to find out later in my time at NASA, you really look for people that are adaptable and have a more diverse background, because the way you operate in the office, you have assigned tasks for six months, nine months, a year, and these are technical tasks that you do when you're not training or flying a mission. And you get switched to another one and [so] you try and develop a corps of astronauts that really have a very broad base of experience and knowledge that covers the wide spectrum of the space program.

"That was part of the reason for going to all the centers, because you got to learn what they did at all the other centers, so you got a better understanding. You went to the contractors, where they were building the Shuttle, and got to understand that a little bit more. So it gave you a really good, broad experience base. That's what I liked about the job. You weren't stuck in a small, narrow area. It kind of goes back to my nature. The Navy is kind of that way, too. They move you around in jobs every nine months or so, and that's what I always liked. I liked new experiences and learning more, and having a more diverse-type job as opposed to a very narrow focused-type job."

KC-135 Vomit Comet

As a taster of what was to come, each of the TFNG completed familiarization flights on the KSC-135, in small groups, giving them short bursts of simulated weightlessness training. "It's probably a little bit of a reach to call that training," recalled Steve Hawley, "but it's good exposure to what it might be like to be in zero-*g*. Of course, you don't get it for more than 20 or 30 seconds at a time, and how pure a zero-*g* it is, is a function of how well the pilots up front can fly the trajectory. But it's a lot of fun. Everybody enjoys doing it, and I probably only did it a couple of times, but, yes, that was fun."

Fig. 6.9: Group 8 Ascans experiencing short periods of "zero-g" aboard the NASA KC-135 'Vomit Comet' during familiarization flights.

Spacesuits and EVA

All the MS astronauts were trained in spacesuit operations in the NASA Neutral Buoyancy Laboratory (NBL) facility in Building 29 at JSC. This was a large water tank where astronauts could practice and develop their Extra-Vehicular Activity (EVA) skills. In order to work in the NBL, the candidates had to qualify from scuba diving training and were required to maintain their scuba proficiency as MS.

"The scuba diving was challenging for me because I'm not a swimmer," confessed Rhea Seddon. "I had to work hard at being able to do that; to be comfortable in the water, to be comfortable under water, to be able to tread water for 20 minutes. I had to do some additional work. It was frustrating, but I was just determined that that was not going to stop me from being an astronaut. Then they found out that I was too small for the spacesuit, so I really didn't have to do that training anyway. The suit training – especially for small people – that was physically demanding."

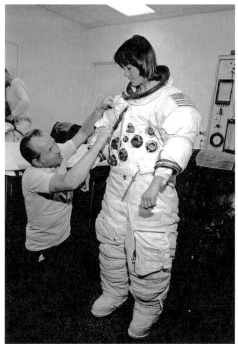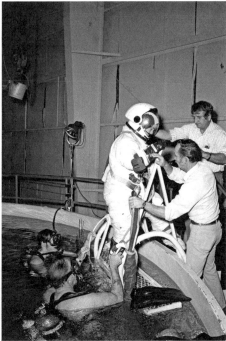

Fig. 6.10: The TFNG were all trained in spacesuit operations and had to pass a scuba course prior to wearing the full suit. Here, Anna Fisher is seen during 1979/1980 in underwater simulations of EVA activities.

Apart from attitude changes brought about by the selection of female astronauts, there were other factors to consider. For instance, the Shuttle's cargo bay doors were made easier to operate in case the female astronauts had to suit up and venture out into space on EVA to close them if they jammed. The Shuttle spacesuits had been made with smaller fittings to accommodate women, while the flight and middeck seats had been modified to allow them to move up and down to make it easier for the women to reach switches. "Women are just not as strong as we are in the hands and arms," Alan Bean stated, "so we modified the spacesuit and made some changes in levers and things so their hand and arm strength can do the job."

Dealing with the media
One of the often overlooked aspects of astronaut training is learning how to handle being interviewed by the media. As the 1978 group were the first to be selected for some years, with the added spotlight of the first females and minority candidates, the press were naturally eager to record and photograph everything. Such attention is not helpful when trying to master some of the training devices, especially being seen making errors while learning the skills.

"There were press everywhere we went, particularly since we were the first Shuttle astronauts," Jeff Hoffman recalled. "I think it was much more of a unique reception probably than Group 12 or Group 13 who went ten years later. Everywhere we went, we were followed by the media. I remember we went for water survival training. We were supposed to be by ourselves in little one-person life rafts. Ten feet away were these big boats with the photographers clicking away, mostly at the women of course, but my raft was right next to Judy Resnik's. We were surrounded by photographers. It was really quite something." NASA provided media training to help handle the press or television reporters, the so-called 'Charm School'. "I don't honestly remember how far into our training they did that," said Hoffman, "but they contracted with one of these training organizations that comes in and gives you an intense two-day crash course in how to deal with the media. It was quite good. I think most everybody who went through it felt that they had a lot of useful hints about preparing certain talking points in advance and how to fend off microphones that are being shoved in your face. There's a lot of little tricks of the trade."

NOW FOR THE REAL WORK

In April 1979, nine months into their two-year Ascan training program, the preparations for the first Shuttle flight were gearing up. The OFT crews were undergoing intensive training in the Shuttle simulators and controllers were staffing the updated (from Apollo) console layout in Mission Control. The training for the 35 Ascans would have new emphasis because they would be leaving the classroom, as explained by Senior Scientist-Astronaut Ed Gibson, the coordinator of the candidates' technical assignments: "They're being put straight to work like the rest of us. The first months, they were in more of an observer mode. Now they'll be assuming responsibility, the same as anybody else in the office." [11]

Clearly, the group were progressing at a better rate than originally envisaged and they were now being given their first technical assignments, to work on concurrently with the rest of the Ascan program. Steve Hawley, for example, investigated how software could support payloads to be flown on the OFTs. Kathy Sullivan was working in a similar role, but specifically for the second flight, while Fred Gregory was involved in enhancing the Orbiter cockpit layout, and Pinky Nelson evaluated methods of getting into and out of the Shuttle's new EVA spacesuit. All 35 candidates were excited with this new development, trying to absorb a tremendous amount of information and to learn every facet of their roles. They would continue flying the T-38, working one-to-one with veteran astronauts, keeping up to speed in their specialist field and keeping fit. As the first Shuttle crews were beginning training in the Shuttle Mission Simulator, the 35 Ascans were assigned as observers. As Steve Hawley commented at the time, "The best way to learn how to do it is to go out and do it."

THE LONG WAIT

By 1980, the group had been in training for two years, and the Shuttle was still a year away from its first flight. "A couple of them have complained to me that they're just as far from flight as they were when they came aboard two years ago," commented the group's coordinator Al Bean in 1980. "Well, I know some guys, like Bob Crippen, who've waited more than ten years without a flight, so these new guys are going to have to learn a little patience." [12]

Social bonding

In the interim, the training continued and the technical assignments increased, but they all managed to partake in social activities together. "In a lot of ways it was like a Navy squadron," said John Creighton. "You socialize together on the weekends and do a lot of things together, travel as a group to go to visit these things, so you built sort of a close-knit group and a camaraderie among the crews. That lasted throughout the initial year or 14 months of training, and up until the selection of the first crews out of our group, and then the group tended to fragment a little bit. Then you started socializing more within the crews, once you got assigned to a crew to go fly."

As the *Los Angeles Times* reported in September 1978, the Ascans were spending so much time together that they had become a particularly tight-knit group. The newspaper reported that they gathered together on Friday nights for 'Happy Hour', having Texas-style barbecues at the homes of the married members, and divided into Red and Blue teams for softball and touch football games. The married astronauts spent more time with each other than they spent with the single astronauts, and vice versa. An exception to the rule was Shannon Lucid, a mother of three, whose husband left his job and followed her to Houston when she was selected. "I don't have a lot of time to do things on the side," she said. "You can't do everything [although] I expect to fly a lot." The newspaper also gleefully reported that there had been at least one romance developing within Group 8, stating that Rhea Seddon was dating one of the pilots. As history records, she and Hoot Gibson were later married.

"These guys see the astronaut job as a career; you know, 20 years with the Space Shuttle," Al Bean said of the new group. "Everybody here figures they'll fly two or three times and some of them probably figure they'll fly 20 times."

While there was overt attention given to the six women candidates, with continuing speculation as to which of them would be the first to fly, it was a similar story for the three African-American candidates. "It may be better to be second or third, because then you can enjoy the experience a little more," said Guion Bluford. "It [being first] isn't something I'm running after." He may not have been going after it, but he would indeed become the first of the three to fly, on the STS-8 mission. Fred Gregory agreed it was not something he was after either, but added philosophically: "Let's get there first. I'm so glad I'm here, I wouldn't trade this for anything."

"We thought the Shuttle would be flying within a year of when we got down here," recalled Mike Coats. "As it turned out, it was three years before the first Shuttle flew, but it was a really good time. In fact, they told us that first year, during that astronaut candidacy period, if you had something else to do, like I did finishing up a master's degree – I had to take some night classes to do that, because I'd been in graduate school – they said this is a good time to do it, because after the year is up you're going to get awfully busy. The hours will get very long. So the first year was actually fairly relaxing, even though it was 'fire hose' treatment. It wasn't nearly as intense as later, when you got assigned to a crew and you're in the simulator all the time learning. That first year was enjoyable."

End of the Ascan probation

On August 31, 1979, NASA issued a press release stating that the 35 astronaut candidates had completed their Ascan training and evaluation period and were now "eligible for selection to flight crews." The two-year period had been cut down based on the experience with this group, and all future candidates selected for the astronaut program would now undergo a one-year training and evaluation period instead. "We are pleased with this newest group of astronauts' performance and their adaptation to the Space Shuttle training program," reported George W.S. Abbey, Director of Flight Operations. With this decision, the total strength of the Astronaut Office eligible for selection to a Shuttle flight crew suddenly shot up from 27, where it had remained for some years, to 62, with another selection process already announced on August 1, 1979, for a new class of Ascans to arrive during 1980. [13]

The TFNG found out about the change in their status only a short time before the official announcement later the same day. Chris Kraft and George Abbey had called an "all-hands" meeting and told the 35 that they could drop the word "candidate" from their job titles, as their first technical assignments had produced results beyond expectations. [14] "All of a sudden, we were here for about one year," recalled Anna Fisher, "and John [Young, Chief Astronaut] announced, 'Okay, that's it. Training is over. You're all astronauts now'. We were supposed to have a two-year period. [We thought] 'No, we've got a lot of work to do now. We don't have time for all this foolishness'." They immediately became eligible for even more technical assignments and Public Relations activities.

"I didn't have any expectations other than flying Space Shuttle," explained Brewster Shaw, "and I soon learned that the percentage of the time you got to fly the Space Shuttle was pretty miniscule, relative to the percentage of the time that you were here working for the agency, and that there were a lot of other things you were going to do that would take up all your time… That was made clear to us pretty soon. So we all got assigned various technical assignments."

John Creighton thought the group learned a lot about NASA. "Originally, they were talking about it being an 18-month to two-year training program," he said, "but they cut it shorter to about 14 months, and then the subsequent classes were limited to a year."

Reflecting on the first Ascan course

"They had a course laid out for us," said Loren Shriver, "but there was, I think, a lot of real-time adjustment. And of course, each time we finished a phase or a course, then they would take the lessons learned from our experience, and I think by the time three or four more classes came through, they had it pretty well ironed out the way they wanted it. But it was a fairly new process [with us]. But we had a lot of great training… and… all kinds of lectures and information that, had it not been for that [the Ascan program], I never would have been exposed to them. And it was just all so interesting. That was an aspect of our Ascan training that was completely different than my technical background as a test pilot in the aero-astro engineering field that I went through school in. This was completely different, and I just really enjoyed that. I really liked that a lot. To get a little bit of exposure [to different aspects]. You might have to try to use some of that training in the future on looking down at Earth and seeing the various geology that was represented there… I would never be proficient at that, but at least I was exposed to some of the terms and some of the thought processes that they were going to be using, so it was great."

A year into their career with NASA, the group were already working on their first technical assignments in the Astronaut Office, supporting the OFT missions, and primarily focusing on the first launch of the Shuttle system, designated STS-1, to be flown by John Young and Bob Crippen. These varied, important and critical roles represented the next step in their personal journey to space and are described in the following chapter.

References[3]

1. **The Astronaut Maker**, Michael Cassutt, Chicago Review Press Inc., 2018, p. 192.
2. **Riding Rockets, The Outrageous Tales of a Space Shuttle Astronaut**, Mike Mullane, Scribner, New York, 2006, p. 28.

---

[3] Unless specifically stated, all quotes used extensively in this and following sections are taken from the official NASA JSC Oral History transcripts detailed in the Bibliography.

3. **Go for Orbit, One of America's First Women Astronauts Finds Her Space**, Rhea Seddon, Your Space Press, Murfreesboro, Tennessee, 2015, pp. 42–43.
4. NASA News Release, JSC 78-27, June 29, 1978.
5. *Sextet for Space*, Peter Gwynne/Holly Morris, *Newsweek* magazine, August 14, 1978, p. 5.
6. *Astronaut candidates to take parasail training at Vance AFB, August 28*, NASA JSC News 78-39, August 23, 1978; NASA JSC Houston *Roundup* news magazine, September 1, 1978, Vol. 17, No. 17, p. 1.
7. NASA JSC News Release 78-37; **Astronautics and Aeronautics** 1978, NASA SP-4023,1986, p. 191.
8. Paraphrased from *Space Lures New Astronauts,* unaccredited article, *Los Angeles Times*, issue 3, September 1980, p. 128.
9. NASA JSC News Release, 79-20, April 8, 1979.
10. *Astronaut Candidate Kathryn Sullivan Sets Altitude Record*, NASA JSC News Release 79-47, July 2, 1979.
11. *Schooling of Astronauts, 35 New Candidates is Varied, Exciting*, NASA News Release 79-20, April 8, 1979.
12. *Astronauts of the '80s*, Thomas O'Toole, *Washington Post*, issue 20, July 1980.
13. *35 Astronaut Candidates Complete Training and Evaluation Period*, NASA JSC News Release, 79-53, August 31, 1979.
14. Reference 1, p. 210.

7

Silver Pin Astronauts

> "We actually have more jobs that we support
> than any one individual can do in a career.
> All of them have training value of some kind.
> All of them are important to the program in some way;
> some of them probably a little more fun than others."
> *Steve Hawley, JSC Oral History*

As Steve Hawley explained, the astronauts would not be able to rotate through every support position in their careers, but some assignments were clearly more popular than others. "I never got to be a Capcom [Capsule Communicator], for example, but that's a job that lots of people get rotated through," Hawley observed. By the fall of 1979, with their Ascan training program completed a year earlier than planned, the rookie astronauts were assigned to technical and support roles in the Shuttle program, which they hoped would lead to at least one flight into space on the Space Shuttle.

What Do Astronauts Do?

This question was posed in an unsigned Astronaut Office handout dated April 3, 1986, as part of the briefing material provided to the Rogers Commission investigating the *Challenger* Accident. [1] The answer was given as:

- Provide flight crews for NASA space vehicles and stations
- Participate in the design, development and operations of space vehicles and space stations

© Springer Nature Switzerland AG 2020   194
D. J. Shayler, C. Burgess, *NASA's First Space Shuttle Astronaut Selection*,
Springer Praxis Books, https://doi.org/10.1007/978-3-030-45742-6_7

- Provide crew input into the design and development of space vehicle and station operating technologies and procedures
- Perform crew evaluations supporting design, text and checkout of space machines, and
- Perform other duties as assigned.

In order to achieve this, all astronauts are required to complete a range of technical and support assignments prior to and following their time as part of a flight crew; in training, in flight, or during post-flight periods. Many astronauts completed more than one assignment, and in fact some had concurrent triple or quadruple assignments or were in various stages of training having been assigned to more than one mission at one time. As the memo stated, the bottom line was that "many astronauts do more than one job."

TECHNICAL AND SUPPORT ASSIGNMENTS

Following on from a huge ground test program and the 1977 series of Approach and Landing Tests (ALT), the first orbital missions of the Shuttle system were flown under the Orbital Test Flight (OFT) program. OFT was designed to test and evaluate Shuttle hardware, systems and procedures from pre-launch to post-landing, prior to the system being declared 'operational.' Originally, the OFT program envisaged six missions, designated STS-1 through STS-6 and crewed by six, two-man teams of veteran astronauts chosen from earlier selections. Each flight initially had a back-up pair of astronauts, who would serve as the prime crew on a later OFT flight. By 1982, with the Thirty-Five New Guys (TFNG) four years into their careers at NASA and now eligible for crew assignment, the OFT program was cut to just four missions. The back-up positions were terminated after the third flight. NASA reasoned that there was a pool of suitably trained astronauts (though not all were flight experienced) to draw from to replace any ailing, injured or reassigned crewmember as required. Though the next two missions, STS-5 and STS-6, were no longer officially part of the OFT program, the assigned veterans were already in training as four-man crews, including the first astronauts to hold the Mission Specialist (MS) designation. As earlier documents suggested, the opportunity to assign the first Group 8 astronauts would therefore not arise until STS-7 onwards. In the meantime, there was plenty of work for them to do in support of these first six missions.

The flight crew on any space flight is the visible element of the mission, but that is only part of the story. There are those who work the consoles at the Mission Control Center (MCC) in Houston, and the hundreds of workers down at the Cape, at other NASA field sites, or at the dozens of contractors who process the hardware for each mission. During the Shuttle program, there was another group who were often overlooked in the reporting of each mission: the astronaut support team.

As the program developed, the size of the astronaut team increased, and while the Shuttle flew between 5–7 persons on most missions, there were normally scores of other astronauts in the office, with many in training for forthcoming flights, or debriefing following a recent mission. Others were fulfilling their technical assignments, serving in managerial roles, on sick leave, or simply on vacation or a day off. As dedicated, busy and sought after as the astronauts naturally are, they remain human and need time away from 'the office', an opportunity to unwind with the family and home life.

Astronaut Silver Pin

Though not eligible for these first six missions, all 35 members of the 1978 selection were now termed 'astronauts' having completed their Ascan training program. Though this was their official title, they still prized the real goal of making a space flight and, when in front of the public, the chance to describe their experiences from a real mission rather than how they hoped it would be on a forthcoming flight. They had the title and the blue suit to wear, but the TFNG would remain space flight rookies for some time to come. "Once you go from astronaut candidate to astronaut, you get an astronaut pin that's silver," recalled Mike Mullane in his 2003 Oral History. **[2]** "Then, when you fly, you [are eligible for] a gold one. You're not really an astronaut until you're wearing that gold one, and I remember I never wore my silver one. I don't think I ever put it on, because I remember telling somebody [that] to wear that, to consider yourself an astronaut, to me it would be like a stewardess wearing flight attendant wings and saying you're a pilot... To me… you're not an astronaut until those SRBs ignite."

Though the experience of those Solid Rocket Boosters (SRB) igniting was still a few years away, there remained plenty of work to be done in supporting the OFT and the Shuttle program through their technical support assignments, well before any of them were announced to a crew. As Guion Bluford said in 2004: "After a year of training, and a strong demand for our talents in the Astronaut Office, John Young decided to put the Ascans to work. Several jobs were parceled out to us as the Johnson Space Center [JSC] prepared to fly the Space Shuttle for the first time." **[3]**

But what exactly are these 'technical or support assignments'? It is the work that usually goes under the radar with the public, but forms the bulk of the astronaut profession. Very little time, percentage wise, in any space explorer's career is spent actually 'in space' flying the mission. An astronaut's time is mostly spent on the ground, attending countless meetings, briefings, simulations, training sessions, survival courses, and public or press appearances.

Having expected to receive their first technical assignments once they had completed the two-year training program, or at least at the end of their first 12 months at NASA, it came as a bit of a surprise to the TFNG to learn that they would begin their initial technical assignments in April 1979, just nine months after joining the agency. They would be deemed fully-fledged astronauts after just 12 months – such was the pace of the program and the need to get the Shuttle in orbit – primarily in recognition of the way the whole group had performed in their short time at JSC. Whatever it was that the selection board had seen in these 35 candidates had shone through in their Ascan training, as each and every one of them progressed to earn their silver astronaut pin months before they had expected to. The objective now, over what was to prove to be three years, was to ensure that they were capable and ready for crew assignment on the Space Shuttle.

Fig. 7.1: STS-1 astronauts John Young (2nd from left) and Bob Crippen (2nd from right) pose inside the Vehicle Assembly Building at KSC during the Shuttle Interface Test prior to rollout to the launch pad. With them are support astronauts Dick Scobee (left) and Fred Gregory. Space Shuttle *Columbia* is visible behind them, mated to its External Tank and Solid Rocket Boosters (Image courtesy of Ed Hengeveld).

The Pilot Pool

While the new class of astronauts were now eligible to wear the silver pin, they were still some way away from the coveted gold version. As well as receiving their first technical assignments, the group were also transferred administratively to the Astronaut Pilot Pool, under the Mission Operations Directorate, Flight and Systems Branch at JSC, for their Advanced Training Program. This next phase of their preparations for space flight featured some 16 courses over the next twelve months or so, totaling approximately 375 hours, and were completed prior to their first crew assignment in addition to various technical assignments. It would provide them with the skills to operate the complex Space Shuttle, by completing a number of generic simulations and training programs (see Chapter 8).

An undated document, probably circa 1979/1980, researched by the authors reveals the initial technical assignments of the TFNG to be conducted over the next two to three years, prior to assignment to their first spaceflights. These are listed in Table 7.1.

TABLE 7.1: CLASS OF 1978 INITIAL ASTRONAUT OFFICE TECHNICAL ASSIGNMENTS 1979-1983

Bluford (MS): RMS; Spacelab 3; Shuttle systems; SAIL; FSL; DOD payloads.

Brandenstein (PLT): Support crew STS-1 & 2 (Capcom / cue card development / crew equipment); vehicle integration.

Buchli (MS): T-38 MS training development; support crew STS-1 & 2 (SMS / initial Orbiter EVA development / Capcom).

Coats (PLT): Landing and rollout; SAIL; training; support crew STS-4.

Covey (PLT): Back-up flight software; board representative / CRT (Cathode Ray Tube) verification) SAIL / FSL / STS-2 & STS-3 chase plane; prior STS-1 astronaut support Orbiter engineering development and testing.

Creighton (PLT): Initial SAIL crew; flight software (Version 18 coordinator / enhancement coordinator) T&O board and OAS CB representative / malfunction Caution & Display (C&D) system study; Contingency On-scene Director (COD).

Fabian (MS): Skylab re-boost support; RMS; proximity operations and rendezvous; crew equipment; leader of Operations Support Group; DOD payloads.

Fisher A. (MS): RMS; EVA; tile repair; medical operations STS-1 & 2; SAIL.

Gardner D. (MS): T-38 MS training development; Astronaut Project Manager for flight software in Shuttle onboard computers leading up to STS-1; flight software (Version 16 Coordinator / Enhancement Coordinator / T&O Board & ASCB Representative) / displays and controls; Flight Data File coordinator; STS-4 support crew; DOD payloads.

Gibson R. (PLT): Skylab re-boost; CRT display verification; STA HUD evaluation pilot; STS-1 & STS-2 chase pilot; SAIL.

Gregory F. (PLT): Crew station enhancement; vehicle integration; Flight Data File coordinator (Manager); CB representative KSC initial Orbiter checkout and launch support STS-1 & STS-2.

(continued)

TABLE 7.1: (continued)

Griggs   (MS): TPS repair; development and testing of HUD; support crew STS-3 (Capcom); approach and landing avionics system; development of MMU; requirements definition and verification of on-orbit rendezvous, and entry flight phase software and procedures.

Hart (MS): Spacelab 1; Skylab re-boost support; support crew STS-1 & 2 (SSME liaison / Capcom / HP-41).

Hauck   (PLT): Support crew STS-1 & 2 (Flight Data File crew coordinator / Orbiter GCA development / Capcom); Leader of Development Group; Orbiter night landing project officer (test pilot) for development of techniques and landing aids; Navy Administrative coordinator for miscellaneous activities.

Hawley (MS): Payload software; SAIL (simulation pilot prior to STS-1); vehicle integration.

Hoffman (MS): Spacelab 2; orbit navigation; guidance and control (orbit digital autopilot insertion/deorbit targeting); proximity operations and rendezvous; STS-6 support; During preparation for OFT he worked in the Flight Systems Laboratory (FSL) at Downey California; testing guidance, navigation and flight control systems.

Lucid (MS): Spacelab 1; crew training; SAIL; FSL (Downey, California); Space Telescope; proximity operations and rendezvous group; SPAS; FSL at Downey working with the prox ops and rendezvous Group.

McBride (PLT): STS-1 (lead) chase pilot; SAIL.

McNair (MS): Solar Maximum Mission retrieval; payload and RMS retention; IECM, PDP, REM; Space Telescope; Orbiter systems.

Mullane (MS): Spacelab 1; IUS, Syncom IV and TDRS; SAIL; Support crew STS-6; DOD payloads.

Nagel (PLT): Software; SAIL; STS-1 Chase; Support crew STS-1 & 2 (entry maneuvers & Capcom).

Nelson G. (MS): EMU / EVA development; Space Telescope; WB-57F crewman (scientific instrument operator); initial Malfunction Book development; STS-1 Chase (prime chase photographer); support crew STS-3 (Capcom).

Onizuka (MS): SSUS, IUS; TPS repair; vehicle integration; crew equipment; RMS; Shuttle Student Involvement Program; DOD payloads; member Orbiter Test and Checkout Team & Launch Support Team at KSC for STS-1 & STS-2.

Resnik   (MS): Spacelab software; payload software; RMS software; RMS; Power Extension Package (PEP); 25 KW Power Module; training techniques; Orbiter development including experiment software.

Ride (MS): RMS; support crew STS-2 & 3 (on-orbit Capcom /visual operation I/F / RMS procedures / Orbit Flight Data File); STS-1 Chase

Scobee   (PLT): Spacelab 1; SPS engineering simulation evaluations; SMS verification; 747 (SCA) Pilot / IP; vehicle integration; STS-1 chase; Leader of Verification Group; Instructor Pilot SCA Boeing 747.

Seddon (MS): Food systems; Spacelab 3 & 4; Orbiter & payload software; CRT display verification; orbital medical kit; medical checklists; medical ops STS-1, 2 & 3; SAIL; FDF; launch and landing rescue helicopter physician; support crew STS-6.

Shaw (PLT): Skylab re-boost support; SAIL initial crew; auto land; Optional TAEM Targeting (OTT); Shuttle Training Aircraft; STS-3 support; support crew & Entry Capcom STS-3 & 4.

Shriver (PLT): Vehicle integration; STS-2 chase; Vandenberg launch site crew interface.

(continued)

TABLE 7.1: (continued)

Stewart (MS): Entry Flight Control (EFSIG Representative); FSL; entry training; MOCR
aerodynamic console STS-1 & 2; support crew STS-4; testing & evaluation of entry
flight control system for STS-1; ascent orbit procedures development; payload coordination; Ascent / Orbit Capcom STS-5.

Sullivan (MS): OASTA-1 and OSS-1; WB-57F Crewman (qualified as a systems engineer
in 1978) participated in several remote sensing projects in Alaska; flight software
(version 16 and 18); STS-2 chase; vehicle integration; co-investigator SIR-B experiment
(flown later on STS-41G); software development STS-1 & 2; lead chase photographer
for STS-2 launch and landing; Orbiter cargo test, checkout and launch support KSC
STS-3, 4, 5 & 6.

Thagard (MS): RMS; SAIL; proximity operations and rendezvous; SPAS.

Van Hoften (MS): Lead Astronaut Support Team (AST), KSC; responsible for Orbiter
turnaround, testing and flight preparations.

Walker D. (PLT): Deputy (Acting) Chief, JSC Aircraft Operations Division; CB Safety
Officer; STS-1 chase; SAIL; Mission Support Group leader STS-5 & 6.

Williams D. (PLT): SPS engineering development simulations; crew equipment; SAIL (test
pilot); vehicle integration; participated in Orbiter test, checkout launch and landing
operations support at KSC.

*Adapted from an undated NASA document (certainly originating from @ 1980-1982). Copy
on file, AIS Archive.*

ASTRONAUT SUPPORT ROLES EXPLAINED

For each Shuttle flight, a team comprising a couple of dozen astronauts were
assigned to various support roles at JSC in Houston, at the Kennedy Space Center
(KSC) in Florida, at key sites across the United States, or at remote locations on
the eastern side of the Atlantic Ocean. Some of these were single-mission assignments, others were short tours of duty, but all offered that valuable experience of
being part of the 'Shuttle system' and were a key element in the individual's preparation for assignment to a flight crew. Perhaps the most prominent of these, in the
eyes of the public, were those astronauts who served at the Capcom console in
Mission Control at JSC. However, this was just one of the myriad roles that the
astronauts were required to fulfil before progressing to crew assignments, most of
which were overlooked in written accounts. They were performed in between
their training and flights, and were as important to their flight preparations as the
mission training itself.

Below is a selection of these major assignments, described with additional
comments from some of the Group 8 astronauts who participated in these roles
during the early years of their careers. Between 1966 and 1975, the astronauts had
formed a third tier *support crew* for each Apollo mission. They also fulfilled the
Capcom roles for Apollo/Skylab and Apollo-Soyuz (ASTP), but with the more

complex Shuttle intended to fly more frequent missions, this new support network had to be much more in-depth than the generally single-mission support of the Apollo-era crews. For Shuttle support, the 'tours' had to be more program orientated rather than flight-specific simply because of the frequency and variation of the missions on offer.

LAUNCH SUPPORT ROLES (KSC FLORIDA)

All 135 Shuttle missions were launched from KSC in Florida and required a small team of astronauts to support the larger workforce at the Cape in preparing each mission and its associated hardware for launch, with the pace increasing as the launch date approached. These roles were developed by the members of the TFNG from the very first Shuttle missions in 1981 and 1982.

Fig. 7.2: Loren Shriver, the first Astronaut Support Person, and suit tech Al Rochford on the aft flight deck of *Columbia* as they observe STS-1 astronauts Young and Crippen. *Columbia* was in the OPF when this photo was taken on October 13, 1979 (Image courtesy of Ed Hengeveld).

Cape Crusaders

For each Shuttle mission, a team of about 5−8 astronauts served as the flight crew's point of contact between JSC and KSC, flying their T-38s between the centers on 90-minute trips each way. Known officially as the Astronaut Support Personnel (ASP), they were also the eyes and ears for the Shuttle vehicle the crew were about to fly. Better known collectively as the "Cape Crusaders" (also known as C-Squared or the C²s, see sidebar: *Origin of the Cape Crusaders*), their role was a busy one during a nominal year of flight operations, with between 25−35 trips to KSC a year. They were present for the Terminal Countdown Demonstration Test full dress rehearsal, the launch and the landing. The astronauts were assigned to the Crusaders for a 'tour' which could vary depending on flight operations and other assignments or requirements. A typical stay at the Cape could vary between a single night or four days away from home, especially when there were delays to a launch.

"There always was a team of from three to six astronauts assigned," recalled Dan Brandenstein. [4] "It was one of the technical assignments, and they were assigned for support down at the Cape. They worked crew-related issues as far as getting the vehicle processed, getting experiments integrated, following the various payloads as they were getting checked out [because] the crew couldn't be down there all the time. We switched from having a support crew supporting one crew. I mean, it was kind of like a support crew for everybody in a particular area, and it was down at the Cape. They're also the ones that helped strap the crew in when they launched, and during the countdown they did all the switch positioning, the prepositioning of the switches and everything for the crew. So it was a real interesting job, in that you got to work on the real hardware, and that's always fun. You got to see a lot of Shuttles launched because you were always down there for the launches and the like. So it was another one of the interesting jobs."

One of Fred Gregory's first technical assignments was to be assigned to KSC as an astronaut liaison. "We had a name; we were called the Cape Crusaders, the C²s. Every Sunday evening or Monday morning, four of us would fly down to the Cape [Canaveral] and we would stay all week. We would attend meetings. We were always in our blue suit, and one of the privileges was that you would spend a lot of time sitting in the cockpit of *Columbia*. So I was exposed to the hardware in '79 or probably a year, year and a half after I got there, and I stayed through the second launch of the Orbiter, STS-1 and -2. I was there for both of those." [5]

Fig. 7.3: Just one version of a Cape Crusader cartoon (Image from the collection of Steve Hawley, used with permission. Artist unknown).

ORIGIN OF THE CAPE CRUSADERS

According to the dictionary, the term 'crusader' can mean "a vigorous campaigner in favor of a cause." [6] The exact origin of the term 'Cape Crusader' in NASA-speak has been lost in time, but appears to come from the early 1980s and the first Shuttle missions. While capturing the Oral Histories for NASA, the JSC team was unable to uncover whether the term had existed during the Apollo era as nobody ever mentioned it.

Unlike Apollo, the Shuttle required a greater astronaut presence at the Cape over a longer timescale, and when the first astronaut support team assigned to STS-1 went to the Cape, they found they were indeed embarking on a "vigorous campaign in favor of a defined cause", namely the first launch of *Columbia*. It was probably Loren Shriver, the first Astronaut Support Person who, recalling the 'caped' superheroes of his youth, suggested a connection to this new astronaut 'crusade' and the location at which they were working – the Cape – hence "*Caped Crusaders*."

Rhea Seddon suggested in her Oral History that "*Batman* was the Caped Crusader, maybe that's where it came from. I recall that there was a little cartoon character [and] that was either their nickname or that's what they were called – *Crusader* somebody [*Crusader Rabbit*, a cartoon which aired through the 1950s]. It came out of comic books or TV, and it was a silly thing. Not that we all took ourselves very seriously." Whatever the origin, the name was certainly apt, and it has stuck ever since.

ASP duties also included responsibility for the *Switch List* positions, which involved ensuring that the Shuttle Orbiter cockpit switches were set correctly for launch, or participating in the communications checks (*Comm Checks*) between the Orbiter, the Launch Control ground control team at the Cape and Mission Control in Houston. They also assisted the ground closeout crew, the handful of technicians who were trained to oversee the pre-launch preparation in the White Room (which was positioned next to the crew side access hatch on top of the launch access tower), assisted the flight crew into their seats, set the switches to launch positions, and set up the seats for each crewmember. After working a few missions, a Cape Crusader could qualify to become the Prime Astronaut Support Person (with at least one back-up ASP), taking the lead for that particular mission, helping the flight crew into their seats, and being one of the last people to see them before the hatches were closed for launch.

As the program unfolded, this became a much practiced and very polished operation, and though the closeout crew knew their jobs inside out, the astronauts, as members of the support personnel, helped with hands-on activities; mainly ensuring that the crew were strapped in, that none of the switches were knocked in the process of getting them into the launch position, and ensuring that the detailed process did not miss a step. All this was made more difficult by the fact that the Shuttle Orbiter was standing on its tail on launch day, with the crew compartment tiled 90 degrees vertically and the flight crew lying on their backs. The seats that normally would be viewed conventionally on the vehicle flight and mid-deck floors now appeared to be hanging on the wall. This meant that the ASPs, technicians and flight crew had to be careful where they placed their hands and feet as they climbed through the vehicle to help the crew into their seats.

Being an Astronaut Support Person

Steve Hawley, one of the first ASPs, explained his role on some of the early Shuttle missions: "The ASP was the astronaut who was part of the closeout crew and strapped in the flight crew (along with the suit tech). I was the backup ASP on STS-2 (Ellison Onizuka was prime) and I was prime for STS-3 and STS-4. The ASP was also responsible for representing the crew at KSC meetings or in tests on the vehicle, and for coordinating the astronaut support to the practice and actual countdowns. For example, there was a rule that once the prelaunch switch list was complete, there needed to be an astronaut on board continuously until launch. This was to ensure that no switches were changed without instruction from the LCC [Launch Control Center] and that if any switches were moved, as part of the normal countdown or as part of troubleshooting, they would only be moved by an astronaut. I was also responsible, as part of the closeout crew, to go extract the crew in case of a scrub. I would stay on board in that case until one or two other astronauts relieved me. Those assignments were designated as part of my launch planning. To be clear, the astronauts assigned to KSC were determined by management, but the ASP was responsible for organizing how those guys would be used to staff the different tasks." Hawley recalled that during STS-3 he thought that the unpainted (to save weight) External Tank (ET) looked "kind of weird" in orange instead of the white-painted tanks of STS-1 and STS-2. As the ASP for STS-4, he had to contend with added security as *Columbia* carried a classified payload. "To protect the secrecy of what was in the payload bay," he continued, "there were metal covers locked in place over the aft windows so that no one could look into the payload bay. One of my jobs was to unlock and remove the covers as part of prime crew ingress. I had the key and I was really worried about losing it, which would have meant we couldn't launch. Fortunately, I didn't." **[7]**

Ascent Abort Support Roles

During the early years of the Shuttle program, astronauts were dispatched to remote sites on the west coast of North Africa or in Spain,[1] to support and assist in any abort situation during the ascent to orbit which could have resulted in a

---

[1] As the odds of actually landing at one of these site were slim, the equipment available there was kept to the minimum. Early sites included Morón (southern Spain), Banjul (Gambia, West Africa) and Ben Guerir (Morocco, North Africa). Dakar (Senegal, West Africa) was also an early site for TAL aborts, as was Zaragoza Air Base (eastern Spain).

landing at one of those sites. Fortunately, these sites were never called upon to be used in an emergency throughout the history of the program.

TAL Sites: Described as a safety net for the Shuttle launch as it arced over the Atlantic, a number of European and African locations were designated Transoceanic Abort Landing (TAL) sites. Here, it would have been possible to land the Orbiter in the event of an abort during the ascent if it had been impossible either to Return To the Launch Site (RTLS – never a preferred option for the crew), or Abort To Orbit (ATO, as in the case of STS-51F in July 1985). Fortunately, it was never necessary to use these sites, but it was always good to have them in reserve. One of the astronaut support roles was to be stationed at the TAL sites in case the stricken Shuttle needed to land there. The role was called TALCOM, for TAL Communicator, and their job was to confirm the weather and landing conditions around the site and communicate directly with the crew, as Capcom in a rapidly changing situation minutes after the mission left the launch pad in Florida. As some of these sites were remote this was not one of the preferred roles, especially during the early days of the program.

AOA Sites: If the Shuttle had been unable to reach a stable orbit but had acquired sufficient velocity to orbit the Earth just once and return to a landing site in the continental USA, then the Abort Once Around (AOA) mode would have been the preferred option. While this was never actually performed during the program, each launch required astronauts to be dispatched to AOA sites such as Edwards or Northrup, in case such an emergency did occur. **[8]**

Launch and Landing Helicopter Physician

For the first few Shuttle flights under the OFT program, it was decided that as many members of the Astronaut Office as possible should be involved in supporting the missions, especially the first launch. Part of this support included assigning medically trained members of the Astronaut Office to the search and rescue teams on standby during launch and landing. One of those assigned was Rhea Seddon, who explained the assignment in 2010: "I guess because of my emergency room work, and [because] Anna [Fisher] was doing something else, [I was asked] to look into that. When we looked at how many helicopters were going to be deployed both at the Cape [Canaveral, Florida], at Edwards [Air Force Base, California], and at Northrup Strip in New Mexico, I think they needed six docs. We had five in the office [Seddon, Fisher, Norman Thagard and, from the 1980 selection, Jim Bagian and William Fisher, Anna Fisher's husband]. Craig [L.] Fischer was head of medical operations at the time [and] one of the flight surgeons we involved. We talked to the SAR [Search and Rescue] forces and yes, if they were told to do so, they would put us on the helicopters. And yes, they would appreciate our input because they didn't really know much about the Space Shuttle or about what the suits were that the astronauts would wear or how the escape systems worked, that sort of thing. They needed a liaison with our office so I mapped it all out.

"I said one of the problems is [that] I have done trauma life support, but the other physicians that we're working with were not used to dealing much with trauma. I think Bill and Anna probably had. It was Bill, Anna, me, Norm, Jim, and Craig Fischer. That was six of us. Jim, Norm, and Craig had not had any trauma experience, and Bill, Anna, and I were a few years away from the experience we had. I asked if we could do some additional trauma training. There was a good course in San Diego [California], and then there's an advanced trauma life support course. I said, 'If you want us to do this, we want to be capable of doing it'.

"[They] let us do some additional training, which all of us enjoyed, getting back into medicine. Craig was a pathologist, so he was interested in 'How do I handle emergencies', because pathologists don't have an awful lot of emergencies. We did some additional training, but then we began working. We were each assigned to a different launch site for the first flight and began to work with the crews that were going to man those sites, [to] learn about search and rescue. I didn't know anything about helicopters, about SAR forces, about PJs, the para-jumpers that jump in the water with people. It was fascinating because the helicopter crew that I worked with out at the Cape had been doing search and rescue for years down at Homestead Air Force Base. They had all fought in Vietnam; they were the crazy people that went into the jungle and pulled people out in Vietnam, handled helicopters that were under fire. Getting to know those people and getting to hear what they had done; they were just awesome folks.

"I was assigned a PJ named John Smith, who had absolutely no hair. Big burly guy, and just one of the nicest people around. To look at them you would think they were animals of some sort. The things that they did were just physically awesome. Just very, very nice people. I got some time flying a helicopter, and just had a great time being a part of that, learning the on-scene rescue, evaluation, and treatment of patients.

"I was used to working in an emergency room, where patients all came in with their IVs and their tubes down their throat; their limbs bandaged and splinted, and laid on a table. Now I had to deal with what we do if we get out there and they drag an astronaut in a pressure suit out of a Space Shuttle. At first, John [Smith] deferred to me… 'You do your stuff and tell me what you need'. I came to realize that he was much better at doing the initial evaluation and treatment because that's what he had been doing for years. He was in charge at first, and then I was going to take over on the things that I knew how to do. I could do the follow-up part. We worked really well as a team, and I learned a lot of stuff. I got to see the first launch from the skid strip at the Cape and watched it with those guys. It was a really nice experience.

"When I first got into it I thought, 'This is crazy', but it certainly got us very involved in what was going on. I think we added value because we could say, 'If we get people in this condition we need to take them to this facility. Here's what we're going to have to do if we have to take somebody's helmet off; here's how we would do this, and here's how we would do that. Here's what the Shuttle can do, here's what all these words meant', so that the rescue forces understood what was going on. We had a number of practices. I think we all learned a lot. Luckily we didn't have to use that, but we were prepared.

"I did have the opportunity, as one of my assignments, to help with the development of the medical checklist that flew and the equipment that was carried in the medical kits. We certainly developed what we were going to do and probably somebody wrote it down as a checklist and put down, 'Here's what you need to do to get people out of the Shuttle'. They had a variety of ways. You could take them out the overhead windows; you could take them out the side hatch; you could saw a hole in the side. There were all kinds of permutations and combinations of how it could happen, and I'm sure they documented all of that. I wasn't in charge of writing the checklist *per se*. The PJs had to wear the SCAPE [Self-Contained Atmosphere Protective Ensemble] suits, the big aluminum suits, because of leaking fumes and stuff like that. I thought they were going to kill themselves in Florida, trying to climb up on the Shuttle model and hoisting people out on a backboard through the overhead windows. John would get back to the helicopter after they'd carried them, and he would just be drenched with sweat. I thought, 'They're going to kill themselves just with the rescue part. I'm going to have to rescue the rescuers'. It was amazing what they could do and what I learned from them." **[9]**

Weather Support

The *Weather Coordinator* provided the link between the Astronaut Office, the Mission Management Team (MMT), the Lead Flight Director (FD) in MCC Houston and the Launch Director at the Cape. In addition, the *Weather Pilots* (known as WX – pronounced 'wex') were senior pilot astronauts who flew the Shuttle Training Aircraft (STA) from KSC in Florida or Edwards Air Force Base (AFB) in California, to report on current weather conditions at altitude immediately prior to a Shuttle launch or commitment for entry and landing at a primary landing site. Information provided by the WX allowed the MMT to decide whether to proceed with the launch or order a delay. The WX pilots were also utilized to provide up-to-date information either to permit or waive-off a landing at primary or alternative sites, prior to committing to the de-orbit burn.

Family Escort

A member of the Astronaut Office was detailed to the immediate families or dependents of the prime crew. Their role was to provide advice, guidance, and support to the families throughout the mission, from preparation for launch at the Cape through the launch and orbital phases to entry and landing. Occasionally, an astronaut was also assigned to the prime crewmembers' extended family groups.

In the early years, according to Mike Mullane, there were no formal criteria for selecting one or more Family Escorts, with the spouses of the crewmembers putting forward a few names. The Family Escort, assisted by members of the Public Affairs Office (PAO), would be on hand during the day of launch as the families viewed the ascent from the roof of the LCC. In the event of a serious mishap, the family could be isolated from the press and looked after respectfully. Here, the unspoken role of the Family Escort as Casualty Assistance Officer would come into play. The accidents with STS-51L *Challenger* at the Cape in January 1986, and the return of STS-107 *Columbia* in February 2003 proved the most challenging for the Family Escorts, who had to provide immediate support for the families through these tragic losses.

Mike Coats was one of the TFNG who pioneered the Family Escort role. "We were preparing for the first Shuttle mission, and I was personally fortunate to be asked to be the Family Escort. I escorted John Young's and Bob [Robert] Crippen's families during the activities for their Shuttle mission – both for launch and then during the mission, and for landing out at Edwards Air Force Base in California. This was precedent-setting, if you will, and I really enjoyed that. It was fun to be right in the middle of all the 'firsts' that were going on."

Brief HQ Guests

At most launches and landings, there were a (sometimes quite sizable) number of VIP guests invited to attend by NASA Headquarters. As a result, at least one astronaut was designated as 'a briefer', whose role was to respond to any questions these guests might have about the flight or program. Rhea Seddon recalled her unexpected experience in this role: "The end of STS-7 had an interesting story, because they were meant to land at the Cape, and there was this immense crowd of VIPs who were going to be the very first special people to get to meet the now-famous first American woman to fly in space [Sally Ride]. But the STS-7 crew landed at Edwards. As soon as it became clear that they were going to send the vehicle to Edwards, NASA recognized that there was a big problem. There's 2,000 [or so] VIPs in Florida expecting to meet someone really neat and interesting, and she won't be there. Someone went patrolling the hallways in Building 4 and grabbed me. 'We need a pilot, and we need another one of the women, and we need to go placate these people'. They grabbed P. J. [Paul J.] Weitz and they grabbed me. They said, 'Go to the Cape. You're the replacement. You're the designated hitters. Faint substitutes, to be sure, but too bad, you're it'. We flew down to the Cape and walked into this immense sea of people. It was a really interesting moment. [One] thought instantly went through my head, [which was] 'I am really happy that Sally's in California and gets the six or eight hours to just digest what she's just done, absorb it, let it be hers, because it ain't much going to be hers. It's quickly going to become everybody else's'. I was just instantly really happy for her that she had this little hiatus to make her own initial sense of the flight – to enjoy it and bask in the moment."

Crew Transport Vehicle
At KSC, either the Chief of the Astronaut Office or his deputy were *authorized* to ride with the crew in the Crew Transport Vehicle (CTV) as far as the LCC.

MISSION SUPPORT ROLES (JSC HOUSTON)

With the vehicle safely in orbit, the support priority switched from the Cape to Houston for the duration of the mission. The roles here included:

Fig. 7.4: Dan Brandenstein (nearest camera) and Terry Hart at the Capcom console in Mission Control on April 10, 1981, the first launch attempt for STS-1 (Image courtesy of Ed Hengeveld).

Capcom
The term 'Capcom' derived from the early days of the space program during Project Mercury, where those in the control center *COMmunicated* with the lone astronaut in his *CAPsule*, hence 'Capsule Communicator' or 'Capcom'. Though the astronauts never liked using the word 'capsule' to describe their spacecraft, the name stuck and continues to this day in the International Space Station (ISS) Mission Control. Soon after the TFNG completed their Ascan training, a small group of them was assigned to train as Capcoms for the first few Shuttle missions. The first

Shuttle Capcom, for the launch of STS-1, was Dan Brandenstein, with many (though not all) of the group fulfilling the role over the next 30 years. Shannon Lucid, the last TFNG to be stationed at the Capcom console in MCC, was on duty during STS-135, the final Shuttle mission in July 2011. Assigned to a Flight Crew under a Flight Director (FD), the teams were assigned for launch and entry in eight-hour shifts around the clock as Orbit 1, 2, 3 and occasionally Orbit 4, or to the Planning Shift, with members (prime and back-up) manning the Capcom console.

Mike Mullane said that working as a Capcom was a real privilege, "and it's also something that gives you an insight into the best team on the Earth, and that's MCC. I had the greatest admiration for those people. To see them working together. There are times when you're on teams, and usually they're smaller than that, that click really well, but to have a team that large that is that trained and that good, I really felt proud of being part of it, and every astronaut should be a Capcom. Every astronaut should get in there and be part of the Capcom because it is just such a neat view of what NASA's all about. That's the heart of NASA in that MCC."

Steve Nagel recalled the early days at the Shuttle Capcom console for STS-2. "I was sitting next to Rick [Frederick H.] Hauck, who was the entry Capcom. We had it different in those days. Nowadays [he was speaking in 2002], there is a flight control team that works both ascent and entry, and then you have your orbit teams. [Back] Then, they had a separate team for ascent, a separate team for entry, and a separate team for orbit. So I was on the entry team with Rick. I don't know if I ever really even talked on the radio to the Shuttle. I think I just rode sidesaddle with him, learning from him on that one and assisting him. Then, on the third flight, I did it. I was the [prime] Capcom. And I only did it for one flight. Nowadays, these guys go over for assignment and the Shuttle flies often enough that they'll do several flights while they're working over there a couple years. I wound up being there just for two flights, but there was a longer time between flights then. Really, I don't remember much about STS-2 except that I was kind of Rick's assistant and learning from him. It was like OJT [On-the-Job Training]."

"Working as a Capcom during that time, I thought, was just a kick," recalled Pinky Nelson, "I mean, it was an incredible challenge, because we didn't have the TDRSS [Tracking and Data Relay Satellite System] satellites, so we only had the ground sites. So the time that you could communicate was very limited. You'd get a three-minute pass over Hawaii and a two-minute pass over Botswana or something, so you had to plan. Unlike now, when you can talk pretty much anytime, you had to plan very carefully and prioritize what you were going to say, and the data came down in spurts, so the folks in the back rooms had to really plan for looking at their data and analyzing it and being able to make decisions based on spurts of data rather than continuous data. So it was kind of a different way to operate.

"I really liked being a Capcom during that, having to organize what you were going to say; [being] able to say things in a really succinct and precise way and make sure that the language you used was just what they were expecting to hear so that you wouldn't have to repeat things, and to be able to listen. When you went AOS – Acquisition Of Signal – over a site, you would call up and say, '*Columbia*, Houston through Hawaii for two and a half', or something like that, and then you could just tell by the tone of their voice in the answer whether they were up to their ears or whether they were ready to listen. So there was a lot of judgment that had to be made, just in terms of [always having] a pile of stuff to get up. How much of this should I attempt to get up? What has to go up? Do I need to listen instead of talk?

"I found that to be just an interesting experience, a challenging job, and I really liked it. I liked that idea that it's a really high-tech machine. Really, the key communication was so subtle, the voice communication was really subtle and interesting, because there were times during those missions where there are always little things going wrong, and with just two people, it's just incredibly taxed. There were times when you could just tell by the tone of their voice it was like, 'Just knock it off for a while here. We're busy up here'.

"There were a few run-ins. I remember Neil [B.] Hutchinson, the Flight Director, was trying to get me to get a message up and I just wouldn't do it, because I knew that they just weren't ready to act on it. It was important, but wasn't critical or anything. And Neil was ready to kill me, and I just kind of sat there and just said, 'No. They're busy. They don't need to do this now'. So that was fun."

Rhea Seddon worked the Capcom console between her second and third flights, "and it was one of the best jobs that you could have," she recalled. "It really kept you in the loop of what was going on currently with the flying world. I had sort of been out in the science part of the universe and had not been able to keep up with what other people were doing, and what the Shuttle was doing, and engineering changes and all of that, so it was wonderful to be the Capcom. You had to do shift work again, and it was very busy. There was a lot going on, and you would do these sims [simulations] that would last all day, or have to come in at midnight, so that was somewhat difficult."

Seddon was not one of those originally selected for Capcom assignments in 1979 and only later realized that the assignment offered experience in understanding what went on during a flight and how the ground team supported it. "I made a flight or two and I'd worked with people on the ground, but I didn't really fully understand what they did and how they did it. So this was a wonderful experience to get to see that and be part of it, when things went wrong, how people worked on the ground. Of course, I'd been the beneficiary of that, because on our first flight the ground had an awful lot of work to do to put together a flyswatter and an unplanned spacewalk and rendezvous. So it was fun to see how that was done in Mission Control and to be a part of it. As a Capcom, you really got involved. You were representing the crew, and you had to pull in resources. You couldn't do it while you were sitting there as Capcom in Mission Control, so you had to call in people and think about who the expert was and who could help. Then you were in charge of explaining everything to the crew or explaining the evolution of what was going on. So I gained a lot of appreciation for what went on, on the ground, and I really enjoyed the job."

Anna Fisher's tour as a Capcom gave her experience at both ends of her astronaut career. "When I was a Capcom, it was for STS-9 [in 1983] and it was a very, very different environment," Fisher recalled. "We had all-paper procedures, and everything was done by documenting and paper." Interviewed in May 2011, Fisher noted how things were changing for the ISS program as the Shuttle fleet retired: "I'm just in the middle of my flow for ISS training, so I haven't actually started working on console yet. I'll probably start doing that in about the July time frame. But just from my exposure to what I've seen, it's just a totally different environment, both in the Shuttle and in Station, because everything is so electronic now. Everything's available electronically, and you can be in Mission Control and have your laptop there and be doing e-mail and work. When I was a Capcom [for STS-9] you were in Mission Control, and there was no way other than by the phone to communicate with other people or do anything about your office job. You were just there. That's a pretty big difference, actually, because [now] – just like in everything – you can pretty much stay connected no matter what you're doing.

"The difference in a Shuttle Capcom and an ISS Capcom is [that] a Shuttle mission is just like a sprint, and ISS, as a Capcom, as a crew, is more like a marathon. When you're working a Shuttle mission, everything has to be done rapidly. The biggest difference is you're manned around the clock. You're there for one week or two weeks, and then that flight is over. So what I've seen on the ISS side so far is [that] it's a much slower process. If the crew asks a question, you might not have the right person there to get an answer right away. The habitability aspects, the stowage aspects, keeping track of where things are, just the sheer amount of knowledge you have to have about the systems… The Capcom flow for ISS systems just scratches the surface just so you'll be familiar with the vocabulary and generally how the Station is organized. There's just so much to know. At the same time, other than on certain days, like docking days, undocking days, when a Progress is coming or a Soyuz, other than that, it's a slower pace, I would imagine. That's just kind of my guess, as opposed to when you're in the Shuttle environment."

Flight Data File

The Flight Data File (FDF) was a small library of printed documents (originally) stored aboard the Orbiter, which provided the flight crew with a back-up for checklists and procedures for the different phases of their mission should there be problems in communications or onboard systems. These included nominal and contingency guidelines for ascent, entry, Extra-Vehicular Activity (EVA), flight plan, orbit operations, Payload Deployment and Retrieval Subsystem (PDRS) operations, photo/TV, post-insertion and rendezvous, and so on. These documents had to be updated and maintained, so one of the early roles for a new astronaut was to track these changes and ensure that the flight crew had the most up-to-date versions.

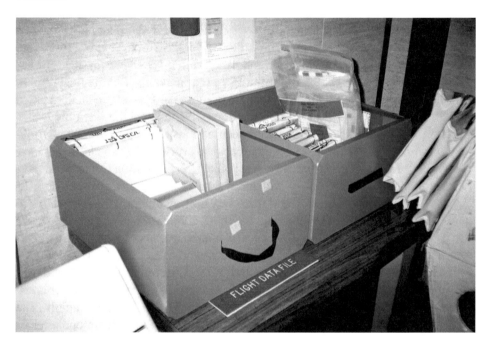

Fig. 7.5: Training copies of the Flight Data File from a later Space Shuttle mission, located in the mock-up of the Shuttle middeck, Building 5 at JSC (Image from the Astro Info Service Ltd archive).

Flight Simulation Laboratory

Guion Bluford recalled that "In addition to working in the SAIL [Shuttle Avionics Integration Laboratory], I was also assigned to work in the Flight Systems Laboratory [FSL] at the Rockwell International Corporation facility in Downey, California. This facility was used to verify the flight software for deorbit burns, entry, and landing. As part of that job, I was also checked out to fly simulated Shuttle approaches with T-38s on the White Sands Test Facility [WSTF] range in New Mexico. NASA put large speed brakes on the T-38s to simulate Shuttle approaches. This was done to help train pilot astronauts. This was an exciting time for me, because it gave me an opportunity to see and verify the flight software for all flight phases of space operations. I would spend a week in Houston flying

Shuttle ascents in the SAIL and then the following week I would fly a T-38 out to El Paso, Texas, fly simulated Shuttle approaches on the White Sands Test Facility range, and then fly to Downey, California, to fly Space Shuttle approaches in the FSL. I did this for several years as we prepared for the first four flights of the Space Shuttle."

Launch Systems Evaluation Advisory Team

This team, which included an astronaut representative, looked at the loads on the vehicle during launch and the aerodynamic loading and structural limits prior to each launch. A final recommendation was made by the JSC Launch Systems Evaluation Advisory Team (LSEAT) to the KSC launch director at Launch minus 30 minutes. Steve Hawley worked on LSEAT and recalled the sometimes early starts. "We were there for instance at two in the morning, before Shuttle launch at eight."

PAO Support

Astronauts were detailed to support the PAO staff, which included assignments to the large media companies of the United States. In the early 1980s, this meant working with the radio stations or TV broadcasters as the mission progressed (this was in the days long before social media and websites).

"For the early missions, NASA assigned astronauts to the broadcast networks," Steve Hawley explained. "I remember Joe Kerwin worked for one of the networks (I think NBC) for STS-1 and I was really impressed at what he brought to the coverage. I'm not sure that every network wanted someone, but for STS-5, NBC did. I was assigned, although it wasn't a job I wanted. I worked with Jane Pauley on the *Today Show*. We did both launch and landing coverage. My job was to be on air with her and to explain what was going on during launch and landing. We were at KSC together and then we were together at EDW [Edwards Air Force Base, California]. I was on call while the mission was on orbit in case something happened, but I don't recall if they used me. It actually turned out to be a fun assignment and I stayed friends with Jane and her husband for a number of years. I don't recall when we stopped offering astronauts to the networks." **[10]**

Kathy Sullivan was at the Cape during STS-1, assigned as a technical advisor and commentary support to the American Broadcasting Company (ABC). "ABC News, as it turns out, was trying to really boost their competitive advantage on the TV coverage," she recalled in her Oral History. "Gene [Eugene A.] Cernan was ABC's main on-camera commentator for the Space Shuttle Program. They were intrigued with the bookends: last man on the Moon, long hiatus, and first flight of the Space Shuttle. You know, 'A handsome Navy test pilot, this is very cool. He'll be our on-camera presence'.

"I show up, twenty-nine-year-old girl astronaut. I haven't ever flown in space, but I'm friendly. They said, 'Lovely. Glad to have you. Go talk to the radio guys'. So they banished me to radio. 'We paid for *this* guy [Cernan]'. The radio folks were lots of fun. I actually had a really good time with them. We did some work-ups in Houston before everything shifted down, and they flew me down to the Cape with the radio guys. Their plan was to just bring me back to Houston, while Gene would go out and cover the landing, and I was kind of disappointed at that. I thought it would be cool to see a landing, but, you know, I had figured out what the food chain was and where I had been placed on the ABC food chain, so that was kind of understandable.

"We're down at the Cape, and we're in the countdown. They've suited up. Gene is bantering with whoever the commentator was. We get down to the twenty-minute hold when you take the computers over to flight mode, and the radio guys are not trying to fill every single second contiguously, so I'm actually having time to listen to the technical loops. I know where we are, I know what's supposed to happen. I'm following the countdown flow [and] I hear 'Crip' [Robert L. Crippen, STS-1 Pilot, inside *Columbia* on the Pad] say, 'BFS [Backup Flight Software] didn't follow the pass'. I know right away we're not going anywhere. There's no way. The very first launch, when the crew took the BFS to Ops 1, if it didn't synch up and track the primary flight set, I just absolutely knew there was no way [we would launch]. I couldn't imagine that anyone was going to come up with enough of a diagnosis in ten minutes that we're going to launch STS-1 when the BFS didn't follow the pass. This was just not going to happen. We're going to stop and really look deeply at this. I'm convinced.

"We go into a hold. I'm listening to the chatter on the loops and it's clear that the way people are thinking this through, the launch will be scrubbed. I can just tell from the way the Launch Director and the NASA Test Director are talking; their heads are about where my head is. There's no reason to hurry up and push through this. Twenty-four hours doesn't hurt a thing. Let's really know what we're doing here. Forty-eight hours doesn't even really hurt a thing; let's know what we're doing.

"The hold just keeps extending and extending, and the agency's not saying anything yet about what's going on or what the intentions are, because it's more important to stay focused on the technical stuff. I tell the radio guys, 'This ain't going today' [and they reply] 'No, they just said they're in a hold'. [I said] 'Well, I know, but they've only said that because…', and I'm explaining to them. 'But trust me; we're not going anywhere today'. Gene, meanwhile, is up there, and they're getting 'Stretch it, stretch it. We need more fill. We don't know when it's launching. Come on', from the editors. And they start asking Gene more and more detailed questions about what exactly is happening with the computers here, and he starts giving answers. They might have been Apollo system answers, but they weren't accurate Shuttle system answers. So I listened to a couple of these, and he kept getting further off the ranch. At some point, I tugged the radio producer's sleeve and said, 'Does it matter that he's not giving correct answers?'

"Whoom! The whole booth goes sort of nuts, and the next thing I know I'm on camera. It's radio, so I've got grubby jeans on. Nobody's seeing me. The next thing I know, they're going, 'Get that girl up here', and I'm yanked up and plunked onto the TV desk. Now the challenge is, how do you start giving answers without discrediting an accomplished astronaut? I didn't know Gene Cernan at all then. I still don't know him super well, but, you know, this is a really accomplished guy and he flew to the Moon and all that cool stuff. I'll confess I probably had my feelings hurt a little bit that I didn't get any attention by the TV guys.

"But I had no desire to come across as, 'What an idiot! Let me tell you what really happened'. So how do I finesse the transition? 'What the captain meant to say was…', and then start giving the better answers. So I sort of helped them out of that. [I] Ran up to New York and gave a speech at the Explorers Club of New York the next day. I was one of first female members of the Explorers Club. [I] Raced back and skidded out to the launch site again to cover for radio the next morning. Well, then they flew us back to Houston, and we're covering a little bit of the mission. The fun part of all of that was what I had done, and how I had done it. The radio guys and the TV guy I had bailed out went to bat with the executive producer and said, 'Take her to the landing site. We want her at Edwards Air Force Base'. I'm good on my feet. I follow things. I'm a good commentator. I'm a great explainer… they've now seen all of that. But I haven't been to the Moon; I'm not a Navy test pilot. 'But she actually speaks well and makes it understandable in human terms. Get her to the landing site'. So I did end up going out to the landing site for STS-1. That was a fun story, and Gene never said anything. I think I must have finessed that little transition reasonably well. As far as I could tell, he didn't have any cause to take offense."

By 1984, in-depth coverage of each Shuttle launch was being curtailed by the major U.S. networks, as Ox van Hoften recalled during the lead up to STS-41C: "Pinky and I had dinner with the ABC anchorwoman Lynn Shearer, who sat there and said, 'Well, I just want you guys to know that this is the last flight we're going to cover live. All the rest of them [TV networks] are just going to have a death watch here'. That's exactly what she told us. We said, 'Well, that's great, Lynn. Thank you so much for that'. That was a pretty amazing comment, I thought. And they did. After that, they didn't have quite the same hype that we had down there."

Spacecraft Analysis

In addition to conducting mission simulations and operations with flight controllers, provisions were made for key NASA engineering and scientific personnel, along with representatives of the major contractors, to support each mission.

This increased presence strengthened the problem-solving capabilities of the MCC team. The spacecraft program office support team occupied what became known as the Spacecraft Analysis Room, or 'SPAN'. JSC and industry engineering teams supported missions in this little room, just off to the side of the Flight Control Room (FCR) in the MCC.

This arrangement enabled immediate contact with key JSC engineering and industry representatives when any assistance was required in resolving technical anomalies that might arise during missions. Throughout a mission, SPAN was operated in consecutive 12-hour shifts by a group of astronauts and other support personal assigned to a particular phase or flight day. The group supported the mission by analyzing data returned from the Orbiter, and offered real-time support in the event of malfunctions and/or equipment failures.

Shuttle Mission Simulator

There were two Shuttle Mission Simulators (SMS) in Building 5 at JSC. They were supported by the Guidance and Navigator Simulator (GNS) in Building 35. Support astronauts used these devices to replicate flight issues, and to explore contingency and workaround tasks, assisting the crew on orbit in real-time. Many of the astronauts rotated to assignment in the SMS at some point in their careers before flying in space. "After that initial orientation of NASA acronyms and locations and things, then we began some initial training in the Single-System Trainers [SST] that they have and in the Shuttle Mission Simulators," recalled Fred Gregory. "We did not interface with [the simulator engineers] daily, though they were kind of our super dads down there to make sure that everything was prepared for us."

Hoot Gibson recalled how little simulator training the group received at first. "We were being trained on Space Shuttle systems and Space Shuttle flight dynamics, although not deeply into it. The training organization over at JSC was pretty heavily involved in training for the first four STS flights. Of course, you just had two-person crews, so each of those two guys on those first four flights had to learn everything. They had to learn how to do spacewalks; they had to be able to go outside and winch the doors closed if you couldn't close the doors electrically. We weren't seeing much at all in the way of simulator training. At that time, I think we had the second simulator, what we called the mission simulator. We had the Motion-Base [simulator], and then the mission simulator also got called the Fixed-Base Simulator. I think that one came on a little later, so we really weren't doing a whole lot of training in the simulators. They brought us in when they did, I've been told, because they believed we were within six months of the first launch, and it actually turned out to be almost three years. We didn't train a whole lot after our initial training."

"We [eventually] got simulator training in several different Space Shuttle simulators," explained Guion Bluford. "These included the Single-System Trainer, and the Shuttle Motion-Base and Fixed-Base Simulators. We learned how to use the Space Shuttle Flight Data File. This included the various checklists and cue cards used by the astronauts to fly the Shuttle. Simulator training gave us valuable exposure to how the Shuttle flies in space and how the Space Shuttle systems work. Mission Specialist candidates were also trained in the RMS [Remote Manipulator System] simulator. This trainer was in Building 9 and it was used for procedure development and camera coordination. The best RMS training occurred in the SMS, in which the RMS was simulated by computer graphics."

"Back then, we didn't get much simulator time. They do nowadays," recalled John Creighton. "When they get in, as a part of that, they'll start getting in the simulator fairly soon, but back then, all the simulator time was being taken up by the first couple of crews that were getting ready to fly the Space Shuttle, John [W.] Young and [Robert L.] Crippen, and [Richard H.] Truly and [Joe H.] Engle."

Dan Brandenstein explained that "[we] got a full set of briefs on each system on the Shuttle, so you knew how the electrical system worked and how the hydraulic system worked and how the computer worked, and you got some time in simulators. You didn't get time in the upscale simulators, the moving base or the fixed-base. They had what they called Single-System Trainers, where you kind of go in and you just learn one system at a time. The cockpit didn't move, but it had the basic displays and things. A lot of the switches, in a Single-System Trainer… that weren't used in the level of classes you were getting, were just pictures of the switches that you needed to operate to learn the system you were working with, [that] you actually operated. So, yes, you got quite a bit of time doing that. Once again, it was the first time through, and you got a pretty good understanding of it, but it isn't until you really got further down the line in the real mission training that you really get to understand a lot of the subtleties of the Shuttle."

Shuttle Avionics Integration Laboratory

Located in Building 16, the Shuttle Avionics Integration Laboratory (SAIL) was a mock-up of the Shuttle cockpit and payload bay, designated OV-095. All the instruments and controls were linked to flight-standard cabling, and used the same software that the General Purpose Computers (GPC) used on the operational vehicles. Several computer systems were linked up to simulate the operational environment without ever leaving the ground. One computer simulated the atmosphere, a second simulated the engines, and a third simulated the aerodynamics and equations of motion. Astronauts then 'flew' simulated ascent and entry profiles, deliberately stressing the systems to ensure that they would work as designed on a real flight, with the simulator staff inserting different abort and contingency scenarios into nominal operations.

Fig. 7.6: [main] The cockpit of the Shuttle Avionics Integration Laboratory (SAIL) later in the Shuttle Program (showing the upgraded glass cockpit displays. [inset left] An exterior view of the SAIL. [inset right] The wiring feeds in the mock-up payload bay.

Several astronauts worked in SAIL as a round-the-clock activity. "We'd work a different shift every week, which made us exhausted a lot of the time, but back then we were young and we could do that," Mike Coats recalled in his Oral History. "Steve Hawley and I seemed to be on the same shift a lot, and it was fun because Steve was a scientist. In the simulator there in the SAIL facility, which was a cockpit of the Shuttle, they have a technician in there all the time as well. We were supposed to be the crewmen.

"The technician was Hispanic, and he'd be talking Spanish to other technicians that were on the headset listening in, and unbeknown to them Steve spoke Spanish. He'd just come from Chile, where he'd been an astronomer down there observing, and he spoke good Spanish. We were in there probably a month, and this guy would be joking about us or saying things about us in Spanish, and Steve would

Fig. 7.7: A youthful Steve Hawley looks in on the SAIL facility circa 1979/1980 (Image from the collection of Steve Hawley, used with permission).

tell me later what he was saying, and we were enjoying that. Then one day Steve lets out a sentence in perfect Spanish, and this guy's face just turned white. You could see all the blood drained out of his face, and he's trying to remember all the things he'd said about these two young astronauts. He never said another word on the simulator. Steve enjoyed that a lot."

Working in SAIL was disruptive to a normal routine at times, as Coats explained: "You were supposed to work three shifts at SAIL, so you'd work a shift, but you'd have to get there an hour ahead of time for a brief and then a debrief afterwards. So it was 10 hours there, and then you'd go over to the office and try to get whatever work done there. You might have a class or something like that. Then you'd be working Saturdays and Sundays. It was about 12 hours a day, sometimes seven days a week. You thought it was important because you were getting ready for that first Shuttle flight, but it was hard on the families."

Don Williams said that he learned far more about the Shuttle systems from SAIL than from other training devices. "It was actually pretty exciting. I learned a lot about the systems, particularly the computer systems and data systems in the Shuttle, during that time, a lot more than perhaps I learned later on. But it gave me a good foundation and a lot of familiarity with how the software is put together and how it interacted with the hardware, and how the crew interacted with the computer systems on the Shuttle." He also noted that the work could be demanding: "Working midnight to 8:00 in the morning every third week wasn't too much fun, and the other two you'd work from 8:00 in the morning until 4:00 [in the afternoon, or] 4:00 [pm] until midnight. Every week you'd switch to a different shift. It was seven days a week for a while. It was a tough schedule."

Steve Nagel went further, stating that working in SAIL "was kind of dog work… the average number of formal runs per day was seven. So [it's] not like [you're running a lot], there [was] a lot of preparation time for each run. Then you make the run, then there's a lot of time afterwards. So things moved pretty slowly, but you really got to know the software, from a personal standpoint, and how it all played together with the avionics there. You learned a lot. It just took a long time to learn it. But, yes, we were doing good work there. It was a good place to work. I wound up working there three times, so three different assignments. I think I paid my dues in SAIL."

"One job that we had to do was to review all the test procedures that would get printed," recalled Robert Gibson, "and virtually every day in our in-basket there would show up what was called TCPs, Test and Checkout Procedures. So one astronaut would be assigned to TCP review for the whole week. It was your job to get over there, and sometimes it could be a mountain of paper that was in that inbox. You had to go through and review it for accuracy. We always wound up making little changes to it and corrections to it. Another astronaut would be assigned to the TCP review meetings, because they would get together when they had one and everybody would provide their inputs to it, and then from there it would go to a final TCP. Then you could run it in the simulator. The really wonderful thing about the SAIL lab was that it wasn't simulated equipment, it was real hardware. You had the real Space Shuttle computers, you had real rate gyros, and you had real accelerometers. You even had the cable trays, so you had the cables that ran from the nose section of the Orbiter. We had an Orbiter cargo bay with all the electrical wires and things running back to the aft avionics compartment as well, which is where a lot of the electronics that controlled the main engines and the fuel lines and a lot of stuff would be located. We wanted to have all the cable lengths because that can make a difference in the electric signals getting through. It was a fascinating place to work."

Not everyone in the office looked forward to working in SAIL, mainly due to the hours and the schedule. Steve Hawley originally did not want to do it, "but off I went, and I found [that] you really learn a lot [there]. So that was really helpful to me because I learned a whole lot about the software. This is before STS-1 when I worked over there, and by helping develop the test cases and running the test cases in the simulator and evaluating the results that came from the test cases, you learned a lot about how the software works. I didn't realize how important that would be later to me going through training as a crewmember, but for five flights one of the things that really helped me was I had a good solid understanding of how the software works, which is critical to how the whole Shuttle performs.

"It was a kind of an inefficient way to spend your time. You'd sit around and wait for two or three hours, and then you'd hurry up and run a test case, and maybe then go off and wait two or three more hours before you ran another test case…

My experience over the twenty-five years has always been [like] that. I was in the second group of astronauts that ever went over to do that job. The first group of astronauts [who worked SAIL], all I remember them talking about is 'What a waste of time this is. It's so inefficient'. So all of us in the second bunch that went over were telling people 'Man, this is great. You learn a lot about the software. You learn how the Shuttle works'. [Suddenly] everybody wants to go to SAIL.

"I was assigned to SAIL, the software verification facility, for STS-1. As it turned out, I had the job of trying to reproduce the BFS tracking problem that caused the scrub of STS-1 on April 10 [1981]. It took loading the software into the GPCs something like 175 times before we could get the problem to recur. I probably still hold the record for most times 'IPL-ing the GPCs [Initial Program Loading the General Purpose Computers]'. One other thing I remember is that the day before launch, the facility allowed some media to visit. One of the local TV reporters talked to me while I was sitting in the simulator. I remember he asked me 'When is your flight?' I told him 'Tomorrow – we're all flying tomorrow'." **[10]**

Weightless Environment Training Facility
The Weightless Environment Training Facility (WETF) in Building 29 at JSC was operated between 1980 and 1998. Building 29 originally housed the centrifuges used during project Apollo, but for the Shuttle era it was repurposed as a water tank to train the astronauts for Shuttle-based EVAs. Measuring 78 ft (24 m) by 33 ft (10 m) and 25 ft (7.6 m) deep, the WETF pool proved too small to hold the mock-up Hubble Space Telescope (HST) or elements of the ISS and was replaced by the Neutral Buoyancy Laboratory (NBL) in the Sonny Carter facility. Originally built to support EVA training for Space Station *Freedom*, the NBL began operational use in 1997 to support ISS operations.

For the TFNG, the original WETF was the main water tank for both EVA training and water egress training from the Shuttle. During each mission, whether an EVA was planned or not, an EVA astronaut support team stood by for any contingency EVA requirements. Such contingency occurred during STS-41C, 51A, 51D and STS-49, where real-time situations necessitated a quick simulation of any new procedures, prior to the flight crew conducting the EVA in space.

Because of the limited dimensions of the WETF pool, it became necessary to utilize the tank in the Neutral Buoyancy Simulator (NBS) building at Marshall Space Flight Center (MSFC) in Huntsville, Alabama to train the astronauts for servicing HST, and later to demonstrate space station assembly techniques, until the NBL facility was ready in the late 1990s. The NBS had been built in 1967 to support Apollo Applications Orbital Workshop (AAP, later Skylab) EVA development. **[11]**

Chase Plane Team
During the first few Shuttle missions, several of the TFNG crewed T-38 jets to support the final stages of a mission. One or more aircraft would be airborne as the Orbiter descended towards the landing site, providing additional guidance to the

flight deck crew regarding the descent, landing conditions and the state of the Orbiter, in particular confirming that the landing gear had lowered, the condition of the aerodynamic surfaces and the integrity of the thermal protection system after entry. They also positioned the aircraft high enough to view and film the ascent of the Shuttle from the pad at launch.

"I just thought it was just an amazing experience, being on the chase team," said Pinky Nelson. "We did a lot of training. I got a tremendous amount of terrific airplane-flying experience, going out to the range at Edwards and practicing rendezvousing with the Shuttle and chasing it down."

"I was a chase pilot for STS-1 and STS-2," recalled Hoot Gibson. "I was Chase-1 for STS-2, so I did what they did. I was down at the Cape for launch and then made my way out to Edwards. It was supposed to be a five-day mission. We were really looking forward to five days out at Edwards because we would [only] have been on alert during the crew's awake time. When the crew went to bed for the night, we weren't going to be on alert anymore, so we didn't have to hang around out at the flight line. We were going to have a great time running because Edwards is a great place to go running. We were going to get to enjoy the gym and go running and all of those great things. [Then] they had a fuel cell failure, and they came down in two days. I only got to enjoy Edwards for two days, once again."

The chase teams were referred to as the Chase Air Force, with their own embroidered patch, "Yes, we did have a patch," Gibson explained. "In fact, I still have one of my jackets that has a patch that Dick Gray [of the Aircraft Directorate] actually developed. It has a T-38 joined up on a Shuttle, and it says Shuttle Chase Team on the patch. We got a little bit notorious on STS-1 because Jon McBride and Dave Walker really liked to be off practicing chase, so we did a lot of it. Since we were support for the upcoming mission, we had the highest priority for getting T-38s, and we wound up doing probably a whole lot more flying than we needed to.

"We got really hammered by the office and because we did so much practicing, we had over-flown our T-38 budget. We the pilots, the individual pilots, all four of whom were Navy pilots. Jon McBride, Dick Gray (former Navy), Dave Walker, and myself, were Chase-1, -2, -3, and -4. All former Navy and we got beat up. In fact, George Abbey sent out a note to all of us that said, 'You guys are walking until you make your time for the six months come out to your allocated flight time', which was normally 15 hours a month of T-38 time. I think because we were mission support for STS-1, we could have 20 hours a month. We over-flew that. We went well beyond our 20 hours a month, so he sent us out a memo that said, 'All of you boys are walking', basically. 'Make your time for the six months come out to…' six months times whatever our number of hours was supposed to be. I think I was the only one that actually abided by that. I think the rest of the boys said, 'Ah, phooey, I'm not doing that'. I think I was the only one that did that."

Tower

During the early stages of the program, an astronaut was assigned to the control tower (TWR) at the landing sites, though this assignment did not continue for long once the first few landings had provided confidence in the system. "If I remember correctly, Dave Walker was in the tower at the SLF [Shuttle Landing Facility, KSC] for the STS-7 landing," recalled Steve Hawley. "I was waiting for the landing in the O&C [Operations & Checkout] building listening to the chatter on the loops. MCC kept asking if the STA was airborne yet since it was time to make the weather observations according to the timeline. I think it was Dave [Walker] who told them the STA couldn't take off because it was too foggy. MCC kept asking for status reports since they needed the weather observations. I recall being amused that somehow it didn't register that if it was too foggy for the STA to take off, it was too foggy to land the Shuttle. [So] STS-7 went to Edwards." [12]

Center Operations Directorate [COD]

The Center Operations Directorate (COD) is responsible for the day-to-day running of JSC. One of the roles of the COD during a mission was to be in contact with the local police department that provided security for each crew member's family and home when the mission was in flight. During a mission, one member of the Astronaut Office was given a concurrent assignment to be the point of contact between the Office and the COD, representing the interests of the flight crew.

EOM Exchange Crew

The End Of Mission (EOM) Exchange Crew was formed occasionally by members of the Cape Crusaders, to provide an 'astronaut crew' on hand at the end of a mission to de-configure the Orbiter after its flight when the prime crew had departed the Orbiter. The ground crew would secure the vehicle and supervise its transport to the Orbiter Processing Facility (OPF) if it landed at the Cape, or prepare it for ferry flight if it landed elsewhere.

"Sometimes, but not always," explained Steve Hawley, "the ASP/Cape Crusaders were also assigned as the EOM exchange crew. Those were the guys who relieved the flight crew after landing and stayed with the vehicle through the power down. Dan Brandenstein and I were the exchange crew for STS-3. We were at White Sands Space Harbor [New Mexico] for the only landing there in the history of the program. That was a memorable experience for many reasons. The weather was marginal due to winds and the gypsum dust was blowing everywhere. Dan and I were cleaning dust out of the Orbiter the whole time we were on board. I was told that the KSC guys were also trying to clean up gypsum after the vehicle got back to Florida." [7]

Crew Recovery Team

In the event of a Shuttle landing outside of the continental United States, the Flight Crew Operations Directorate (FCOD) at JSC would have dispatched a KC-135 aircraft from Ellington Field to the landing location with the aim of returning the flight crew to the United States as soon as possible. In the event of injury to a member of the flight crew, they would have been evacuated by appropriate means to the nearest U.S. military base for treatment prior to repatriation to the U.S. Fortunately, this type of contingency was not required during the Shuttle program, but members of the senior management at NASA (including positions at the FCOD fulfilled by senior Group 8 management astronauts) remained on standby during a mission if required.

Fig. 7.8: Members of the JSC astronaut corps, Vehicle Integration Test Team (VITT) and other personnel at the completion of a countdown demonstration test (CDDT) at Launch Pad 39A, Kennedy Space Center. The participants are (from left): Wilbur J. Etbauer, engineer with the VITT; MS and TFNG astronaut James D. van Hoften; Terri Stanford, engineer from JSC's Flight Operations Directorate; MS and TFNG astronaut Steven A. Hawley; astronaut Richard N. Richards (Class of 1980); astronaut Michael J. Smith (Class of 1980); Richard W. Nygren, head of the VITT; MS and TFNG astronaut Kathryn D. Sullivan; astronaut Henry W. Hartsfield Jr., STS-4 Pilot; Mark Haynes, a co-op student participating with the VITT; astronaut Thomas K. Mattingly II, STS-4 CDR; and TFNG astronaut Donald E. Williams. (Image courtesy Ed Hengeveld.)

MANAGEMENT SUPPORT ROLES

During the extensive 1986 Presidential Commission investigation into the *Challenger* accident, they found that there had been a departure from the philosophy set up in the 1960s and 1970s which saw astronauts assigned to management positions. These astronauts brought their experience and a keen appreciation of flight safety and operational procedures to the role. The Commission recommended that "NASA should encourage the transition of qualified astronauts into agency management positions," and that "the function of the Flight Crew Operations Director should be elevated in the NASA organization structure." It was also recommended that a representative from the Astronaut Office should be assigned to the Shuttle Safety Panel, and that the mission commander (CDR), or their designated representative, should "attend the Flight Readiness Review, participate in the acceptance of the vehicle for flight, and certify that the crew is properly ready for flight." [13] This was seen to be implemented during the 1990s, when several former Group 8 astronauts moved from the Astronaut Office to more senior managerial positions in the Shuttle program.

FCOD Management Support
The FCOD were members of the senior management chain at NASA JSC, and were on hand to assist the FD at MCC in making decisions that had no mission safety consequences but which could have cost or public perception consequences. They could not overrule the FD during a mission, but merely advise them. During a mission, an astronaut was assigned to the Operations Support Room (OSR), which was an annex to the MCC, or down at KSC, as Steve Hawley explained: "OSR was staffed by technical people and program representatives. [It] often did some leg work for the MCC [such as] notifications, coordination, perhaps setting up some testing or sims if necessary. We had [an astronaut] rep [there] to make sure that crew interests were represented, who could find astronauts to do tasks [such as flying a part to a contractor, or to KSC, or running a sim], and for the purpose of making sure FCOD management was aware of what was going on." [14]

Contingency Action Center
In the event of a mishap occurring during a mission, the manning of action centers and communication networks followed predetermined guidelines and included a representative of the MMT. After the *Challenger* accident, this would often, but not always, mean a veteran or senior managerial astronaut. These centers were located at NASA HQ, MSFC, KSC, and JSC.

Mishap Representative
This was a member of the FCOD senior management, on hand to assist the Mishap Investigation Team dispatched to a contingency site near to where a mishap occurred. Usually, a senior astronaut was on hand to support this position.

READY TO EARN THEIR ASTRONAUT WINGS

From 1979, members of the 1978 astronaut selection provided the various support roles for the early Shuttle missions flown by *Columbia* under the OFT, as well as the first two 'operational missions' by *Columbia* and the inaugural space flight of *Challenger*. With four years of NASA astronaut experience and training behind them, they were now eligible to be assigned to their first missions and undertake the next step in their career. The Shuttle crew training program would finally lead each of them into space. With only 5–7 flight seats to fill, the TFNG could not all possibly fly on the first Shuttle mission available, which was STS-7, nor fully crew the next five missions as rookies. NASA systems demanded that a veteran CDR should lead each mission, but as many as possible from the 1978 group would be assigned over the next 14 missions. Some of these seats would be taken by veteran astronauts as CDR and MS, as well as the first Payload Specialist (PS) selected for the program. For the TFNG, the next step was to qualify from the Shuttle mission training program, another challenge in their new chosen career. This is explored in the next chapter, mainly in the context of the early to mid-1980s, during which all 35 flew their first missions.

References

1. *What do Astronauts Do?* unsigned Astronaut Office handout dated April 3, 1986, on file AIS Archive.
2. Mike Mullane, NASA Oral History, January 24, 2003.
3. Guion Bluford, NASA Oral History August 3, 2004.
4. Dan Brandenstein, NASA Oral History, January 19, 1999.
5. Fred Gregory, NASA Oral History, April 29, 2004.
6. **The Concise Oxford Dictionary**, 1991 edition.
7. Email to author David J. Shayler from Steve Hawley, January 28, 2019.
8. *NASA Shuttle Abort Modes*, in **Space Rescue, Ensuring the Safety of Manned Spaceflight**, David J. Shayler, Springer-Praxis, 2009, pp. 216–224.
9. Rhea Seddon, NASA Oral History, May 20, 2010
10. Email to author David J. Shayler from Steve Hawley, January 29, 2019.
11. **Walking in Space**, David J. Shayler, Springer-Praxis, 2004; **Hubble Space Telescope** and **Enhancing Hubble's Vision**, David J. Shayler with David M. Harland, Springer-Praxis, 2016, and **Linking the Space Shuttle and Space Stations**, and **Assembling and Supplying the ISS**, David J. Shayler, Springer-Praxis, 2017.
12. Email to author David J. Shayler from Steve Hawley, January 22, 2020.
13. *Report of the Presidential Commission on the Space Shuttle* Challenger *Accident*, June 1986, Recommendation II, p. 199, and Recommendation V, p. 200.
14. Email to author David J. Shayler from Steve Hawley, January 23, 2020.

8

Preparing to fly

> *Today, the [training] facility computers and*
> *equipment are out of date and obsolete.*
> *Plans exist to update the equipment [but]*
> *over a 10-year period and are minimal.*
> Robert K. Holkan, Chief,
> DG6 JSC Training Division.
> From a memo dated April 11, 1986.

Despite years of planning and 60 months of flight operations, the comment above illustrates the frustration that was being felt in preparing Shuttle crews for space flight using equipment which should have been state of the art, but was instead already falling behind the requirements for a rapidly expanding program.

Training for space flight always presents a challenge for those who undertake the program and those who prepare and supervise their courses. Schedules were always tight during the Mercury, Gemini and Apollo programs, even when there were relatively few active astronauts vying for "sim time." By 1986, almost two decades later, it was becoming even more of a challenge during the early years of the Shuttle program to ensure that several multi-person crews remained at the peak of performance and ready to fly, while simultaneously implementing refinements to the system based on debriefings of recently-returned crews, initiating new training programs for future crews, and always pushing for new equipment and software programs within a limited budget.

Between 1959 and 1969, NASA had chosen 73 men in seven small groups of between six and 19 members. This cadre trained for or supported 37 missions across a 22-year period under the Mercury, Gemini, Apollo (including Skylab and the Apollo-Soyuz Test Project – ASTP), a number of ground and airborne

© Springer Nature Switzerland AG 2020
D. J. Shayler, C. Burgess, *NASA's First Space Shuttle Astronaut Selection*,
Springer Praxis Books, https://doi.org/10.1007/978-3-030-45742-6_8

simulation, and the early Shuttle Approach and Landing Test (ALT) and Orbital Flight Test (OFT) programs.

In contrast, 84 men and women had been chosen in just four new astronauts groups in the seven years between 1978 and 1985. There were plans for more to arrive at regular intervals to meet the expected Shuttle flight rate (at least 26 missions a year from two different launch sites), crewing requirements of up to seven persons on one mission, and natural attrition rates. Clearly, the training program for such an influx of crews and missions would be put to the test, but it was a problem which NASA had addressed, or so they thought given the headline above, long before any astronauts were chosen to train for Shuttle missions or new astronaut groups were specifically selected for that program.

A NEW TRAINING PROTOCOL

Twenty years after NASA had chosen its first seven astronauts to train for the one-man Project Mercury missions of up to 24 hours in Earth orbit, the agency was faced with a group five times as large for flights on the multi-seated Space Shuttle, on Earth orbital missions of between seven to ten days. As NASA's programs progressed from Mercury to the two-man Gemini and then to the three-man Apollo spacecraft, the training syllabus had focused on a small crew mastering the workings of their spacecraft, acquiring the techniques of rendezvous and docking, and gaining new experiences in walking in space, lunar geology and space navigation, with a little science thrown in[1]. For those who had not previously qualified as a jet pilot in the U.S. armed forces (the civilian members of the 1965 and 1967 Scientist-Astronaut selections), there was also the rather daunting challenge of a 53-week U.S. Air Force (USAF) military jet pilot training course to graduate from, prior to tackling the astronaut basic, survival and academic training program.

A change in direction

Having moved on from the one-shot ballistic spacecraft that had featured in the 1960s and 1970s, the Space Shuttle necessitated a whole new approach to training for space flight, and the 35 members of NASA's latest group of astronauts would be the first to experience this transition. The era of 'hot-shot' test pilots with what became known as the "*Right Stuff*" aura and a 'can-do' attitude was over, to be

---

[1] Significantly more scientific research was conducted by the three Skylab crews during their 28, 59 and 84-day missions in 1973/4. This included extensive studies in astrophysics, solar physics, Earth observations and resources, materials science and space manufacturing, engineering and technology, life sciences, and student experiments and science demonstrations.

replaced by astronauts with more academic and engineering skills rather than test flying credentials.

Early 1970s documentation researched by the authors from the NASA archives helps to explain the protocol in developing this new training program. While the experience of past programs would be useful, far more aspects would be radically new. At this point, there was no presidential decree to achieve new objectives (such as a space station) "within a decade," but there was a desire to push rapidly and safely towards the first orbital test flights of the Shuttle system, qualifying the design and format for much more ambitious missions in content and frequency. By the mid-1980s, the program had evolved to achieve 24 relatively successful Shuttle missions, but it was a system already under stress as a production line procession of crews and missions rotated through the program. A 'routine' process was emerging, not with the Shuttle missions themselves, but in assigning crews to train and fly those missions. But that process came to an abrupt halt on a cold January day in 1986, when *Challenger* and her crew of seven, including four members of the 1978 astronaut selection, were lost just seconds after leaving the launch pad, in full view of the gathered officials, family members and onlookers, and the world's media.

Reviewing Shuttle crew training 1986
The subsequent post-*Challenger* inquiry and recovery program would see the Shuttle return to flight in 1988, with a very different emphasis to its mandate. Before that, however, the causes and roots of the *Challenger* accident had to be scrutinized by a Presidential Commission, which turned the American space agency upside down as it examined all aspects of preparing for and flying the Shuttle in detail. Part of this in-depth assessment included the normally restricted process of training the crews, trying to uncover any issues which may have contributed to the loss of the astronauts and the vehicle.

Though training did not directly contribute to the tragedy, the 1986 *Report of the Presidential Commission on the Space Shuttle* Challenger *Accident* stated: "An assessment of the system's overall performance is best made by studying the process at the end of the production chain: crew training. Analysis of training schedules for previous flights and projected training schedules for flights in the spring and summer of 1986 reveal a clear trend: less and less time was going to be available for crew members to accomplish their required training." [1]

A number of issues were highlighted which directly or indirectly affected crew training, pressurizing the training process. These were summarized as late changes in the manifest from hardware problems, customer requests, operational constraints and external factors. In its recommendations, the Commission noted (Item 8) that reliance on the Shuttle as the nation's principle space launch capability created pressure on NASA to increase the flight rate. [2] In the future, NASA was advised to set a flight rate that was consistent with its resources, including establishing a firm payload assignment policy for more rigorous control of cargo manifest changes, to "limit the pressures such changes exert on schedules and crew training."

By late 1985, the system in place for producing crew training materials was struggling to keep up with the increasing and changing flight rate. The number of training simulators was another limiting factor, and the capacity of the two available simulators could not sustain training more than 12–15 crews each year (which the program had yet to attain, but which was planned for the next few years). Another limiting factor was when flights followed each other in rapid succession, where critical anomalies occurring on one mission could not be fully identified and corrected before the next mission flew. As Group 7 astronaut Henry Hartsfield testified, "Had we not had the accident, we were going to be up against a wall. STS-61H… would have to average 31 hours in the simulator to accomplish their required training, and STS-61K would have to average 33 hours… That's ridiculous. For the first time, somebody was going to have to stand up and say we have got to slip the launch because we are not going to have the crew trained." **[3]**

Clearly, the loss of the *Challenger* and her crew brought to the fore issues which had been getting worse for some time, as reflected in the statement at the start of this chapter. To enable the Shuttle to resume flying, these and many other recommendations, findings and issues had to be addressed.

The documentation released on Shuttle crew training, as part of the investigation into the loss of *Challenger*, provided a rare insight into how crews were trained for Shuttle flights in the 1980s, at the height of the active involvement of the 'Thirty-Five New Guys' (TFNG) in the program. Using this information as the basis for this chapter, and following a review of crew positions below, we present an overview of the Space Shuttle crew training process circa 1980s, which all of the TFNG completed on the path to their first space flights.

SPACE SHUTTLE CREWING NOMENCLATURE

Before reviewing the training program that all members of the Class of 1978 passed through as either a Shuttle Pilot (PLT) or Mission Specialist (MS)[2], it is useful to review the different positions to which a crewmember on the Shuttle could be assigned.

---

[2]The Ascan training program instigated in 1978 continued, with a few refinements, through to the Class of 2004 (Group 19 "The Peacocks"), who were the final group to receive Shuttle mission training. From the Class of 2009 (Class 20 "The Chumps"), the emphasis shifted to the ISS and future space exploration programs.

Prime Crew: The group of astronauts who fly the mission, which for Space Shuttle missions was as little as two and occasionally as many as eight. The Shuttle did have the capacity to fly ten crew members in an emergency situation, but this was never called upon operationally.

Back-up Crew: During the Mercury, Gemini and Apollo era, in the event of injury or illness during training, NASA assigned a team of astronauts to 'back-up' the prime crew. By shadowing their training program, the back-up team could step in at short notice to replace part or all of the prime crew, allowing the mission to proceed with little delay.

Though this system worked well and was utilized on some of those pioneering missions, it was not fully adopted for the Shuttle program after the OFT missions were completed, because it was felt that there would be an adequate pool of suitably prepared crewmembers capable of stepping in should the need arise. It was also thought that adding a second 'crew' to each flight would simply clog up the training syllabus and simulator time, which was already tight. However, there were a few occasions – mostly later in the program – where a sole NASA astronaut was assigned as back-up to a crew for a specific time or purpose. In general, from STS-5 in 1982 to the end of the program 30 years later, relatively few 'official' back-up NASA crewmembers were assigned other than for most of the Payload Specialist (PS) positions.

The system that had been in place between 1965 and 1972 (during Gemini and Apollo), in which a complete back-up crew could expect to skip the next two flights and be prime crew on the third to utilize their training to the maximum, was no longer viable for the expected frequent flights and rapid turnaround of the Shuttle program. The "back-up one mission, skip two, fly the third" system had been created in the early 1960s by former Mercury astronaut and Director of the Flight Operations Directorate, Deke Slayton, and worked extremely well. However, where the Gemini or Apollo flight profiles and objectives were relatively similar within the two programs, the expected frequency of the Shuttle missions and the multiplicity of operations on orbit between the ascent and entry phases meant that specific mission training for more than one crew per mission would have been difficult. As virtually no mission would be a duplicate of the previous flight, creating an on-going repetitive system from one mission to the next would have been almost impossible, and that was without considering hardware delays and changes to the launch manifest.

While no formal back-up system was adopted for the Shuttle's operational missions, some crews during the period 1982–83 were prepared to 'support' a mission with a similar payload, such as a Tracking and Data Relay Satellite (TDRS) deployment crew supporting an earlier TDRS flight, or a commercial satellite deployment crew supported by another team preparing for a similar mission themselves a few months down the line. Though not officially a 'back-up' team, this capability was available if required (which it was not) to replace a crewmember or even a whole crew. It offered additional and early standalone training in the Shuttle Mission Simulators (SMS) for the later crew and useful support for the earlier flight crew, until the astronaut pool included several flown Shuttle PLT and MS. The original assignments (prior to 1983 problems with the Inertial Upper Stage – IUS – deploying TDRS satellites) were:

- STS-5 crew (Comsat deployment mission) – supported by the STS-7 crew
- STS-6 crew (TDRS deployment mission) – supported by the STS-8 crew
- STS-7 crew (Comsat deployment mission) – supported by the STS-11 crew
- STS-8 crew (TDRS deployment mission) – supported by the STS-12 crew

In 2019, Steve Hawley explained the philosophy of these 'support' assignments: "There were often more simulations for training the control center than the crew might be able to support given their other obligations, and so what we could do in order to support the training and the Flight Directors and the control center people would be to use another crew further down the line that had a similar payload – it would still be good training for them, not negative training – and they would support what we would call the integrated sims if the prime crew wasn't available. We did that more commonly earlier in the program than we did later." Hawley could not remember when this was stopped, but by 1984 the manifest and payloads began to change more frequently, making it difficult to match crews with similar payloads in 'support' roles for simulations.

THE ASTRONAUT ROLES ON A SHUTTLE CREW

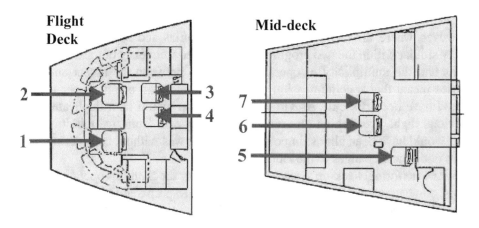

Fig. 8.1: A plan view showing the Space Shuttle Flight Deck (left) and Mid-deck (right) seating (Image courtesy of Joachim Becker, SpaceFacts.de).

The specific crew roles were defined as:

Commander: (CDR – Seat 1, the left-side front position on the flight deck). Each CDR was responsible for the vehicle in which they were flying, their crew, the success of the mission and overall safety. The position was analogous to the duties of a ship's captain.

Pilot: (PLT – Seat 2, right-side front position on the flight deck). Each PLT assisted their CDR and was formally second in command. Their primary responsibilities included operating and flying the Orbiter. During the orbital phase of the mission, both the CDR and the PLT assisted the rest of the crew in aspects of the flight, notably the deployment or retrieval of payloads, operation of the robotic arm, or in support of scientific objectives.

Mission Specialist: (MS#). According to NASA documentation, future MS were: "required to have a detailed knowledge of the Shuttle systems and operational characteristics, mission requirements and objectivities, and supporting systems for each of the experiments to be conducted on the assigned mission. By working closely with the Commander and Pilot, the MS are also responsible for the coordination of onboard operations involving crew activity planning, the use and monitoring of onboard consumables (fuel, food, water etc.), and for conducting experiments and payload activities. MS also perform experiments, spacewalks [Extra-Vehicular Activity, or EVA] and payload handling functions involving the [Remote Manipulator System] RMS arm."

The MS were not required to 'fly' the vehicle or land it, which established two paths of mission training for the NASA crew – one for the flight deck crew (CDR and PLT) and one for the MS – that could be merged when required. Each took on specific but mostly interchangeable responsibilities as a crewmember. The number of MS assigned to a flight crew depended upon on the mission flown.

On all Shuttle flights between STS-5 in 1982 and the end of the program in 2011, MS were assigned to each flight crew. These could number as few as two or as many as five. With the inclusion of foreign nationals in the NASA astronaut training program from 1980, each new astronaut selection (though not all) after that date could include representatives from partner agencies in the Space Station Program who had completed the NASA MS training syllabus, (Canada, Europe, Japan, and even Russia from the early 1990s, whose cosmonauts needed to complete only an abbreviated MS training program due to their previous experience).

While this seemed outwardly sensible in a growing international program, these international crewmembers, as well as American or overseas PS and USAF Manned Spaceflight Engineer (MSE) candidates, took up some of the limited number of flight seats on a mission given their importance to that mission or payload, which in turn restricted the seats available to NASA's career astronauts and the frequency in which they flew. Outwardly, the inclusion of such non-NASA personnel into a crew never seemed to cause major crew integration problems, though there were some initial doubts and some adapted to their temporary astronaut role better than others. For today's International Space Station (ISS) Program, such multinational crewing or 'non-career' crewmembers are now familiar, but in the early 1980s this was still a relatively new phenomenon and in some cases took a while to get used to by the parties involved.

Which Mission Specialist does what?
Once assigned to a Shuttle crew, the MS received a numerical designation which did not necessarily reflect their previous medical or scientific, technical or engineering experience. Though taken into account on some missions, it often meant that an astronomer would not *always* fly on an astrophysical research flight, nor a doctor on a medically-orientated mission, or an engineer on a deployment mission.

In his NASA Oral History, Mike Mullane explained the differences in these roles: "At various times, different people – Judy [Resnik] and I, for example – [laid] out the roles and responsibilities for the [three] Mission Specialists, [such as] their roles and responsibilities, generically, so you don't have to reinvent a training program every time you name a crew[3]. So we came up with a scheme in which MS-1 had overall responsibility for payloads and experiments in orbit, and MS-2 had primary responsibility for flight engineering and helping the Pilot and Commander during ascent and entry, and back-up for payloads. MS-3 really had responsibilities for independent experiments and also an EVA responsibility. **[4]**

"So that's kind of the way we laid it out. MS-3 would be typically the most junior and the lowest training requirement but heavy on EVA. MS-1 would have the largest overall responsibility and, in principle, ought to be the most experienced member of the astronaut Mission Specialist crew. And MS-2 had the most simulation time, [spending] an enormous amount of time in the simulator… We split it that way in order to recognize the fact that the flight engineering role was the dominant training requirement for one of the Mission Specialists, and therefore that person shouldn't be burdened with overall responsibility for the satellites."

Mission Specialist 1: Usually occupied Flight Deck Seat 3 located behind the PLT. Normally 'ascent trained', their primary role was to assist the flight deck crew in operating Shuttle systems, by monitoring displays of checklists on the way to orbit. On some flights, MS-1 would swap places for the descent with another crewmember, usually MS-3 who had flown the ascent on the middeck.

---

[3] Here, Mullane is referring to a nominal five-person Shuttle crew, consisting of CDR, PLT and three MS.

Mission Specialist 2: [see *Flight Engineer* below]

Mission Specialist 3: Normally occupied Seat 5 near the left-side hatch window on the middeck for launch. As they could be 'descent trained', they could swap with MS-1 in the Flight Deck Seat 3 seat position to assist the flight deck crew during the entry and landing phase. From 1988, in response to upgrades following the loss of *Challenger*, the MS-3 position was given added responsibility to activate the escape slide pole, designed for rapid crew exit from a stricken Orbiter. Fortunately, though they were well trained for this role, it was never called upon in a real emergency. From STS-26 in 1988, the MS-3 role also encompassed the responsibility of helping crewmembers get out of their launch-entry suits, as well as stowing the middeck MS seats and ensuring that equipment on the middeck was ready for orbital operations. At the end of a mission, MS-3 also led the crew's efforts to unpack the seats and assisted in donning launch-entry suits in preparation for landing. Often overlooked, this role could be quite involved and at times very busy.

Mission Specialist 4 and 5: Launched and landed on the middeck, taking Seats 6 and 7 facing the middeck lockers and having minimal roles during ascent or entry. During Shuttle-Mir and when returning ISS crewmembers, one or more of these seats would be reconfigured into a recumbent seating position, allowing the crewmember to lie prone on the middeck floor with their feet in vacant locker spaces to ease their adaptation back to Earth's gravity after a lengthy stay on a space station. Yet again a member of the TFNG, Shannon Lucid, pioneered this system when returning from her six-month Mir mission in 1996. It became the standard on Shuttle missions returning ISS resident crewmembers through to 2009.

Flight Engineer (FE): Always allocated to MS-2, occupying Flight Deck Seat 4 between and behind the CDR and PLT. This position was developed during STS-5 (by Class of 1967 astronauts William B. Lenoir during ascent and Joe Allen during descent) and evaluated on STS-6 (by Donald H. Peterson (class of 1969) for both the ascent and descent phases). The MS-2 role included the responsibility of assisting the flight deck crew during the ascent to orbit, for deorbit preparations, and during entry and landing, as well as sometimes assisting with complex orbital maneuvers of the Shuttle during science operations. They also monitored potential abort modes during ascent and contingency planning in the event of a mishap during landing, keeping track of real-time information via the Capcom and calling out milestones during the ascent and descent to the flight deck crew. Essentially, MS-2 was also a third set of eyes and ears on the flight deck during critical phases of the mission. For one member of the Class of '78, this proved very useful in his rise 'up the ranks' on the Shuttle flight deck.

As Steve Nagel, the Group 8 astronaut who flew initially as MS-2 (despite being selected as a pilot) before going on to fly as PLT and then commanding two subsequent missions[4], explained: "The MS-2 is like the Flight Engineer, so you learn all the same Shuttle systems that you'd have to learn as the Pilot or the Commander. So the switch from MS-2 seat to the Pilot seat wasn't that hard. It was just that I had all the head knowledge, and now I just had to be able to put it into practice. I had done a little bit of that, but at a certain point I had stopped training for the second flight [to fly his first mission], and then I got right back with it after the first flight. The missions were totally different, but what I had to learn for the Shuttle systems and all that was almost a one-to-one carryover, except for the Spacelab… The Spacelab was something totally different, and I had had some classes on it and learned some about it before the first flight, even. Then I had to stop [for the first flight], and [then] pick it up and really hit it hard before the second flight. So the Spacelab was a big difference."

Developing the roles on the mission was an on-going process as the program matured, as Steve Hawley explained in a 2019 interview. A professional astronomer by trade, Hawley was determined to master the new skills required for sitting in flight deck seat 4. "It was challenging, but it was awfully exciting, [and] one of the things that motivated me [was that] I really wanted to understand how everything worked. So I learned how the software worked and then I learned how the other systems worked. As a matter of fact, I think it was prior to 41D [in 1984], I wrote a little handbook that explained why the malfunction procedures for aborts were written the way they were, and I did it for the benefit of the others in training. I had a good time understanding why we do it, and a lot of the time I found out that people thought they knew why these steps were being done and they were wrong." **[5]**

Payload Commander (PC): This was announced as a new crew role by NASA in January 1990, with the change from commercial satellite deployment to more scientific payloads. It required a flight-experienced MS to be assigned but was not a dual role as with MS-2/FE. The PC was given additional responsibility for managing the major science or other payload assigned to the mission. The role was to provide long-range leadership in the development and planning of payload crew science activities. The PC had overall responsibility for the planning, integration and on-orbit coordination of payload/Shuttle activities on their assigned mission, while the mission CDR retained overall responsibility for crew and mission safety. The initial allocations of this role were made after the respective crewmembers had been assigned to their missions. Of the four initial PC assigned, three were from Group 8 (Norman Thagard on STS-42/IML-1, Kathy Sullivan on STS-45/

---

[4]Dave Griggs was another pilot candidate from the Class of 1978 who flew his first mission, STS-51D in 1985, as MS-2. At the time of his death in an off-duty plane crash in 1989, Griggs was in training as PLT for STS-33, which would have been his second space flight. It was widely expected that he would also have transitioned to the command seat on his third mission, had he lived.

Atlas 1, and Jeff Hoffman on STS-46/Eureka/TSS-1), reflecting the importance of this new position as the science became more complex on Shuttle missions. As with EVA crew assignments in the 1990s for complex spacewalking objectives requiring longer training time, subsequent PC assignments would be made in advance of the remainder of the flight crew in order to help identify and resolve training issues and operational constraints prior to crew training. In addition, it was envisaged that the role of PC would serve as a foundation for the development of the Space Station *Freedom*/ISS Increment Commander concept, another example of the significant role that members of Group 8 provided both in the development of Shuttle operations and also in planning and preparing the initial ISS crewing regimes.

In her 2009 Oral History, Kathy Sullivan explained that the role of PC had more than one element to it. [6] "Payload Commander was the NASA Mission Specialist who would oversee and organize typically two Mission Specialists and two Payload Specialists who work back-to-back shifts operating complex, multi-experiment Spacelab flights. I think several factors went into conceiving the Payload Commander role.

"One is [that] you've got a very complex Spacelab mission with a dozen, to three or four dozen experiments. The training time that the responsible Mission Specialist should put into that needs to be longer than the Pilot and Commander probably need to put into the basic orbital operations. So you're going to want to slot in Mission Specialists 18, maybe 24 or 36 months in advance so they can build the relationships that are necessary with the scientific team and the payload operations team, get out to the factories, laboratories, engineering facilities, and see the flight hardware [to ensure] that the mission simulations are going to represent those payloads. Sometimes it's built into the Shuttle Mission Simulator. Sometimes the preparation is done differently. The payload crew [played] a really substantial role in helping the simulation teams know how to model the payloads correctly."

Sullivan also explained the authority given to the PC in these more complex missions, "to provide a long lead time for the mission crew, to be sure that one of the NASA Mission Specialists is considered and recognized as authoritative in all those early planning decisions. You want that group to be able to make effective decisions and move the flight preparations forward, not have various people on the team saying someone doesn't have the authority to do this or [that] we need the [mission] Commander to make this final decision. Naming a Mission Specialist as Payload Commander gave that authority. You're also counting on that person to use smart judgment when different payload operating or crew issues do impinge on larger flight ops constraints that do eventually need concurrence by the Flight Director or by the Commander. You're not going to step in and overrule those or supplant those. You need someone who's representing CB [Astronaut Office] and JSC [Johnson Space Center, Houston, Texas] and able to keep the ball moving in those early lead phases. Helping the Payload Specialist folks, who sometimes have had prior flight experience – one of our Payload Specialists had flown before – but often this is their first time. Giving them some guidance about 'This is how it's going to get done,' or, 'This is the way things normally go.' Being that training voice about life aboard the Shuttle, with some authority backing that up.

"That was the whole idea. Later, closer to launch, when you combine the payload crew with the Shuttle crew, that balance shifts around. The Shuttle Commander is the Shuttle Commander, there's no two ways around that. You help the Commander in that sense, because you know the mission teams. You know the experiment teams. You have a little more insight about the personalities, cultures, backgrounds, mindsets that the payload team brings to bear and can help jumpstart the overall crew's understanding of that by the time investment that you've made. That was the idea behind it."

Payload Specialist (PS): This position was created specifically for the Shuttle program. They were professionals in the physical or life sciences, or technicians trained in Shuttle-specific requirements or hardware. The PS were chosen by sponsoring organizations, customers, or investigator working groups approved by NASA. They could also be politicians with links to the space program or guest observers flying on behalf of major suppliers. Though not career astronauts, they had to meet NASA's health and fitness standards and pass security vetting, but they did not have to be U.S. citizens. In addition to preparations at a company facility, university or government agency, the PS also had to complete a comprehensive flight training program to become familiar with Shuttle systems, support equipment, crew operations, housekeeping duties and emergency procedures. This specific training at NASA centers usually took place as much as two years prior to a flight (sometimes a much shorter time period), though each mission required some individual adjustments to this schedule. Several Group 8 astronauts flew in a crew that included PS.

Manned Spaceflight Engineer (MSE): Despite the cancellation of the USAF Manned Orbiting Laboratory (MOL) program in 1969, without a single crewed mission flown, the USAF sought to be involved in the development of the Space Shuttle for military objectives, and were adamant that the size of the Orbiter's payload bay should fit their requirements for planned classified payload deployments, rather than NASAs smaller bay size for more scientific operations. At its creation in the late 1960s, the Shuttle was intended to have a 'launch-all' capability, handling not only civilian payloads but all of America's military and classified payloads as well. There were plans for the USAF to purchase up to three Shuttle Orbiters of its own (known as "Blue Shuttles") and associated hardware (engines/ fuel tank/boosters, etc.) and operate them with all-military crews into strategically advantageous polar orbits from a specially converted facility, Space Launch

Complex 6 (known as 'Slick 6'), at Vandenberg Air Force Base (AFB) in California. Hoping for up to 12−14 Department of Defense (DOD) Shuttle missions each year, there were plans to create a classified Mission Control Center in Colorado, but USAF budget restrictions, problems in reaching the required annual Shuttle launch rate, hardware difficulties and the loss of *Challenger* contributed to the USAF pulling out of the Shuttle program after 1986.

The MSE designation was therefore created in the late 1970s to train and fly serving military personnel to accompany classified payloads on dedicated Shuttle missions. Though plans for its own fleet of Orbiters were eventually abandoned, the USAF still desired to have its own personnel on-hand for classified military payloads, and as a result chose a total of 32 military candidates in three groups (13 in 1979, 14 in 1982 and 5 in 1985) to train as MSE. While military observers and specialist crewmembers were trained and assigned to crews, in the end only two MSE flew, on two separate Shuttle DOD missions during 1985, the crews of which both included representatives from Group 8 (Shriver, Onizuka and Buchli (with Gary E. Payton, USAF) on STS-51C and Bob Stewart (with William A. "Bill" Pailes, USAF) on STS-51J). Following the loss of *Challenger* and subsequent grounding for nearly three years, the USAF budget restrictions and a move to expendable launch vehicles signaled the end of the MSE program by the late 1980s, with no further flights completed. A third military crewmember, Thomas Hennen, was not an MSE but flew on STS-44 in 1991 under the command of Group 8 astronaut Fred Gregory.

Crew responsibilities

With the crew assigned and 'seated', each crew member received specific crew responsibilities as well as key roles in preparing for their missions.

Key crew roles

RMS Operator: On most, but not all Shuttle flights, the Canadian-built RMS robotic arm was carried on the port longeron of the payload bay. The operation of the RMS, from the aft flight deck station, was normally undertaken by a specially trained MS, though some CDRs were also RMS qualified. As well as learning to lift, deploy, capture, maneuver and lower payloads around the payload bay safely and securely, RMS operators became an integral part of Shuttle EVA operations, relocating an EVA crewmember positioned on the end of the arm around the payload bay, or positioning equipment in support of their spacewalk.

EVA Crew: For safety reasons and to achieve mission objectives, at least two astronauts in each crew were trained as EVA crewmembers, using the designations EV1 and EV2 (plus EV3 or EV4 etc., as required), with suit identification marks distinguishing each astronaut out on EVA. All EVA teams featured a lead crewmember (usually EV1). As the Shuttle program developed, more extensive EVA programs were devised in preparation for assembling the space station and for servicing large payloads. Several members of the Class of 1978 were instrumental in developing the Shuttle EVA system that was so critical to the success of missions such as servicing the Hubble Space Telescope (HST) and much later during the assembly of the ISS. When multiple EVAs were planned, a gap day was normally inserted into the flight plan between the EVA days to allow the EVA crew to rest, clean and prepare their equipment for the next sojourn and to review plans for the next spacewalk. Another crewmember (normally the PLT) was assigned to the EVA 'crew' as an Intra-vehicular crewmember (or IV1), to help the EVA crew put on and remove their pressure suits and associated equipment. Usually, an EVA-trained crewmember also acted as chorographer for the spacewalk taking place by observing and directing from the flight deck windows, while other members of the crew controlled the Orbiter or took still photography and film during the spacewalk. Shuttle-based EVAs, like RMS operations and subsequent rendezvous, docking and logistics transfers to space stations, required a team effort that often lasted several days.

Rendezvous and Proximity Operations: Rendezvous with another large object and keeping the Shuttle in close proximity (Prox Ops) only became a major crew objective of several missions towards the end of the Group 8 era. The techniques were pioneered during the early Shuttle missions, with Group 8 astronauts heavily involved during retrieval missions such as Solar Max, Westar, Palapa and Leasat, but with the advent of HST servicing missions and docking to Mir and ISS, it became even more important to practice and master these skills, as no American astronaut had rendezvoused and docked with a second object since the mid-1970s. During the mission, the CDR and PLT handled maneuvering the Orbiter, while the MS handled the RMS, performed visual cueing roles, operated the laser ranging equipment and conducted photo documentation. The key was to have as many eyes as possible on this tricky and intensive operation.

Docking/Space Station Operations/Logistics Transfer. Unfortunately, despite being part of the original planning, the use of the Shuttle with a space station did not reach flight stage until the mid-1990s. By then, most of the TFNG had relinquished their active flight status to seek new goals. As a result, only three members of the selection (Gibson, Lucid and Thagard) visited a space station, the Russian Mir. In 1995, Hoot Gibson performed the first docking between two spacecraft by an American astronaut since Tom Stafford had linked his Apollo spacecraft with Aleksey Leonov's Soyuz in 1975. Already on Mir and scheduled to come home with Gibson and his crew on *Atlantis* was fellow Group 8 astronaut Norman Thagard, who had become the first American to live and work on a space station since Skylab over 20 years before. The following year, Shannon Lucid lived aboard

Mir for six months, initiating a continual presence on the aging Russian station by a series of five American astronauts (working rotational residencies over the next three years) in advance of ISS assembly. These missions to Mir spearheaded a cooperative program with the Russians that resulted in the creation of the ISS, which is still in orbit and operating over 20 years after its assembly began.

Science support: With no space station to build or visit, the early Shuttle missions between 1983 and 1995 (the period in which many of the TFNG completed their missions) had the primary objectives of satellite deployment, retrieval and repair, or science-focused missions using research payloads in the cargo bay. On these missions, many MS worked as part of the 'science' crew supporting the collection of data, while the 'Orbiter' crew (the CDR, PLT and MS-2/FE) looked after the vehicle's systems and maneuvers, as well as sometimes participating in certain biomedical experiments, albeit reluctantly in some cases. During this period, the PC role was created, which would in turn evolve into the Science Officer position assigned to early NASA ISS resident crew members.

TRAINING THE SHUTTLE ASTRONAUTS

The following are summaries of Shuttle mission training programs gleaned from NASA documents from two periods. [7]

The first selection is from 1980, at the very start of the TFNG assignments following Ascan training and the year prior to the Shuttle reaching orbit. The second selection originated from documents made available during the 1986 *Challenger* enquiries, investigating the roles and depth of NASA astronaut participation in the Shuttle program at that time. This was after all 35 of the Group 8 astronauts had flown at least one Shuttle mission each, and while the Shuttle training program evolved over the ensuing years, these documents offer a good representation of the type of training conducted in the mid- to late-1980s when most of the group were active.

NOVEMBER 1980 TRAINING STATUS

In 1980, Shuttle *mission* training was categorized in four areas:

- *Basic training:* This was generic in nature and aimed mostly at recently selected astronauts. It was designed to familiarize them with the NASA infrastructure and the Space Transportation System (STS) program, and to prepare them for further training in more individual roles.
- *Advanced training:* Under this phase, the individual gained the generic knowledge and skills necessary to perform STS operations at a proficiency level required by the position (PLT/MS) for a "typical" flight.

- *Flight-Specific Training:* Based around recurring simulator training, this phase provided the individual with the knowledge and necessary skills unique to their flight. It also included practicing generic STS operations in a flight-specific environment to prepare them for a mission.
- *Recurring training:* This phase was used to maintain the proficiency of experienced astronauts in critical areas of STS operations which were seldom used, as well as incorporating any new information and techniques to a particular field.

In summary, new astronauts would undergo the 'basic training' program as part of the Ascan phase, prior to moving on to advanced training as qualified astronauts who could be included in the 'pilot pool' of available personnel. Potential flight crew members would be selected from this pool to receive flight-specific training, to a level that would deem them to be 'flight ready' and able to perform their assigned mission. After their flight and any post-mission requirements and recovery, the crew would be broken up and returned to the pool of available personnel, where they would occasionally complete any recurring training required prior to their next flight assignment, be reassigned to administrative roles, or eventually step down as an active astronaut.

At the time (1980), this was all theoretical of course, as the Shuttle had yet to fly. However, with a new group of 19 Ascans joining the team (Group 9), plus the 35 Group 8 astronauts and over 20 members of the earlier groups still active, an effective system had to be devised to balance those in training preparing for missions alongside post-mission activity, support roles, natural attrition, medical illness and vacations. It must be remembered that though astronaut training is always intense and involved due to the requirements of the role, the members of the astronaut team (as with flight controllers, trainers, managers and workers) male or female, could become ill, take vacations, travel on official business and request more family time away from constant assignments. Also to be taken into consideration were their responsibilities outside of NASA (especially with serving military astronauts), and times when members of the office were preparing to leave and their absence had to be covered. Just one year before Shuttle started flying, as the 1980 document underlined, the training program was still immature. In the late 1970s, to help develop the Shuttle training program at a quicker pace, it had been decided that the first six flights of the Shuttle (all originally part of the OFT program, subsequently cut to just four missions) would have members of earlier selection groups assigned to the crews. The same 1980 document also revealed that "new Pilots and Mission Specialists [from Group 8 initially] would be assigned to flights beginning with STS-7," which is exactly the way it transpired.

Projection (1980)

As mentioned, by the time the first members of Group 8 began flying in 1983, it had already been decided that no back-up crews would be assigned after the OFT series had been completed. Instead, crewmembers with suitable skills and/or similar flight experience would be available to fulfill a given role as required, or could be assigned to a crew to replace an ill or removed crewmember.

In attempting to make future projections, certain assumptions clearly had to be made. As a guide, the 1980 document focused upon the pilots rather than the MS, suggesting that the re-fly rate for pilots could be an average of three flights per year. Training would be as a crew rather than individual simulator time, and with several 'older astronauts' passing the age of 50 by 1983, the attrition from the office was expected to be heavy during the early operational years; therefore a lower average re-fly rate would be acceptable during those initial years (Financial Year (FY) 1985 and earlier) in order to allow more pilots to be trained "as insurance against attrition losses." Another interesting point made in this 1980 document was the recognition that the serving military astronauts from the Class of 1978, who were just two years into their assignment at NASA and recently out of Ascan training, were scheduled to return to the DOD by 1985/1986, suggesting that the Flight Operations Directorate at JSC was already considering looking for replacements

The 1980 projections suggested that each crew would be spending at least 25 hours per week in training by the flight of STS-7, with an average of 16 hours spent in the simulators by crews training across a 50-hour working week. The unknown variables at the time were the 'classified' developments in the DOD Shuttle Operations and Planning Center (SOPC) objectives, especially as all crew training was planned and agreed to occur at JSC. Some of the basic requirements to schedule these classified objectives remained secret to all but a small group.

Five months prior to the first Shuttle flight, sufficient development had been made to the training program to be able to propose some forward planning as the new group of astronauts were completing their basic training and beginning to receive mission support assignments, pending allocation to their first flights.

In 1980, the main objective of this early Shuttle training program was to develop a list of minimum requirements for a small number of *pilots*. After two years of experience from the STS-1 training flow, significant progress had been made in the simulation of a Shuttle mission model. Based upon the experience of STS-1, a refinement of these definitions followed, and in some cases they were redefined, to take full advantage of the cheaper part-task facilities available within the training program to make the best use of this resource on a restricted budget.

The predictions in 1980 suggested that standardization of the ascent and entry phases would be possible as early as STS-10, or possibly sooner. Repetition was felt to be key at this time, with a program of "standard orbit training" being introduced after two or three missions had flown similar profiles (which would later be defined as TDRS or Comsat deployments, as well as payload bay science missions such as Space Radar Laboratory). Part of this prediction was to factor in the possibility of pilots flying an average of three missions per year. This did not fully come to fruition, although it was evaluated several times by flying the same astronaut on a second mission a few months after his previous flight. This included: Bob Crippen commanding STS-41C in April 1984 and STS-41G six months later in October; Karol Bobko commanding STS-51D in April 1985 and STS-51J that October; and TFNG Steve Nagel flying as MS on STS-51G in June 1985 and, after a record four-month turnaround, returning to space as PLT on STS-61A that October.

Another factor which had to be taken into consideration was the size of the Orbiter 'fleet.' At the start of the program, only a small fleet of Orbiters had been formally authorized, limiting flight seats even further and, as a result, requiring crew performance standards to be maintained at a high level for longer periods between flights. By early 1980, only the first operational orbiter, OV-102 *Columbia*, had been delivered to the Cape and it would not be ready to fly on a mission for a further year. After flying for a fifth time, *Columbia* would be taken 'offline' to undergo a period of upgrades before returning to fly the program's ninth mission. The second Orbiter manifested to fly in space was the former Structural Test Article (STA-099). This had been converted into the operational Orbiter OV-099 *Challenger* and followed *Columbia* into orbit, on the sixth mission in 1983. The other Orbiter, OV-101 *Enterprise* used in the 1977 ALT program, was to have joined them, but it was deemed too costly to convert the atmospheric test vehicle to orbital standards, so *Enterprise* was relegated to further ground testing and destined never to fly into space. At this time, there were just *two* authorized vehicles left in the Rockwell production flow at Palmdale, OV-103 *Discovery* and OV-104 *Atlantis*. They would not be operationally ready until 1984 and 1985 respectively, so *Columbia* and *Challenger* would bear the brunt of missions between 1981 and 1984. While Shuttle missions had been planned to last for 7–10 days each, it still required a considerable effort to maintain and process each vehicle for flight, a factor which had not been fully appreciated in the early projections that described a larger fleet of Orbiters flying one mission a week. Each of the Orbiter vehicles also required occasional down time for maintenance, effectively taking them out of the flight schedule for critical servicing and upgrades throughout their operational life. Taking just one Orbiter out of the schedule for a short period was a serious detriment to mission planning. Actually losing one to accident or permanent grounding was unthinkable.

There was also the question of OV-105. In the early 1980s, the oft-proposed fifth Orbiter was not intended to be a flight vehicle, but merely authorized for fabrication as a set of structural spares in the event of the loss or severe damage of one of the other Orbiters. As for further vehicles joining the fleet, any prospect of an OV-106 or OV-107 and beyond never progressed further than initial discussions.

As the program matured, factors affecting the training focused upon simplifying the flight software and creating workaround and waiver options. Critical to all of this was the availability of the SMS, which the 1980 report noted was the "single most critical element of the training program." The SMS was already overloaded, and was projected to be even more so by 1984. Even this early in the program, it was already deemed inefficient to conduct part-task Guidance and Navigation (G&N) training in the SMS, warranting a separate (and costly) simulator for that role. The document indicated that the Shuttle program was "currently too immature to commit to a more extensive trainer [i.e. a third SMS]", but if the Guidance and Navigation Simulator (GNS) could "grow gracefully toward a third SMS capability" it would be beneficial. As Carl Shelly noted in his presentation, "[The] GNS is deliverable earlier than a third SMS station. We need it now for its part-task training application, but not later than one year preceding a flight rate of 22–25 flights per year to relieve SMS training load (currently early 1984)."

Locating the GNS in Building 35 at JSC provided a level of redundancy in the training capacity, in the event of the SMS being lost due to hurricane damage, flooding or accident. Consideration was also being given to moving the SMS up one floor to reduce this risk.

A different approach
The change from Apollo era training to the Shuttle was significant. There were some similarities, such as in crew performance standards and crew training for nominal, abnormal (e.g. aborts) and malfunction situations, but the differences between the programs were notable.

Generic training for Shuttle focused more upon operational procedures (or "how to work it"), as opposed to understanding the technology *plus* operational procedures (or "how it works, plus how to work it") as it had been in past programs. Standard procedures and tasks were now listed in a Flight Data File (FDF), with approximately 80 percent of the program classed as 'generic'. Flight-specific training had been reduced substantially by limiting recurring rehearsals of specific plans or procedures, while part-task trainers saw greater use. For example, the Orbiter Single System Trainer (SST) saved 63 hours of SMS time per crewmember, while the Orbiter GNS was expected to save a further 113 crew hours in the SMS once implemented. Though the use of computers was expanding, in 1980 it was still expected that Shuttle crews would spend a significant amount of time utilizing self-study workbooks (then termed "paper training"). Overall, Shuttle training offered a far more structured program based upon formal task analyses of the onboard job.

FLIGHT CREW TASKS CIRCA 1980

In the 1980 document, STS-1 PLT Robert Crippen listed the tasks the flight crew were expected to handle during a nominal Shuttle mission:

Monitor/mode vehicle

For this function, the flight deck astronauts monitored the dynamics of the vehicle by observing both the ascent and entry trajectories, essentially serving as a back-up to the vehicle's computers and being ready to take over manual control in the event of contingency or off-nominal situations. They were trained to monitor the sequence of automatic events, including the separation of the twin Solid Rocket Boosters (SRB) and the External Tank (ET), the cut-off of the three main engines and activation of the vehicle's cooling system. They would also monitor Main Propulsion System (MPS) propellant dumps during nominal and abort modes.

On orbit, at the beginning and end of orbital operations, the CDR, PLT and FE (MS-2) astronauts would be busy. For a nominal mission, the astronauts would become familiar with activation and deactivation of the heaters, the Auxiliary Power Unit (APU) and hydraulics; MPS idling and re-pressurization; reconfiguring the Reaction Control Systems (RCS) and Orbital Maneuvering Systems (OMS); closing the ET door on the belly of the Orbiter and reconfiguring the Data Processing Software (DPS). They would initiate opening and closing the payload bay doors and vent doors, deploying and stowing the radiators, and reconfiguring the vehicle's electrical systems, the Environmental Control and Life Support System (ECLSS), and the Guidance, Navigation and Control (GN&C) subsystems.

On orbit, they would perform maneuvers by adjusting the attitude of the vehicle and initiating orbit adjustment burns as required. They were also required to update the vehicle's attitude and the state of the navigation platform by performing alignments of the Inertial Measuring Unit (IMU), monitoring the star tracking acquisitions and maintaining the correct navigation state.

Malfunction safing/reconfiguration

A major element of Ascan training was learning how to cope with and overcome system failures or problems. The future astronauts were trained during simulations to respond to various malfunction subsystems and, where possible, reconfigure the vehicle accordingly to address issues by means of prescribed actions.

For problems with the MPS, training addressed the failure of one or more of the Space Shuttle Main Engines (SSME), problems with the flow of data or communication with the MPS, leaks of the He (Helium), high LO_2/LH_2 (Liquid Oxygen/Liquid Hydrogen) pressure, and low LH_2 ullage pressure.

The Ascans trained in responsive actions to signals of over speed or under speed of the APUs and hydraulics, excess Exhaust Gas Temperature (EGT) and oil temperatures, leaking APUs, and leaking or low pressure hydraulics. For problems encountered with the communications systems, the astronauts were trained to be ready to reconfigure the system and manage the back-up antennas.

For the ECLSS, they simulated sealing cryogenics leakages and reconfiguring the system, including the fans, the cabin pressure system, the flash evaporator, the H_2O (water) and Freon loops, and the heaters.

With problems in the OMS or RCS, they were taught how to identity and isolate a leak and reconfigure the systems around the problem. If a thruster failed again, the astronauts were instructed how to isolate the faulty jet and reconfigure the remaining thrusters to compensate for the loss. They also learnt how to regulate and manage the propellant valves.

For GN&C issues, the crews were instructed how to manage the flight control system channel, the failure of aero surfaces, and controls to isolate IMU failures and TACAN (TACtical Air Navigation) failures.

The DPS training included identifying component failures, corrective actions and reconfiguration, as well as the interaction of the system with the GN&C components and increased task complexity. For electrical problems, instruction was given on AC Bus and invertor management, and DC Bus and fuel cell management.

Crew system operations

Training was given on how to use, clean and maintain the waste management system (the Shuttle toilet), the food system (including selecting their own meals from the available menu and tasting samples of the selection to help them choose). This training also included operation, maintenance and housekeeping of the onboard galley. They were instructed on the range of cameras carried in the crew compartment (TV, still and movie) and in the stowage onboard the vehicle. Medical emergency training was included, as were familiarization sessions for the range of crewmembers' personal equipment, the Earth terrain maps, navigation devices, stowing and setting up the middeck seats, and various escape procedures.

As it took about a year to train a Shuttle crew, depending on the mission objectives and the past experience of the assigned crew, the total training hours required for all these tasks varied mission by mission.

EVA operations

The MS (and some pilots) were instructed and qualified for the Extravehicular Mobility Unit (EMU) EVA suit, operating the airlock, and a number of contingency EVA tasks such as closing the payload bay doors, stowing a failed Ku-band antenna, possible thermal protection system tile repairs, lowering the payload support platforms and so on. This took place initially in bench tests of equipment, followed by unsuited and suited training in $1g$ simulations, in aircraft flying multiple parabolic curves to give up to 30-seconds of 'weightlessness' on each parabola, and finally over many hours spent suited in the huge water tanks at JSC and the Marshall Space Flight Center (MSFC), Huntsville, Alabama.

FLIGHT DATA FILE

The FDF featured volumes of *paper* checklists that were either flight-specific or generic, carried on board for reference. It covered ascent, entry, post-insertion and de-orbit preparation for both nominal and contingency situations; EVA; the Crew Activity Plan/Flight Plan; orbital operations, Payload Deployment and Retrieval Subsystem (PDRS); photo/TV; and rendezvous.

In the days prior to laptops and personal computers, there were also smaller pocket checklists for ascent, orbit, and entry, as well as an Orbiter systems data book, a DPS dictionary, malfunction procedures books, orbit operations schedules, crew systems checklist, photo/TV checklist, medical checklist, EVA checklist, egress procedures for both nominal and contingency situations, and an ever-changing 'updates' book. As up-to-date as the Shuttle was, it still required a veritable library of documents to back-up the onboard computers, as well as teams of controllers in Launch and Mission Control, scores of contractors located across the country, hundreds of hours of training, and the capabilities of the astronauts to retain the information presented.

CREW RESPONSIBILITIES

Once a crew was assigned, the enormity of what lay before them became evident. There was so much to prepare for and keep track of, and with seemingly very little time to do so. With no formal back-up or support crews as in the Apollo days, the flight crew was entrusted with deciding which of the team would be responsible for what. Therefore, in addition to understanding the major elements of the mission, there was a division of labor to enable the crew to follow and report on developments (or in some cases the lack of developments) in key areas, and for each to be the crewmember responsible for certain items (primary) or to support (back-up) a fellow crewmember in other areas. Such specialization changed flight to flight depending on how the CDR and his crew split up the list, but could typically include:

- For the CDR, PLT and MS-2: The DPS, the MPS (SSME/ET/SRB), OMS and RCS; the APU and hydraulics; the Electrical Power Distribution System (EPDS); and the ECLSS.

- For the MS: Responsibilities included supporting the flight deck crew with some of their tasks; communications and instrumentation; opening and closing the payload bay doors; the photo, TV and cam recorder equipment; serving as crew medic; crew personal equipment; keeping the FDF up to date; in-flight maintenance checks; being aware of changes to the Flight Rules; the Text And Graphics Systems (TAGS) on the early flights; primary and secondary payload hardware and systems; scientific experiments and investigations; Detailed Test Objectives (DTO), Detailed Supplementary Objectives (DSO) and the RMS; Earth observation objectives including geography, meteorology and oceanography; middeck and student experiments; the EVA/EMU and associated equipment; and even inviting guests to the launch, designing the crew T-shirts, and the occasional comic crew photo. All this had to be scheduled, assigned and completed before launch, keeping the crew training diary very busy as the clock ticked down towards lift-off.

THE TRAINING DIVISION (1986)

At the time of the *Challenger* accident in 1986, at least 25 'crews' had progressed through the Shuttle mission training flow to flight status, with a further ten complete crews (over 60 crew seats), including about 20 Group 8 astronauts, in various stages of training for missions in the remaining months of that year.

The "just to be sure" syndrome

During the investigation into the *Challenger* accident, part of the Presidential Commission's remit was to evaluate the training that the STS-51L crew had received and to determine whether this had any bearing on the accident. Two main points were discussed in detail: their workload in the weeks leading up to the accent; and their workload and frequency of use of the training facilities.

By 1986, the Shuttle had been flying for five years, but crews had been 'in training' for almost a decade, including the crews who flew the ALT series of flights in 1977, those who had supported the original OFT program, the first operational mission, and the maiden launch of *Challenger*, the second Orbiter. Members of the 1978 group had been in various stages of preparation for crew assignments or specific mission assignments for seven years, and three more astronaut selections had followed in 1980, 1984 and 1985. The training methodology would evolve following the *Challenger* accident as the program moved into the 1990s, but by then most of the 1978 class had moved on from active flight training to managerial roles or had left the program. At the time of STS-51L, all 35 members of the 1978 group had flown into space at least once, with several having logged two missions. Therefore, reviewing the 1986 accounts of Shuttle crew training is relevant both to the peak of Group 8 participation and to the level of Shuttle mission training in the mid-1980s.

In his memo to the Commission, Robert Holkan, Chief of the JSC Training Division, wrote: "Training workload generally peaks in the last 10 weeks before flight due to the arrival of the specific software products used during the flight. These products are installed in the simulators and the final training is conducted. However, late delivery of these products can cause training compression and increase the crew workload." It was also pointed out in the memo that there was a certain amount of self-generated loading on each crew: "This load comes from the legitimate concerns felt by each crew that they need just one more simulator run or just one more meeting with the checklist people, or some similar group, before the flight 'just to be sure'."

According to Holkan, the SMS, the main training facility, "has been a constant source of problems through the entire program." He cited that it had been less than capable since the start, and that the system had provided adequate training only through the constant efforts of the personnel utilizing it. "It is basically not a good teaching environment," Holkan continued, underlining, just five years into the flight program, that "the facility computers and equipment are out of date and obsolete." On a brighter note, he observed that there were plans to update the equipment with the aim of increasing the capabilities of the SMS as "an effective teaching machine", but added that "the funds for those modifications are programmed out over a 10-year period and are minimal."

Shuttle Training Division

From the late 1970s, a number of the veteran astronauts were pathfinders for the new Shuttle training regime, eventually providing the crewing for ALT program and the first six operational missions. Their experience proved useful in defining the techniques, equipment and procedures, but it was the TFNG who were the first large group to progress through this training cycle with no previous space flight experience and only a few years' experience of working at NASA. It is worth exploring the training division circa 1986, as a background to the processes that each of these 35 astronauts – and many that followed them – experienced prior to taking their first flight into space.

During the mid-1980s, at the peak of Group 8 involvement in Shuttle flight operations, the JSC Training Division had a staff of 270 who were responsible for training both the flight crews and the flight controllers. Within this group were approximately 150 instructors who specialized in the various media, with a further 70 who were more computer orientated and who were dedicated to programming and developing the training facility, adding new lessons into the syllabus.

The Training Division was responsible for developing training plans and maintaining a training calendar, which included standalone sessions, integrated simulations, joint simulations, and simulator and trainer requirements. This covered the ascent and reentry phase including various abort modes (Return To Launch Site (RTLS), Transoceanic Abort Landing (TAL), Abort To Orbit (ATO), Abort Once

Around (AOA)), the orbital phase including systems operations, payload deployment, and rendezvous and proximity operations (there was no docking training at this stage). Science and EVA training were not mentioned, while space station training was limited to concepts, requirements and facilities.

The 'charter' for the Training Division was to "provide [a] training program for flight crews and flight controllers." To meet this objective, the functions of the Training Division were listed as: To define training requirements; formulate a training plan; develop a training course, material and scripts; conduct training sessions; integrate, schedule and record training programs; define the required training facilities; negotiate user agreements for outside facilities; develop and operate selected training facilities; and provide real-time support for EVA and crew systems. To achieve this, in the days before laptops and digital programs, the Training Division generated a significant amount of printed products, including training catalogs, training plans and schedules; Shuttle Flight Operations Manuals (SFOM); workbooks; training lessons; and scripts and training records to provide a historical database of how each 'crew' progressed through the training cycle.

Training Hierarchy

Preparing a crew for a flight on the Shuttle relied upon a range of techniques, which included (but was not limited to):

- Paper media workbooks, training manuals and briefing notes.
- Computer-aided instruction which, in the 1980s, came from "Regency" small computers that provided a number of lessons to the student crewmember/controller on how particular instruments worked[5].
- Single System Trainer (SST). This was a medium-complexity machine that allowed the student to learn about one system of the Shuttle in depth, such as the electrical system, prior to moving on to another system.
- Water Emersion Training Facility (WETF), the water tank used to train EVA crews for spacewalking using neutral buoyancy. At this time, the former centrifuge building (Bldg. 29) had been adopted post-Apollo to provide a large water tank on site at JSC. While the WETF was capable of supporting water egress training for Shuttle crews, it was unsuitable for supporting EVA training to service larger payloads such as HST and Solar Max. Since the late 1970s, the larger Neutral Buoyancy Simulator (NBS) 'pool' at MSFC had been used for EVA and RMS crew training, but in 1997 a new, much larger Neutral Buoyancy Laboratory (NBL) was completed at the Sonny Carter Training Facility near to JSC. This was used for Hubble servicing training and for developing and practicing ISS assembly tasks. **[8]**

---

[5] Regency provided a programmable 64 x 64 spot touch screen that displayed switches and indicators, and component schematics. Using this, the trainers could communicate with the teaching software by touching the screen in the appropriate place.

- Shuttle Mission Simulator. This complex machine was able to duplicate all phases of a Shuttle flight profile, with simulated views out of the windows, motions, and working cockpit instruments.

A building block approach was followed for Shuttle training, in the three distinct phases of basic training, advanced training and finally flight-specific training prior to launch.

Self-study formed part of the basic training phase, with workbooks, texts and videotapes supplementing the training before the students progressed to the advanced phase that incorporated computer-aided instruction. The advanced training also incorporated one-to-one instruction, using flight hardware trainers and SST to focus upon the avionics system, Orbiter systems, crew systems, and payload and carriers.

As the advance training progressed, it inevitably increased in both complexity and overall cost. Once flight-specific training came to the fore, the team instruction became more involved, with advanced and flight-specific training requiring more use of the SMS, together with phase training for ascent and aborts, orbital operations, entry and landing, rendezvous, payload deployments, Prox Ops and EVA (there was still no reference to docking at this stage of the Shuttle program). Integrated simulations developed communications skills, trained the flight controllers, crew and payload customer, and verified procedures.

PLANNING A TRAINING CYCLE (Circa 1986)

With 25 crews successfully progressing through the training cycle and several more in various stages of preparation for their missions, the assigned 'training-flow' appeared to be working, for the time being. In developing the training plan system, the team had factored in a smooth flow for new crew assignments as the flight rate increased. That helped reduce the workload on the training management. Parts of this system had been automated, but there were always unforeseen incidents requiring a short-term change to the plan, and such incidents, as well as changes in the manifests, triggered the *unplanned requirements cycle*. If a manifest change impacted crew assignments, the extra work this entailed was amplified. Any change in the manifest usually resulted in unutilized training, but this was not always the case. The crew of STS-51D had received rendezvous proximity training as part of an early manifest, but when that manifest changed it appeared their training would be wasted. On their flight, however, during the attempt to rendezvous with and activate the stranded Leasat satellite, this early training came in handy, even though they had not planned to use it on their flown mission. The rule to remember in planning a training cycle was that "schedule slips cause inefficiency."

The start of the process generated a training calendar listing requirements vs. tasks, a flight manifest that included the flight-specific requirements of that mission, and the training records from previous crews. This provided a summary of crew training requirements for that mission around the time the flight crew was named. From this, a crew training guide was produced listing the training sessions to be completed and the time before launch to complete them. Typically, as many as *ten* crews could be in training at this point. Unassigned crews in the pilot pool received about 16 hours per week of generic training, but only after prime crews had been scheduled, taking advantage of schedule inefficiency at this time. A series of training interfaces and reviews usually began prior to Launch [L]-40 weeks, with the crew being named and their tasks defined at L-38 weeks. A cargo review was completed at L-36 weeks, followed by a training guide review at L-32 weeks. The first training status report was issued around L-29 weeks, with flight-specific training commencing at L-27 weeks. The training team met to review the training (termed Team Tag Up or TTU) on weeks L-20, L-16, L-12, L-10, L-8, L-6, and L-4, with the second status report reviewed during L-14 and the third and final report during L-4. About two weeks after launch, a post-flight training report was completed and published.

Cataloguing the training

Having recognized the need to establish a database of training experiences since before the first Shuttle flight, the JSC Training Division had maintained a training catalog, which detailed exactly what training each crew person required for a specific flight. In 1985, a similar catalog recorded the training of Shuttle flight controllers. Any changes to the training flow were instigated by the Training Division Office, with the cooperation and input from the Astronaut Office as the system developed.

Standalone training flows

Standalone training began in 1979 and was completed in the SMS. For this simulation, the instructors acted as the Capcom and in all positions of Mission Control. Initially, the crews were instructed and tested on their actions and procedures in nominal situations. Once they had demonstrated their ability to do this, the training teams introduced malfunctions to tax them and teach them the correct responses. Some limited contingency training was also completed, but it did include near-catastrophic situations. This type of training evolved over the ensuing years, resulting in fewer training hours across all areas. All changes were recorded in the training catalog.

The original training flows for Shuttle OFT crews were developed in 1978, based upon experiences from earlier programs and "guesswork" about the Shuttle. Each of the four OFT crews followed this flow during their training and each pair commented upon it afterward. This was all collated and led to the first major revisions for the STS-5 crew (1982). Logically, the experience base would grow as more missions were flown, enabling further refinements to the training flow. Over time, lessons were added, some removed, and the content or sequence was changed, with successive reviews further defining the amount and type of training required to prepare a crew to fly the Space Shuttle. By 1986, the training team felt that the Orbiter part of the training catalog was very mature and required minimal changes. However, further refinement was necessary regarding payloads, RMS operations, and rendezvous and Prox Ops (see Table 8.1).

TABLE 8.1: SHUTTLE TRAINING CATALOG REQUIRED TRAINING HOURS

Training Area	Nov 1982	Sep 1984	Jan 1986
Ascent	97	91	89
Orbit	129	93	89
DPS	61	43	43
GNC	86	104	98
Support Sys	89	126	126
Crew Sys	76	79	88
EVA	133	154	152
Deorbit/Entry	183	140	141
PDRS	146	141	143
Prox Ops	71	65	63
PAM	47	48	51
Spacelab	179	146	146
IUS	40	63	58
Total	*1317*	*1293*	*1287*

Adapted from the Transmittal of Official STS-51L Training Presentation [& attachments], compiled by Frank E. Hughes, Training Group Lead, Mission Planning and Operations Team (DG6), NASA JSC, Houston, Texas, dated April 11, 1986, Ref DG6-86-107. Presented to the Presidential Commission on the STS-51L accident, at JSC on Tuesday April 1, 1986. Copy on file AIS Archive.

In the three years between STS-5 and STS-51L, Shuttle training hours had reduced from 129 hours in November 1982, to 93 hours by September 1984, and to 89 hours by January 1986. Several factors led to the reduction between 1982 and 1984, most notably that the CDR was recently flown and therefore did not require advanced training. Several of the lessons were combined to make better use of the simulator time. As Part-Task Trainers (PTT) developed, some lessons were dropped because the material was covered in new lessons developed for the PTT. Dedicated lessons for the MS were combined with the same lessons taught to the CDR and PLT. Timeline and Crew Activity Plan (CAP) lessons were dropped as the material was being duplicated in new payload-specific lessons. As the training flow evolved, new Flight Procedure Handbooks were introduced.

Interestingly, the refinements introduced between 1984 and 1986 included a change in the philosophy regarding the CDR taking advanced lessons. Such advanced sessions now became a requirement for the CDR, based on when they had last flown. Several of the Flight Procedure Handbooks were deleted and new Handbooks introduced into the system, while Shuttle Portable Onboard Computer lessons were introduced. Orbit Skills and Orbit Timeline lessons (totaling eight hours) were dropped, as these were now covered in the payload lessons.

Integrated Training

These sessions were designed to develop coordination between the flight crew and Mission Control teams. For these sessions, the Flight Director (FD) and his team were located in Mission Control, while the crew entered the SMS. Radio links between the two were simulated using data and phone links. The integrated training process featured a script that the training team followed, which detailed exactly what malfunctions were to be used and when in the session they were to be implemented. The responses from the crew and controllers were monitored and any incorrect actions were discussed and rectified after the session. In reality, the integrated sessions probably trained the flight controllers more than the flight crew, but the development of a coordinated crew/controller team was paramount. A schematic of integrated and joint integrated sessions involving the Payload Operations Center is shown in Fig. 8.2.

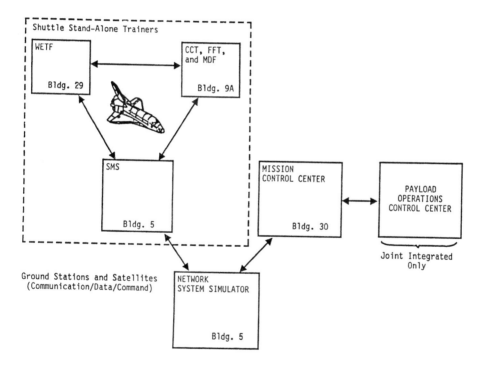

Fig. 8.2: The Integrated and Joint Integrated Simulation Network circa 1986. Taken from the Official STS-51L Training Presentation, April 1, 1986.

Towards the 1990s

As previously mentioned, during the review into the STS-51L accident the JSC training department acknowledged that the Shuttle crew training facilities were out of date or obsolete. As concerning as this was, just five years into the operational program, there were plans to address this situation. Reading between the lines, it appears that long-requested funds would be channeled directly into upgrading the trainers, as the scope of the program had already exceeded the capabilities of the trainers as constructed in the late 1970s. These upgrades included improved fidelity displays for the MPS, and visuals out of the windows. There was also a desire to provide a GNC trainer to support the training flow of pilots at an improved rate. Additionally, there had been issues with the delivery of software upgrades, which had caused compression in the training program. Changes to the manifest led to challenges in delivering new upgrades in a timely way, and as a result, the end-of-the-line training loads were delivered later than desired. The best-case scenario was to have a stabilized manifest cycle, but this became an ongoing issue through the Shuttle era and affected a variety of support areas across the program. It was also found that increasing the crew workloads quite naturally created a situation in which the crews requested more time with the final flight software product, to ensure that everything worked as expected before they needed to use them during the mission.

The loss of the training skill base, from the retirement of experienced astronauts but also from the pool of crew trainers or flight controllers, also affected the smooth training flow, as any new person joining the team required 6–12 months to train up, although new methods were always being sought to improve each training position. Even by 1986, the Shuttle training program was still being described as "evolutionary." However, the new format of not assigning dedicated back-up crews, with replacement crewmembers available as required from a pilot pool, was working well. Recent studies into re-flying a CDR and crew had also suggested major benefits. Still requiring further work was the amount and timing of Ascent Integrated Training for each flight, and there was an urgent need to upgrade the RMS and Prox Ops training philosophy.

Looking back almost 40 years, the participation of the TFNG in supporting the earlier veteran astronauts helped to create the Shuttle training program which, with regular upgrades and refinements, operated successfully for three decades. As a group, they were instrumental in ensuring the smooth transition from the earlier one-shot missions of the Mercury-Gemini-Apollo era to the more routine operations of the Shuttle. The legacy to this has been in supporting the assembly and resupply of the ISS, although most of the TFNG had long since retired by then. Forty years after the selection of the eighth astronaut group, as NASA once again prepares for a new generation of human spacecraft to replace the Space Shuttle, its current astronaut team, like the TFNG before them, provide the vanguard of crews as America embarks into its next era of human space flight.

SHUTTLE CREW TRAINING CYCLE

All of the above resulted in a definitive program to prepare selected crewmembers for their flights on the Shuttle system, the contingency procedures they hoped they would never need, and the specific mission they had been selected for. A detailed account of crew training is beyond the scope of this book, but for the 1978 selection, a summary of the Shuttle training flow is presented here for the reader's reference.

Upon selection to a crew, their training became a priority, and as the earlier flights progressed so their training moved toward the top of the queue for simulator time. Flight-specific training generally consisted of the following:

- The CDR and PLT logged flying hours in the Shuttle Training Aircraft (STA), practicing the approach and landing profiles until just a few days before launch. They also maintained their flight proficiency level by flying their T-38 jets.
- The whole crew participated in the crew systems refresher course, whether they were rookies on their first mission or space flight veterans, as these systems were constantly updated and changed as the program matured.
- Refresher courses and flight-specific profiles were included in Flight Operations training.
- The flight-specific training changed depending on the missions, but could encompass some of the following

  o PDRS, which included RMS training, the various deployment systems for different payloads and any retrieval operations specific to their mission.
  o Carriers, which were a range of hardware devised for the STS system and designed to 'carry' various payloads, experiments and hardware. They were interchangeable between missions as required. These included: the European-built Spacelab pressurized long module and unpressurized pallets, instrument pointing system and Igloo system unit; the Payload Assist Module (PAM), which was phased out by the 1990s; the IUS; the Centaur upper stage, which was deleted from the manifest after the loss of *Challenger*; and, from 1990, the pressurized SpaceHab augmentation module.
  o Attached payloads that could be coupled, which included the Get Away Special (GAS) experiment canisters, the Hitchhiker payload attachment devices and any mission-specific packages
  o Middeck experiments, which included small experiments that could fit in or on the front of middeck lockers, such as the educational student experiments and flight-specific experiments such as the SAREX Ham radio or IMAX camera.
  o Prox Ops/Rendezvous, which included any joint operations with specific communication satellites, the Shuttle free-flying Pallet Satellite (SPAS), larger science satellites such as Solar Max, the Long Duration Exposure Facility (LDEF) and HST, and, from 1995, rendezvous and docking training for the Russian Mir space station[6].

---

[6] By the fall of 1998 and the start of ISS assembly, none of the remaining TFNG in the Astronaut Office were directly involved as crewmembers in assembly mission training.

- ○ EVA training, which included familiarization with the associated EVA hardware (pressure suit, tools, tethers, restraints) and procedures (donning and doffing, pre-breathing, airlock operations); contingency operations and scenarios; and planned EVAs for specific missions (such as Solar Max retrieval and repair, or HST servicing).
- ○ Manned Maneuvering Units (MMU) until the mid-1980s, and the smaller SAFER rescue unit from the early 1990s.
- ○ Virtual Reality (VR) training was just coming into the mainstream at JSC as the STS-61 crew was preparing for the first Hubble servicing mission. Now commonplace for ISS EVA training, this was a new tool in the EVA training portfolio in the early 1990s.
- ○ Space station construction techniques and simulations (none of the TFNG were assigned to the ISS assembly missions).
- ○ DTO, DSO and Risk Mitigation Experiments (RME), which consisted of a variety of new investigations aimed at evaluating procedures, hardware and systems that may or may not be used operationally on future missions. They could be investigations on the characteristics of space flight, the operation of the Orbiter, the environment the Shuttle was flying in, or research into reducing potential risks. The DTO were usually more engineering based, the DSO were experiment based, and the RME were risk based.

Crew training could vary from printed workbooks and classroom studies, to bench reviews, 1*g* simulations, reduced gravity simulations, and visits to contractors or other NASA field centers, such as KSC for launch training. There were also programs of public affairs duties, such as interviews with the media and press conferences, as well as specific tasks such as designing the mission logo and choosing the food to be included in the mission menu. A variety of training devices were developed to support such pre-flight crew preparations.

SPACE SHUTTLE FLIGHT CREW TRAINING FACILITIES

The facilities for training Shuttle crews were far more extensive and widespread than were available during the Mercury, Gemini and Apollo eras, but still limited in comparison those for the current ISS resident crew training. Below is just a brief summary of the different facilities and locations used by the TFNG in their preparations for the various missions they were assigned to or flew.

Johnson Space Center, Houston, Texas

Building 1: Headquarters Administration Building
Center Management and support administration offices

Building 2: Public Affairs Office
Includes the media briefing room for press conferences, and interview rooms

Building 4 (North (the original Astronaut Office location) and South): Mission Operations Support Offices
Incudes the Astronaut Office and the Mission Control Center (MCC) FD Offices. Over the years, Building 4 also housed a number of smaller training devices and aids, including:

- Single System Trainer for the Orbiter and Spacelab science lab
- Crew Software Trainer (CST) part of the SST system
- Part Task Trainers also linked to the SST
- Shuttle Procedures Simulator (SPS) [also known as the 'Spare Parts Simulator']. In the early days of the program, parts from this trainer were often cannibalized to get the more critical SMS working. By the early 1980s, the SPS was scrapped and a GNS was constructed for part-task training from the remaining parts.

Fig. 8.3: (top pair) Space Shuttle Mission Simulator (SMS), Building 5, JSC. (bottom left) SMS Forward Flight Deck (Glass Cockpit). (right) SMS Aft Flight Deck (All images from the AIS archive).

Building 5: Space Mission Simulation Facility (now the Jake Garn Mission Simulation and Training Facility)

Used in series of simulations (or "sims") during training. Integrated simulations were linked to MCC but standalone simulations were not. The Joint Integration Sims saw the crew in the simulation patched into MCC and MSFC, the NBL, or Goddard Space Flight Center (GSFC, Greenbelt, Maryland). These simulations were conducted at varying levels of intensity, with the less complex ones staffed by simulator instructors or members of the training staff standing in for flight controllers (and throwing in unexpected failures and contingencies to push the crew responses and experience). More complex sims with MCC involvement required a dedicated console in MCC called the Simulated Control Area.

- *Shuttle Mission Simulator-Fixed Base (SMS-FB)*
  This simulator included high-fidelity mock-ups of the Orbiter flight deck and a low-fidelity mock-up of the middeck. Computer-generated realistic views were provided out of the forward, aft and overhead windows.
- *Shuttle Mission Simulator-Motion Base (SMS-MB)*
  This comprised the forward part of the flight deck of the Orbiter, using a six-axis hexapod motion system with additional extended pitch axis to generate motion cuing for all phases of the flight. This simulation only gave the crews accurate visual scenes outside the forward compartment windows.

Building 7 Crew and Thermal Systems Division

- *EVA Mobility Unit Malfunction Simulator (EMU MALF SIM)*
  Familiarized an EVA crewmember with the various potential failures while wearing the EMU.
- *Caution and Warning Simulator (CWS)*
  Similar training sessions about the various cautions and warnings generated from the EMU.
- *EVA Vacuum Chamber (EVA VC)*
  Pressurized simulations wearing the Shuttle EVA suit

In addition to training sessions, the vacuum chambers were used to qualify the suits prior to the first Shuttle missions. One of George 'Pinky' Nelson's first technical assignments was to support the development of the Shuttle EMU prior to STS-1. "I sought that out, actually, because it really looked like fun to be able to work in the suit, [to] go outside," he recalled in 2004. **[9]** "Story Musgrave at the time was the EMU person. So I started working with Story, and he helped check me out in the suit. There were three or four of us who were working EVA-related issues. Anna Fisher and Jim Buchli were working EVA issues, closing the payload bay doors and tools and things like that, and I was working the suit side of things, so we overlapped quite a bit. Story was a fabulous mentor in terms of just physically learning how to use the suit. His depth of knowledge of the suit and the way he operated in terms of really digging in and getting to the bottom of every system, really knowing everything inside out, was a great example of how to work, so I learned a lot from just being around Story and watching him work, and then getting checked out in the A7LBs, in the Apollo suit. [We] Had this little water tank in Houston and I did some work in the tank at Marshall [MSFC], in Huntsville [Alabama].

"I spent a lot of time going to design reviews and some trips up to Hamilton Standard [Inc.], where the suit was being designed. I don't think I ever went to [International Latex Corp. in] Dover [Delaware] during that time. I might have once to see where the fabric part of the suit was being put together. But the suit was one of the long poles in getting the Shuttle ready to fly. The folks in Houston who were in charge of it, [Walter W.] Guy and his group, were really working hard, and it was a difficult task to get it pulled together. The suit actually blew up shortly before STS-1. I was home working in my garden. I was playing hooky one afternoon, and I got a call from George Abbey. He said, 'Where the hell are you?' [I replied] 'Well, I'm home working in the garden'. He said, 'Okay. Get in here. We just had an accident with the spacesuit'. They [the technicians] were doing some testing in one of the vacuum chambers in Building 7, and… they had the suit unmanned, pressurized, in the vacuum chamber. They were going to do some tests and they were going through the procedures of donning the suit and flipping all the switches in the right order and going through the checklist.

"There's a point, when you get in the suit, that you move a valve. There's a slider valve on the front of the suit, and you move this slider valve over, and it pushes a lever inside a regulator, and opens up a line that brings the high-pressure emergency [oxygen] tanks on line. You do that just before you go outside. You don't need them when you're in the cabin, because you can always repressurize the airlock. When you're going to go outside, you need these high-pressure tanks. They're two little stainless-steel tanks about six inches in diameter, maybe seven. And it turned out that when this technician did that, he threw that switch and the suit basically blew up. I mean not just pneumatically, but burst into flames, [and he] got severely burned. It was pure oxygen in there. The backpack is made basically out of a big block of aluminum, and aluminum is flammable in pure oxygen. So this thing just went up in smoke. And they reacted very well. So then I was put on the Investigation Board for that, and spent a couple of months at least just focusing on what had caused this and [whether we could] identify it and fix it and get it ready before STS-1. So I learned even more about the design and manufacturing and materials and all of that in the suit during that process. It was fascinating. The [NASA] system for handling that kind of an incident really is very good. We've seen it with the big accidents we've had. They really can get to the bottom of a problem very well."

Building 9 Space Vehicle Mock-up Facility (SVMF)
This facility developed, operated and maintained the various mock-ups and training facilities to support crew training and engineering activities.

- *Crew Compartment Trainer (CCT)*
  This was a high-fidelity replica of the Orbiter crew station and was used for crew training and engineering evaluations of on-orbit procedures. The astronauts followed over 20 different classes to learn how to operate all the Orbiter subsystems. Capable of tilting to nose up configuration, the CCT permitted training in pre-launch activities at JSC in advance of progressing to the Cape, thus saving the crew travelling time and training time in Florida. The crew module was an accurate representation of the flight deck and mid-deck but with non-functioning switches, connections, guards and protection devices. It did have exactly the same physical characteristics and movement as the real vehicle. The fabrication closely replicated the actual vehicle and included fully functional flight-like CCTV systems.
- *Forward Fuselage Trainer (FFT)*
  This was a full-scale mock-up of the Shuttle Orbiter, minus its wings. It was used as a test-bed for various upgrades to the fleet, for astronaut training such as a 1*g* walkthrough of EVA airlock and payload bay operations (including lighting and CCTV systems), and for emergency egress by means of a Sky Genie™ from the crew overhead windows or via a functional (inflatable aircraft-like) escape slide from the side hatch when platforms and ladders were not available. Fabricated at JSC in the 1970s, this was the oldest mockup in the SMVF.

On May 3, 1993, Rhea Seddon, in training as PC/MS-1 for STS-58/Spacelab 2, was participating in a simulated emergency evacuation of the Orbiter in Building 9, wearing the heavy and bulky orange launch and entry suit. [10] The crewmembers seated on the middeck successfully evacuated the Orbiter from the side hatch using the inflatable escape slide. However, Seddon and her colleagues on the flight deck had to unstrap, carefully descend the ladder to the middeck and then exit from the side hatch. This was not as easy as it might sound, as with the faceplate of the suit closed their vision was impaired, and getting entangled in the leads and cables was common in such tests. As MS-1, seated behind the PLT on the right of the flight deck, it was Seddon's responsibility to be the 'spotter' for the CDR, MS-2 and PLT, ensuring they exited their seats before leaving the flight deck herself as the last crewmember to vacate the vehicle.

As she wrote in 2015 "I kept my spot behind Rick [Searfoss, the PLT] and pulled all their straps and lines clear as they left the flight deck. Then I huffed and puffed my way across Bill's (McArthur, MS-2) seat and down the ladder. 'I hate this stuff. I hate this stuff,' I panted under the cumbersome load [of the launch and entry suit]." Having managed to get downstairs to the middeck, she sat on the edge of the hatch and prepared to go down the inflatable slide, thinking "this is the easy part." The slide traverse was fine until she neared the bottom, when she felt "a horrendous pain in my left foot. Trying to stand up at the bottom of the slide, my left

Fig. 8.4: [top left] Space Vehicle Mock-up Facility, Building 9, JSC. [top right] Crew Compartment Trainer. [bottom left] Forward Fuselage Trainer, [bottom right] 1G Full Fuselage Trainer (All images from the AIS archive).

foot collapsed under me." Sprawled on the floor with trainers and medics around her, she was shouting "left foot, left foot," but with the faceplate closed no one could hear her. It was not until she opened the visor that she could shout where the pain was coming from. After she was painfully extracted from the boot and then the flight suit, she was examined. "After a brief discussion [with the duty flight surgeon in the Flight Medical Clinic at JSC] and even briefer exam of my foot and ankle, he assured me it was a sprain, but we'd take an x-ray. As he turned to leave

Fig. 8.5: [top left] 1G Flight Deck Commander Seat. [top right] 1G Payload Bay Trainer. [bottom pair] PDRS (RMS) Trainer (All images from the AIS archive).

the room I glared at his departing back, thinking 'I'm a doctor, you turkey, and something is BROKEN'!"

On subsequent examination, it was discovered that Seddon had broken four metatarsal bones in her left foot. After further consultation she was advised to have surgery to align the bones with screws that could be removed later, which would mean she had to be in a cast for six weeks after surgery and a walking boot for a further six weeks. An investigation into the incident determined that the slide had slightly deflated each time a crewmember descended, and buckled a little when Seddon made her ride down, causing her left foot to catch at the bottom and twist backwards. It was also determined that she had inadvertently pointed her toes to the left, so future crews were reminded to ensure their toes pointed up before descending the slide. Fortunately, her flight was delayed for a couple of months for other reasons, which gave her more time to recover and continue training. To her surprise, Seddon was able to fly the mission **[11]**

- *One-g trainer for the Orbiter cabin (01G-CAB)*
  <u>Flight Deck</u>: This trainer included Orbiter displays and controls, representative of the flight article, which were limited in use or, like the hand controller, functional for positional adjustment only. The four flight deck seats replicated the flight vehicle, and the Sky Genie™ device was used for emergency crew egress training. The CCTV systems supported systems training and featured two 10-inch color monitors, allowing the crew to view live video from any of the CCTV around the FFT payload bay.
  <u>Middeck</u>: Up to three MS and one instructor could be seated in this area and there was capacity for additional seats, a treadmill and biomedical attachments if required. The switches on the middeck were replicas of those in the flight vehicles, and the middeck could be fitted with an airlock featuring working hatches and systems, the post-*Challenger* crew escape system (slide-pole), a functional side hatch, middeck accommodation rack (the galley), and stowage for TV equipment, locker trays, cameras, etc. Three or four sleep stations could also be fitted, but the Waste Management Compartment (WMC) included was not a functional model. Middeck lockers were configured according to the mission in the forward and aft positions, with fully operable doors, latches, hinges, and trays and padding cushions, giving the crew an accurate representation of the locker layout and content they would find on orbit. It also gave an accurate representation of the locker numbering systems used on the vehicle to locate specific items from the crew manifest documents. **[12]**
- *One-g trainer for payload (01G-PLB)*
  <u>Payload Bay:</u> This was used for fit checks and orientation simulations of specific payload configurations. With flight-like thermal blankets, a mock-up winch and lighting, it could also be used for crew familiarization. A non-functional RMS was mounted on the port side, while the airlock could be mounted externally in the payload bay and interfaced with tunnel adapters. Other mock-ups that could be fitted into the payload bay simulator included the SpaceHab or Spacelab pressurized modules, unpressurized pallets and, from the mid-1990s, the Orbiter Docking System (ODS)
- *Precision Air-Bearing Floor (PABF)*
  Replicating EVA on Earth is impossible without assistance. A number of simulators have been developed over the years to facilitate this. The PABF provided a two-dimensional simulation (and three degrees of freedom) of a microgravity environment. A polished metal surface measuring 32 x 24 ft. (10 x 7 m) was used to teach and develop mass handing techniques by using a thin cushion of air, similar to an air-hockey table, to 'float' heavy objects over its polished surface.

- *Partial Gravity Simulator (PGS, also known as POGO)*
  This simulation combined servos, air-bearing and gimbals to simulate a reduction in gravity and was used by the astronauts to evaluate their ability to overcome tasks in simulated partial or microgravity conditions.
- *PDRS Deployment/Berthing Trainer (RMS TRNR)*
  Working from a mockup of the Aft Flight Deck and RMS work station, crews could simulate using the RMS to move objects (including air-filled balloon mock-ups of payload) around the payload bay to rehearse the deployment or retrieval of payloads.
- *Spacelab Simulator (SLS)*
  Functional mock-ups of the Spacelab pressurized laboratory module.

Building 16 (and 16A) Shuttle Avionics Integration Laboratory

- *Shuttle Avionics Laboratory (SAIL)*
  This was the only facility in which actual Orbiter hardware and flight software could be integrated and tested in a simulated flight environment. The set up consisted of an avionics mock-up, designated OV-095, contained within a basic skeleton shape of a real Orbiter. However, its electronics were identical to the flight vehicles, so it was sometimes preferred to the dedicated training simulator as a training device. Operated for the whole of the Shuttle program up to 2011, many of the 1978 astronauts were the first to work in this facility, both prior to STS-1 and in support of other missions.
- *Shuttle Engineering Simulator (SES, now Systems Engineering Simulator)*
  The SES has been in continuous operation since it was installed in 1968 and supported real-time man-in-the-loop simulations for Shuttle for many years. The Orbiter forward cockpit mock-up was located in East High Bay of Building 16 and was used for the two main areas of operation entry and on-orbit simulation, using a range of digital computers. The other mock-ups in this area included an Orbiter aft section and an MMU station. On-orbit operation could support station docking and berthing, payload handling and deployment, MMU operations and several other devices. Accurate representations out of the window, thanks to ever-improving computer programs and graphics, linked the SES to SAIL.
- *Orbiter Guidance and Navigation Simulator (GNS)*
  This was linked to the fixed- and motion-base simulators in Building 5. Originally, this was to be used for guidance and navigation issues, but following the 1986 Rodgers commission enquiry into the *Challenger* accident, the GNS was upgraded to a fully-fixed-base simulator as a part-task trainer to assist pilots in mastering the navigation training or flight techniques as required.

Building 17 Space Food Systems Laboratory

- *Selection and preparation of food for the crew menus.*
  Inside this simulator is a test kitchen, a food processing laboratory, a food packing laboratory and an analytical laboratory. In this building, a dietitian and nutrition team helped the astronauts to select items from the menu, which utilized eight types of food processing techniques: rehydration, thermo stabilization, irradiation, intermediate moisture, natural form, fresh, refrigerated and frozen.

Building 29 Weightless Environment Training Facility

- Originally called the Neutral Buoyancy Trainer (NBT, later WETF), this included mock-ups of the Orbiter middeck crew compartments, an EVA airlock, the scientific airlock (Spacelab), Instrument Pointing Systems, IUS tilt cradle, and mock-up payload bay doors and Ku band antenna. The round portion of this building was used for centrifuge training for Gemini and Apollo crews. The WETF was superseded by the much larger NBL at the Sonny Carter facility just north and offsite of JSC.

Fig. 8.6: [left images] WETF, Building 29, JSC (Image from the AIS archive). [right] STS-41D EVA crew and TFNG Mike Mullane (EV1, red stripes on suit) and Steve Hawley (EV2, no stripes) conducting contingency EVA training in the WETF assisted by support divers.

Building 30 Mission Control Center Houston

- *Mission Control Center (MCC)*
  Created in 1965, this facility became famous through its radio callsign of 'Houston', although that city is about 30 miles (48 km) north of the JSC site where MCC is located. The original Mission Operations Control Rooms (MOCR, pronounced "mohker") used during the Gemini and Apollo era were renamed Flight Control Rooms (FCR, pronounced "flicker") for the Shuttle program. FCR-1 was used for the first Shuttle missions, while FCR-2 was used during classified DOD Shuttle missions. From the mid-1990s, a new five-story extension was built, with two new control rooms ("White" and "Blue") developed for the space station made operational. The older FCR-2 was used in tandem with the White control room until 1996 and then the "White" control room was used by itself until the end of the Shuttle program. Teams of controllers manned the consoles around the clock during each Shuttle mission and one of the assignments for both new and veteran astronauts was to man the famous Capcom console, talking directly to the flight crew. Capcoms were assigned to an FD team for Ascent/Entry/Orbit 1, 2, 3 or Orbit 4 (the planning shifts). Capcoms from the 1978 selection on console during key events in the program included: Dan Brandenstein (1981 launch Capcom, STS-1 *Columbia*); Dick Covey (1986 launch Capcom, STS-51L *Challenger*); and Shannon Lucid (2011 Lead Capcom Planning Shift STS-135 (*Atlantis*), the final mission in the program).

- *Payload Operations Control Center (POCC (Spacelab) until 1990 when it moved to MSFC)*
  The POCC was operated by flight controllers and researchers during Spacelab missions (except for Spacelab-D1 in 1985, see below), where they received and analyzed data from the experiments onboard the Orbiter, directed the science operations during the mission, and liaised with controllers in the adjacent MCC. From 1990, this facility was moved to MSFC, in Huntsville, Alabama.

Building 33 Space Environment Simulation Test Facility

- *ECLSS Test Article, JSC.*
  This facility provided a real-time, crew-in-the-loop engineering simulator. It was used to test changes with existing systems and for engineering analysis, in this case for the ECLSS of the Shuttle Orbiter.

Building 35 Guidance and Navigation Simulation Facility

- *Orbiter Guidance and Navigation Simulator (GNS)*
  Believed to have housed a fixed-base crew station of the SMS.

Building 45 Project Engineering

- *Virtual Reality*
  An immersive environment facility that provides real-time, integrated EVA/
  robotics procedures and development training. This was just coming online
  during the training for the STS-61 crew, including TFNG Jeff Hoffman, but
  grew in importance as the assembly of the ISS approached.

Building 259 Astronaut Selection and Isolation Quarters

In 1967, this building located at the rear of JSC was used as a warehouse, but
was modified in the 1980s to support the Shuttle program. Many astronauts looked
upon this inconspicuous building, separated from the more well-known buildings
on the site, as a very special place in which they shared many unforgettable
moments together as a crew, including sharing a dinner with their spouses prior to
departing to KSC for the launch. History may have echoed around the halls and
rooms in Building 259, including its use as the selection office that processed their
applications to join the program, but time caught up with the aging building and a
new facility (Building 27) was constructed in 2004.

OTHER NASA LOCATIONS

Not all the Shuttle training could be accomplished on site at JSC, requiring the
astronauts and crew to travel to other NASA field centers or to locations across the
United States.

Ellington Field, Houston

- *KC-135.* This military version of the Boeing 707 was NASA's reduced gravity aircraft (known as the "Vomit Comet") from 1973. There were two stationed at Ellington Field (N930NA and N931NA), both of which were
  retired in 2004.
- *T-38.* Since the 1960s, NASA has operated a fleet of (updated) T-38s, which
  are used for astronaut transport between sites, as a proficiency trainer, and
  as chase aircraft.

NASA Forward Operating Location, El Paso, Texas

- *Gulfstream II Shuttle Training Aircraft (STA).* Four of these modified aircraft were used to duplicate the Shuttle's approach profile and handling
  qualities, giving Shuttle pilots accurate simulations of the Orbiter's descent
  prior to attempting the task on a live mission. Operations were completed at
  White Sands Space Harbor, New Mexico and at the Shuttle Landing Facility

(SLF) at KSC in Florida. These aircraft (tail numbers N944NA (s/n 144), N945NA (s/n 118), N946NA (s/n 146), and N947NA (s/n 147) were also used to assess weather conditions prior to Shuttle launches and landings. All four aircraft were finally retired in August 2011.

Sonny Carter Training Facility (SCTF)

- Site of the Neutral Buoyancy Laboratory, featuring a 202 ft (62 m) long by 102 ft (31 m) wide and 40 ft 6 in (12.34 m) deep diving tank holding 6.2 million gallons (23.5 million liters) of water. This is large enough to contain full-sized replicas of the ISS modules, payload and visiting vehicles, but not the full-sized truss structure. When this facility was opened in 1995, the second Hubble Service Mission crew (STS-82), including TFNG Steve Hawley, were the last to train heavily in the water tank at MSFC and conducted their final sims in the new tank at the SCTF just north of JSC.

Goddard Space Flight Center (GSFC), Greenbelt, Maryland

- *Hubble servicing training (lead center)*
- *Spartan free flying platform (lead center).* The POCC at Goddard monitored all free-flying (satellite) systems delivered, retrieved or serviced by the Shuttle.
- *Compton Gamma Ray Observatory (lead center)*

Kennedy Space Flight Center (KSC), Florida

- *Operations and Checkout Building KSC* (O & C Building)
- *Launch Complex 39, KSC* (launch simulations, and escape and fire training)
- *Launch Pad 39A* (simulated pad ingress/egress training; simulated countdown)
- *Launch Pad 39B* (as above)
- *Shuttle Landing Facility (SLF)* From 1984, the SLF became the preferred and primary landing site for most (but not all) missions.

Marshall Space Flight Center (MSFC), Huntsville, Alabama

- *Neutral Buoyancy Simulator (NBS)* Constructed in 1968 in Building 4705, this 75 ft (22.86 m) diameter, 40 ft (12.19 m) deep water tank held 1.4 million gallons (5.2 million liters) of crystal clear water and was the primary training facility for large-scale EVA by engineers and astronauts until the Neutral Buoyancy Simulator (NBS) was opened at the Sonny Carter Training Facility near JSC in the mid 1990s.

- *Payload Operations Control Center (POCC).* Moved from JSC in 1990 and was renamed the Spacelab Mission Operations Center to control all Spacelab missions (Except Spacelab-D1 and D2 which used the Space Operations Center of the German Institute of Aviation and Spaceflight Research and Development (DFVLR), at Oberpfaffenhofen near Munich in Germany.
- *Shuttle ET; SSME, IUS payload and related crew training (lead center)*

Additional locations

- *Ellington AFB/KSC/Edwards AFB.* Shuttle Training Aircraft training locations.
- *Dryden Flight Research Center, Edwards AFB, California.* Primary landing facility for the early missions. From 1984, it became the primary alternate landing site in the event of a non-return to KSC.
- *White Sands Space Harbor, White Sands, New Mexico.* A back-up landing site, used only once in the program (STS-3, March 1982)
- *Martin Marietta Plant, Denver, Colorado.* Prime contractor for the MMU, where a simulator was provided for training on the MMU flying and operational techniques. Astronauts wearing the simulated MMU were able to 'fly missions' against a full-scale mock-up of a portion of the Orbiter, using controls similar to the flight MMUs to maneuver the unit in three straight-line directions and in pitch, yaw and roll.
- *Hamilton Sunstrand facilities, Nassau Bay near JSC.* Hamilton Sunstrand of Windsor Locke, Connecticut, were the prime contractor for the Shuttle spacesuit and PLSS. At this facility, Shuttle EMU units were stored, repaired, tested and prepared for flight.
- *SpaceHab Inc., Webster, Clear Lake.* On Gemini Street, offsite but close to JSC. This provided a location to develop and produce the pressurized SpaceHab middeck augmentation module flown from 1993 (STS-57). A SpaceHab mock-up was also available at Ames Research Center.

DEPARTMENT OF DEFENSE (DOD)

The DOD originally intended to fly its own Shuttle missions from both KSC and Vandenberg AFB. The first DOD payload was carried on STS-4 in June 1982, the fourth and final OFT. Though many more were planned, in the end only ten dedicated DOD Shuttle missions were flown, all within the NASA STS program, between January 1985 and December 1992. The interesting point here is that at least one serving military member of the 1978 selection was assigned to each of the ten mostly classified missions, creating yet another unique fact in Space Shuttle history.

TABLE 8.2: GROUP 8 DOD SHUTTLE MISSION ASSIGNMENTS 1985-1992

Mission DOD	STS	Year flown	Primary DoD payload	Commander	Pilot	Mission Specialists
1	51C	1985	Magnum (USA-8)		Shriver	Onizuka, Buchli
2	51J		2x DSCS III (USA-11 &-12)			Stewart
Planned	*62A[1]*	*1986*	*Teal Ruby*			*Mullane, Gardner, D.*
3	27	1988	Lacross/Onyx (USA-34)	Gibson, R.		Mullane
4	28	1989	SDS-2 (USA-40)	Shaw		
5	33		Magnum (USA-48)	Gregory, F.		
6	36	1990	Misty (USA-53/AFP-731)	Creighton		Mullane
7	38		SDS-2 (USA-67)	Covey		
8	39	1991	AFP-675 (various)	Coats		Bluford
9	44		DSP	Gregory, F.		
10	53	1992	SDS-2 (USA-89)	Walker, D.		Bluford

[1]STS-62A (planned July 1986 launch) was to be the first polar orbit Shuttle mission, and the maiden Shuttle launch from Space Launch Complex 6 (Slick 6), Vandenberg AFB. It was canceled in the wake of the *Challenger* accident earlier that year.

As all the crews for the Shuttle DOD missions comprised current or former members of the U.S. military, it is obvious that details of their assignments, missions and experiences have been limited over the years since they flew, and will probably remain so for some time to come. When the authors researched their previous volume of astronaut selections in 2015, which included the selections for the USAF MOL program, some details of that program, crew training and hardware had recently been declassified, 50 years after the program was terminated. Significant gaps still remain in the information about the MOL program, and primarily the activities of the astronauts assigned to it. **[13]** The same is true for the astronauts assigned to the ten DOD Shuttle missions and most of their activities on those flights.

From what the authors can determine, most of the Shuttle Orbiter training for the DOD missions was conducted at JSC, which is sensible since most of the training hardware and software was based there. However, there were other training sessions in 'other locations' around the country. There were also visits to the various payload manufacturers to become familiar with the hardware they were to carry, as well as training sessions with the flight controller teams at Sunnyvale as part of the classified preparations for each DOD mission.

- *SCF Satellite Control Facility (DOD missions), Sunnyvale, California* (USAF)
- *Defense Language Institute (DLI), Monterey, California (U.S. Army).* Russian language studies for training in Russia for Mir residency missions

Contractor locations for DOD shuttle training and familiarization included:

- *Lockheed Martin (Lacrosse or Onyx radar imaging satellite deployed from STS-27 and Misty recon satellite deployed during STS-36).*
- *TRW (contractor for Magnum SIGINT spy satellite launched on STS-51C and STS-33).*
- *General Electric/Hughes Aircraft Company [subsequently Lockheed Martin Space Systems]. Contractor for the Defense Satellite Communication Systems (DSCS III) 2 launched on STS-51J.*
- *Rockwell International Space Division (Teal Ruby, intended payload for STS-62A).*
- *Hughes Aircraft (contractor for second generation Satellite Data System (SDS) military comsats deployed from STS-28, 38 and 53).*
- *TRW/Northup Grumman Aerospace Systems (Defense Support Program recon satellite, deployed from STS-44)*
- *STS-39 carried a range of six payload 'packages' designated AFP-675, with instruments provide by the Phillips Laboratory; USN Research Laboratory; Los Alamos National Laboratory; and University of Florida*

FOREIGN LANDS

The members of the 1978 selection were as much pioneers of many of the astronaut roles and assignments as were their predecessors. In particular, they expanded the international astronaut training program, following on from the groundbreaking work done during ASTP in the early 1970s. ASTP was a single mission that involved the final Apollo spacecraft joining up in Earth orbit with cosmonauts of the Soviet Union flying a solo Soyuz, though many more joint missions were proposed and planned. The ASTP program was seen as a period of détente, but changes in the wider world away from the space programs of both nations put paid to follow up missions for two decades. Despite this, the premise of the Shuttle program and major cooperative ventures with Canada (the supplier of the RMS), Europe (ESA/Spacelab) and Japan (Spacelab J), together with their partnership within the *Freedom* (subsequently International) Space Station Program, meant that crews assigned to missions flying associated hardware would travel to foreign countries for at least some training and familiarization sessions. The inclusion of international PS, and later the inclusion of the new Russia that emerged from the breakup of the Soviet Union into the ISS program, expanded this activity globally. Today, many astronauts travel the world to train with colleagues across the globe.

Canada

- *Spar Aerospace Ltd., Weston, Ontario, Canada.* This was the primary location for early RMS training.

Europe

- *Bristol Aerospace Systems (BAe) Bristol, England.* Visited by the STS-61 crew for familiarization with the replacement Hubble solar arrays they were to fit on the first servicing mission.
- *European Astronaut Center, Cologne, Germany.* The mission management for the 1985 Spacelab D mission was handled by the Space Operations Center of the DFVLR, at Oberpfaffenhofen, near Munich, Germany (now the DLR Columbus control center).

Japan

The payload crew of Mark Lee (PC), Jan Davis and Mae Jemison visited Japan for STS-47/Spacelab J training multiple times over three years prior to the mission. The Orbiter crew of CDR Hoot Gibson, PLT Curt Brown and MS-2/FE Jay Apt were assigned to the flight about 12 months prior to launch and, as a result, "trained in Japan from May 2 to 11, 1992," as Gibson explained. "We trained in Tokyo and Tsukuba, but had some time for sightseeing. We rode the multiple trains that went to Mount Fuji and we rode the subways around Tokyo. There were also evening receptions because this was the first time the Japanese had met an orbiter crew (CDR/PLT/MS-2). [We] also debriefed in Japan after the mission during November 14 to 20, 1992, staying in Tokyo again and travelling on the Bullet Train to Kobe for the debriefing, before spending a few days relaxation prior to heading back home." **[14]**

Russia

- *Cosmonaut Training Center named for Yuri Gagarin (TsPK), Moscow.* This was used for Mir resident crew training and Shuttle visiting crew familiarization training. **[15]**

Contractors

In the course of their time with NASA, members of the 1978 class of astronauts followed the tradition of earlier groups in visiting major and smaller contractors across the United States. They met workers as representatives of the Astronaut Office and the "public face of the space program," or as members of a specific crew both during training and after the mission was completed, to offer a personal thank you to the workforce for their efforts and dedication.

Major Shuttle contractors visited included:

- Rockwell International, (now Boeing) Downey, California (Orbiter vehicle)
- Rocketdyne Division of Rockwell International, Canoga Park California (SSME)

- Martin Marietta Corp., Michoud Aerospace, New Orleans, Louisiana (ET)
- Morton Thiokol Chemical Corp., Brigham City, Utah (SRB)
- Spar Aerospace, Canada (RMS robotic arm)
- Hamilton Standard (now Hamilton Sundstrand), Windsor Locke, Connecticut (Shuttle EMU)
- Martin Marietta Denver plant, Denver, Colorado (prime contractor of the MMU).

This represents just a small selection of the dozens of contractors stretched across the United States, big and small, visited by members of the TFNG in their preparations for their missions on the Space Shuttle.

PUTTING THEORY INTO PRACTICE

So what did all this mean? To give just one early example of Group 8 mission training, we need only to turn to the STS-7 Post-Flight Training Report, which recorded the first four TFNG to go through the system and complete a mission. **[16]**

Standalone training was conducted in the SMS with the objective of supporting the earlier STS-5 integrated sims. "The CDR (Crippen) and PLT (Hauck) attended all non-payload related lessons and the majority of the payload lessons. MS-2 (Ride) functioned as the Flight Engineer and attended essentially all lessons. MS-1 (Fabian) and MS-3 (Thagard) attended only payload and orbit timeline lessons." Crew systems training began in July 1982 and was completed on June 8, 1983 (just ten days before launch). Contingency EVA training had to be adjusted halfway through the training program due to Norman Thagard joining the crew in December 1982 and replacing Robert Crippen as the original STS-7 EV1 crewmember. Thagard's preparation was the shortest EVA training time period (six months) of any crewmember at the time.

The STS-7 crew supported the STS-5 deployment integrated sims, allowing them to acquire additional knowledge about the PAM and Hughes satellite systems and operations. This meant that their training for STS-7 could be further developed and refined. The STS-7 crew were also the first to require Prox Ops training. Despite not having the luxury of previously developed on-orbit procedures, their training was a success, enabling advancements in generic Prox Ops training by using the STS-7 training as a guide in redesigning the program for future flights.

The crew was also the first to receive PDRS training. There were several problems with the training loads in the SMS for STS-7, due in part to the slip in the launch of STS-6. This necessitated a number of recommendations to correct the issue for future missions. In payload briefings, conducted together with Bob Crippen, Hauck, Ride and Fabian received four hours training each for PAM-D briefing, Palapa B1, Anik D, and OSTA 2, and two hours each for the Monodisperse

Latex Reactor (MLR) experiment and the seven GAS experiments, as well as 16 hours training each on CFES.

During their year of mission training, Rick Hauck would log 1,039 hours training for STS-7, Fabian accumulated 719 hours, and Ride 988 hours, with Thagard acquiring 334 hours in six months. A total of 162 hours for Hauck, 132 hours for Fabian, 154 hours for Ride and 97 hours for Thagard were logged in the series of Integrated Simulations (split into ascent aborts; orbit procedures Day 1, 2, 3, 4 and 5; de-orbit prep; entry; a 58-hour simulation; and a contingency orbit Day 6).

T- MINUS AND COUNTING

As shown in this chapter, the training for a Shuttle crew, especially in the early years when the first members of the 1978 group flew, was both challenging and demanding, from the time that they arrived at NASA, through the Ascan training program, the technical and support assignments, being assigned to a crew, and preparing and flying a mission. Of course, the work did not stop there, because invariably following the mission came the post-flight debriefings, medical examinations, press conferences, report writing, interviews, public tours, homecoming celebrations and finally, after all the dedication and intensity of preparing for and flying the mission in the media spotlight, the reality of returning to Earth and catching up on all the household chores that had remained untouched while the celebrated family member was 'out of town' for a while!

With their mission training behind them, it was at last time to put all the theory and practice into action and go fly the mission. It was time for the TFNG to earn their astronaut gold pins.

References

1. *Report of the Presidential Commission on the Space Shuttle Challenger Accident*, June 1986, p. 166.
2. Reference 1, p. 201.
3. Commission Work Session, Mission Planning and Operations Panel, JSC, April 1, 1986, p. 198.
4. Mike Mullane, NASA Oral History, January 24, 2003.
5. Steve Hawley interview with author David J. Shayler, October 2019.
6. Kathy Sullivan, NASA Oral History, May 28, 2009.
7. **Shuttle Flight Crew Training**, Carl B. Shelly, JSC Flight Operations, November 1980; also a Memo from Robert K. Holkan, Chief of Training Division (mail code DG6) to the Lead, Mission Planning and Operations Team/STS-51-L Data Design and Analysis Task Force, April 11, 1986 (initiated by Frank Hughes, DG6). Copies on file, AIS Archives.
8. **The Hubble Space Telescope: From Concept to Success**, David J. Shayler with David M. Harland, Springer-Praxis, 2016, pp. 213–224.
9. George Nelson, NASA Oral History May 6, 2004.

10. NASA News Release 93-032, May 4, 1993.

11. **Go for Orbit, One of America's First Women Astronauts Finds Her Space**, Rhea Seddon, Your Space Press, 2015, pp. 413–416.

12. For an explanation of the middeck locker numbering systems see **Linking the Space Shuttle and Space Station**, David J. Shayler, Springer-Praxis, 2017, pp. 176-181.

13. **Last of NASA's Original Pilot Astronauts: Expanding the Space Frontier in the Late Sixties**, David J. Shayler and Colin Burgess, Springer-Praxis, 2017.

14. Emails to David J. Shayler from Robert "Hoot" Gibson, August 20 and 22, 2019.

15. For details of on Mir training by NASA astronauts, see *Cosmonaut Training for Astronauts*, in **Russia's Cosmonauts: Inside the Yuri Gagarin Training Center**, Rex D. Hall, David J. Shayler and Bert Vis, Springer Praxis 2005 pp. 274–289; also for Shuttle-Mir docking missions, see **Linking the Space Shuttle and Space Stations: Early Docking Technologies from Concept to Implementation**, David J. Shayler, Springer-Praxis, 2017.

16. **STS-7 Post-Flight Training Report**, Mission Operations Directorate Training Division, TD225/A192, NASA LBJ Space Center, October 13, 1983. Copy in file AIS archives.

9

NASA's All-Electric Flying Machine

"Looks real good over here... Everything looks real good...
They're down, pick up your feet.
Five, four, three, two, one, touchdown...
Welcome home Columbia. Beautiful, beautiful."
Jon McBride, T-38 pilot Chase 1,
April 14, 1981, during the landing of *Columbia* (STS-1). [1]

For the sake of emphasis, it is worth repeating the well-chosen words of Mike Coats when he spoke about the heterogeneous mix within the TFNG class. "Most of the previous classes had either been pilot astronauts or scientist astronauts, including medical doctors. In our class, we had 15 pilot astronauts, and we had six women. We had engineers, scientists, medical doctors – a real mixture of folks. And the age difference was pretty large too. It was fascinating to see the interaction of that class of 35 people, because nobody was really senior to anybody else, and we outnumbered all the astronauts that were already here. There were only 29 astronauts when we got here, and we were 35 more.

"Suddenly we dominated, but they were glad to see us because they had a lot of work they needed us to do. It was really fun for us because there were enough Apollo astronauts still left over that they were able to mentor us, and we really enjoyed that. Of course we were still developing the Space Shuttle and learning how the Orbiter was going to operate. We got involved right away in developing operational procedures and flight rules for this amazing new vehicle; it was really a special time." [2] Though their number initially made a difference in the Astronaut Office, the Thirty-Five New Guys (TFNG) would not remain the latest additions for long, because another selection was being organized even as the Class of 1978 were completing their Ascan program (see sidebar: *Nineteen More "New Guys"*).

© Springer Nature Switzerland AG 2020
D. J. Shayler, C. Burgess, *NASA's First Space Shuttle Astronaut Selection*,
Springer Praxis Books, https://doi.org/10.1007/978-3-030-45742-6_9

Years later, Ox van Hoften recalled the challenges of those early years, as they were waiting for a flight assignment with new astronauts already coming in behind them. "As the program wore on, we got put into groups working for various people throughout the thing. When you don't know where it is and you don't know how you're graded and whatnot, it got to be very, very awkward, because some people got what looked like good jobs and other people got what looked like not-so-good jobs, and that was hard."

NINETEEN MORE "NEW GUYS"
While still unflown by the summer of 1980, the TFNG were not lacking for work as they supported the activities necessary to get the Shuttle off the ground and through the first half-a-dozen missions to qualify the system, prior to the more extensive missions in which they were expected to be involved. Though technically still rookies, the TFNG would soon be followed by a new group of Ascans, who were scheduled to arrive shortly. In anticipation of a greater flight rate, as well as the natural attrition of veteran astronauts once the Shuttle started flying, the call for the ninth astronaut selection was issued on August 1, 1979. Later that same month, the Ascan program for the TFNG ended officially and earlier than originally expected, making them 'Silver Pin Astronauts'. [3] Desiring to expand the diversity of the group even further, NASA issued an additional call just one month later, encouraging those of Hispanic origin to apply for astronaut training. [4] By December 1979, a total of 3,278 applications had been received for consideration for the agency's ninth group of astronauts. [5] Following the selection process, and with the first Shuttle flight less than a year away, 19 new candidates were named by NASA on May 29, 1980. [6] The new group would report to the Johnson Space Center (JSC) in Houston to begin their year-long Ascan training program on June 30. [7]

In addition, as part of an agreement with the European Space Agency (ESA), who would be providing the Spacelab science laboratory for the Shuttle, two ESA astronauts also joined Group 9 for Mission Specialist (MS) training, the first non-American candidates to undergo NASA astronaut training. [8] This trend continued for the next 24 years, with international astronauts selected to train as MS for future Space Shuttle crew assignments as part of the Group 14 (1992), 15 (1994), 16 (1996), 17 (1998) and 19 (2004, the last group selected for flights on the Space Shuttle) NASA astronaut selections, supporting future Space Shuttle crewing through to the end of the program in 2011.

With their Ascan training now behind them, technical assignments allocated and preparations towards the first Shuttle launch progressing, the TFNG began to immerse themselves in their various assignments as they supported efforts to get the Shuttle into orbit. As the flight planning evolved, the complexity of each individual mission clearly demonstrated the wisdom of the decision to expand the Astronaut Office substantially, not only to fulfil the flight crew positions, but also to assign a significant number of astronauts to each flight in a variety of supporting roles (though without formally identifying them as a 'support crew'). These assignments helped the new astronauts to expand their understanding of the Shuttle system and gave each of them the opportunity to get up close and personal with the hardware and procedures they would expect to be using on their own missions.

As the Shuttle's first orbital mission drew closer, the TFNG waited patiently for news of their own flight assignment. By 1981, three years after arriving at JSC and with only a limited number of veteran senior astronauts remaining active, they all hoped it would not be long before they received the news they were all eager to hear. But first, the Shuttle had to prove it could fly well, fly safely, and then do so again and again. It was time to put all those years of design, debate and development into practice and begin the Orbital Flight Test (OFT) program.

The real role of an astronaut
Presenting detailed accounts of every assignment or mission by the members of the 1978 selection is beyond the scope of this current work. Indeed, it would take a library of titles to do this justice. Therefore, readers are respectfully directed to the Bibliography for further reading concerning the missions of the TFNG. Here, and in the following three chapters, we summarize their assignments as a group, in both the often-overlooked mission support roles and in the subsequent flight crew assignments for each of the missions that included one or more of the group. In fact, this begins with the very first mission in 1981 and ends three decades later in 2011 at the close of the Shuttle program, demonstrating the key role that the TFNG played as a group in encompassing the whole Shuttle orbital flight program from start to finish. These summaries present the significant assignments and departures from within the group, as well as mission milestones.

The image that most people have of an astronaut is of someone who spends a considerable amount of time in space, on mission after mission. This could not be further from the truth. History tells us that, as a group, the 35 members of the 1978 selection collectively spent approximately 978 days in space in total (with some flying together on the same mission), spread across 16 years and one month (or 5,870 days) between June 1983 (STS-7) and July 1999 (STS-93). That works out at less than 17 percent of group's total NASA careers actually spent in space. Shannon Lucid, the longest serving and most experienced member of the group in terms of flight time, accumulated 33 years and six months of service in her NASA career between July 1978 and January 2012. Her flight record of 224 days was achieved across five missions, over an 11-year period between June 1985 and September 1996. Clearly, the time astronauts (or cosmonauts) actually spend in space is only a fraction of that spent on the ground, and the bulk of their role is taken up with training, simulations, endless meetings, outreach activities and both domestic and (more recently) international travel. Examining the records of each of the TFNG underlines this, revealing a wealth of assignments and roles spread across their years at NASA in between their occasional assignment to a flight crew, starting with their assignments in the fall of 1979 in support of the very first Shuttle mission.

SUPPORTING THE FLIGHT TESTS

The main support assignments for STS-1 were announced in an Astronaut Office memo (CB-79-054) dated October 23, 1979. Assigned to the prime crew for this important and historic mission were John W. Young (Commander, or CDR) and Robert L. Crippen (Pilot, or PLT). They would be backed up by Joe H. Engle (CDR) and Richard H. Truly (PLT), who were also in training to fly the second test flight, STS-2, a few months later.

Quite naturally, Young was the primary point of contact within the Astronaut Office for all matters related to the first flight. but seven members of the TFNG were assigned in support roles for this crucial mission. Detailed to the Capcom group in the Mission Control Center (MCC) at Houston, together with some of the veteran astronauts, were Dan Brandenstein ('Silver' team Capcom for launch, replacing the retired Ed Gibson), Rick Hauck (back-up entry Capcom 'Crimson' team) and Jim Buchli (back-up orbit Capcom 'Bronze' team). Detailed to the support team down at the Cape were Terry Hart, Jon McBride, Steve Nagel and Loren Shriver (the first Astronaut Support Person, ASP).

The primary T-38 chase plane crews for STS-1 were subsequently identified as pilot Jon McBride (in a dual assignment) with photo/observer George 'Pinky' Nelson (prime Chase 1), and pilot Dave Walker with photo/observer Mike Mullane (back-up Chase). The Capcom on duty for the chase plane crews during the glide and landing phase at the end of the mission was Rick Hauck (Group 6 astronaut Joe Allen was assigned to the Capcom console in MCC, talking directly to Young and Crippen in the returning *Columbia*). Hauck would communicate with the T-38 chase planes as they rendezvoused with *Columbia* at 16,000 ft (4,900 m). **[9]**

The same October 1979 Astronaut Office memo detailed another group of astronauts who were assigned to a team headed by Paul 'PJ' Weitz, the Primary Point of Contact (PPOC) for creating and developing a generic support role for the whole OFT program. Included in this list were: Mike Coats, responsible for approach, landing and rollout issues; Guion Bluford, on the Configuration Control Board (CCB) and for the Technical Status Review (TSR); Don Williams and Anna Fisher, handling crew equipment, crew station and crew station improvement issues, with Williams given a concurrent assignment related to the Displays & Controls (D&C); and Ox van Hoften, responsible for issues regarding engineering simulation coordination. Due to the importance of the first mission to both the program and the future of American human space flight, other members of the TFNG would be assigned to support roles closer to the mission, either for contingency situations or actively participating during the flown mission.

"We Want to Dust it Off First"
STS-1/OFT-1 Columbia. *Launched April 12, 1981; landed April 14, 1981*

Following four years of construction, the first operational Orbiter, OV-102 *Columbia*, had arrived from primary contractor Rockwell on top of the Shuttle Carrier Aircraft (SCA) on March 25, 1979, just a few weeks after the TFNG had celebrated the first anniversary of their selection to NASA. Delays in qualifying the Space Shuttle Main Engines (SSME) and completing the application of 30,000 insulation tiles on *Columbia*'s external surfaces[1], together with on-going issues with the SSME, meant that the intended first launch of the Shuttle had to be pushed back from 1979 to 1980, and eventually to 1981. By the time the vehicle was ready to fly, she had spent 750 days at the Cape (610 days in the Orbiter Processing Facility (OPF), a further 35 days in the Vehicle Assembly Building (VAB) and 105 days out on Pad 39A) before finally lifting off. These delays had a knock-on effect on the maiden flights of the TFNG, but conversely offered the opportunity for the 35 rookie astronauts to acquire further experience in supporting the initial Shuttle flights. They were at the forefront of preparing the vehicle to fly and developing its systems for subsequent flights (including their own), and gained valuable knowledge which would serve them well in their later careers.

Originally, the ascent Capcoms would have been Ed Gibson as prime and Dan Brandenstein as his back-up, but as Brandenstein recalled in 1999, reflecting on his good fortune to be launch Capcom on that first ascent, "when Ed Gibson retired, as opposed to pulling another experienced astronaut in to be the ascent Capcom, I inherited that position, and Terry Hart moved up to be my back-up. So that was kind of a thrill for me. To this day, I think I was more excited being Capcom on the first mission than I was actually flying my own mission. I still look back and think about it, listen to the tapes. I mean, I settled down, but you make

---

[1]When *Columbia* arrived at the Cape, approximately 8,000 individual tiles were still to be attached.

calls back and forth just to kind of check to make sure everything was working all right, and the first couple, it's real obvious that… I was pretty excited." **[10]**

Brandenstein went on to recall experiencing the differences between being a launch Capcom back in Houston, watching a launch live as it left the Cape, and participating in one. "I never saw a launch until STS-3, because I was Capcom also for ascent on STS-2. I still remember a bunch of the wives went down to the launch of STS-1, my wife being one of them, so all I saw of it was on TV. When she came back about two days later after the launch… she was still ricocheting off the ceiling. In all the years we'd been married, I'd never seen my wife half that excited. She was just something else. 'You wouldn't believe it. You just can't believe it. What you see on TV is nothing! You ought to be there. You won't believe it'.

"But I had to wait till STS-3 before I got to see one in person. And it is [exciting]. I think it's significant that I think it's more exciting watching one as a spectator than being on one. People always look at me like I'm smoking something. But really, I always [have] explanations. I call them pilot explanations for things. I can explain medical things in pilot talk, and it's probably not right, but it at least is a way of explaining it that satisfies me. But when you're watching one [a launch], you have no real responsibility, and it is noisy. You hear the popping and the cracking and the big long flames shooting out and everything like that, and you have no responsibility. I get a lump in my throat and chills up and down the spine and all that. But when you're on board, you're responsible for that baby, so you're checking instruments and you're making sure everything is working all right. You're not there to take it in; you're there to make it work. That's certainly a different perspective. I mean, don't get me wrong, I would never turn down a launch, the opportunity to go fly, to go watch one, but from a pure spectacle standpoint, the spectator point of view is more thrilling than the flying point of view."

After a two-day delay caused by a computer synchronization issue, *Columbia* finally left the pad for the first time on April 12, heading into space on the first of four planned test flights. Eight minutes later, *Columbia* was in orbit. For the next two days the Orbiter was put through its paces by a jubilant Young and Crippen, as they evaluated living and working in this new spacious (for now, with only two crewmembers on board) American spacecraft[2]. *Columbia* had launched on the 20th anniversary of Yuri Gagarin's first human space flight, and so much had been achieved in the two short decades separating the flight of the small, single-seat Vostok 'sharik' sphere and the huge, multi-seat Shuttle Orbiter. The prospects for the Shuttle, and for the TFNG, looked promising, even after just a two-day test flight.

---

[2] Crippen was on his first space flight, but for Young, who had experienced the cramped confines of Gemini and Apollo on his previous four journeys into space, the internal volume of the Shuttle – especially the virtually empty middeck – was an absolute joy.

Pinky Nelson was part of the Chase Team for the returning *Columbia* on that first mission, "Jon McBride and I were Chase 1, so it was our job to rendezvous with *Columbia* on reentry and then follow it down to the runway. My job was to photograph the tiles and to make the calls as it came down, in case the radar altimeter didn't work and things like that, [to] be able to call air speed and altitude. The training for that was amazing. Jon McBride and Dave Walker and then a couple of the staff pilots, Dick [Richard E.] Gray, especially, built this kind of empire around the chase program. There were just a ton of us, and every once in a while we'd be out at El Paso [Texas] and there'd be eleven T-38s lined up on the ramp. A lot of flying. So that experience was just great.

"So I got to see the launch of STS-1 from above the launch pad in a T-38. Jon and I were up orbiting… south of the launch pad and got to see the first launch, see it come off and go right by us. That was just really cool. I remember my response at that, as it went by, was 'I'll be damned! It worked'. We had so many problems getting STS-1 ready to go in the first place. So that was a really interesting experience. The chase team was terrific. McBride and the whole team did a good job.

"We almost blew it [though]. I mean, we'd been practicing with the radar folks from both Edwards [Air Force Base (AFB), California] and the radar organization in L.A. [Los Angeles, California], and the Shuttle was coming in and it was approaching and we were getting ready to go intercept it. We took off at Edwards and the first thing we did, we called the radar folks, and they said 'You're not going to believe this. We just lost all our power'. They were sitting in a dark room. They lost everything. They couldn't guide us. So then the folks at Edwards took over, and so the rendezvous was really kind of grab-ass, but we saw the Shuttle and Jon is a great pilot [and he] got us up to it. I had this camera in the back and Jon's flying around and I'm taking pictures. But we got the whole surface of the Orbiter photographed and Jon called out the landing and all that."

There were always contingency plans in place during Shuttle flights, especially for the first couple of missions. "I was Chase-4 on STS-1," said Robert "Hoot" Gibson, "so when *Columbia* launched on April 12, I was actually sitting in El Paso [Texas]… Chase-3 and Chase-4 [were] in El Paso. White Sands [Northrup Strip, New Mexico] was the Abort Once Around [AOA] strip, if they had had a problem that required an AOA. Dave Walker was Chase-3, so he was lead, and I was Chase-4. I had the TV cameraman in the back seat, and so we were sitting in El Paso, watching the launch. If they had declared an AOA, we would have hopped right in our T-38s and gotten right up on station over White Sands to chase them when they came back in. As soon as they got to MECO, Main Engine Cut-Off, we knew that they weren't going Abort Once Around. We hopped in our airplanes and flew to Edwards to be in place to cover them, just in case they had to land on the first three or four orbits. Jon McBride and Dickie [Richard E.] Gray were down at

Cape Canaveral for the launch, to chase them if they had to do an RTLS, Return to Launch Site. So then I wound up staying at Edwards. It was only a two-day mission, so we were at Edwards for two days. I was airborne over the alternate runway aim point, and Jon McBride and Dick Gray got to chase STS-1 when it came down to land."

As *Columbia* landed, Young joked with Capcom Joe Allen that he wanted the Orbiter rolled straight into the hanger, to which Allen replied "No, we want to dust it off first." Clearly, the first orbital flight of the Shuttle system had been an out-standing success. The landing of *Columbia* signaled the end of only the 77th trip into orbit since 1961, 32 of them by American astronauts. Witnessing this latest step in that adventure were 35 eager rookies who were supporting the develop-ments and hoping to take a similar ride themselves very soon.

This first flight had been planned as a 'minimum mission', sufficient to qualify the whole Space Transportation System (STS) from launch processing, through countdown, ascent, orbital operations, entry, landing and post-flight activities. Through it all, *Columbia* had performed magnificently, so much so that John

Fig. 9.1: Astronauts Sally Ride and Jim Buchli at the Capcom console in Mission Control during an STS-2 simulation (Image courtesy of Ed Hengeveld).

Young commented just after landing STS-1 that: "This is the world's greatest all-electric flying machine." A new era of human space flight beckoned and the TFNG were standing in line to participate, but as *Columbia* was taken back to the Cape on top of the SCA, it was time to prove it could be done all over again.

"First the Good News…"
STS-2/OFT-2 Columbia. *Launched November 12, 1981; landed November 14, 1981*

For the second flight of *Columbia*, all the Capcoms were from the TFNG, with Brandenstein again taking the prime role for ascent on the 'Silver' team backed up by Hart. Buchli was prime for the orbit 'Bronze' team backed up by Sally Ride (who was also identified as the primary Capcom during planned Remote Manipulator System (RMS) Flight Test Objectives), and Hauck was prime for the 'Crimson' entry team backed up by Nagel. [11]

An Astronaut Office memo noted that members of the STS-2 support crew had also been assigned dual technical responsibilities, in areas that the prime crew thought needed additional attention prior to their mission. [12] Terry Hart was assigned to the Flight Data File (FDF) for ascent/post-insertion and contingency Extra-Vehicular Activity (EVA), flight techniques for ascent, contingency EVA activities, and development of the Hewlett-Packard HP-41C™ hand-held alpha-numeric display calculator. Steve Nagel was assigned to FDF issues opposite to Hart covering deorbit preparations and entry, as well as entry flight techniques, entry data maneuvers and supporting the potential of contingency aborts to the Trans-Atlantic landing site in Rota, Spain, though he did not actually go there. Sally Ride was assigned to cover the overall coordination of STS-2 FDF inputs, specializing in on-orbit issues. She was also point of contact for the RMS (the robot arm) and the Office of Space and Terrestrial Applications 1 (OSTA-1) payload carried on the flight. Ellison Onizuka was prime ASP on this flight, with Steve Hawley as his back-up.

Two other TFNG were assigned as prime Chase 1 crew, though this was not mentioned in the Astronaut Office memo. Hoot Gibson was assigned as pilot, with Rhea Seddon (Gibson's future wife) flying in the back seat as photo/observer.

The primary objective of the second Shuttle mission was to continue the orbital evaluation of the Shuttle and its systems. Since this flight marked the first occasion that a manned spacecraft had been launched into orbit for the second time with a crew aboard, it was also a milestone in space flight history, demonstrating the Shuttle system's long-advertised capability to fly its vehicles repeatedly. Unfortunately, a problem with one of the three onboard fuel cells shortly after entering orbit meant that the planned five-day flight had to be curtailed to just two days. As a result, the crew's workload was reorganized to fit as much of the planned work into two days as possible. This included limited tests of the new Canadian-built RMS, which would become one of the most beneficial working aids of the Shuttle program and a specialization area for several members of the 1978 selection.

In fact, it was one of those who had worked so hard on developing the RMS systems that informed the crew of their shortened mission. On November 13, 1981, Bronze Team Capcom Sally Ride had to tell the STS-2 crew that they were coming home early: "First the bad news. Our plan is that we're running a minimum mission and you'll be coming home tomorrow... Think of that you got all of the good OSTA [science package] data and all of the RMS data and you just did a good job. We're going to bring you in early." [13] Joe Engle later commented during the post-flight press conference that while STS-2 may have been a minimum mission, it gave maximum mission accomplishment, achieving over 90 percent of the pre-mission objectives.

Columbia soaks in space
STS-3/OFT-3 Columbia. *Launched March 22, 1982; landed March 30, 1982*

The initial support assignments for the third mission were listed early, in an Astronaut Office memo dated July 21, 1981. The support crew was to be Dave Griggs, Pinky Nelson and Brewster Shaw, who would also be the back-up Capcoms. After the STS-2 mission had been completed they would be joined by Terry Hart, Sally Ride and Steve Nagel, who would occupy the prime positions on the Capcom console. Hart and his back-up Griggs were assigned to the ascent 'Ivory' team, Ride and her back-up Nelson to the 'Silver' orbit team, and Nagel and his back-up Shaw to the 'Crystal' entry team. [14] The Astronaut Office memo also noted that Dale Gardner was to assume management of the FDF for the Astronaut Office from STS-3 and subsequent missions, until he was assigned to a flight. Steve Hawley was prime ASP and he and Dan Brandenstein formed the End Of Mission (EOM) exchange crew. The T-38 Chase team members included TFNG astronauts Dick Covey and Ron McNair, the latter returning to light duties after his automobile accident in September 1981 (see sidebar: *McNair Injured*).

The STS-3 mission extended the duration of a Shuttle on orbit to eight days, the nominal flight time for the early missions. The crew was able to put the RMS through its paces in a series of hot and cold soaks in space, in temperatures of between -66 degrees C and +93 degrees C. As with most flights, the mission encountered small, niggling problems that prevented it from being considered a 100 percent success, but the OFT program was nevertheless progressing well with just one flight left.

Fig. 9.2: Astronaut Judith A. Resnik confers with Flight Directors Harold M. Draughon and Jay H. Greene at their console during Day 5 of the STS-3 mission. Another huddle in the background includes astronauts George D. Nelson and Sally K. Ride (Image courtesy of Ed Hengeveld).

THE FIRST FLIGHT-SUITABLE TFNG CANDIDATES

Confidence in the Shuttle system was growing following the success of STS-3 and, as it required about a year to train a flight crew, it was now time to select the crews for missions beyond STS-6. The Advisory Group on Shuttle Crew Selection proposed that Bob Crippen and Dick Truly should be promoted to command STS-7 and STS-8 respectfully, based upon their distinguished work on the first two missions as Shuttle PLT. As these would be satellite deployment missions and not dedicated science flights, the other seats would be open to the first members of the 1978 select. But who would be chosen?

A memo from NASA Associate Director Henry E. ("Pete") Clements listed the MS candidates for the two flights, and also stated that proposed PLT for the two missions were Rick Hauck and Dan Brandenstein respectively, with no other candidates identified. [15] Their selection ahead of the other 13 TFNG pilot astronauts once again highlighted a trend that had become normal practice in the release of crew assignments since the Gemini days and which continues to this day: No one ever explained why or how one person was chosen over a colleague.

In 2018, an insight to their assignment was provided by American author Michael Cassutt in his book on NASA engineer and manager George W. S. Abbey, who in 1982 was serving as JSC's Director of Flight Operations with responsibility for selecting the Shuttle crews. [16] In selecting the first of the TFNG to flight crews, Abbey had set clear goals. Firstly, he wanted to choose MS with clear leadership skills, looking to reassign them quickly to far more complex missions further downstream. Next, he recognized the need to fly the first American woman and the first African-American as soon as possible, and from the start his leading contenders were Sally Ride and Guion Bluford. When Abbey presented his selections to Chris Kraft, the Director of JSC, Kraft decided that he wanted to discuss Abbey's choices further. To justify his decisions, Abbey worked with Pete Clements, with added contributions from STS-7 CDR Bob Crippen, Leonard S. Nicholson (then Acting Associate Director, JSC), and Samuel L. Pool (Space and Life Sciences Directorate). Dr. Carolyn Huntoon (Deputy Chief of Personnel Development, and later the first female Director of JSC) had also been consulted by the Advisory Group with regard to the readiness of the six eligible female candidates

In April 1982, a shortlist of suitable candidates for STS-7 and STS-8 was assembled and their suitability assessed based upon training experiences. The memo noted that there was little data on susceptibility to 'space sickness' (later called Space Adaptation Syndrome, or SAS) at that time, but the medical team was producing a protocol to gather this data on all astronauts. To decide who should fly, the Advisory Group met and reviewed all 35 astronauts from the Class of 1978 and found that all were technically qualified and each had good public presence. This latter point was an important factor as the first flights of the new astronauts were bound to attract the attention of the media, especially when the first female and ethnic astronauts were named to a flight. Those assignments would naturally be a 'news story', but would also have the potential to be an irritating distraction to the overall objectives of the missions and the abilities of the whole crew. The ten shortlisted candidates included all six female astronauts, Judy Resnik, Sally Ride, Rhea Seddon, Shannon Lucid, Kathy Sullivan and Anna Fisher, indicating that either STS-7 or STS-8 would see the first American female astronaut in orbit. Considered alongside the women were Dale Gardner, Ron McNair, Guion Bluford and John Fabian. From these, the six women and Fabian were considered for STS-7, while Bluford, Gardner and McNair were shortlisted for STS-8[3].

---

[3] At this point, only one female had flown in space, with Soviet cosmonaut Valentina Tereshkova having spent three days in orbit aboard the single-seat Vostok 6 in 1963.

For STS-7, the April 1982 review summarized each candidate's experience and skills:

- The only male MS candidate, John Fabian, had extensive RMS, Proximity Operations (Prox Ops) and rendezvous experience. He had worked on Crew Equipment assignments, was a supervisor of training and the FDF, and was working on Department of Defense (DOD) payloads. He was recognized as a leader, well organized and extremely well qualified in Prox Ops and RMS.
- Anna Fisher had gained experience in the RMS, EVA, tile repairs, the Shuttle Avionics Integration Laboratory (SAIL) and medical support operations, and had an outstanding public presence.
- Judy Resnik had extensive RMS experience and was extremely well qualified on the systems. She had also worked on Spacelab, payload software, the Power Extension Package (PEP) and Power Module concepts, and training issues.
- Sally Ride had worked as a Chase plane crewmember on STS-1 and as support crew and Capcom on STS-2 and 3, and was also extremely well qualified on the RMS.
- Kathy Sullivan was experienced in the OSTA-1 and OSS-1 payloads, had flown as a crewmember on the WB-57 high-altitude research aircraft and as a Chase crewmember for STS-2, and had worked on flight software and vehicle integration down at the Cape.
- Shannon Lucid had experienced assignments in SAIL and the Flight Simulation Laboratory (FSL), and had worked issues on Spacelab, crew training and the Space Telescope as well as Prox Ops and rendezvous.
- Rhea Seddon had gained experience on the Shuttle's food systems, Spacelab, displays and payload software, and the Orbiter medical kit and check list. She had supported the first three Shuttle flights, medical operations and in SAIL, but was not RMS trained and had only limited experience with the payload.

Though she was on the list, Seddon was not under consideration for STS-7 having also just informed her superiors in March that she was pregnant. "This was the first time an astronaut had become pregnant, and we wanted everyone to know [that] I intended to have a baby and come right back to my career… I didn't want to be held back on jobs or flight assignments," she wrote in 2015. She informed Chief Astronaut John Young, who "didn't seem to know what to say except congratulations. We talked to Mr. [George] Abbey, Head of Flight Operations, and got his usual taciturn response." They had a better response from Center Director Chris Kraft, who was pleased for Seddon and comfortable with her decision to continue her current career path. Seddon continued her assignments for a while, but was grounded from making T-38 flights and had to fly commercially to her assignments as helio-physician for STS-3. She was too advanced in her pregnancy to fulfil that role for the STS-4 landing on July 4. Three weeks later, on July 26, 1982, Seddon and Gibson became the proud parents of their first son. **[17]**

Bluford and McNair were on a shorter list to become the first black American astronaut in orbit on STS-8. McNair's recovery from injuries sustained in his car accident possibly restricted his consideration for that flight, although the April 1982 matrix stated that there was "no apparent problem related to the automobile accident" so he remained under consideration. The shortlisted MS candidates for STS-8, therefore, were:

- Ron McNair, who had worked Solar Max issues. He had limited RMS training but had experience in deployable experiments and Space Telescope, as well as the various systems onboard the Orbiter. He was also said to be "outgoing," with an outstanding public presence.
- Guion Bluford also had limited RMS experience, though more than McNair, and had worked on Spacelab and Shuttle systems, in SAIL and in FSL, and on DOD payloads
- Dale Gardner had extensive experience in flight software and was considered extremely well qualified in that area. He had worked on the displays and controls, and supported STS-4 and DOD payloads.

Interestingly, Fred Gregory, the remaining African-American from the 1978 selection, was not even in the matrix for consideration for a position as either PLT or MS-2 on the STS-8 crew, though the reasons why remain unclear.

For STS-7, Ride apparently had a slight advantage in orbital systems over Resnik, the other leading candidate, but for Abbey it was Ride's skills in handling the RMS that swung it in her favor. For STS-8, Bluford had the slight edge over McNair in RMS experience. In early April, the individuals concerned were informed of their assignments but were told not to say anything to the others, as the official announcement would be made at the next Monday morning pilots' meeting.

MCNAIR INJURED
On the evening of Sunday, September 13, 1981, five months after the triumphant return of *Columbia* from STS-1, Ron McNair was seriously injured in an automobile accident on Interstate 45 in Houston, Texas. While his wife had received only minor injuries, McNair had sustained eight broken ribs and a bruised lung and was admitted to the emergency room of Southeast Memorial Hospital in Clear Lake. He was subsequently placed into intensive care. The next day, he was transferred to a room on the orthopedic floor and was listed as in a stable condition. McNair was later sent home to recuperate from his injuries. He expressed eagerness to return to his duties at JSC and expected to do so "within weeks." In the attachment to the Press Release

(continued)

issued by JSC, he was thankful that both he and his wife were not more seriously hurt and that they both expected to recover fully from their injuries.
Medical privacy naturally precluded further information, but what is clear is
that McNair was off work for some months while he recovered. Some reports
have indicated that the incident almost cost him his astronaut career. He
would not appear in a support role for six months, until serving on the T-38
Chase team as a rear-seat crewmember during STS-3 and on STS-4, where
he was listed on the Public Affairs Office (PAO) support team. [18]

The First Assignments
On April 19, NASA released the names of the crew for STS-7, which was planned
as a six-day flight for *Challenger* in April 1983 to deploy two comsats (Telesat-F
and Palapa B1), a German Shuttle Pallet Satellite (SPAS) and carry the second
Office of Space and Terrestrial Applications instrument package (OSTA-2). [19]
In addition to veteran STS-1 PLT Bob Crippen, now assigned as mission CDR,
were the first three members from the 1978 selection to be named to a flight crew:
PLT Rick Hauck, and MS John Fabian (MS-1) and Sally Ride (MS-2).

In the April 19 Monday pilots' meeting, George Abbey had simply informed the
group that new crew assignments had been made, and then read out the list of names
for STS-7, 8 and 9. In the room, there was a mixture of congratulations from the
other TFNG that their colleagues were finally going to fly, and disappointment that
they were not among the first themselves. Rhea Seddon later wrote that she was
naturally a little disappointed when Sally Ride was named as the first U.S. female
astronaut to fly in space, but Seddon was pregnant with her first child, having made
the decision, with her husband Hoot Gibson, to start a family. She reasoned that she
would rather have children and no space flights than lots of space flights but no children. In addition, she wrote, "Sally would have to go through the rest of her life as
the FAWIS, the First American Woman In Space. Maybe that would not be such a
good deal. Maybe it would be easier to be just *one* of the first." [20]

In the same Press Release, NASA also issued the names of the crew for STS-8
under CDR Dick Truly, which included flight assignments for another three
Group 8 astronauts: Dan Brandenstein as PLT, and MS Guion Bluford (MS-1)
and Dale Gardner (MS-2). For STS-9, John Young was announced as CDR, with
Group 8's Brewster Shaw as PLT. STS-9 would fly the first Spacelab long module
science mission, with MS Owen Garriott and Robert Parker already assigned.
Two Payload Specialists (PS), one from the U.S. and one from Europe, were still
to be announced.

At last, after four years of preparations, seven of the 35 TFNG were to enter the
far more demanding mission training process. They were still at least a year away
from flying, but a lot closer than the remaining 28.

Fig. 9.3: Astronauts Ken Mattingly (front) and Hank Hartsfield, the back-up crew for STS-2, watch an unmanned demonstration of the slide wire crew escape system at the Launch Pad 39A service tower. A sandbag-laden gondola basket rolled down a cable to ground level at approximately 60 mph. At right is support crewmember Don Williams. (Image courtesy Ed Hengeveld.)

America's 206th Birthday Present
STS-4/OFT-4 Columbia. *Launched June 27, 1982; landed July 4, 1982*

An Astronaut Office Memo dated May 20, 1982 (the first the authors discovered that detailed far more support roles in the office, including the first from the 1980 selection) indicated that several of the Astronaut Office personnel would "support the STS-4 flight as needed." Dave Griggs, George Nelson and Brewster Shaw were moved up to the prime Capcom positions for launch ('Ivory' team), orbit ('Bronze' team) and entry ('Crystal' team) respectively, with TFNG colleagues Bob Stewart '(Ivory') and Mike Coats ('Bronze') serving in the back-up roles on the console. [21]

The memo also added that manning of the verification facilities (SAIL, FSL etc.) during the mission would be completed according to real-time requirements, with assignments made as the need arose. This was an indication of the pool of experienced astronauts that was now available to support the missions, offering far more flexibility in manning levels and giving the astronauts a broad range of useful experiences in preparation for their first flights. Other TFNG assignments in support of the last OFT mission were also noted in the memo, with Dick Scobee and Dale Gardner assigned to Mission Control Support but not Capcom duties, and a dedicated team of astronauts down at the Cape designated as the KSC Launch Support Team.

The launch support team for STS-4 consisted of Steve Hawley as ASP and Ox van Hoften as his back-up. Kathy Sullivan provided countdown support and Anna Fisher was on duty as medical helio-physician (possibly replacing the pregnant Rhea Seddon). Norman Thagard occupied the helio-physician position out at Northrup Strip where STS-3 had landed (later known as White Sands Space Harbor), for a potential emergency or alternative landing. Over at Dakar in Senegal, Dave Walker was ready to handle communications if *Columbia* had to conduct a Transoceanic Abort Landing (TAL). It is believed that after the launch – and now back in Houston – van Hoften became the liaison for the Center Operations Directorate (COD). At the end of the mission, a scheduled landing at Edwards AFB would be covered by Dick Covey (COD), who was also assigned as Chase 1 prime pilot. The EOM Exchange Crew was Loren Shriver and Kathy Sullivan. Don Williams was assigned to the Control Tower (TWR) for the landing at Edwards. Detailed to support PAO operations for the final OFT were Brandenstein, Hart, Hauck, Jeff Hoffman, McNair (now back in the office following his accident), Nagel and Judy Resnik.

STS-4 was the final OFT mission and the first to carry a classified DOD payload. Launched on June 27, 1982 and landing eight days later, this was the first American mission not to have a formal back up crew. The primary objectives included continuing the various orbital tests on the vehicle, as well as a range of 20 scientific studies along with the classified payload DOD-82-1 (later identified as the Cirrus cryogenic infrared radiance instrument designed to obtain spectral data on the exhaust of rockets and aircraft), and a UV horizon scanner. STS-4 was the first of ten classified Shuttle missions, all of which included at least one member of the TFNG. Other highlights included Ken Mattingly successfully demonstrating donning and doffing the new Shuttle EVA suit, and airlock operations that stopped short of actually cracking the hatch and going outside. That activity was planned for the next mission.

On July 4, 1982, President Ronald Reagan and his wife Nancy witnessed the landing of *Columbia* at Edwards and declared the STS program 'operational', signaling the end of the formal testing phase of the Shuttle program. That same day, a new Orbiter, *Challenger* (OV-099), left California bound for the Kennedy Space Center (KSC) in Florida. *Challenger* was the former Structural Test Article (STA-099), which had been upgraded to orbital flight capability to replace

Enterprise, whose reconfiguration would have been too costly following its work on the 1977 Approach and Landing Tests (ALT) and subsequent ground test programs. *Challenger* was to take on the majority of the missions over the next two years, with *Columbia* due to be refitted after STS-5 and the next Orbiter, OV-103 *Discovery*, still under construction. *Challenger* was also the Orbiter which would have both joyous and tragic connections for the TFNG.

Fig. 9.4: STS-4 CDR Ken Mattingly and PLT Hank Hartsfield discuss mission events with astronauts and administrators during a post-flight crew debriefing held in a JSC conference room. Seated around the conference table [clockwise from lower left] are: Dick Truly [back to camera]; astronaut William B. Lenoir (blue shirt), Hartsfield, Mattingly, astronaut Robert F. Overmyer, TFNG astronaut S. David Griggs, astronaut Joe Allen, Chief Astronaut John W. Young, NASA Administrator George W. Abbey, and astronaut Vance D. Brand. On the perimeter of the room are TFNG astronauts George D. Nelson (left) and Francis (Dick) Scobee (right). (Image courtesy Ed Hengeveld.)

AN OPERATIONAL SHUTTLE

With the completion of the four OFT missions and the delivery of the second Orbiter, NASA considered the program operational. Regular missions would begin to deliver commercial and military satellites into orbit, deploy planetary probes, operate scientific platforms and perform research missions from facilities in the payload bay. This claim of operational capability after only four OFT missions proved premature in the long run and generated a false sense of security with the system, at least for the public, media and politicians, if not to everyone within the program. It is worth recalling that even after a decade of flying three X-15 rocket planes (1959–1968), with 199 successful drop flights from their carrier aircraft (and several more canceled drops), the program was never classed as operational and remained merely a research program, a stepping stone in the challenge to master hypersonic flight.

As the Shuttle began this operational phase, one of the first members of the TFNG was assigned to an administrative role to broaden their experiences within the NASA system. In September 1981, Don Williams was named as Deputy Manager, Operations Integration at the National Space Transportation System Program Office, an appointment which extended to July 1983.

The following month, on October 20, 1982, NASA named the first Shuttle crew to be assigned to a fully-classified military payload. [22] Designated STS-10 (the tenth planned mission in the series) and due to be launched from KSC, the mission was under the command of Apollo and Shuttle veteran Thomas 'TK' Mattingly II. He was joined by three more representatives of the TFNG: Loren Shriver as PLT, Ellison Onizuka (the first Asian-American named to a flight crew) as MS-1 and Jim Buchli as MS-2. A fifth crew member, an unidentified USAF Manned Spaceflight Engineer (MSE, later identified as Gary Payton) would be announced shortly before the mission. For the first time on an American mission, virtually no other details were released due to its classified nature, other than its scheduled launch in the last quarter of 1983 and that it would be flown by *Challenger* on her planned fourth mission.

"We Deliver"
STS-5 Columbia. *Launched November 11, 1982; landed November 16, 1982*

The Astronaut Office memo for STS-5, issued on October 8, 1982, reported that the criteria for this flight would be the same as for STS-4. [23] The TFNG supporting the mission at MCC were Dave Walker, Dick Scobee and Jeff Hoffman. Capcom assignments were mainly filled by the newer Group 9 astronauts, although Mike Coats was prime Capcom on the 'Granite' orbit team and Dick Covey was prime on the 'Gray' planning team, backed up by Jon McBride. Other TFNG support assignments included the KSC launch support team of Ox van Hoften as ASP, Kathy Sullivan as back-up ASP and Anna Fisher as countdown support. EOM (post-landing) support assignments at Edwards showed Brewster Shaw and Shannon Lucid as the Exchange Crew and Norm Thagard as the Landing Support Doctor. Four TFNG were assigned to PAO support duties: Fred Gregory (CNN); Pinky Nelson (ABC); Steve Hawley (NBC) and Rhea Seddon (CBS).

STS-5 launched on November 11, 1982 the first *operational* mission in the STS program. During the mission, the crew successfully deployed the first two commercial satellites from a Shuttle payload bay. Unfortunately, the planned and much-anticipated first Shuttle-based EVA by Joe Allen and Bill Lenoir had to be abandoned due to problems with both EVA suits and was rescheduled for STS-6. The STS-5 mission was notable for other reasons. It was the first four-person launch and the first to fly astronauts occupying the new position of MS. Further, the role of MS-2 as Flight Engineer (FE), assisting the CDR (Vance Brand) and PLT (Bob Overmyer) during the ascent, was a position pioneered by Bill Lenoir in flight (helped in its development by Joe Allen, who served in the role for the descent), a role many of the TFNG would make their own.

STS-5 landed on November 16, 1982, and following the mission de-processing period, *Columbia* was taken off the flight program for the next year to be modified to fly a large crew and carry the Spacelab pressurized module in the cargo bay on its next mission, STS-9. With the success of STS-5 in deploying the comsats, but disappointment in not being able to complete the first Shuttle EVA and qualify the new EMU suits for operational use, NASA turned to the growing issue of the individual crewmember's personal adaptation to space flight conditions.

Adapting to space flight… or not
Space sickness affected a percentage of crews during their first days on orbit (see sidebar: *Space Adaptation Syndrome*). In order to gather information early, and to investigate the conditions and possible preventative measures, NASA named two physician astronauts to the crews of STS-7 and STS-8 on December 21, 1982, to "assist in the accomplishment of additional mission objectives." TFNG Norman Thagard was assigned to STS-7 and veteran Group 6 Scientist-Astronaut Bill Thornton to STS-8. Both would conduct medical tests to gather data on the physiological changes associated with SAS. The tests would focus upon the neurological systems and continued a program of studies which began on STS-4, aimed at better understanding the process and developing an effective countermeasure. [24]

SPACE ADAPTATION SYNDROME
On March 23, 1983, NASA management at JSC issued a statement [Release 83-010] regarding the formal policy for the public release of information relating to the health of the astronaut crews during space flights. This was followed by a Fact Sheet [Release 83-024, June 22, 1983] detailing the historical perspective of Space Adaptation Syndrome (SAS) and its implications on the current research into the phenomenon during Shuttle operations.

(continued)

In humans, the neuro-vestibular system supplies information about the direction and magnitude of gravitational forces acting on the body, to maintain equilibrium and spatial orientation. Once in space, gravitational forces are neutralized, causing significant realignment of the body's sensory system and initiating a range of clinical symptoms. The most noticeable is SAS and includes dizziness, vertigo and vomiting. [25]

Early American space flights did not reveal this condition, mainly due to the lack of room available in the habitable volume of the spacecraft, although such effects were first recorded in 1961 by Soviet cosmonaut Gherman Titov aboard Vostok 2. With the introduction of 'roomier' vehicles such as Skylab and the Space Shuttle, significant movements of the head or body position could induce such symptoms, though not in every astronaut. As many as half of all space explorers can experience symptoms of SAS during their initial adaptation to microgravity. It can last a couple of days as the body adapts to the conditions of space flight, and can return after landing as the subject readjusts back to Earth's gravitational environment.

Aware of the condition, NASA initiated the collection of data from individuals who flew on the Shuttle during the first six missions. [26] A Detailed Supplementary Objective (DSO) was developed (initially DSO 401, but expanded to include more than 25 investigations over a five-year period) to begin this process by conducting in-flight observations, supported by a series of pre- and post-flight data collection procedures, in order to develop ground-based tests to deduce susceptibility to SAS. This was also combined with the development of space motion sickness medication and a crew testing program. It was found that adjusting the mission timeline to allow susceptible crewmembers time to adjust to space flight was advantageous where necessary and possible, such as not scheduling a planned EVA for early in the mission, for example.

This early in the Shuttle program, there was still only limited data available to provide an accurate account of the occurrence of SAS, but these tests continued as the program evolved. One early link discovered was between disorientation and motion sickness. During the initial phases of space flight, crewmembers reported experiencing nausea and/or spatial disorientation in the first few hours or days of a mission when they moved their head or body too rapidly.

The research into adapting to space flight (and to life back on Earth) continued throughout the Shuttle program and featured dedicated Spacelab life sciences missions. Research programs continued in the Shuttle-Mir and International Space Station (ISS) programs, to better understand space adaptation during short-, medium- and long-duration space flights.

(continued)

Naturally, NASA's policy of being open with the press and public had to be balanced against the privacy of the individual astronauts, hence the March 1983 memo. The medical condition of each astronaut would be kept private unless it was determined to affect the mission adversely. This raised some initial concerns during some of the press conferences held in 1983, when reporters did not receive the in-depth answers they wanted on the health of the crews. Reported data were presented in an 'average population' format, without identifying individuals.

Six years later in October 1989, following the *Challenger* tragedy and the Shuttle's return to flight, NASA issued another press release on the policy of talking about the crews' medical condition during a mission. [27] It was stated that during future space flights (beginning with STS-34), medical consultations between the crew and NASA physicians would be routine, both to improve the understanding of SAS symptoms and to provide a timely treatment of the condition. Flight surgeons in Mission Control would hold a Private Medical Communication (PMC) with crewmembers in the pre-sleep periods on the first two days on a mission, when the condition is most prevalent. Additional consultation could be requested by the Flight Surgeon or crewmember as required. The release went on to state that "The consultations will be confidential because of the physician-patient relationship and privacy laws. If a crew health problem is determined to affect a mission adversely, the Flight Surgeon will prepare a statement for public release which will address the nature, gravity and prognosis of the situation. Information beyond that required to understand the mission impact will not be released."

Events gather pace

Barely a month into 1983, NASA announced new crew assignments. [28] On February 4, the STS-11 crew were identified, with veteran Vance Brand (Class of 1966) as CDR and Hoot Gibson as PLT. The MS included veteran Bruce McCandless (Class of 1966) and rookies Ron McNair (MS-1) and Robert Stewart (MS-2). STS-11 was scheduled as a seven-day mission and the fifth flight of *Challenger* and was planned for a January 1984 launch. It would feature the deployment of an Indonesian communications satellite and a test structure called the Payload Deployment and Retrieval System Test Article (PDRSTA), which would be used in conjunction with further evaluation of the performance and capabilities of the RMS.

In the same announcement, the crew for STS-12 were identified as CDR Hank Hartsfield (Class of 1969) and four more members of the 1978 selection: Mike Coats (PLT) Mike Mullane (MS-1), Steve Hawley (MS-2) and the second female from their selection to receive a flight assignment, Judy Resnik (MS-3). STS-12 was planned for March 1984 as the maiden flight of OV-103 *Discovery*, with the primary payload being the third in the Tracking and Data Relay Satellite System (TDRS-C). A commercial PS from McDonnell (Charles Walker) was named to the crew four months later, to operate the company's Continuous Flow Electrophoresis (CFES) equipment to demonstrate the efficiency with which electrically-charged biological cells could separate in microgravity. The CFES equipment had already flown on STS-4 and was also scheduled to fly on STS-6, 7 and 8, operated by NASA crewmembers.

Two weeks later, on February 18, the pace stepped up as the crew of STS-13 was identified. **[29]** Bob Crippen (Class of 1969) was set to command his third mission, although at this point he was still to fly his second (STS-7). Crippen would be joined by PLT Dick Scobee, with MS Terry Hart (MS-1), Ox van Hoften (MS-2) and Pinky Nelson (MS-3) completing the crew. STS-13 was planned as the sixth flight of *Challenger*, a five-day mission in April 1984 to repair the malfunctioning Solar Maximum Mission satellite and deploy the Long Duration Exposure Facility (LDEF).

The same news release also revealed the Shuttle crew for STS-18/Spacelab 3. Robert F. Overmyer (Class of 1969) was assigned as CDR, with Fred Gregory assigned as PLT. The MS were Norm Thagard as MS-2, Bill Thornton (Class of 1967) and Don Lind (Class of 1966). Thagard and Thornton were first scheduled to fly on STS-7 and 8 respectively, later in 1983. The PS for STS-18 would be named to the flight in due course. STS-18 was the first operational Spacelab mission, featuring experiments in material processing, space technology and life sciences. Spacelab 3 was to have flown after Spacelab 2, a developmental flight with 13 major experiments in plasma physics, infrared astronomy and solar physics, but a delay in the development of the Instrument Pointing System necessary for Spacelab 2 science operations necessitated switching the two missions. Two MS for Spacelab 2 (then manifested as STS-24) were named in the press release as Karl Henize and Anthony England (both class of 1967).

Assigning crewmembers early to science flights would become standard for subsequent missions, to provide a long lead time for training and association with the payload prior to the assignment of the Orbiter crew. The exchange of the flight order for the two Spacelab missions was an early indication that keeping the manifest on target was already becoming a challenge, as many of the TFNG would personally experience throughout their careers.

Moving Office

In his NASA Oral History, Hoot Gibson was asked to recall his experiences of being assigned his first office space in Building 4 (the Astronaut Office) and again once he was assigned to his first crew.

Fig. 9.5: John Fabian (left) and Fred Gregory at their desks in one room (with windows) of the Astronaut Office, Building 4, JSC. (Image courtesy Ed Hengeveld.)

"Ellison [Onizuka] and I got put in one of the few interior rooms that didn't have a window. Everybody else had a window. I think there were only two offices that were on this internal wall over by what was Rick Nygren's office [Richard W. Nygren of the Flight Control Division], so we didn't have a window. It was Ellison and me in one, and then the other one… was Mike Coats and maybe somebody else. I'm not sure about that, but I think that's what I found out the first day, was where my office was and that Ellison and I were going to be officemates. When the '80 class came in, we got Woody [Sherwood C.] Spring in addition in our office. Ellison and I were in that office together all the way up until he got assigned to STS-10. It was supposed to be the tenth mission to launch. What we always did when you got assigned to a crew was you moved into a crew office. I guess on a five-person crew, it was three in one office and two in another office, right next to each other. We were constantly moving offices around. Every time we'd have another crew selection, they'd put out a new office assignment list, and you'd have to move. As I remember, it was four-and-a-half years that Ellison got assigned to STS-10 and moved from that office. I, at some point after that, got assigned to the next mission, STS-11, which turned into [STS]-41B. When we were first assigned, it was STS-11. At that point, I moved out of that office and actually got an office with a window." **[30]**

Challenger takes up the challenge
STS-6 Challenger. *Maiden mission. Launched April 4, 1983; landed April 9, 1983*

The sixth flight of the program was originally manifested as the final OFT mission, but when that series was cut to four test flights, STS-6 became the first mission for OV-099 *Challenger*, the second operational Orbiter.

Though the wider astronaut support roles were missing from the files researched by the authors[4], at least the TFNG who supported the Capcom assignments were known. Dick Covey was prime for ascent on the 'Emerald' team, Jon McBride was prime for Orbit 1 on the 'Ivory' team and Covey was also assigned to the 'Stone' planning team. All the other Capcom assignments were fulfilled by Group 9 astronauts, reflecting the shift in staffing as the Group 8 astronauts were increasingly 'promoted' to flight crew training and the Group 9 astronauts moved in to gain experience of the ground support roles that the TFNG had filled since the start of the program. This pattern was reflected in Mission Support assignments throughout the 30-year Shuttle program.

The maiden flight of *Challenger* was highly successful, with the high point being the first EVA conducted directly from the Space Shuttle on April 8, nine years after the previous U.S. EVA from Skylab. The 4 hr 17 min Shuttle demonstration EVA by Story Musgrave and Don Peterson qualified the EVA/EMU (Extravehicular Mobility Unit, the spacesuit) hardware for operational use, initiating an EVA program directly from an airlock on the Shuttle that would continue to 2005[5]. Operation of the Inertial Upper Stage (IUS) after a flawless deployment by the crew was less successful, with the two-stage IUS tumbling out of control and placing the TDRS in a low elliptical orbit instead of its planned geostationary one. Over the next three months, the excess onboard fuel for the attitude control thrusters was used to move the satellite gradually to its required position. This mishap led to the creation of an investigation board on April 7 to examine the performance of the IUS following its deployment from *Challenger*. [**32**] It was subsequently determined that a collapsed second-stage nozzle seal had caused the failure. As a result, it was announced on May 27 that the second TDRS (B) would be deleted from the STS-8 manifest. [**33**]

---

[4]The authors found little in the way of support information for STS-6 and would welcome any information to fill the gaps.

[5]With the delivery of the Quest airlock to ISS during STS-104 in July 2001, Shuttle crewmembers now had easier access to the station's exterior as the construction expanded. The first EVA from Quest was the third EVA of STS-104 and was followed by three more Shuttle-based EVAs during STS-105 and 108 before the Quest was put into regular service by ISS assembly crews from 2002. The final EVAs directly from a Shuttle airlock were conducted during the STS-114 Return-to-Flight mission in 2005. All the subsequent EVAs by Shuttle crews were conducted from Quest until the end of the program in 2011.

On May 16, 1983, NASA announced a call for its tenth group of astronauts to fill an expected 12 vacancies (six PLT and six MS). **[34]** The successful candidates would be announced in the spring of 1984. This was the first of what was intended to be an annual selection of Space Shuttle astronauts. The statement went on to explain that "The number of candidates to be recruited in subsequent selection periods will vary depending upon mission requirements and the rate of attrition to the existing astronaut corps." With the Shuttle now flying, many of the remaining astronauts selected in the 1960s were close to retirement, leaving the TFNG as the senior Shuttle-era class in the agency, though they were all yet to fly.

By early June 1983, over two years into Shuttle flights and a year after the end of the orbital test phase, the new era of American manned space flight was on the brink of expansion. Up to now, all of the six Shuttle missions flown had been crewed by veterans from the groups selected in the 1960s. But with STS-7, this was all about to change. Of the 35 candidates chosen in January 1978, there were now 23 (or 65.7 percent of the group) assigned to mission training, with the remainder soon to be named to their own maiden flights. The first TFNG were about to earn their astronaut wings and gold pins.

References

1. STS-1 Mission Commentary transcript, Tape 0201, April 14, 1981, p. 1.
2. Mike Coats, NASA Oral History, August 5, 2015.
3. *NASA to recruit more Shuttle astronauts*, NASA JSC News 79-050, August 1, 1979.
4. *Hispanics are encouraged to apply for the astronaut program*, NASA JSC News 79-056. September 12, 1979
5. *NASA receives 3,278 applicants for astronaut program*, NASA JSC News 79-074, December 11, 1979.
6. *NASA selects 19 astronaut candidates*, NASA JSC News 80-038, May 29, 1980.
7. *New astronaut candidates report for training*, NASA JSC News 80-041, June 30, 1980.
8. *Two Europeans accepted for Space Shuttle Mission Specialist training*, NASA JSC News 80-045, July 7, 1980.
9. *Flight Control of STS-1*, JSC News 81-012 April 3, 1981; and STS-1 Mission Commentary, Tape 0201, April 14, 1981.
10. Dan Brandenstein, NASA Oral History, January 19, 1999.
11. *Flight Control of STS-2*, NASA JSC News 81-044, October 27, 1981.
12. *STS-2 Support Crew*, Astronaut Office memo, January 30, 1981.
13. STS-2 Mission Commentary, Tape M0075, November 13, 1981.
14. *Flight Control of STS-3*, NASA JSC News Release 82-014, March 15, 1982.
15. Memo from Henry E. Clements to Dr. Chris Kraft and Cliff Charlesworth, dated April 14, 1982, with attached '*Matrix of mission specialists being considered for the STS-7 and STS-8 flights*'. Copy on File AIS Archives.
16. **Astronaut Maker, How one NASA engineer ran human spaceflight for a generation**, Michael Cassutt, Chicago Review Press Inc., 2010, pp. 239–240.
17. **Go for Orbit, One of America's First Women Astronauts Finds Her Space**, Rhea Seddon, Your Space Press, 2015, pp. 130–135.
18. *Astronaut Injured in Traffic Accident*, NASA JSC news release 81-035, September 14, 1981.

19. *Three Crews Announced*, NASA JSC News Release 82-023, April 19, 1982.
20. Reference 17, pp. 134–135.
21. *Flight Control of STS-4*, NASA JSC News Release 82-034, June 17, 1982.
22. *NASA Names STS-10 Astronaut Crew*, NASA JSC News 82-046, October 20, 1982.
23. *Flight Control of STS-5*, NASA JSC News 82-051, November 10, 1982.
24. *Fifth Crewmember Named to STS-7 and STS-8*, NASA JSC News 82-190, December 21, 1982.
25. **Space Physiology and Medicine**, Arnauld E. Nicogossian and James F. Parker Jr., NASA SP-447, 1982, pp. 141–155.
26. **Results of the Life Sciences DSO conducted onboard the Space Shuttle 1981-1986**, NASA May 1987 [no ID number], specifically Section Five: *Space Motion Sickness*, pp. 117–170.
27. *Private Medical Talk Policy Set for Space missions*, NASA JSC News Release 89-035, October 12, 1989.
28. *STS-11 and STS-12 Crews Named*, NASA JSC News, 83-003, February 4, 1983.
29. *Crewmembers Named for STS-13, Spacelab 2 and Spacelab 3*, NASA JSC News 83-004, February 18, 1983.
30. Robert Gibson, NASA Oral History, November 1, 2013.
31. *Flight Control of STS-6*, NASA JSC News 83-009, March 21, 1983.
32. *IUS Investigation Board members named*, NASA JSC News 83-012, April 7, 1983.
33. *Second Relay Satellite deleted from Shuttle Flight 8 manifest*, NASA JSC News 83-020, May 27, 1983.
34. *Astronaut Recruitment*, NASA JSC News 83-015, May 16, 1983.

10

The TFNG take wings

"We are go for main engine start...
we have main engine start...
and ignition... and liftoff... liftoff of STS-7
and America's first woman astronaut...
and the Shuttle has cleared the tower."
STS-7 Air-to-Ground Transcript
June 18, 1983.

The most productive era for the Thirty-Five New Guys (TFNG) as a group, in terms of missions flown, was from the middle of June 1983 through to the beginning of January 1986. In this period, all 35 members of the Class of 1978 flew in space for the first time, with some able to complete a second mission. Following the four Orbital Test Flights (OFT) and two 'operational' missions, NASA flew a total of 18 successful missions during that period, all but one of which (STS-51F in July 1985) included at least one of the Class of 1978 among the flight crew. This phase of their story began over a decade after the authorization to develop the Space Shuttle program and five years after their selection to train to fly on that vehicle. The formal Shuttle testing phase was complete and two operational Orbiters (*Columbia* and *Challenger*) were available, with a third (*Discovery*) soon to be delivered. As the first members from the Class of 1978 finally seized the opportunity to become 'real' astronauts and fly into orbit, the whole group now had the chance to make their mark on space history, exploring and expanding the new possibilities that the Space Shuttle offered.

© Springer Nature Switzerland AG 2020   307
D. J. Shayler, C. Burgess, *NASA's First Space Shuttle Astronaut Selection*,
Springer Praxis Books, https://doi.org/10.1007/978-3-030-45742-6_10

ASTRONAUTS AT LAST

It had taken five years of training and preparation, interspersed with delays in developing and qualifying the hardware, but finally on June 18, 1983, the first four representatives of the TFNG left the pad and headed for orbit, immediately making history as the world's first five-person crew to fly. Under the command of former STS-1 Pilot (PLT) Robert Crippen on his second mission, the other four astronauts represented the transition to the ethos of the 1978 selection, though this was not emphasized. Veteran Commander (CDR) Crippen and PLT Rick Hauck, a Vietnam veteran and the first of the new generation of pilot astronauts to fly, had old-school NASA backgrounds, having both been seconded from the U.S. Navy (USN). Joining them were the three Mission Specialists (MS), who represented a cross-section of experience in both civilian and military achievements. Sally Ride, who held a PhD in physics and had acquired teaching experience, became the much anticipated and publicized first American female to reach space. She was joined by United States Air Force (USAF) pilot John Fabian, another Vietnam combat veteran who had also found time to earn a PhD in Aeronautics and Astronautics, and experience in academia at the USAF Academy. The third MS was Dr. Norman Thagard, an electrical engineer, pilot and the third crewmember who was a Vietnam combat veteran − with the United States Marine Corps (USMC) − who more recently had studied for a Medical Doctorate (MD). Thagard had been assigned to the flight just six months earlier with the specific remit to study the effects of Space Adaptation Syndrome (SAS) on himself and his crewmates. So, on that first flight of the TFNG, there was a mix of veterans in space flight, combat and flying, together with talented scientists, engineers, a doctor and an academic. This was precisely what NASA had been searching for when it announced in 1976 that its astronaut selection process would be broadened, and chose its first astronauts specifically to crew the Space Shuttle.

Ride Sally Ride

In her 2002 NASA Oral History, Sally Ride was asked about the pre-flight media attention on her and the flight, and whether that attention had impacted the whole crew or their training: "Actually it didn't. NASA did a very good job protecting me and protecting the rest of the crew. I did very few interviews from the time that we entered training until our crew press conference and the interviews afterward. Then we did no more interviews until our pre-flight press conference about a month before the flight. Right after that press conference, we did a day of solid interviews. NASA protected me while we were in training, and even the day that we did all our interviews, we did them in pairs. I did most of my interviews with Rick Hauck or Bob Crippen. NASA's attitude was 'She's going to get all the attention, and we need to help her', and they did. They did a really good job shielding me from the media so that I could train with the rest of the crew and not be singled out. We also tried to get across that space flight really is a team thing." She was then asked about her awareness of all the attention she was receiving in the newspaper articles, and all the special attention from the different organizations: "I was only vaguely aware of it. The training, particularly in the last couple of months before a flight, is very intense. It was early in the Shuttle program. Four of us on the flight had never flown in space before. We had a lot to learn and there was a lot of information coming at us. The training really accelerated and intensified during those two months before the flight. I was spending virtually all my time trying to learn things; what I'd learned, practice, and just stuff that one last fact into my brain. I was barely watching the news at night and really wasn't aware of all the attention. Of course, I was a little bit aware of it – I couldn't help but be – but it wasn't impacting my training at all." [1]

At the Kennedy Space Center (KSC), an estimated crowd of 500,000 people witnessed the launch of the first American woman into orbit, including an estimated 1,628 journalists, the fourth largest press turnout in KSC history [2]. Advised to "have a ball" in the last message relayed to the crew cabin prior to lift off, Sally Ride's enthusiasm in making it to orbit was evident in her explanation of ascent as "definitely a [Disney] premier E ticket."

STS-7 (June 18 – 24, 1983)
Flight Crew: Robert L. Crippen (CDR), Frederick H. HAUCK (PLT), John M. FABIAN (MS-1), Sally K. RIDE (MS-2), Norman E. THAGARD (MS-3)
Spacecraft: Challenger (OV-099) 2nd mission
Objective: 7th Shuttle mission; commercial satellite deployment; SAS medical investigations
Duration: 6 days 2 hours 23 minutes 59 seconds
Support Assignments: Capcom teams for STS-7 were identified in an Astronaut Office memo dated April 22, 1983 and confirmed in a NASA News Release a few weeks later. [3] This also reflected the constantly evolving developments in the Astronaut Office, as members of the TFNG were being supplanted in the majority of the support team roles by Group 9 astronauts. For STS-7, only two members of the 1978 group were assigned to Capcom duties. Prime Capcom on the Orbit 1 'Ivory' team was Jon McBRIDE, backed up by Terry HART. The latter was also assigned to Team 4 duties (an offline, on-call, troubleshooting team)[1].

---

[1] For these mission summaries, members of the TFNG are identified by their surname in CAPITALS. The authors were unable to find details of other astronaut support assignments for STS-6, 7 and 9 during their research, and would welcome any information from readers regarding such assignments to complete their records.

During their mission, the STS-7 astronauts deployed satellites for Canada (ANIK-C2) and Indonesia (Palapa B-1) and operated the Canadian-built Remote Manipulator System (RMS) to perform the first deployment and retrieval exercise with the Shuttle Pallet Satellite (SPAS-01). Along with Crippen, Hauck conducted the first close proximity piloting of the Orbiter, alongside the SPAS-01 free-flying satellite. Though he was a late addition to the crew, and with no formal planning to do so, Thagard completed one of the SPAS retrievals using the RMS, but for most of the mission he was occupied with conducting various medical tests, collecting data on the physiological changes associated with the body's adaptation to space.

In 2006, John Fabian recalled this division of work between the MS: "We trained as a crew of four, and then relatively late in the program, Norm came on, and we tried really hard to integrate Norm into the rest of the crew activities. Sally, for example, was lead on our electrophoresis experiment, doing some fluid work, trying to get some very pure substances out of this process. And I was the lead on doing the deployment and the retrieval using the Remote Manipulator System. But Sally was going to do essentially as much of that as I was. But remember, Norm Thagard [had worked] on the RMS, so he was no stranger to this system. So we worked hard to try to find a way to work Norm into that, and we ended up having him do one of the retrievals without it being written in the checklist that way." [4]

Fig. 10.1: (left) A new generation of astronaut role models: Sally Ride, MS on STS-7 and America's first woman in space; and Guion Bluford, MS on STS-8 and the first African-American to fly in space. (right) Norm Thagard conducting important SAS studies during the STS-7 mission.

Taking your turn in the barrel

Fabian also recalled the post-flight Public Relations (PR) tour after the missions: "Rick and I went to Indonesia [a PR tour following the deployment of the Palapa satellite during the mission], and had a grand time. We went to a meeting of the congress or parliament or whatever it's called in Indonesia; went to the president's house and visited with the president; talked to a lot of school kids, here and there; went to their equivalent of the Fourth of July, a national independence holiday. Had a really nice time, treated very nicely. I got to go back to my hometown and was treated like a hero, got the key to the city for what it's worth. We went to New York City and went to the major's house, Mayor [Edward I.] Koch, and met Floyd Patterson, the famous boxer. It was fun. It really was fun. We all laugh about it, you know, 'It's your turn in the barrel', and everybody has to go through this post-flight, but it's one of the things that you do in return for the reward of flying in space. But it was really fun. I didn't do very many things that I thought were distasteful. There were a few, frankly, and you can get that – if you're not careful – you can get to feeling that you're being utilized. In other ways, it was something like being wallpaper.

"When we went to Washington we went to the state dinner, and that was very nice. There was a lot of media attention to the fact that Sally was there, and Norm Thagard was knocked into the wall by a photographer, in trying to get to Sally. And they're not courteous people. I mean, there's something of the vulture that's going on there [going after the story] because Sally is here. Sally was married at the time to Steve Hawley, and we went someplace, and Steve was with us. And we went through the door, and someone [said] 'Stop. You can't come here. This is for the astronauts'. And Steve says, 'We *are* the astronauts'. Really funny."

Becoming America's First Woman in Space

After the mission, attention naturally focused on Sally Ride, who recalled her experiences in her 2002 Oral History. "I remember being disappointed that we weren't going to land in Florida, but I grew up in California, and we'd spent a lot of time at Edwards Air Force Base [AFB]. The pilots had done a lot of approach-and-landing practice at Edwards, so it almost felt like a second home. But there weren't many people there waiting for us!" There were still crew activities to perform and a significant amount of PR for Ride, because 20 years after Valentina Tereshkova became the first female to enter orbit, Ride would forever be known as the first *American* female to fly in space. "I think that's when all the attention really hit me. While I was in training, I had been protected from it all. I had the world's best excuse: 'I've got to train, because I have this job to do'. NASA was very, very supportive of that. So my training wasn't affected at all. But the moment we landed, that protective shield was gone. I came face-to-face with a flurry of media activity. There was a lot more attention on us than there was on previous crews, probably even more than the STS-1 crew."

She received some assistance from NASA for this increase in activities and attention after her flight: "[I had] a lot of help with fielding the requests. All the requests went through NASA, through the Public Affairs Office [PAO] and through the Astronaut Office, but very little help in preparing to talk with either the press or to make public appearances. Now, of course, all astronauts learn 'on the job' how to give a talk and how to work with the public and with various organizations.

"I'd done my share of public appearances and speeches before I'd gone into training, so I knew how to talk to the press and I knew how to go and show my slides and give a good speech. But just the sheer volume of it was something that was completely different for me, and people reacted much differently to me after my flight than they did before my flight. Everybody wanted a piece of me after the flight." Occasionally, there were times when she felt that some of the events she attended or some of the questions were not necessarily NASA-related, but not too much. "Most often, people were just interested in the flight, in my experience, in my view of the historical nature of the flight, that sort of thing. Not too much on my personal life. That really wasn't as much of a problem for me as the sheer volume of things that I had to do. It was just incessant for months." Fortunately, not long after she returned on STS-7, Ride learned that she would be flying again, and thought "Thank goodness. Back into training, safe again." **[1]**

STS-8 (August 30 – September 5, 1983)
Flight Crew: Richard H. Truly (CDR), Daniel C. BRANDENSTEIN (PLT), Dale A. GARDNER (MS-1), Guion S. BLUFORD (MS-2), William E. Thornton (MS-3)
Spacecraft: Challenger (OV-099) 3rd mission
Objective: 8th Shuttle mission; satellite deployment; RMS load evaluation tests; continued SAS medical investigations
Duration: 6 days 1 hour 8 minutes 43 seconds
Support Assignments: The support assignments for this mission fulfilled by the TFNG were identified in an Astronaut Office memo dated July 20, 1983. Mission Control Center (MCC) Support Astronaut Office Representative was Loren SHRIVER; Family escorts were Steve NAGEL, Rick HAUCK and Jim BUCHLI; Jeff HOFFMAN was listed as Capcom. However, when a news release was issued on August 23, HOFFMAN was not assigned to Capcom duties. **[5]** This demonstrates the difficulty in detailing support assignments and the often-unreported changes to the original lists with regard to who actually fulfilled an assignment real-time. Down at KSC, the Launch Support Team for STS-8 included Dave WALKER as Weather Pilot (WX), Shannon LUCID as Astronaut Support Person (ASP), and John CREIGHTON.

Both the launch and landing of this flight took place at night, recording more firsts for the Space Shuttle program and the TFNG. It was later revealed that NASA had dodged a bullet on this flight, having come perilously close to recording the first catastrophic loss of a Space Shuttle and her crew[2].

After the removal of the second Tracking and Data Relay Satellite (TDRS), the rather empty payload bay of *Challenger* was partially filled by the dumbbell-shaped Payload Flight Test Article (PFTA), brought forward from the STS-11 mission. Unberthed but not released by the RMS, the aluminum and stainless steel PFTA measured 15 x 16 ft (4.5 x 4.8 m) with a mass of 8,500 lbs. (3,855 kg). It was fitted with four RMS grapple fixtures and was used to simulate a large mass for flight testing of the robot arm. During the mission, the crew gained experience in evaluating the reaction of the RMS shoulder, elbow and wrist joints to higher loads than previously flown.

By flying on STS-8, Guion Bluford became the first African-American to fly in space though not the first black astronaut. Three years earlier, the Soviets had launched Cuban Air Force pilot Arnaldo Tamayo-Mendez aboard Soyuz 37 for a week-long visiting mission to Salyut 6. Despite this, Bluford's mission was a significant step forward for the diversity of American space flight crewing, and he was fully aware, as with Sally Ride before him, about the role model he now became to the youth of America. "Although there was a lot of interest in my participation on the mission, I focused my attention on making the mission a success and stopped doing PR events during the last six months of training," he said in 2004. Bluford went on to comment on his first experiences with microgravity, or zero-g: "Once we completed our OMS [Orbital Maneuvering System] burns, I unstrapped from my seat and started floating on the top of the cockpit. I remember saying to myself 'Oh, my goodness, zero-g', and like all the other astronauts before me I fumbled around in zero-g for quite a while before I got my space legs. However, it was a great feeling, and I knew right away that I was going to enjoy this experience." **[6].** Bluford and Gardner were both responsible for operating the Continuous Flow Electrophoresis System (CFES) package during the mission, and while Gardner helped Truly with the PFTA operations, Bluford took photos to document the event.

---

[2] A post-flight inspection carried out on September 27 revealed the shocking discovery of corrosion, particularly in the left-hand Solid Rocket Booster (SRB) nozzle, which breached the 1.5 inch (4 cm) limit to just 0.5 inches (1.3 cm). Had the nozzle burned through on ascent, it would have caused a side-thrusting escape of superhot gas exhaust, resulting in the total loss of control and break-up of *Challenger*. Later estimates suggested that at SRB separation a cataclysmic burn-through, similar to that on STS-51L (mission 25), would have taken place just 14 seconds later. All of this came as a complete shock to the crew.

Bluford spoke of his post-flight agenda in his 2004 Oral History. After the mission, NASA Headquarters assigned Mary Weatherspoon to work with Bluford on his PR agenda. "Mary was a public relations specialist from NASA Headquarters who had lots of experience doing PR support for the NASA Administrator," Bluford explained. "We worked together to determine which events we should do and how best to support all the speaking requests. She handled all the transportation and logistics for each PR trip and she served as my escort at many PR functions. We worked well together as a team. From October to December of 1983, we made three to four trips a month to various parts of the country. We tried not to spend a lot of time crisscrossing the country, but tried to focus on a particular area of the country on each trip. In several cases, we convinced people to change the date of their events in order to best accommodate my schedule. Between trips, I would spend a lot of time answering the mail and preparing for the next trip. On each trip, I talked about my experiences of flying on STS-8, the importance of the space program, the need for more scientists and engineers in this country, and I tried to acknowledge the role of teachers, parents, and role models in my life. I used the PR trips to thank the American people for giving me the opportunity to fly in space and tried to show my appreciation to those organizations that helped me the most in life. I particularly focused my gratitude on Penn State University, the City of Philadelphia, the United States Air Force and the Tuskegee Airman. It was a wonderful three months.

"I went back home to Philadelphia for four days in November and rode in the Thanksgiving Day Parade. I met with Mayor Wilson Goode of Philadelphia and Governor Richard Thornburgh of Pennsylvania. I visited the University of Pennsylvania's Children's Hospital and several schools in Philadelphia, including Overbrook Senior High, my alma mater. I spent time at the Franklin Institute in downtown Philadelphia talking with school kids about the importance of studying math and science and I participated in numerous press conferences. It was a busy four days. I went to Hollywood [California] and joined up with Bob Crippen to do a TV special on the 25th anniversary of NASA. Bob Hope hosted the event, as we highlighted some of the many accomplishments of the Agency. I also attended an awards program for the NAACP [National Association for the Advancement of Colored People], and received the NAACP Image Award. It was an exciting experience.

"In October, my wife and I went to Washington D.C. to attend several events. We attended a ceremony in the Pentagon, hosted by the Chief of Staff of the Air Force, General Charles A. Gabriel, who presented me my Air Force Command Pilot Astronaut Wings. John Fabian also received his Air Force Astronaut Wings at the same event. There was a small reception, after the ceremony, with quite a few flag officers. From there, the wife and I went to the Smithsonian National Air

& Space Museum to join President [Ronald] Reagan. The President gave a speech recognizing NASA on its 25th anniversary. I participated in that event with Sally Ride, several other astronauts, and with the NASA Administrator. By the end of the year, I decided to get off the PR circuit and return to my normal duties in the Astronaut Office. Although I enjoyed my experience giving speeches and signing autographs, I felt it was time for me to support some of the other astronauts who were getting ready to fly. However, I had one more surprise that occurred after the Christmas and New Year's holidays. Among the mail that I had received during the holidays there was a letter from the Undersecretary of Defense for Personnel. In the letter, he congratulated me on my accomplishments and officially notified me that I was promoted to full colonel. The Department of Defense [DOD] had decided to re-initiate an old policy of promoting astronauts when they flew in space. I was authorized to wear the new rank, as the Air Force got approval from Congress for my promotion. It was a great gift from an organization that I felt very proud of."

The 'Eyes' patch

Personal crew patches for American missions had been common since 1965 and Gemini 5. Continuing the tradition, each Shuttle crew designed their own personal emblem which became synonymous with their flight, and naturally became a collectable item in patch, decal, lapel pin and memorabilia form. The official STS-8 emblem depicted the dramatic separation of the twin Solid Rocket Boosters (SRB) as the Orbiter streaks to orbit, but there was a second patch, known as the "eyes" patch, that was not worn by the crew but was present as a cake on the table during the pre-launch meal.

In his Oral History, Bluford commented on the comic patch. "The 'joke patch' was shaped like an eight-ball with two of the Shuttle windows depicted on it. Dick Truly was depicted as half asleep looking out one of the windows while the rest of the crew was shown wide-eyed looking out of the other window. An *Aviation Week* article on our flight highlighted the 'joke, or eight ball' patch when they described our mission." Designed by pilot Dan Brandenstein, the 'unofficial' STS-8 patch was given a slightly different explanation in *JSC Roundup*: "Richard Truly [in the right hand window], steely eyed and bespectacled veteran of STS-2 and now Commander of STS-8, will lead four space rookies who are shown here [in the left window, hanging on] slightly awed by their participation." [7] There is, however, another explanation for the symbolism. The bespectacled eyes in the right hand window represented Dr. Bill Thornton, chasing after the blood of his four concerned colleagues in the left window, who are frantically trying to get away from participating in his space sickness tests during the mission.

Fig. 10.2: The STS-8 'Eyes' joke patch (Image courtesy of Joachim Becker, SpaceFacts.de)

A new and confusing numbering system

On September 9, 1983, NASA issued a memo outlining changes to both the schedule and the designations of future Shuttle missions. This was partly due to delays with some payloads and the potential confusion of missions flying out of numerical sequence, as well as identifying the proposed use of the Shuttle launch facility at the Space Launch Complex (SLC-6 or "Slick 6"), at the USAF Vandenberg AFB in California. As some reordering of the flights was expected, crews would now be announced by *payload assignment* rather than a specific STS number, beginning in 1984.

The designation of a mission would now consist of a numerical prefix to indicate the U.S. Fiscal Year (FY)[3] in which the launch would occur, a numeral to identify the launch site (either "1" for KSC or "2" for Vandenberg), and a letter suffix to reflect the original sequence of launch.

---

[3] The U.S. Financial, or Fiscal Year, which defines the U.S. federal government's budget, runs annually from October 1 through September 30.

The first seven missions already flown would not be amended retrospectively, and as the next mission (STS-8) would fly within FY 1983, it too would not be given a new designation. The next flight, manifested to carry Spacelab 1, was scheduled to be the first mission launched in FY 1984, but as it was well advanced in its processing schedule it was decided to leave it officially as STS-9, though it also became known administratively as STS-41A. The STS-9/STS-41A designation can be used as an example of the coding in NASA's new system: *STS* indicated a Space Transportation System flight; the *4* meant that the planned launch was in FY 198*4*; the *1* indicated a launch from KSC; and the *A* showed that this was the first mission assigned to FY 1984. With 26 letters of the alphabet available, this somewhat over-optimistically suggested up to 26 missions could be launched within a year, or one every two weeks, from *both* KSC and Vandenberg, giving a maximum on paper of 52 missions in 12 months! The resulting adjustments to the crews already assigned can be seen in Table 10.1

The former STS-10 mission became another designation casualty, due to the delays caused by difficulties with the Inertial Upper Stage (IUS) following its deployment from STS-6. The classified STS-10 payload was to have deployed on an IUS, which meant that a postponement was necessary to allow the problem to be resolved. As a result, the mission designated STS-10 was permanently deleted from the manifest, with the crew stood down pending reassignment to a new mission.

As logical as the new system looked on paper, in reality it was subject to change almost immediately, as the crews and assigned Orbiters changed over the months while the basic payload remained with the same designation code. This also meant that the missions frequently flew out of numerical sequence over the next three years, making it even more confusing than before. The idea was sound in trying to track a payload through the system, assign the hardware and not get bogged down in numerical identification too early, but due to the complexity of the payloads to be flown and the uncertainty that they would remain on time, or that the assigned Orbiter would be ready to fly, it simply did not work as intended. It also gave rise to the Astronaut Office lore: "Don't fall in love with your Orbiter or payload, as it will change."

It would probably have been far easier to keep the lettering system pre-flight to track the main payload, then assign the next STS-Arabic number to the mission once it had left the pad. Thus the tenth mission, STS-41B, would have become STS-10 on launch, the next one, STS-41C, would be STS-11, and so on. That way, even if the payload letter designations went awry, at least the STS launch sequence would remain intact. This method had been adopted by NASA before in unmanned programs, where the letter code assigned prior to launch was replaced by the next number in the sequence at lift-off, but it was decided this would be impractical for the Shuttle. As author Michael Cassutt explained: "After the IUS failure [in April

TABLE 10.1: STS CREW ASSIGNMENTS (1983/1984) NASA ASTRONAUTS ONLY [September 9, 1983, JSC]

Old designator	New designator	Primary payload	Planned Launch date	Commander	Pilot	Mission Specialists		
STS-9	41-A[1]	Spacelab 1	Oct. 1983	Young	SHAW	Garriott	Parker	
STS-10	-	DOD	Cancelled	Mattingly	SHRIVER	ONIZUKA	BUCHLI	
STS MISSIONS FOR CALENDAR YEAR 1984								
STS-11	41-B	Palapa; Westar	Jan. 1984	Brand	GIBSON R.	McCandless	STEWART	McNAIR
STS-13	41-C	Solar Max; LDEF	Apr. 1984	Crippen	SCOBEE	NELSON G.	VAN HOFTEN	HART
STS-12	41-D	Telesat; Syncom	Jun. 1984	Hartsfield	COATS	RESNIK	HAWLEY	MULLANE
STS-15	41-E	DOD	Jul. 1984	Crew not assigned				
STS-14	41-F		Aug. 1984	Crew not assigned				
STS-17	41-G		Aug. 1984	Crew not assigned				
STS-16	41-H		Sep. 1984	Crew not assigned				
STS-18[2]	51-A		Oct. 1984	Crew not assigned				
STS-20	51-B	Spacelab 3	Nov. 1984	Overmyer	GREGORY F.	Lind	THAGARD	Thornton W.
STS-19	51-C		Dec. 1984	Crew not assigned				

[1]The STS-9 designator remained in place but the mission was also unofficially known as STS-41A. Group 8 astronauts are shown in CAPITALS.
[2]Although STS-51A and the following two missions were planned for launch in 1984, they were the first three missions manifested after the start of U.S. Fiscal Year 1985 (October 1) and were therefore designated 51A, 51B and 51C instead of 41I, 41J and 41K.

1983], STS-10 had been postponed for at least a year and was likely to be launched after STS-14. STS-12 was slipping behind STS-13… In the program office [they] were facing a flight order that went 11, 13, 12, maybe 14, 10. Missions beyond that were in total chaos." **[8]**

The origins of this new system came from a team in the program office that required payloads to be planned in advance, in some cases as much as a year or two. But there was an ulterior motive as well, as apparently the new center director, Gerry Griffin, who had been a Flight Director involved with Apollo 13 in 1970, did not want to fly an STS-13. Thus, the original STS-13 would be re-designated and fly as STS-41C in 1984, while the thirteenth Shuttle mission launched would now be known as STS-41G.

On September 21, NASA confirmed two new crews for missions designated STS-41E and STS-41G, as well as the new assignment for the former STS-10 crew. The STS-41E mission was planned for a June 1984 launch and the second flight of the new *Discovery* Orbiter. Its payload was listed as two commercial satellites (Telesat and Syncom IV-1), the Large Format Camera, and a large experiment support structure provided by NASA's Office of Aeronautics and Space Technology (OAST). The STS-41E crew, commanded by Karol Bobko, included four more TFNG: Don Williams as PLT, following his return to the Astronaut Office after completing his temporary management role; and MS Rhea Seddon (MS-1), Dave Griggs (MS-2), and Jeff Hoffman (MS-3). Griggs had been selected in 1978 as a pilot astronaut but was taking the role of MS-2/Flight Engineer (FE) for this mission, giving him the chance to gain experience prior to future assignment as PLT then CDR. **[9]**

The selection of Griggs as an MS on this crew is interesting. By the fall of 1983, 22 of the 35 TFNG had received a flight assignment or had flown, and with another third of the group still awaiting assignment, it seemed for a while that several of the MS would fly before all of the pilots, some perhaps twice. To counteract this, with the Group 9 astronauts also becoming eligible for crew assignment and with Group 10 Ascans now in training, George Abbey decided to assign two of the pilots as MS-2 for their first missions. Thus, Griggs was assigned to the Bobko crew and Steve Nagel was named as MS-2 to the Brandenstein crew two months later.

In the same announcement, it was revealed that the payload for STS-41F would be classified and the mission flown by the astronauts previously assigned to the cancelled STS-10 mission: Mattingly, Shriver, Onizuka and Buchli. STS-41G was to be commanded by STS-7 veteran Rick Hauck, the first TFNG pilot to receive a command seat which, after just one mission and only five years at NASA, was a significant achievement. Fellow TFNG Dave Walker would be his PLT, with two others, Anna Fisher and Dale Gardner, as MS. Rounding out the crew was MS Joe Allen (Class of 1967). This mission was planned for August 1984 on *Discovery*, with the payload manifested to include three commercial satellite deployments: Telstar 3-C, Satellite Business Systems-D and Syncom IV-2, and an astronomy free-flyer called SPARTAN (Shuttle Pointed Autonomous Research Tool for AstroNomy).

Fig. 10.3: The October 16, 1983 rollout ceremony for OV-103 *Discovery* at the Rockwell factory in Palmdale, California, with the STS-12 (41D) crew present (on the podium in blue flight suits from far left: Hawley, Mullane, Resnik, and at far right, Coats and Hartsfield), as well as NASA dignitaries, employees and visitors. The Orbiter is positioned behind a reviewing stand, with Don Beall of Rockwell (at lectern) introducing Dr. Rocco A. Petrone, the president of the company. Members of the TFNG flew on each of the five operational Orbiters and were represented on the maiden flight crews of *Discovery*, *Atlantis* and *Endeavour*.

Delivering *Discovery*
Over in California, on October 16, 1983, following four years and nine months of construction and testing, Rockwell held the rollout ceremonies for the latest Orbiter, OV-103 *Discovery*, witnessed by the five astronauts of STS-41D (formerly STS-12) who were to crew this new vehicle on her maiden flight. "I remember the event pretty well, particularly as it rolled into public view for the first time," recalled Steve Hawley 36 years later, "I recalled thinking that it looked clean and new. I had seen it before in the plant while it was being worked on and I'm sure I participated in some of the testing. The rollout made it seem real that we might actually get to fly it." The crew was there to wave to the crowd and thank the

Rockwell workforce, and Hawley recalled that, at the time, there may have been a subset of the "Cape Crusaders" at Palmdale to participate in the activities, keeping the crew informed as the vehicle was completed and prepared for its move to the Cape. Hawley also reflected on the end of *Discovery*'s career following its retirement in 2011 after 39 missions over 27 years, and its relocation the following year to public display at the Steven F. Udvar-Hazy Center, Dulles, Virginia. "I have been to Dulles to see *Discovery*. My thought was how sad it was to see it as a display. I always felt a connection to *Discovery*, since I got to fly it for the first time. It was a great vehicle." **[10]**

On November 5, the new Orbiter was taken overland to Edwards AFB for transportation on the Shuttle Carrier Aircraft (SCA) to KSC, arriving on November 9. *Discovery* was lighter than the previous Orbiters, as much as 3.6 tons less than *Columbia*, benefitting from lessons learned during the construction and testing of the three previous vehicles. The weight-saving measures included using quilted Advanced Flexible Reusable Surface Insulation (AFRSI) blankets rather than white Low temperature Reusable Surface Insulation (LRSI) tiles, and the use of graphite epoxy instead of aluminum for the payload bay roofs as well as in some of wing spars and beams. After it arrived at KSC, workers began preparing the Orbiter for its first mission (STS-41D) and completed modifications (along with *Challenger*) to enable it to carry and deploy the liquid-fueled Centaur G upper stage.

Even more assignments

By November 1983, with an increase in the planned launch manifest, NASA announced the crews for STS-41G, 51A, 51C and 51F, as well as a launch ready crew, bringing the total to 12 crews in various stages of mission training for flights planned over the next 18 months. **[11]** The schedule drawn up at this time can be seen in Table 10.2

In this schedule, Rick Hauck's original STS-41G crew now moved to 41H, the Bobko crew from 41E to 41F and Mattingly's DOD crew moved again, this time from 41F to 41E. A new crew was named to STS-41G, with Bob Crippen as CDR (though he was yet to fly as CDR for STS-41C), Jon McBride as PLT, and Kathy Sullivan and Sally Ride as MS, together with David C. Leestma, the first of the Group 9 astronauts to receive an assignment. Crippen would trial a short-duration training profile intended for later CDR and PLT to follow, to evaluate whether this could be a viable option to turnaround and re-fly the same flight deck crew (CDR, PLT and MS-2/FE) quickly in the future. It was also becoming increasingly evident that as the frequency of missions increased, so the time between them would diminish, meaning a CDR could fly a second mission much sooner, ideally after a shorter training turnaround, using their experience to compensate for the lack of training hours. This was exactly what Crippen would evaluate between STS-41C and 41G. The 41G crew announcement also meant that Sally Ride had the

TABLE 10.2: STS CREW ASSIGNMENTS (1984-85) NASA ASTRONAUTS ONLY [September 9, 1983, JSC]

STS flight	Primary payload	Planned Launch date	Commander	Pilot	Mission Specialists			
SPACE SHUTTLE FLIGHTS CALENDAR YEAR 1984								
41B	Palapa; Westar	January 29, 1984	Brand	GIBSON R.	McCandless	STEWART	McNAIR	
41C	Solar Max; LDEF	April 4, 1984	Crippen	SCOBEE	NELSON G.	VAN HOFTEN	HART	
41D	Telesat; Syncom	June 4, 1984	Hartsfield	COATS	RESNIK	HAWLEY	MULLANE	
41E	DOD	Classified	Mattingly	SHRIVER	ONIZUKA	BUCHLI		
41F	Telstar; SBS, Syncom	August 9, 1984	Bobko	WILLIAMS D.	SEDDON	HOFFMAN	GRIGGS	
41G	OSTA; ERBS	August 30, 1984	Crippen	McBRIDE	SULLIVAN	RIDE	Leestma	
41H	DOD or TDRS-B	To Be Determined	HAUCK	WALKER D.	FISHER A.	GARDNER D.	Allen J.	
51A	MSL; Telesat	October 24, 1984	BRANDENSTEIN	CREIGHTON	LUCID	FABIAN	NAGEL	
51B	Spacelab 3	November 22, 1984	Overmyer	GREGORY F.	Lind	THAGARD	Thornton W.	
51C	TDRS B or TDRS C	December 20, 1984	Engle	COVEY	BUCHLI	Lounge	Fisher W.	
SPACE SHUTTLE FLIGHTS CALENDAR YEAR 1985								
51F	Spacelab 2	March 29,1985	Fullerton	GRIGGS	Musgrave	England	Henize	
-	Stand-by Crew	TBD	Bobko	Grabe	MULLANE	STEWART	Hilmers	

Group 8 astronauts are shown in CAPITALS.

opportunity to become the first woman to fly two missions, while Kathy Sullivan was scheduled to perform an Extra-Vehicular Activity (EVA, or spacewalk) with Dave Leestma, becoming the first American female to do so, and perhaps the first in the world.

In the same announcement, the STS-51A crew was named under the command of Dan Brandenstein, with Steve Nagel becoming the second pilot to fly as MS on his first mission. This was a notable crew for another reason, as the first NASA crew comprised entirely of Group 8 astronauts, (though Payload Specialists (PS) would be assigned later). In addition, Dave Griggs received his second assignment (though he was yet to fly his first), this time as PLT for STS-51F, while Bob Stewart was to be assigned to the first DOD Launch Ready Standby Crew[4], under the command of Karol Bobko, ready to take a military payload into orbit at short notice.

STS-9 [STS-41A] (November 28 – December 8, 1983)
Flight Crew: John W. Young (CDR), Brewster H. SHAW (PLT), Owen K. Garriott (MS-1), Robert A. R. Parker (MS-2), Ulf Merbold (PS-1, ESA, Germany), Byron K. Lichtenberg (PS-2, U.S.)
Spacecraft: Columbia (OV-102) 6th mission
Objective: 9th Shuttle mission; 1st flight of European-built Spacelab pressurized scientific laboratory (Spacelab 1/Long Module unit #1)
Duration: 10 days 7 hours 47 minutes 23 seconds
Support Assignments: The Astronaut Office support assignment list was missing from the source referenced by the authors. The mission was the first not to include members of the Class of 1978 in Capcom assignments, as all of them were in various stages of mission training, had recently flown, or were about to do so.

STS-9 became the first extended U.S. scientific research mission since Skylab a decade before. A total of 73 separate investigations were carried out in the Spacelab 1 module or Orbiter payload bay, in the fields of astronomy, physics, atmosphere physics, Earth observation, life sciences, material science, space plasma physics and technology. The crew divided into two 12-hour shifts, with the sole TFNG representative on the crew, Brewster Shaw, assigned to the Blue Shift along with Garriott and Lichtenberg, and the remaining crewmembers forming the Red Shift.

The crew of six became the largest so far to fly aboard a single spacecraft. It was also the first international Shuttle crew (with a crewmember from West Germany), and the first to carry Payload Specialists (from the U.S. and the European Space Agency, ESA). After ten days of Spacelab hardware verification and around-the-clock scientific operations, *Columbia* and its laboratory cargo (the

---

[4] It transpired that this would be the *only* DOD standby crew assigned in the program.

heaviest payload to be returned to Earth in the Shuttle's cargo bay) returned to land on the dry lakebed at Edwards AFB, California, completing the fourth mission of that year. NASA was confident that the next year would be much busier.

SATELLITES, SPACEWALKS AND SCHEDULES

Fig. 10.4: (top) Ron McNair plays his saxophone aboard *Challenger* during the STS-41B mission. (bottom) "Cecil B. McNair" taking a leading role.

On February 2, 1984, the day prior to the launch of STS-41B, NASA announced the crew assignments for two further missions. **[12]** Brewster Shaw, recently returned from STS-9, would next command the STS-51D mission scheduled for launch in February 1985. His four colleagues would be members of the 1980 selection. The mission would be the 21st of the Shuttle series and the ninth flight of *Challenger*, and would include the deployment of a Syncom communication satellite and the retrieval of the free-flying Long Duration Exposure Facility (LDEF) that was planned to be released during STS-41C in April 1984.

STS-61D was scheduled for a January 1986 launch, carrying the Spacelab 4 on the ninth mission of *Columbia* and focusing upon experiments in the fields of life sciences. John Fabian was named as MS-2 and Rhea Seddon as MS-3, with another of the Group 9 astronauts (Dr. Jim Bagian) as MS-1. This announcement also revealed NASA's policy on assigning crews to future Spacelab-type missions, with the intention of having three of the crew "share flight deck responsibilities." The CDR and PLT would be announced to join MS-2 Fabian at a later date. The announcement also stated that MS were "frequently selected earlier than flight crews, since their [science] training is more specialized and requires more time."

STS-41B (February 3–11, 1984)
Flight Crew: Vance D. Brand (CDR), Robert L. GIBSON (PLT), Ronald E. McNAIR (MS-1), Robert L. STEWART (MS-2), Bruce McCandless (MS-3)
Spacecraft: Challenger (OV-099) 4th mission
Objective: 10th Shuttle mission; commercial satellite deployment; 1st tests of Manned Maneuvering Unit (MMU)
Duration: 7 days 23 hours 15 minutes 55 seconds
Support Assignments: In the Astronaut Office memo dated January 4, 1984, the Group 8 mission support assignments for STS-41B included Brewster SHAW (CB Rep, MCC Support). KSC launch support was Dave WALKER as WX coordinator and Shannon LUCID as back-up ASP. LUCID was also on the Exchange crew for landing support, with WALKER again as WX coordinator. Jon McBRIDE would fly the Edwards Landing Pattern (ELP) WX T-38, while Anna FISHER led the PAO support team working with the NBS network and Jim BUCHLI worked with CBS. The Capcoms for this mission were all assigned from the 1980 Group 9 astronauts.

Unfortunately, while the mission was completed successfully, the flight was plagued by equipment failures. During their eight-day mission, the crew correctly deployed two Hughes 376 communications satellites. Westar 6 was for America's Western Union, while Palapa B-2 was for Indonesia, but both failed to reach their desired synchronous orbits due to upper stage rocket failures. Nevertheless, the flight was notable as the first in which astronauts tested the MMU, with both Bruce McCandless and Robert Stewart flying untethered up to several hundred feet from the Orbiter. During the first EVA (February 7), as McCandless ventured out 320 feet (98 meters) from the Orbiter, Stewart tested the 'workstation' foot restraint at the end of the RMS. On the seventh day of the mission (February 8), both astronauts performed another EVA, again test-flying the MMU, to practice capture procedures for the Solar Maximum Mission satellite retrieval and repair operation which was planned for the next mission, STS-41C. These EVAs were the first untethered human operations from a spacecraft in flight. As well as undertaking the flight testing of rendezvous sensors and computer programs, the Canadian-built robot arm, operated by McNair, helped position the two EVA crewmen around *Challenger*'s payload bay by moving the platform on which the astronaut was standing, another first. This method of placing an astronaut in a specified position using the robotic arm was used on subsequent Shuttle missions to repair satellites and assemble the International Space Station (ISS). While orbiting the Earth, accomplished saxophonist McNair fulfilled an ambition to become the first person to play the instrument in space. He had brought his own saxophone on the mission and managed to play a few numbers while circling the globe. The STS-41B mission ended with a perfect touchdown, making the first Shuttle landing on the purpose-built runway at KSC.

Fig. 10.5:   Bob Stewart flies the untethered MMU during STS-41B.

Three days after the return of STS-41B (February 14), NASA named the PLT and two of the MS for STS-51K (Spacelab-D1). **[13]** Steve Nagel would fly his second mission, this time as PLT, with Guion Bluford assigned as MS-3 and Group 9 astronaut Bonnie Dunbar as MS-1. The CDR, MS-2 and the European crewmembers would be announced at a later date. Spacelab-D1 was a dedicated mission purchased by the Federal Republic of Germany, with a science program focusing on materials and life science experiments. The mission was planned for launch in September 1985 on the third flight of *Atlantis* and the fourth flight of a Spacelab Long Module.

Later that month, the tribulations of the Mattingly DOD crew in trying to remain assigned to a mission, let alone leave the launch pad, persisted when the STS-41E mission was cancelled and the crew (including Shriver, Onizuka and Buchli) stood down again pending reassignment.

STS-41C (April 6 – 13, 1984)
Flight Crew: Robert L. Crippen (CDR), Francis R. SCOBEE (PLT), Terry J. HART (MS-1), George D. NELSON (MS-2), James D. A. VAN HOFTEN (MS-3)
Spacecraft: Challenger (OV-099) 5th mission
Objective: 11th Shuttle mission; deployment of Long Duration Exposure Facility (LDEF); capture, repair and redeployment of Solar Max satellite
Duration: 6 days 23 hours 40 minutes 7 seconds
Support Assignments: The support assignments for this mission were listed in an Astronaut Office memo dated March 13. Assignments for the Group 8 astronauts were: Brewster SHAW (Mission Operations Control Room (MOCR) Representative, MCC Support); Jon McBRIDE (WX T-38 pilot); John CREIGHTON (CBS PAO Support); Dan BRANDENSTEIN and Steve NAGEL (Family Escorts).

This was the most ambitious Shuttle mission to date in the program. During their seven-day mission, the crew successfully deployed the LDEF, retrieved the ailing Solar Maximum (Solar Max) Satellite, repaired it aboard *Challenger*, and replaced it in orbit using the RMS. The retrieval of Solar Max did not go to plan after the intended method of capture – Pinky Nelson flying the MMU and intending to use the chest-mounted T-pad docking device – failed. Nelson's attempt to steady the Solar Max satellite manually resulted in it becoming even more unstable, but fortunately the controllers managed to regain sufficient control of the satellite overnight to enable it to be captured by the RMS and brought into the payload bay for repair. The mission also included flight testing of the MMU over two EVAs (April 8 and 11), operation of the Cinema 360 and IMAX Camera Systems, and the Bee Hive Honeycomb Structures student experiment. Their mission was extended by a day due to problems capturing the Solar Max satellite, with the landing on April 13 taking place at Edwards AFB instead of at KSC as originally planned, due to forecast bad weather at the Cape.

Flight Control Operations Directorate (FCOD) Manpower Requirements
A presentation by the Astronaut Office on April 27, 1984, indicated the extent to which the office was stretched, with crews in training, crews just off a mission (STS-41B including Gibson, McNair and Stewart), and new astronaut support roles that the TFNG were involved with. Brewster Shaw was now Astronaut Office Deputy for Operations, with Joe Engle, and was also assigned to the Operation

and Training Branch with Vance Brand. Also within this branch was Steve Nagel, working on Shuttle Mission Simulator (SMS) issues. Ron McNair was assigned to Science and Technology payloads, Bob Stewart was working in the DOD Coordination Branch, and Loren Shriver had been detailed to the System Development and Test Branch with assignments relating to systems development for improving the Orbiters' General Purpose Computers (GPC). Steve Nagel was detailed to track the Digital Auto Pilot system under development as part of the Orbiter Experiment Program (OEX DAP). John Fabian was detailed to the Mission Development Branch as lead in Deployment Systems and was also working on the Centaur upper stage with Shannon Lucid. Fabian was given further assignments with Payload Accommodations (Payload Data and Retrieval System, PDRS) and Mission Integration, while any Spacelab issues were taken up by Lucid.

Shortly after STS-41C came home, the STS-41H mission was cancelled and the crew, led by Hauck and including Dave Walker, Dale Gardner and Anna Fisher, were stood down temporarily pending reassignment to a new flight.

First to depart
On May 10, a month after flying on STS-41C, Terry Hart announced his intention to retire from NASA, effective June 15, and return to take a position in an engineering management role for the newly-formed Military and Government Systems Division of his previous employer, Bell Laboratories, in Whippany, New Jersey. [14] This division produced large digital communications networks for government applications. Hart thus became the first member of the Shuttle-era selections to depart the agency and the first of the TFNG to hang up his spacesuit.

The busy but frustrating summer of 1984
On June 7, NASA announced the flight crew for the STS-51H/Earth Observation Mission 1. [15] For the first time since STS-6, none of the crew came from the Class of 1978. In the same release, astronomer and TFNG Jeff Hoffman was named to join the crew of STS-61E (Astro-1) in a partial crew announcement of three MS. Set for a March 1986 launch with a crew of six on *Columbia*, the mission included the deployment of the Intelsat-VI Comsat, and the Astro-1 astronomy package designed to view Halley's Comet. Hoffman's experience prior to joining NASA was in high energy astrophysics.

A very low MECO
Originally scheduled for launch on June 22, 1984, the maiden launch of *Discovery* (on STS-41D) was delayed by electrical problems for three days. Then, on June 25, a fault in one of the onboard computers halted the launch at T-6 minutes. The

following day, the count reached the point of starting two of the three Space Shuttle Main Engines (SSME), but they both shut down suddenly just 2.6 seconds later after the main fuel actuator in SSME #3 failed. The result was the first pad abort in the Shuttle program and the first for NASA since Gemini 6 back in 1965. Up on the flight deck, MS-2 Steve Hawley quipped "Gee, I thought we'd be a lot higher at MECO [Main Engine Cut Off]." The abort led to the reassignment of part of the original STS-41D payload, merging it with that of STS-41F. Yet another abort was called on August 29 due to computer software issues, but the next day *Discovery* finally left the pad to become the third operational Orbiter.

The consequences of the aborted STS-41D mission and fate of the 41F crew were reflected in the next manifest, released on August 3, which can be seen in Table 10.3. **[16]**

Things were changing once more, with the 41F (Bobko) crew stood down following the problems in getting STS-41D off the ground and reassigned to STS-51E, which was then scheduled for a launch in February 1985. The original flight of 41F was deleted and some of its cargo placed on the STS-41D flight, delayed due to the June launch abort. The formerly assigned DOD 41E (Mattingly) crew was reassigned yet again, this time to STS-51C, and the 41H (Hauck) crew to 51A. The former 51A (Brandenstein) crew moved to 51D, the previous 51C crew (Engle) now took 51G, and the 51D crew (Shaw) was reassigned to 51L. The STS-51K designation for Spacelab-D1 was erased from the manifest and replaced with STS-61A, while the Launch Ready Standby Crew (Bobko) remained unassigned to a flight.

STS-41D (August 30 – September 5, 1984)
Flight Crew: Henry W. Hartsfield (CDR), Michael COATS (PLT), R. Michael MULLANE (MS-1), Steven A. HAWLEY (MS-2), Judith A. RESNIK (MS-3), Charles D. Walker (PS-1, McDonnell Douglas)
Spacecraft: Discovery (OV-103) 1st mission
Objective: 12th Shuttle mission; maiden flight of Discovery; commercial satellite deployment
Duration: 6 days 0 hours 56 minutes 4 seconds
Support Assignments: The original Astronaut Office support listing for 41D had been released on May 23 prior to the pad aborts and was largely retained without change for the new mission. The assignments for Group 8 astronauts were: Brewster SHAW (MCC support in SPacecraft ANalysis, or SPAN), assisted by Dan BRANDENSTEIN and Bob STEWART; Loren SHRIVER (PAO support with NBC), assisted by Steve NAGEL (ABC) and Norm THAGARD (CBS); and Dick COVEY (Family Escort).

TABLE 10.3:   STS CREW ASSIGNMENTS 1984-86 NASA ASTRONAUTS ONLY [August 3, 1984, JSC 84-036]

STS flight	Primary payload	Orbiter	Planned launch date	Commander	Pilot	Mission Specialists		
STS MISSIONS FOR CALENDAR YEAR 1984								
41D	OAST; SBS; Telstar; Syncom	Discovery	Aug. 24	Hartsfield	COATS	MULLANE	HAWLEY	RESNIK
41G	OSTA; ERBS; LFC	Challenger	Oct. 1	Crippen	McBRIDE	SULLIVAN	RIDE	Leestma
51A	Telesat; Syncom[1]	Discovery	Nov. 2	HAUCK	WALKER D.	FISHER A.	GARDNER D.	Allen J.
51C	DOD	Not listed	Classified	Mattingly[2]	SHRIVER	ONIZUKA	BUCHLI	-
STS MISSIONS FOR CALENDAR YEAR 1985								
51B	Spacelab 3	Discovery	Jan. 17	Overmyer	GREGORY F.	Lind	THAGARD	Thornton W.
51E	Telesat; TDRS-B	Challenger	Feb. 12	Bobko	WILLIAMS D.	SEDDON	HOFFMAN	GRIGGS
51D	LDEF retrieval; Syncom	Discovery	Mar. 18	BRANDENSTEIN	CREIGHTON	LUCID	FABIAN	NAGEL
51F	Spacelab 2	Challenger	Apr. 17	Fullerton	GRIGGS	Musgrave	England	Henize
51G	EASE/ACCESS; Telesat; Arabsat; Morelos	Columbia	May 30	Engle	COVEY	VAN HOFTEN	Lounge	Fisher W.
51L	EOS; TDRS-C; OASIS	Challenger	Jul. 2	SHAW	O'Connor	Cleave	Spring	Ross
61A	Spacelab D-1	Columbia	Oct. 14	Hartsfield	NAGEL	BUCHLI	BLUFORD	Dunbar
51H	Earth Observation Mission	Atlantis	Nov. 27	Brand	Smith M.	Springer	Garriott	Nicollier
STS MISSIONS FOR CALENDAR YEAR 1986								
61D	Spacelab 4	Columbia	Jan. 28	TBA	TBA	FABIAN	Bagian	SEDDON
61E	Astro-1	Columbia	Mar 6	TBA	TBA	Parker	Leestma	HOFFMAN
Standby Crew	-	-	-	Bobko	Grabe	MULLANE	STEWART	Hilmers

[1]STS-51A was listed as a dual option, either a civilian satellite deployment mission or a classified DOD flight.
[2]The 'classified' crew was not named in the memo but Mattingly's team had been reassigned from STS-41E.
Group 8 astronauts are shown in CAPITALS.

On this mission, Judy Resnik became the second American woman, and only the fourth in history, to journey into space. During their seven-day mission, the crew successfully activated the 102-ft (31 m) OAST-1 solar cell wing experiment, deployed the SBS-D, SYNCOM IV-2, and TELSTAR 3-C satellites, and conducted the student crystal growth experiment as well as photography experiments using the IMAX motion picture camera. McDonnell PS Charles Walker spent most of the mission on the middeck operating the CFES-III equipment (which he had to repair). Ironically, Walker was an unsuccessful applicant for the 1978 selection, but was now flying with four fellow Group 8 astronaut-applicants who had made it through the selection process. The 41D crew also earned the name "Icebusters" (the name and associated logo were in parody of the 1984 feature film *Ghostbusters*) after successfully removing hazardous ice particles from the Orbiter using the Shuttle's robotic arm. The problem had been caused by venting water from the Orbiter fuel cell system, which created a large chunk of ice on the exterior of *Discovery*. Using the RMS rendered the proposed contingency EVA by Mullane and Hawley unnecessary.

Refueling and retrieval

Barely a month later, the next Shuttle left the pad. One of the key objectives on this flight was to demonstrate the feasibility of refueling techniques by EVA, a key to future satellite refueling missions and to extending the operational lifetime of a spacecraft already on orbit. Depicted in pictures and articles about satellite servicing, maintenance and space station assembly over the years, these techniques were demonstrated on several Shuttle missions during 1984 and 1985 and became one of the Shuttle's key legacies over the next 25 years. On STS-41G, one of the TFNG, Kathy Sullivan, was once again right at the forefront of this new technique.

STS-41G (October 4 – 13, 1984)

Flight Crew: Robert L. Crippen (CDR), Jon A. McBRIDE (PLT), Sally K. RIDE (MS-1), Kathryn D. SULLIVAN (MS-2), David C. Leestma (MS-3), Paul D. Scully-Power (PS-1, U.S. Navy), Marc Garneau (PS-2, Canadian Space Agency (CSA))

Spacecraft: Challenger (OV-099) 6th mission

Objective: 13th Shuttle mission; scientific satellite deployment; Space Imaging Radar (SIR-B) experiments; satellite refueling demonstration (EVA supported)

Duration: 8 days 5 hours 23 minutes 38 seconds

Support Assignments: The Astronaut Office memo dated September 10, 1984, listed the mission support assignments for 41G, including those for Group 8 astronauts: Brewster SHAW (MCC Support Lead in SPAN Reps); Dan BRANDENSTEIN (SMS Support); Dick COVEY and Bob STEWART

(PAO Support); John FABIAN and Pinky NELSON (Family Escort). In the announced Capcom assignments, just two TFNG were named to the flight control teams, with Ron McNAIR listed on the Orbit 1 'Orion' team and Pinky NELSON on the Orbit 2 'Granite' team for the EVA. [17]

STS-41G carried the largest crew to fly on the same spacecraft to date (seven) and was the first mission to include two females on the same crew. During their eight-day mission, the crew deployed the Earth Radiation Budget Satellite (ERBUS), and conducted scientific observations of the Earth with the OSTA-3 pallet and Large Format Camera. On October 11, Kathy Sullivan became the first U.S. woman to perform an EVA[5] when she successfully conducted a 3 hour 30 minute spacewalk, alongside fellow MS David Leestma, to demonstrate the feasibility of future satellite refueling, using hydrazine fuel with the Orbital Refueling System (ORS) in the payload bay. During the flight, the crew also carried out numerous in-cabin experiments as well as activating eight 'Getaway Special' (GAS) canisters.

Two months after the previous announcement, NASA made further amendments to the manifest by naming a new crew and changing a previously announced assignment. [18] Five astronauts, Robert Gibson (CDR), Charles Bolden (PLT, Class of 1980), and MS Franklin Chang Díaz (also from the Class of 1980), George "Pinky" Nelson and Steve Hawley, were named to the STS-51I crew that was scheduled for launch in August 1985 on *Columbia*. The mission included the deployment of two communications satellites and operation of the Material Science Laboratory (MSL) material processing experiment.

The single crew change announced in the release was made to STS-51F, with PLT Dave Griggs replaced by Group 9 astronaut Roy Bridges due to the proximity of the two missions Griggs was assigned to (STS-51E and 51F), leaving him insufficient time to train adequately for both. Keeping to the schedule and retaining astronauts on their assigned crews and missions was becoming increasingly challenging, just a year into the new system.

STS-51A (November 8 – 16, 1984)
Flight Crew: Frederick H. HAUCK (CDR), David M. WALKER (PLT), Joseph P. Allen (MS-1), Anna L. T. FISHER (MS-2), Dale A. GARDNER (MS-3)
Spacecraft: Discovery (OV-103) 2nd mission
Objective: 14th Shuttle mission; commercial satellite deployment and retrieval
Duration: 7 days 23 hours 44 minutes 56 seconds

---

[5] Sullivan was just beaten to the accolade of being the world's first female to complete an EVA by Soviet cosmonaut Svetlana Savitskaya, who did so three months earlier. Savitskaya also usurped Sally Ride's chance of becoming the first female to make two space flights, having previously flown in 1982.

Support Assignments: Mission support assignments for 51A were listed in an Astronaut Office memo dated October 15, 1984, and included six members of the TFNG: Brewster SHAW (MCC Support Lead in SPAN Reps), supported by Robert GIBSON, Pinky NELSON and Ox VAN HOFTEN; Mike MULLANE (PAO Support with CNN); and Ron McNAIR (Capcom for Orbit 2 'Amber' flight team). **[19]**

Fig. 10.6: (main) Dale Gardner in training with the Stinger/MMU combination. (inset) Gardner puts his training into practice by capturing the Westar satellite using the 'stinger' while flying the MMU.

Anna Fisher had been assigned to the crew just two weeks before delivering her first daughter and was launched 14 months later, making her the first mother to fly in space. During the mission, the crew deployed Canada's Anik D-2 (Telesat H) and the Hughes LEASAT-1 (Syncom IV-1) satellites, and retrieved the Palapa B-2 and Westar VI satellites for their return to Earth, the first space salvage mission in history. Joe Allen and Dale Gardner successfully captured and secured the two satellites, with the two men completing two EVAs in a little more than 12 hours. This included Gardner becoming the sixth and last person, and the fourth Group 8 astronaut, to free-fly the MMU during the second spacewalk, in which he captured the Westar satellite. During both EVAs, Anna Fisher assisted the recovery of the satellites by taking control of the RMS robotic arm from the aft flight deck of *Discovery*.

A BUSY YEAR: 1985

NASA had hoped to fly eight or possibly nine missions during 1985, and as many as 12 the next year. While this would be operationally challenging, there was also the question of training crews to meet those requirements. In 1985, the training system could only handle six crews at a time, so it became necessary to assign the crew as close to launch as possible, no more than about a year in advance. That year was split between six months of general training and then six months dedicated mission training, while all the time keeping up with the constantly changing manifest and schedules. This all made achieving a standardization of crewing almost impossible. The idea from headquarters that a CDR-PLT-MS-2 team could fly more than a couple of missions a year seemed very unlikely, and even assigning MS teams to standardize science missions on Spacelab, or IUS-dependent missions, was likely to consume months of training. On top of this was the ever-present risk factor in flying the Shuttle and the clear indication that each mission was a unique and complicated entity it its own right. There was nothing 'standard' or 'routine' about any Shuttle mission.

By 1985, most of the TFNG had flown in space at least once, a few twice, and all the remainder were in training for their maiden missions. At the same time, members of the 1980 selection were now available for their first assignments, allowing some of the more experienced TFNG to take more senior roles on their next missions and to think about the next moves in their careers.

STS-51C (January 24 – 27, 1985)
Flight Crew: Thomas K. Mattingly II (CDR), Loren J. SHRIVER (PLT), Ellison S. ONIZUKA (MS-1), James F. BUCHLI (MS-2), Gary E. Payton (PS-1, USAF Manned Spaceflight Engineer (MSE))
Spacecraft: Discovery (OV-103) 3rd mission
Objective: 15th Shuttle mission; 1st classified Department of Defense (DOD) mission
Duration: 3 days 1 hour 23 minutes 23 seconds
Support Assignments: Interestingly, despite the secrecy surrounding this flight, an internal Astronaut Office memo dated January 3, 1985 revealed which of the TFNG were fulfilling support roles: Bob STEWART (MCC Support Lead SPAN Rep), assisted by Jon McBRIDE; Mike MULLANE and Ron McNAIR (Capcoms, no Flight team details released); Steven Hawley (Exchange crew, Edwards AFB); Brewster SHAW (Family Escort)

STS-51C was the first fully-classified American space flight with a crew, with many of the Shuttle's objectives and its payload shrouded in deep secrecy, although the flight of the first USAF MSE was announced in advance. The DOD part of the mission was completed successfully and included the deployment of a modified

IUS vehicle from *Discovery*'s payload bay. The payload has subsequently been identified as an Aquacade ELINT (ELectronic signals INTelligence satellite). On this flight, Ellison Onizuka became the first American of Japanese ancestry (and the first Buddhist) to fly in space. For the first time, due to the unprecedented security surrounding the mission, the exact roles or crew assignments each astronaut completed were not divulged, setting a pattern for most of the subsequent classified DOD Shuttle missions.

End of year flight assignments

On January 29, two days after the return of STS-51C, NASA named two crews for end-of-year missions in its first crew announcement of the year. [20] The crew for STS-51L was changed from the Shaw crew to one led by Dick Scobee, with Mike Smith (Class of 1980) as PLT and Judy Resnik, Ellison Onizuka and Ron McNair as the MS. STS-51L was intended to launch in November 1985 on *Atlantis* carrying TDRS-C and was, according to the news releases, "an opportunity to re-launch one of the communications satellites retrieved from orbit during the flight of 51A" in November 1984.

The second crew was assigned to STS-61C, planned for a December 1985 launch on *Columbia*. The mission was manifested to deploy two communication satellites and operate the MSL, as well as conducting the EASE/ACCESS space manufacturing structure EVA experiment. In command of this mission would be Mike Coats, with fellow TFNG Norman Thagard and Anna Fisher as MS. PLT John Blaha and MS Robert Springer (both from the Class of 1980) completed the crew. In the same announcement, NASA identified the flight deck crews to go with two complements of MS announced earlier. Dave Griggs was named as PLT for the January 1986 STS-61D/Spacelab 4 mission along with CDR Vance Brand (Class of 1966), while Jon McBride was named as CDR for STS-61E in March 1986, along with PLT Richard Richards (Class of 1980). The original STS-51L crew led by Brewster Shaw was reassigned to an "unspecified mission."

Two DOD crews were named the following month, on February 15, including the first crew scheduled to fly from the Vandenberg AFB in California. [21]. STS-62A was to be commanded by Shuttle veteran Bob Crippen, with Guy Gardner as PLT and Jerry Ross as MS (both from the Class of 1980), along with TFNG Dale Gardner and Mike Mullane. In the same release, the former DOD Standby Crew was finally assigned to a mission, STS-51J. Karol Bobko would command the mission, with PLT Ron Grabe, MS Dave Hilmers (both of them also from class of 1980) and TFNG Bob Stewart making up the NASA crew.

The Saga of 51E/51D

The month of March 1985 did not start too well for NASA. On the first day, NASA cancelled STS-51E after continued problems with the TDRS satellite were revealed early in the year. The crew was stood down while a revised plan was conceived. Five days later, on March 6, they had their new mission. As the Bobko crew was the priority crew in training, they were reassigned to STS-51D, taking a combination of the 51E and 51D payloads with them and yet again creating a ripple effect in reassigning crews. The Brandenstein crew moved to 51G, which in turn moved the Engle crew to 51I. The ongoing problem with TDRS were the source of a whole sequence of events and manifest changes which followed, including some of the planned DOD missions. There was also a requirement to fly the Bobko crew as soon as possible, to preserve the training schedule for his next mission, 51J, and the remaining flights for the year.

Later that month, in yet another revised and short-lived manifest dated March 25, a new mission briefly appeared as STS-51E (R – for Revised) which was under consideration for a rapid launch between May and September 1985 to deploy the much needed second TDRS (TDRS B) communication satellite prior to the Spacelab-D1 mission that October. This mission was assigned a minimum three-person crew, provisionally identified as Vance Brand (CDR – Class of 1966 and one of the most experienced astronauts in the office), Dave Griggs (PLT) and Rhea Seddon (MS-2/FE). In the end this mission was not pursued, and it disappeared from the manifest as quickly as it had appeared. When asked about this idea, Rhea Seddon said that she had not heard anything about such a plan, though "we were left in limbo for quite some time after our flight was combined with [41D] ... [although it was] interesting to know something was being discussed." **[22]** It seems the suggestion never reached the astronauts concerned and most likely never left the office that created the manifests, underlining the caution that is sometimes necessary when linking paper trails to real events.

March 1985 also saw Rick Hauck assigned as Astronaut Office Point Of Contact (POC) in the project office for integration of the liquid-fueled Centaur upper stage into the Space Shuttle. "After 51A, I was assigned to be the Astronaut Project Officer for Centaur. Centaur was an upper-stage rocket that's very thin-skinned. It has a thin aluminum skin. It's pressure-stabilized, which means if it's not pressurized, it's going to collapse by its own weight. If it was not pressurized but suspended and you pushed on it with your finger, the tank walls would give and you'd see that you're flexing the metal. Its advantage was that it carried liquid oxygen and liquid hydrogen, which, pound for pound, give better propulsion than any other, than a solid-rocket motor. Shuttle was obligated to launch the Ulysses probe and the Galileo probe, both interplanetary probes, and they needed the most powerful rockets they could have, and there was [a] back and forth… could the Inertial Upper Stage launch them or couldn't it? And no, it couldn't.

"In any case, at some point the decision was made… we've got to use the Centaur, which was never meant to be involved in human space flight. And that's important because rockets that are associated with human space flight have certain levels of redundancy and certain design specifications that are supposed to make them more reliable. Clearly, Centaur did not come from that heritage, so, number one, was that going to be an issue in itself, but number two is, if you've got a Return-To-Launch-Site [RTLS] abort or a transatlantic abort [Transoceanic Abort Landing, or TAL] and you've got to land, and you've got a rocket filled with liquid oxygen, liquid hydrogen in the cargo bay, you've got to get rid of the liquid oxygen and liquid hydrogen, so that means you've got to dump it while you're flying through this contingency abort. And to make sure that it can dump safely, you need to have redundant parallel dump valves, helium systems that control the dump valves, [and] software that makes sure that contingencies can be taken care of. Then when you land, here you're sitting with the Shuttle Centaur in the cargo bay that you haven't been able to dump [completely], so you're venting gaseous hydrogen out this side, gaseous oxygen out that side, and this is just not a good idea." **[23]**

On April 5, Steve Hawley and Kathy Sullivan, together with Bruce McCandless (Class of 1966), were named early to the STS-61J mission manifested to deploy the Hubble Space Telescope (HST). **[24]**. The selection reflected the requirement for experienced astronauts to accompany a scientifically important payload, and the extensive EVA program required to back up its successful deployment. That mission, at the time planned for 1986, would not fly until 1990, five years after the astronauts had been named to it.

Another new Orbiter is delivered

April 1985 also saw the arrival at KSC of NASA's fifth and last scheduled Orbiter, OV-104 *Atlantis*. The new vehicle had rolled out from the Palmdale facility on March 6, the same day the former STS-51E crew were reassigned to STS-51D. *Atlantis* had been transported overland to Edwards AFB on April 3 and was then airlifted to the East Coast, arriving at the Cape on April 13. With the delivery of *Atlantis*, NASA now had a fleet of four vehicles available (*Columbia, Challenger, Discovery* and *Atlantis*) to meet its manifest, allowing one to be taken out of the flight line occasionally for its routine maintenance period.

Mission after mission

During the spring and summer of 1985, there was a rapid sequence of four Shuttle missions in just 16 weeks, with nine missions launched in total across the calendar year. This was the most that NASA ever achieved in the Shuttle's 30-year flight history, with the eight launches in 1992 being the only other calendar year to come close to this. It was still far short of the envisaged 26 missions a year.

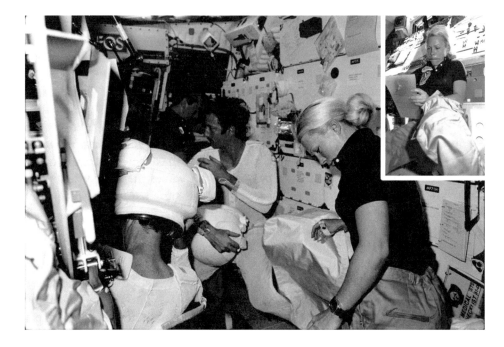

Fig. 10.7: (main) MS Rhea Seddon (right) supports fellow crewmembers Jeff Hoffman (center frame) and Dave Griggs (left, back to camera) as they don their EVA pressure suits for the first unscheduled spacewalk of the Shuttle program. (inset) Seddon had earlier put her surgical skills to good use in fabricating a "fly-swatter" device to be attached to the RMS during the EVA in an attempt to trip the trigger to reactivate the satellite.

STS-51D (April 12 – 19, 1985)
Flight Crew: Karol J. Bobko (CDR), Donald E. WILLIAMS (PLT), S. David GRIGGS (MS-1), Jeffrey A. HOFFMAN (MS-2), M. Rhea SEDDON (MS-3), Edwin J. 'Jake' Garn (PS-1, U.S. Senator); Charles D. Walker (PS-2, McDonnell Douglas)
Spacecraft: Discovery (OV-103) 4th mission
Objective: 16th Shuttle mission; commercial satellite deployment
Duration: 6 days 23 hours 55 minutes 23 seconds
Support Assignments: After the STS-51E flight was scrubbed, the Astronaut Office memo dated January 31, 1985, was simply hand-amended from 51E to 51D. Originally, the only Group 8 astronauts assigned to the flight control team for 51E were Ron McNAIR (Orbit 1 'Altair' team) and Mike MULLANE (Orbit 2 'Amber' team). [25] McNAIR was replaced for 51D, presumably to allow him to train full time for STS-51L, his place taken by Dave Leestma of Group 9. Mike MULLANE was reassigned to the Planning Shift on the 'Sirius' team. [26] Other assignments intended for 51E but reassigned to 51D were: Brewster SHAW (Lead in SPAN), with Pinky NELSON, Judy RESNIK and Sally RIDE; Dave WALKER (Cape Crusaders at KSC, and also KSC Exchange Crew with Mark Brown of

Group 10); Steve HAWLEY (PAO Support for CBS); Jon McBRIDE (Briefer to the guests of the NASA Administrator); Rick HAUCK (PAO Support for NBC back at Johnson Space Center (JSC)). Kathy SULLIVAN was originally assigned to the French CNES team for 51E, but when Patrick Baudry was reassigned to 51G she was taken off the 51D support team. [27]

STS-51D was launched exactly four years after STS-1 and on the 24th Cosmonautics Day celebrations of the 1961 launch of Yuri Gagarin in Vostok. During this week-long mission, Dave Griggs and Jeffrey Hoffman conducted the first unscheduled EVA of the space program (on April 16), in which they tried unsuccessfully to reactivate the malfunctioning Syncom IV-3/LEASAT-3 satellite. To aid in triggering the faulty spring on the side of the rogue satellite, Dr. Rhea Seddon used her surgery skills to fabricate a 'fly-swatter device' from onboard materials. The device was strapped to the end of the RMS, but while it successfully snagged the trigger it also broke, leaving the satellite stranded until another crew could repair it. The crew also conducted several medical experiments, activated two GAS canisters and filmed educational experiments exploring the physics of 'Toys in Space'. Discovery made the fifth Shuttle landing at KSC but suffered extensive brake damage and a ruptured tire during landing. This forced all subsequent Shuttle landings to be made at Edwards AFB, California, until the development and implementation of nose wheel steering once again made landings at KSC feasible.

STS-51B (April 29 – May 6, 1985)
Flight Crew: Robert F. Overmyer (CDR), Frederick D. GREGORY (PLT), Don L. Lind (MS-1), Norman E. THAGARD (MS-2), William E. Thornton (MS-3), Taylor G. Wang (PS-1), Lodewijk van den Berg (PS-2)
Spacecraft: Challenger (OV-099) 7th mission
Objective: 17th Shuttle mission; Spacelab 3 (LM-1) research program
Duration: 7 days 0 hours 8 minutes 46 seconds
Support Assignments: Only one member of the TFNG was assigned as a Capcom for this flight. Mike MULLANE worked console on the Orbit 3 'Polaris' team. [28] Unfortunately, the Astronaut Office Support Team list for STS-51B was missing from the references sourced by the authors.

Norman Thagard acted as MS-2 on this mission, assisting CDR Robert Overmyer and PLT Fred Gregory on ascent and entry. Thagard's duties on orbit included satellite deployment operations with the NUSAT satellite, as well as caring for the 24 rats and two squirrel monkeys contained in the Research Animal Holding Facility, which took considerable time and included clearing up loose floating feces from the cages. Spacelab 3, the second flight with the European-built Long Module, featured 15 primary experiments in the five disciplines of material and life sciences, fluid mechanics, atmosphere physics and astronomy. Once again a two-shift system was followed for the science program, with Thagard assigned to the Gold Shift with Gregory and van den Berg, while the others worked the Silver Shift.

A "Death Star" in the payload bay

On May 31, NASA released the names of two four-man crews assigned to missions of just three days in duration, in close proximity to each other and carrying the controversial Centaur upper stage. [29]. They became known as the Centaur missions. STS-61F would be commanded by Rick Hauck, leading three of the Group 9 astronauts (Roy Bridges as PLT, with MS Dave Hilmers and Mike Lounge) for a launch no earlier than May 15, 1986, to deploy the Ulysses (International Solar Polar) spacecraft. This would be the first mission to use the liquid-fueled Centaur upper stage, and the first deployment of a solar system probe from the Shuttle. The second mission, STS-61G, was to be commanded by Dave Walker, with fellow TFNG John Fabian and James van Hoften as MS, and Group 9 PLT Ron Grabe rounding out the crew. This mission was scheduled for launch on May 21, 1986, just six days after STS-61F, and would also use a Centaur to deploy the Galileo planetary spacecraft on its mission to explore the giant gas planet Jupiter and its system of moons. The close proximity of the missions was necessary to meet the stringent launch windows available to both solar system probes and the four-man crew limit was necessary due to the weight restriction imposed by flying the larger payloads. They would both be flown at just 105 nautical miles (74.97 km) altitude, the lowest a Shuttle had orbited, limited by the mass of the fully-fueled Centaur and its payload.

The Centaur upper stage was a risky payload that concerned the crews from the start, not least because of the fully-fueled stage positioned just behind them in the event of an abort. The Centaur was not a new upper stage, having flown for the first time in 1962, and its combination of liquid oxygen and liquid hydrogen offered far greater power for lifting payloads than either the Payload Assist Module (PAM) or the IUS used previously on the Space Shuttle. The mass of the payloads and the tight launch windows necessitated the use of Centaur, but the Astronaut Office was not happy about it.

"One of the things astronauts do is they start looking at the hardware and so on and so forth," recalled John Fabian. "So I went out to San Diego [California] and watched them working and came back and reported on what I saw." Fabian's concerns were not only about the number of workers in the contractor's clean rooms, but the level of care taken within those rooms. "I wasn't the first to raise concerns about the Centaur. Bill Lenoir had raised concerns about it, because it had a failure mode that, in the event of a return-to-Earth launch abort, if you couldn't vent the propellants overboard, the liquid hydrogen and liquid oxygen overboard, before you came back into the atmosphere for landing, the thing could blow up in the bay. John Young called it 'Death Star', so it had something of a reputation. [4]

"In retrospect," Rick Hauck explained in 2003, "the whole concept of taking something that was never designed to be part of the human space flight mission, one that had many potential failure modes, was not a good idea, because you're always saying, 'Well, I don't want to solve the problems too exhaustively; I'd like to solve them just enough so that I've solved them'. Well, what does that mean? You don't want to spend any more money than you have to, to solve the problem, so you're always trying to figure out, 'Am I compromising too much or not?' And the net result is you're always compromising." **[23]**

Fig. 10.8: The five TFNG who crewed STS-51G are seen on the aft flight deck of *Discovery*. This was the first and only time that all the NASA crewmembers on a Shuttle flight came from the same selection group. (Clockwise from lower left): Brandenstein (CDR), Nagel (MS), Lucid (MS), Fabian (MS) and Creighton (PLT).

STS-51G (June 17 – 24, 1985)
Flight Crew: Daniel C. BRANDENSTEIN (CDR), John O. CREIGHTON (PLT), John M. FABIAN (MS-1), Steven R. NAGEL (MS-2), Shannon W. LUCID (MS-3), Patrick Baudry (PS-1, CNES, France), Prince Sultan Salman A. A. Al-Saud (PS-2, Saudi Arabia)
Spacecraft: Discovery (OV-103) 5th mission
Objective: 18th Shuttle mission; commercial satellite deployment
Duration: 7 days 1 hour 38 minutes 52 seconds
Support Assignments: Group 8 support assignments for STS-51G were identified in an Astronaut Office memo dated May 22, 1985: Loren SHRIVER (Lead in SPAN), with Anna FISHER, Robert GIBSON, Rick HAUCK and Sally RIDE;

Dave WALKER (KSC Launch Support Team/Cape Crusaders); Steve HAWLEY and Dick SCOBEE (Family Escorts). Other support assignments for this mission included Dave WALKER, Dick SCOBEE and Steve HAWLEY deployed to Edwards AFB for End Of Mission (EOM) activities, and Sally RIDE serving as Briefer for guests of the NASA Administrator at the Cape. [30] Two Group 8 astronauts were assigned as Capcoms for STS-51G. Mike COATS worked with Richard Richards (Group 9) on the 'Orion' team for both ascent and entry, while Mike MULLANE worked with Jim Wetherbee (Group 10) on the 'Sirius' team during Orbit 2 shifts.

The international crew, which included French astronaut Patrick Baudry and Saudi Prince Sultan Salman Al-Saud, deployed communications satellites for Mexico (Morelos), the Arab League (Arabsat), and the United States (AT&T Telstar). They used the RMS to deploy and later retrieve the SPARTAN satellite, which performed 17 hours of x-ray astronomy experiments while separated from *Discovery*. In addition, the crew activated the Automated Directional Solidification Furnace (ADSF) and six GAS canisters, participated in biomedical experiments, and conducted a laser tracking experiment as part of the Strategic Defense Initiative (SDI).

On the day NASA launched STS-51G, it also announced a new crew. [31] Loren Shriver was named CDR of STS-61I with PLT Bryan O'Connor and MS William F. Fisher (both from Group 9), together with Sally Ride and the first representative from the Group 10 Class of 1984 astronauts, Mark C. Lee. Scheduled for launch on *Challenger* no earlier than July 15, 1986, the crew was to deploy the Intelsat VI-1 and Insat 1-C communications satellites and carry the Material Science Laboratory-4 (MSL-4) payload.

Abort-to-Orbit
Though the next Shuttle mission, STS-51F, was the first since STS-6 two years and twelve missions earlier not to include a member of the 1978 selection among the crew, it was not originally intended that way. Dave Griggs was initially assigned as PLT for the mission, but as mentioned, he was replaced by Roy Bridges (Class of 1980) due to the close proximity of his earlier mission. Following a successful launch, one of the three main engines failed during the ascent, forcing the crew to follow the Abort-To-Orbit (ATO) profile, the first (and only) such scenario in the entire Shuttle program.

STS-51F (July 29 – August 6, 1985)
Objective: 19th Shuttle flight using *Challenger* (8th mission); 8-day Spacelab 2 research program; verification of Spacelab Igloo/pallet configuration (the pressurized Spacelab Long Module was not carried on this mission).
Support Assignments: Despite the fact that none of the Group 8 members actually flew on this mission, eight of them did serve in various support roles, continuing the TFNG link with each Shuttle mission either in ground support roles or as a

member of the flight crew. **[32]** Loren SHRIVER continued as Lead for the SPAN team, supported by Anna FISHER, Norman THAGARD, Rick HAUCK and Sally RIDE. Dave WALKER continued his role as Cape Crusader with the KSC launch support team and Pinky NELSON served as one of the Family Escorts at the Cape. Mike COATS was the only member of Group 8 to work on MCC. He was again assigned to the 'Orion' team as one of the Capcoms for Ascent and Entry.

During August 1985, Don Williams became Deputy Chief of the Aircraft Operation Division, a role he fulfilled until August 1986.

STS-51I (August 27 – September 3, 1985)
Flight Crew: Joe H. Engle (CDR), Richard O. COVEY (PLT), James D. A. VAN HOFTEN (MS-1), J. Michael Lounge (MS-2), William F. Fisher (MS-3)
Spacecraft: Discovery (OV-103) 6th mission
Objective: 20th Shuttle mission; commercial satellite deployment; satellite capture, repair and redeploy
Duration: 7 days 2 hours 17 minutes 42 seconds
Support Assignments: For this mission, Mike COATS served as Lead Capcom, with fellow Group 8 astronauts Fred GREGORY, Shannon LUCID and Pinky NELSON serving on different MCC shifts during the EVAs. Other support roles listed included: Loren SHRIVER (Lead in SPAN), with Dale GARDNER, Jeff HOFFMAN, Rhea SEDDON and alternates Mike MULLANE and Sally RIDE; Dave WALKER and Dan BRANDENSTEIN (KSC Launch Support/Cape Crusaders); Rick HAUCK (Family Escort); Rhea SEDDON (Briefer for Headquarters guests). **[33]**

Intending to capture, repair and redeploy the faulty LEASAT-3 originally released from STS-51D, this crew first had to deploy its own satellite cargo, the Australian Aussat-1 and ASC-1 owned by the American Satellite Company, on the first day of the mission. A third satellite, another LEASAT, was released later. The first EVA on September 1 saw Ox van Hoften, the largest astronaut in the office, manually grab the satellite and hold it in place while Dr. Bill Fisher attached a grapple bar so that the RMS could be attached. The next day, the two astronauts repaired the satellite by performing 'bypass surgery', before van Hoften became the first human being to launch a satellite by hand, having spun and pushed the repaired communications satellite away from the orbiting *Discovery*.

More assignments, changes and departures.
Several crew assignments were announced on September 19, some of which involved the Group 8 astronauts. **[34]**. Dave Griggs and Bob Stewart were added to the rescheduled EOM/STS-61K mission, now set for a launch in September 1986. Don Williams was named as CDR for STS-61I, which was scheduled to

retrieve the LDEF satellite and deploy the Intelsat V-1 communication satellite. He would be joined by a crew of Group 9 and 10 astronauts. There were also changes made to crews of other flights. Norman Thagard was moved from STS-61H to STS-61G replacing John Fabian, who had announced his intention to retire from NASA effective January 1, 1986, and return to the USAF, while Jim Buchli replaced Thagard on his former mission. The former 61I crew commanded by Loren Shriver was reassigned to STS-61M.

STS-51J (October 3 – 7, 1985)
Flight Crew: Karol J. Bobko (CDR), Ronald J. Grabe (PLT), David C. Hilmers (MS-1), Robert L. STEWART (MS-2), William Pailes (PS-1, USAF MSE)
Spacecraft: Atlantis (OV-104) 1st mission
Objective: 21st Shuttle mission; 2nd classified DOD Shuttle mission; maiden flight of Atlantis
Duration: 4 days 1 hour 44 minutes 38 seconds
Support Assignments: Once again, despite the secret nature of the payload and mission, details of some of the support roles were released in an Astronaut Office memo dated September 19, 1985. Mike COATS again served as Lead Capcom, with Fred GREGORY and Shannon LUCID working other MCC shifts. Other support roles listed included: Loren SHRIVER (Lead in SPAN), with Dave GRIGGS, Ellison ONIZUKA and Norm THAGARD, with alternate John CREIGHTON; Dan BRANDENSTEIN (KSC Cape Crusader); Dave WALKER (Family Escort); Ellison ONIZUKA (Briefer for Headquarters guests).

This mission was the second highly-classified DOD mission, as well as the maiden voyage of *Atlantis*, the final planned Orbiter in the Shuttle fleet. During the flight, the crew is thought to have deployed two DSCS communication satellites.

STS-61A (October 30 – November 6, 1985)
Flight Crew: Henry W. Hartsfield (CDR), Steven R. NAGEL (PLT), Bonnie J. Dunbar (MS-1), James F. BUCHLI (MS-2), Guion S. BLUFORD (MS-3), Ernst W. Messerschmid (PS-1, DFVLR, West Germany), Reinhard Furrer (PS-2, DFVLR, West Germany), Wubbo Ockels (PS-3, ESA, The Netherlands)
Spacecraft: Challenger (OV-099) 9th mission
Objective: 22nd Shuttle mission; Spacelab-D1 (LM-2 unit), West German research program
Duration: 7 days 0 hours 44 minutes 53 seconds
Support Assignments: For this mission, Mike COATS again served as Lead Capcom, assigned with Fred GREGORY on the 'Gray' team for launch and entry phases. Shannon LUCID worked on the (lead) 'Indigo' team for Orbit 1, while Sally RIDE worked Orbit 3 with the 'Rigel' team. [35] Other support roles for this flight included: Loren SHRIVER (Lead in SPAN), with John CREIGHTON, John FABIAN and alternate James VAN HOFTEN; Dan BRANDENSTEIN (KSC

Launch Support/Cape Crusader); Dave GRIGGS (Family Escort). In the Astronaut Office Weekly Report w/e August 28, 1985, Loren SHRIVER was listed as a member of the EOM Support team at Edwards AFB.

This mission was the first to carry eight crewmembers, the largest crew to fly in space, and was a German-financed and dedicated Spacelab mission, the first to be chartered by another nation. The week-long flight was devoted to 76 scientific experiments in the fields of materials processing and life sciences. The Spacelab experiment operations were controlled by the West German DFVLR center, near Munich, using the TDRS-1 (originally known as TDRS-A pre-launch, but then renamed) and Intelsat satellites. With this flight coming ten months after his first mission, 51C, Jim Buchli became the first American astronaut to make two separate flights in a single year[6]. Ironically, a second astronaut on the same mission also achieved this rare event, as Steve Nagel was flying his second mission as PLT just 128 days after flying as MS on STS-51G, setting a new record turnaround. (Buchli was deemed to be the first, having flown his first mission earlier than Nagel.) Again, a two-shift operation was employed on this flight, with Nagel serving as Blue Shift leader and Buchli as Red Shift leader. Guion Bluford was also assigned to the Red Shift.

In November, the POC for aspects of the Flight Data File (FDF) were updated in an Astronaut Office memo, with effect from STS-61B and subsequent missions. Issues related to the Crew Activity Plan (CAP) and the Photo/TV Checklist were the responsibility of the specific crew on each flight, but the relevant Group 8 astronauts assigned (sometimes with members from other astronaut groups) as POC in this memo were: Rick Hauck (CDR, STS-61G) and Dave Walker (CDR, STS-61F) for Centaur Deploy; Ellison Onizuka for IUS deploy; Judy Resnik for PDRS Checklist; and Dick Covey for Rendezvous. **[36]**

STS-61B (November 27 – December 3, 1985)
Flight Crew: Brewster H. SHAW (CDR), Bryan D. O'Connor (PLT), Jerry L. Ross (MS-1), Mary L. Cleave (MS-2), Sherwood C. Spring (MS-3), Charles D. Walker (PS-1, McDonnell Douglas), Rodolfo Neri Vela (PS-2, Mexico)
Spacecraft: Atlantis (OV-104) 2nd mission
Objective: 23rd Shuttle mission; commercial satellite deployment; EVA construction demonstrations
Duration: 6 days 21 hours 4 minutes 49 seconds
Support Assignments: Capcoms for this flight included Fred GREGORY as lead, with Dick COVEY, Shannon LUCID and Sally RIDE. According to an Astronaut

---

[6] The first Soviet cosmonauts to achieve this feat did so in 1969. In January that year, Vladimir Shatalov had flown on Soyuz 4 and was joined during the mission by Alexei Yeliseyev following an EVA from Soyuz 5 on which he had launched. Both then flew together again on Soyuz 8 that October.

Office memo dated November 14, 1985, the Lead in SPAN was now assigned to John CREIGHTON, who was supported by Anna FISHER. Loren SHRIVER was one of the Family Escorts.

During this mission, the crew deployed the communications satellites, conducted two six-hour spacewalks to demonstrate space station construction techniques with the EASE/ACCESS experiments, operated the CFES experiment for McDonnell Douglas and a GAS container for Telesat, Canada, conducted several PS experiments for the Mexican Government and tested the OEX DAP. This was the heaviest payload weight carried to orbit by the Space Shuttle to date.

On December 17, NASA released the final crew announcement of a busy year, which would turn out to be the final such announcement for 13 months though of course this was not known at the time. The recently returned STS-61B CDR, Brewster Shaw, was assigned to command the STS-61N crew, a dedicated DOD mission scheduled for September 1986. The remaining crew consisted of Group 9 MS astronaut Dave Leestma, and Group 10 astronauts Michael McCulley (PLT), Jim Adamson and Mark Brown (both MS). **[37]** During the month, Dave Walker became the new Lead Cape Crusader at KSC.

MISSION IMPOSSIBLE

On January 1, 1986, John Fabian officially left NASA to take up a position as Director of Space, Deputy Chief of Staff, Plans and Operations, Headquarters, USAF, The Pentagon, Washington D.C. At the time of his decision to leave the astronaut program, he was in training as an MS for both STS-61G and SLS-1 and he was subsequently replaced on those crews by other astronauts.

STS-61C (January 12 – 18, 1986)
Flight Crew: Robert L. GIBSON (CDR), Charles F. Bolden (PLT), George D 'Pinky' NELSON (MS-1), Steven A. HAWLEY (MS-2), Franklin R. Chang Díaz (MS-3), Robert J. Cenker (PS-1, RCA), C. William Nelson (PS-2, U.S. Congressman)
Spacecraft: Columbia (OV-102) 7th mission
Objective: 24th Shuttle mission; commercial satellite deployment
Duration: 6 days 2 hours 3 minutes 51 seconds
Support Assignments: Group 8 astronauts assigned to MCC included Fred GREGORY as Lead Capcom on the 'Gray' team for ascent and entry, and Shannon LUCID on the (lead) 'Emerald' team for Orbit 1. Dick COVEY was also listed as Weather (WX) Capcom. **[38]** The Astronaut Office memo dated December 6, 1985 listed other support assignments as: John CREIGHTON (Lead in SPAN), with support from Dave GRIGGS, Loren SHRIVER and Kathy SULLIVAN among others.

Known as the 'end of year clear out flight' due to changes in the manifest, the mission carried only one small satellite for deployment. This flight also received the unfortunate tag of 'Mission Impossible', as it required eight attempts to get into orbit and was delayed twice in getting back on the ground. During the six-day flight, the seven crewmembers deployed the RCA SATCOM KU satellite, as part of a network of three satellites that would provide commercial communications services within the Ku-band of the electromagnetic spectrum. They also conducted experiments in astrophysics and materials processing. Unfortunately, a special camera designed to photograph Comet Halley failed, much to the disappointment of Pinky Nelson and Steve Hawley, the two professional astronomers on-board.

When *Columbia* touched down on Runway 22 at Edwards AFB on January 18, 1986 at the end of STS-61C, a productive six-day mission, the Space Shuttle program had completed 24 missions using NASA's fleet of four Orbiter vehicles. The landing had initially been delayed by a day, and then by bad weather at the Cape, before *Columbia* finally touched down at Edwards on the fifth attempt.

Although successful, the mission had gained a little notoriety in the press, as one of the seven crewmembers was C. William (Bill) Nelson II, who became the second sitting member of Congress (and the first member of the House) to travel into space following the flight of Senator Jake Garn on the STS-51D mission nine months earlier. There were continuing grumblings both within the astronaut corps and the American media about politicians exerting their influence to take precious seats on Shuttle missions, and neither man had been exempt from criticism.

SUMMARY

Flight records indicate that at the end of *Columbia*'s latest mission, all members of the TFNG group had successfully flown into space, amassing a total of 48 flights and with 15 having completed a second mission. Several of them had already been assigned to future Shuttle missions. Then, on the cold, frosty morning of January 28, 1986, just ten days after Hoot Gibson had glided *Columbia* to a safe landing at Edwards, everything would change suddenly and dramatically. On that day, Shuttle *Challenger* was poised to fly into space on what would have been the Orbiter's much-delayed tenth flight.

References

1. Sally Ride, NASA Oral History, October 22, 2002.
2. <u>Florida Today</u>, June 19, 1983.
3. *Flight Control of STS-7*, NASA JSC News 83-017, June 7, 1983.
4. John Fabian, NASA Oral History, February 10, 2006.
5. *Flight Control of STS-8*, NASA JSC News 83-033, August 25, 1983.
6. Guion Bluford, NASA Oral History, August 2, 2004.

7. JSC Space News *Roundup*, Volume 22, Number 7, April 13, 1983, p. 4.
8. **The Astronaut Maker, How One Mysterious Engineer Ran Human Spaceflight for a Generation**, Michael Cassutt, Chicago Review Press, 2018, p. 251.
9. *Shuttle crews selected*, NASA JSC News, 83-036 September 21, 1983.
10. Email to author David J. Shayler from Steve Hawley, November 3, 2019.
11. *STS flight assignments*, NASA JSC News 83-046, November 17, 1983.
12. *STS-51D, 61D Crew Announcements*, NASA JSC News, 84-005, February 2, 1984.
13. *STS-51K Crew Announcement*, NASA JSC News 84-009, February 14, 1984.
14. *Astronaut T. J. Hart to Leave NASA*, NASA News 84-027, May 10, 1984.
15. *NASA Announces Crewmembers for Future Space Shuttle Flights*, NASA JSC News 84-029, June 7, 1984.
16. *NASA Announces Updated Flight Crew Assignments*, NASA JSC News, 84-036, August 3, 1984.
17. *Flight Control of STS-41G*, NASA News 84-046, October 2, 1984.
18. *NASA Announces Flight Assignments and Changes*, NASA JSC News 84-048, October 22, 1984.
19. *Flight Control of STS-51A*, NASA JSC News Release 84-051, November 2, 1984.
20. *NASA Names crews to deploy satellites in year-end flights*, NASA JSC News 85-005, January 29, 1985.
21. *Crews for First Vandenberg Mission, DOD flight named*, NASA JSC News 85-009, February 15, 1985.
22. Email to author David J. Shayler from Rhea Seddon, April 17, 2019.
23. Rick Hauck, NASA Oral History, November 20, 2003.
24. (NASA Announces New Shuttle Manifest) NASA Headquarters News 85-50, April 5, 1985.
25. *Flight Control of STS-51E*, JSC News Release 85-101, February 20, 1985.
26. *Flight Control of Space Shuttle mission 51D*, JSC News 85-015, April 9, 1985.
27. Astronaut Office Weekly Activity Report, w/e April 17, 1985, copy on file AIS Archives.
28. *Flight Control of Space Shuttle Mission 51B*, JSC News 85-018, April 28, 1985.
29. *NASA Names Astronaut Crews for Ulysses, Galileo Missions*, NASA JSC News 85-022, May 31, 1985.
30. Astronaut Office Weekly Activity Report, w/e June 19, 1985, copy on file AIS archives.
31. *NASA Names Astronaut Crew for Space Shuttle Mission 61I*, NASA JSC News, 85-027, June 17, 1985.
32. Astronaut Office Memo dated July 2, 1985; also Astronaut Office Weekly Activity Report w/e July 17, 1985, copy on file, AIS Archives.
33. Astronaut Office Memo August 8, 1985; also Astronaut Office Weekly Activity Report, w/e August 28, 1985, copy on file, AIS Archives.
34. *NASA Names crews for Upcoming Space Shuttle Flights*, NASA JSC News 85-035, September 19, 1985.
35. *Mission Control Teams for Flight 61A/Spacelab-D1*, NASA JSC News 85-042, October 28, 1985.
36. Astronaut Office Memo November 19, 1985, copy on file, AIS Archives.
37. *NASA Names Astronaut Crew for Department of Defense Mission*, NASA JSC News, 85-053, December 17, 1985.
38. *Flight Control of Shuttle Mission 61C*, JSC News 85-052, December 13, 1985.

11

"Go at throttle up"

*"The naming of the crew for the next flight
is a major event in the process
of returning the Shuttle to flight."*
Rear Admiral Richard H. Truly,
NASA Associate Administrator for Spaceflight
STS-26 crew announcement, January 9, 1987.

One year before Dick Truly's statement, the Shuttle manifest for 1986 featured a demanding 15 launches, including 13 from the Kennedy Space Center (KSC) and the first two from Vandenberg Air Force Base (AFB) in California, as shown in Table 11.1.

The highlights included the first deployments of the liquid-fueled Centaur upper stage, the maiden launch from Vandenberg (to send a Shuttle into polar orbit for the first time), and the long awaited deployment of the Hubble Space Telescope (HST). The two delayed Tracking and Data Relay Satellite (TDRS) deployments were also on the manifest, along with four missions carrying commercial satellites, one of which was intended to retrieve the Long Duration Exposure Facility (LDEF) deployed during the STS-41C mission in April 1984. There were three further classified Department of Defense (DOD) missions (two from KSC and the second launch from Vandenberg) and two Spacelab science missions (Astro pallet-only payload, and Earth Observation Mission 1 Small Module and pallet payload).

All 12 of the 1986 missions included at least one member from the Thirty-Five New Guys (TFNG) Class of 1978 on each flight crew. Now a full decade after they responded to the call for new astronauts to train as crewmembers on the Space Shuttle, the remaining 33 members of the TFNG continued to be a strong presence at the forefront of American human space flight.

© Springer Nature Switzerland AG 2020
D. J. Shayler, C. Burgess, *NASA's First Space Shuttle Astronaut Selection*,
Springer Praxis Books, https://doi.org/10.1007/978-3-030-45742-6_11

TABLE 11.1: STS CREW ASSIGNMENTS (1986) NASA ASTRONAUTS ONLY

STS Flight	Primary payload	Orbiter	Planned launch date	Commander	Pilot	Mission Specialists		
61-C	Satellite deployment	Columbia	Jan 12	GIBSON R.	Bolden	NELSON G.	HAWLEY	Chang-Diaz
51-L	TDRS-B	Challenger	Jan 28	SCOBEE	Smith M.	ONIZUKA	RESNIK	McNAIR
61-E	Astro-1	Columbia	Mar 6	McBRIDE	Richard R.	Leestma	HOFFMAN	Parker
61-F	Ulysses solar polar orbiter	Challenger	May 15	HAUCK	Bridges	Hilmers	Lounge	
61-G	Galileo Jupiter probe	Atlantis	May 20	WALKER D.	Grabe	THAGARD	VAN HOFTEN	
61-H	Satellite deployment	Columbia	Jun 24	COATS	Blaha	Springer	FISHER A.	BUCHLI
62-A	DOD	Discovery	Jul 1 1st Vandenberg	Crippen	Gardner S.	MULLANE	Ross	GARDNER D.
61-M	TDRS-C	Challenger	Jul 22	SHRIVER	O'Connor	Lee	RIDE	Fisher W.
61-J	Hubble deployment	Atlantis	Aug 18	Young J.	Bolden	McCandless	HAWLEY	SULLIVAN
61-N	DOD		Sep 4	SHAW	McCulley	Leestma	Brown M.	Adamson
61-I	Intelsat-4; LDEF retrieval		Sep 27	WILLIAMS D.	Smith M.	Bagian	Dunbar	Carter
62-B	DOD		Sep 29 Vandenberg					
61-K	Earth Observation Mission - 1		Oct 1	Brand	GRIGGS	STEWART	Garriott O.	
61-L	Satellite deployment		Nov 1					
71-B	DOD		Dec					

Even taking into account the skills of the astronauts and ground crews, and the success of past missions, it was clearly obvious that suddenly increasing flight operations from nine missions in 1985 to 15 in 1986 – and from two different launch sites on opposite sides of the country – was going to be difficult at best, even without considering potential hardware failures and weather delays. The difficulties experienced each side of the 1985 Christmas holidays in getting STS-61C off the ground, and in returning it to Earth during early January 1986, certainly underlined that, but no one could foretell the enormous and tragic setback the very next mission would bring.

History reminds us that Tuesday, January 28, 1986, would end as a second dark day in NASA's history. The vivid events of that day are etched in the memory of those who were there or watched the event unfold on TV around the world. They would give rise to the desire to honor the sacrifice of those lost that day and, when the time was right, to return astronauts to orbit. Dick Truly's statement which opened this chapter was a step in that direction. The tragic accident which claimed the lives of seven brave astronauts, including four members of the 1978 selection, came 19 years and one day after the horrendous pad fire that claimed the three lives of the first Apollo crew, thereby adding to those still-raw memories of nearly two decades before.

CHALLENGER

The primary goal of STS-51L was to launch the second TDRS (TDRS-B). This was one of the scheduled duties for Mission Specialist (MS) Ellison Onizuka during the six-day *Challenger* flight. He was also due to film Halley's Comet with a hand-held camera. *Challenger* was also to carry the Spartan Halley spacecraft, a small free-flying satellite that MS Ron McNair and Judith Resnik were to release and then pick up two days later using *Challenger*'s Remote Manipulator System (RMS) robotic arm, once Spartan had observed Halley's Comet during its closest approach to the Sun. This intended mission would never be completed.

STS-51L (January 28, 1986)
Flight Crew: Francis R. SCOBEE (CDR), Michael J. Smith (PLT), Ellison S. ONIZUKA (MS-1), Judith A. RESNIK (MS-2), Ronald E. McNAIR (MS-3), Gregory B. Jarvis (PS-1, Hughes Communications Inc.), S. Christa McAuliffe (PS-2, Teacher)
Spacecraft: Challenger (OV-099) 10th mission
Objective: 25th Shuttle mission; deployment of TDRS-B; observations of Comet Halley; Teacher in Space program
Duration: 1 minute 13 seconds (planned as 6 days 34 minutes)

Support Assignments: For this mission, Fred GREGORY served as Lead Capcom, with fellow Group 8 astronaut Dick COVEY, who were both in the Mission Control Center (MCC) as the disaster of 51L unfolded on the TV screens, audio links and computer consoles around them. Had the flight proceeded according to plan, the SPacecraft ANalysis (SPAN) team would have been led by John CREIGHTON, with Steve NAGEL and Don WILLIAMS in support. [1] No other details of planned support roles by the TFNG could be found in the sources researched by the authors.

The STS-51L flight had originally been scheduled for December 1985, but was then rescheduled for launch on January 22, 1986. Operational difficulties further delayed the flight, bumping it daily from January 22 through to Saturday January 25. That morning, a group of senior NASA officials met at the Cape for their "L-1 [Launch minus one] day review." The weather was of principal concern, with the recent cold spell persisting and rain forecast for the next day. For a Shuttle launch, rain would present a no-go situation because even something as innocuous as raindrops could damage the delicate heat tiles on the exterior of the Orbiter, causing pockmarks that might later compromise the integrity of the tiles during the intense heat of re-entry. The group met again later that night, with rain still threatened for the next day, so a decision was made to postpone the launch for another 24 hours.

To the crew's frustration, the weather on Sunday, January 26 was fine. NASA officials were still concerned about a build-up of threatening weather both in the vicinity of the Cape and at the emergency landing field in Africa. However, the expected bad weather never materialized, and in fact conditions would have been perfect for a launch.

Speaking to reporters after the evening meeting on January 26, NASA's Associate Administrator, Jesse W. Moore, issued a sadly prophetic statement: "We're not going to launch this thing and take any kind of risk because we have that schedule pressure. We're going to continue to abide by the flight rules that we've established in this program and we'll sit on the ground until we all believe it's safe to launch." [2]

The following day, Monday 27, the crew were dressed in their flight suits, boarded a transfer bus, and were driven the six miles out to Launch Pad 39B, delighted to hear updates stating that the launch countdown was proceeding and on schedule.

Once at the launch pad, the 51L crew disembarked and entered a small elevator that would take them up to the entry level, where the Shuttle's open hatch awaited them. They gathered in the area outside the hatch where they donned the rest of their flight suits and safety harnesses. Mission Commander (CDR) Dick Scobee and Pilot (PLT) Mike Smith then put on their helmets and safety harnesses before being assisted through the hatch into the Orbiter. They were followed in turn by the two MS who would sit behind them on the flight deck for the launch, Judy Resnik and Ellison Onizuka. Finally, it was the turn of the remaining three crew-members, who would occupy seats on the middeck during the launch phase. They were MS Ron McNair, Payload Specialist (PS) Greg Jarvis from the Hughes Aircraft Company, and a schoolteacher named Christa McAuliffe, who had been selected after a nationwide search among suitable educator applicants.

While the crew were being strapped in, the skies over the Cape were clear, with an air temperature of 4.4°C (40°F). Then, with just nine minutes left on the count-down clock, a hold was announced after it was discovered that an external handle on the Orbiter's hatch was stuck and would not close properly. Technicians were rushed in and tried to remove a bolt causing the problem but were unable to do so. Over the next four hours, the crew waited patiently until the bolt was finally blown, but by then the crosswinds had picked up over the KSC runway to a level that was unacceptable in the event of an emergency Return To Launch Site (RTLS) abort. The decision was made to postpone the flight once again, until 9:38 am the follow-ing morning. The crew was understandably frustrated by yet another delay, espe-cially Judy Resnik. Unfortunately, she had become used to the frustration of technical problems and launch delays, having suffered two such setbacks on her first flight, STS-41D in 1984. STS-51L had now been delayed five times.

It was freezing cold on Launch Pad 39B the following morning, Tuesday, January 28. As night gave way to morning, a dull and wintery sunrise had slowly spread daylight over the Cape. Despite the extreme chill in the air, the early fore-cast was for clear skies, and the crewmembers, woken at 6:30 am, were told this could be their day. Out by the launch pad, where *Challenger* had stood poised for launch over the past 38 days, a battery of 45 massive flood lamps that had lit the steel launch tower during the night were extinguished.

By 7:00 am, the temperature stood at -3°C (27°F), and NASA's "ice team" had completed their morning inspection of the Shuttle's huge external fuel tank, which had been refueled with super-cooled propellants. It was also the team's job to report on the amount of ice at the launch pad in these conditions, in case it reached dangerous levels. Their report was alarming. Sheets of ice and thick, sharp stalac-tites had formed over many of the pad facilities. Given the massive vibration asso-ciated with a Shuttle launch, this ice could be dislodged and tumble down onto the ascending Orbiter. The team managed to break off and remove many of the thicker sections of ice, but it soon began to build up again in the chill, moist air. The situ-ation, as they reported, was far from ideal. Nevertheless, preparations continued.

Once again, the crew were assisted into their flight suits and, following a ten-minute briefing, were driven out to the launch pad. On arrival they stepped down from the van and CDR Scobee chatted briefly with the launch pad crew. "My kind of day," he said. "What a great day for flying."

As she was preparing to enter *Challenger* through the open crew hatch, Christa McAuliffe was handed a last-minute gift by a launch technician that caused her to chuckle. It was a shiny red apple, presented to America's favorite teacher. Although she was unable to take it on board, Christa handed the apple back with a smile, thanking the man for his kindness.

By 8:36 am, all seven crewmembers had clambered through the open hatch and were tightly strapped into their seats aboard *Challenger*, where they began going through their pre-launch checks. During this time, they were informed that liquid hydrogen tanking problems, delays created by hardware interface problems, and a broken water pipe on the pad, had led to the launch being rescheduled for 11:38 am.

NASA officials were still edgy about the cold weather and how it might affect the launch. Up until that time, the average temperature experienced on Shuttle launches had been 23°C (73°F), and the coldest temperature at any lift off was 10.5°C (51°F). By mid-morning, the temperature at the Cape had only crept up to a few degrees above zero. The launch pad crew had earlier been asked to double-check for any problems with ice build-up. Not only could it damage launch equipment, but it had been known on some previous missions that ice falling from the cryogenic-filled External Tank (ET) had hit and damaged the fragile tiles of the Shuttle's exterior during launch. While the pad crew's report at 8:44 am was not altogether encouraging, it did state that the majority of the ice could either be cleared away in time, or even melt to an acceptable level before the revised launch time. They conducted a third inspection at 10:30 am, during which they cleared away some lingering ice from the launch platform. Meanwhile, at T-70 minutes, *Challenger*'s hatch was closed, sealing the crew inside.

As the countdown ticked by, Scobee and Smith ran through all their checks, keeping an eye out for any malfunctions in the Shuttle's systems. They were mindful that they would be completely at the mercy of technology for the first two minutes and eight seconds of the launch, as *Challenger* hurtled ever faster into the Florida skies atop 3,000,000 kg of thrust.

The countdown finally clicked over into the final ten seconds before launch. As the count reached the six seconds mark to lift off, the Shuttle's three main engines burst into life, quickly building up thrust prior to the ignition of the twin Solid Rocket Boosters (SRB). On board, Scobee reported the firing of the main engines to his crew, saying, "There they go, guys!" An excited Judy Resnik cried out, "Alright!" followed by Scobee reporting that all three engines were at 100 percent of their rated power.

At 11:38:00:010 am EST, *Challenger*'s SRBs roared into life, and the winged spacecraft began to rise atop a brilliant pillar of white-hot flame. Clouds of black, orange and white billowed outward as the Shuttle cleared the launch tower and rose into the sky. KSC Public Affairs Officer (PAO) Hugh Harris calmly announced to the onlookers and television audience: "Lift off! Lift off of the twenty-fifth Space Shuttle mission and it has cleared the tower." Steady as a rock, *Challenger* soared into the heavens, trailing a 200-meter geyser of fire and smoke.

Sixteen seconds into the flight, *Challenger* automatically executed a single-axis rotation, and Dick Scobee reported, "Houston, we have a roll program," as the Shuttle arched gracefully backward, assuming a 'heads down' orientation and the correct downrange course for passing through the Earth's atmosphere. This flight configuration ensured that aerodynamic stresses on the Shuttle assembly were tolerable prior to entering orbit.

A further 20 seconds later, *Challenger*'s engines were throttled down to 65 percent of full power as the crew prepared themselves for the Shuttle's passage through the period of highest turbulence, known in the NASA lexicon as Max-Q, or maximum dynamic pressure. At this point, the force of air rushing past the ascending Shuttle assembly could cause severe damage, so the acceleration process is deliberately slowed.

Over the next 14 seconds, the crewmembers were jolted around in their seats as *Challenger* passed through a fierce wind shear. Soon after, the air began to thin as they reached higher altitude, and the outside pressure decreased, at which time the main engines could be brought back up to full thrust. "Throttling up," Scobee announced to ground controllers.

Group 8 astronaut Dick Covey was the Capcom that day, running through the post-launch sequences with Scobee and Smith. All the data before him indicated a perfect flight trajectory as *Challenger* headed east out over the Atlantic, and he responded to Scobee's call with, "*Challenger*, go at throttle up."

Everything was proceeding normally as Dick Scobee pressed the transmit button. It was 70 seconds into the flight. "Roger," he responded. "Go at throttle up." Suddenly the Shuttle assembly began to shudder and sway violently. Pilot Mike Smith realized something was seriously wrong and just had time to utter the words, "Uh oh!" before a massive fireball flashed along the length of *Challenger*, followed by a titanic explosion.

In hindsight, the launch should never have taken place under the prevailing conditions of that freezing cold morning. Even as the final countdown was taking place, two O-rings located in the lowermost field joint of the right-hand SRB had become rock hard due to the extreme cold, which resulted in them losing their designed ability to seal the crucial joint completely. As the Shuttle assembly blazed a path into the Florida skies, a puff of black smoke escaped from the joint at the bottom of the right-hand booster rocket. This then became a thin stream of superhot propellant gas which grew in intensity. Acting like a blowtorch, the white-hot plume of gas then played on one of the steel struts holding the SRB against the huge, hydrogen-filled ET. The strut was burned through and broke apart, at which point immense aerodynamic forces caused the booster to sever and swivel around, slamming into the massive fuel tank, which ruptured. Two million liters of propellant exploded in a huge fireball.

There is an understandable misconception that *Challenger* was consumed by the ensuing fireball, but the Shuttle was actually lost when it was ripped apart by the force of the air, as explained by veteran Shuttle astronaut Don Peterson, a crewmember on STS-6, *Challenger*'s maiden flight into space in April 1983.

"When that vehicle turned sideways, the wind force just ripped it apart," Peterson once told JSC interviewer Jennifer Ross-Nazzal. "When we launch the Shuttle and we light all five engines, we're burning ten and a half tons of fuel per second. That's the weight of three full-sized automobiles every second being burned up. The amount of energy and the force and the power that's in that vehicle is gigantic … and when you're boring through the atmosphere at high speed, the wind force is tremendous. The Shuttle is not designed to stand big side loads. You've got to keep it pointed exactly properly. Once that rocket came loose and pushed the stack sideways, it just came apart. It just literally disintegrated." [3]

There was immediate confusion among those on the ground watching the launch. One moment they were watching a perfect launch and the Shuttle assembly soaring into the blue Florida skies; the next moment, a vast white, orange and red cloud billowed outwards from where *Challenger* had been, and the two boosters unexpectedly emerged from the conflagration, still firing wildly. The cheers of the spectators were quickly choked back amid the confusion, and silence soon fell as the reality of what the massive fireball above them meant hit home. As the seconds ticked by, debris trailing white smoke drifted out of the billowing cloud, while people still clung to the increasingly forlorn hope that *Challenger* had somehow survived the explosion and would appear at any moment, heading back for a landing at the Cape.

Across the United States, millions of people were watching the launch live on television, including many thousands of schoolchildren and teachers who had been following Christa McAuliffe's training, in anticipation of watching the lessons she was going to transmit to them from orbit. Once they realized that something had gone horrifyingly wrong, shocked teachers, many of them crying uncontrollably, ushered their confused and frightened students back into their classrooms. A collective grief would sweep across the country for many days to come, while images of the Shuttle erupting into a fireball and disintegrating on live television will be forever burned into the minds and memories of those who witnessed the tragedy.

As Rhea Seddon later told *Popular Mechanics* magazine: "I happened to be at an off-site building [near the Johnson Space Center, JSC] doing some training for my next mission. The launch was supposed to start around the same time as our meeting, so we found a TV and turned it on. It's always a joyful morning, especially to see friends go to space.

"I said, 'Oh, look, you can see the boosters coming off'. And one of my crewmates – he was looking at his watch – said, 'No, it's too early'. We weren't supposed to jettison the boosters until they were burned out. The camera panned back, and all you could see was a cloud of … stuff. We thought the engines were still going, with the Shuttle attached to the [fuel] tank. Then they showed the ocean and there were pieces coming down – big chunks of something." **[4]**

As Mission Control Capcom on the STS-51L mission, Dick Covey was the designated sole communicator with the crew aboard *Challenger*. "We had been disciplined to watch our data, not to get distracted by watching whatever video might be running in the control center," he recalled, also for *Popular Mechanics* magazine. "So I'm watching my data, and there's nothing unusual through the throttle up. The engine guys confirm that the engines look good, so I make a call: 'Go at throttle up'. Dick [Scobee] responded. And then I'm starting to think about what's the next thing that's coming, if we're going to make a call or whatever, and the data just went all *M*s, which is 'missing'.

"Fred [Gregory] was in charge of weather and so he was able to watch the video, and he almost immediately says, 'Look!' And so I turned and I looked, but I didn't know what I was looking at. I didn't see how it originated, nor did I understand exactly what it was. It just didn't register with me. At best, the Orbiter was separated from everything else and maybe trying to find a way to fly. And the worst thing in an emergency when you are flying is to have someone on the ground trying to talk to you without giving you the help that you need. So I was reluctant to be a distraction unless I had something I could tell them. If they weren't talking to us, then they either didn't have the ability to talk to us, or they were busy enough that they didn't want to talk to us. I asked for emergency procedures, contingency aborts, anything that we could do to help them. And then it became obvious that there wasn't anything we could do, and we were spectators in a tragedy.

"People said, 'Well, the External Tank obviously exploded', because that was what made the big fireball. But what caused it, nobody had a clue. It was days, if not a week, before any of us could say, 'All right, now I'm starting to understand what went wrong'. It was disappointing, angering almost, to find out that discussions had been held relative to [the known issue of joint failure in the SRBs]. Why wasn't that a bigger issue? How did we get to the point of accepting that indicator of a system not working the way it was designed? Why had we been willing to accept that?" **[4]**

Two days after the tragedy, as investigations got underway to determine the cause of the loss of *Challenger* and her seven crewmembers, a somber President Ronald Reagan gave a moving tribute at a memorial service for the seven astronauts at the JSC central mall. In his speech he recalled the lives lost: Richard (Dick) Scobee, Mike Smith, Judy Resnik, Ron McNair, Ellison Onizuka, Greg Jarvis and Christa McAuliffe. He also cited the stirring words of the poem *High Flight* by John Gillespie Magee, and reminded those present that the spirit of the American nation was based on heroism and noble sacrifice.

The cremated remains of the *Challenger* crew were eventually returned to their families for burial, with the final resting place for Dick Scobee and Mike Smith located within the peaceful surroundings of the Arlington National Cemetery in Virginia. Ron McNair's remains were originally buried at the Rest Lawn Memorial Park in Lake City, South Carolina; Ellison Onizuka in the National Memorial Cemetery of the Pacific in Honolulu, Hawaii; and Christa McAuliffe in Calvary Cemetery, Concord, New Hampshire; Greg Jarvis's ashes were returned to Hermosa Beach, California, where they were scattered by his family in the Pacific Ocean. Judy Resnik's family members have never revealed the location of her final resting place. Some sources say that her ashes were buried at sea, others that she is buried in a Jewish cemetery in her home town of Akron, Ohio. What is known is that unidentifiable remains found intermingled within the crushed crew cabin were later cremated together and placed beneath a *Challenger* memorial stone and plaque at Arlington National Cemetery.

The Rogers Commission

On February 3, 1986, six days after the loss of *Challenger*, President Ronald Reagan formed a Commission (Executive Order 12546) to investigate the tragedy. Chaired by attorney William P. Rogers, the commission included: former astronaut Neil A. Armstrong; attorney David C. Acheson; educator and engineer Eugene E. Covert; physicist Richard P. Feynman; editor and publisher Robert B. Holz; Major General Donald J. Kutyna, United States Air Force (USAF); aerospace engineer Robert W. Rummel; aeronautical engineer Joseph F. Sutter; astronomer Dr. Arthur B. C. Walker, Jr.; physicist Albert D. Wheelon; Brigadier General Charles Yeager, USAF Retd.; Dr. Alton G. Keel, Jr.; and TFNG Sally K. Ride.

The commission interviewed over 160 individuals and conducted more than 35 formal panel investigation sessions, generating almost 12,000 pages of transcripts. There were almost 6,300 documents containing over 122,000 pages and hundreds of images, added to over 2,000 pages of hearing transcripts. The resulting report was submitted on June 9, 1986, and offered nine recommendations to improve safety in the Shuttle program. President Reagan then directed NASA to report back within 30 days on how it was going to implement these recommendations.

The recommendations and implementations included: SRB redesign overseen by an independent oversight group; reorganization of the Shuttle management structure and the inclusion of astronauts in management roles; the creation of a new Office of Safety, Reliability and Quality Assurance, headed by an Associate Administrator; improved communications; creation of a crew escape system; revision of the Shuttle flight rate; the addition of a new Orbiter (OV-105); and relocating DOD satellites to Expendable Launch Vehicles (ELV) instead of the Shuttle. In August 1986, it was announced that the Space Shuttle would no longer carry commercial satellite payloads.

In addition to the Commission activities, NASA also put a vast effort into the investigation, including over 1,300 employees from across the NASA field centers, supported by 1,600 employees of other government agencies and more than 3,100 from NASA contractors. In addition, the U.S. military, the Coast Guard and the National Transportation Safety Board were involved in the salvage and analysis of the wreckage. [5]

In 2002, Sally Ride was interviewed about her role in the Commission for the NASA JSC Oral History Project. "The panel, by and large, functioned as a unit. We held hearings, we jointly decided what we should look into, what witnesses should be called before the panel, and where the hearings should be held. We had a large staff so that we could do our own investigative work and conduct our own interviews. The commission worked extensively with the staff throughout the investigation. There was also a large apparatus put in place at NASA to help with the investigation, to analyze data, to look at telemetry, to look through the photographic records, and sift through several years of engineering records. There was a lot of work being done at NASA, under our direction, that was then brought forward to the panel. I participated in all of that. I also chaired a subcommittee on operations that looked into some of the other aspects of the Shuttle flights, like 'Was the astronaut training adequate?' But most of our time was spent on uncovering the root cause of the accident, and the associated organizational and cultural factors that contributed to the accident."

The enormous amount of information the commission had to process was particularly difficult for Ride because the crew were her friends and colleagues. It made the six months on the commission even harder to work through, although she admitted she did not think too much about that at the time. "I was just going from day-to-day and just grinding through all the data that we had to grind through. But, it was a very, very difficult time. It was a difficult time for me and a difficult time for all the other astronauts, for all the reasons that you might expect. It was very, very hard on all of us. You could see it in our faces in the months that followed the accident. Because I was on the commission, I was on TV relatively frequently. They televised our hearings and our visits to the NASA Centers. I looked tired and just kind of gray in the face throughout the months following the accident." [6]

Fig. 11.1: (top) The STS-51L crew in the White Room at KSC Pad 39B during training. [l-r] PS Christa McAuliffe, PS Greg Jarvis, MS and TFNG Judy Resnik, CDR and TFNG Dick Scobee, MS and TFNG Ron McNair, PLT Mike Smith, and MS and TFNG Ellison Onizuka. (bottom) The build-up of ice on Pad 39B is evident in this image. *Challenger*, the ET and the left-hand SRB are in the background.

Fig. 11.2: (top) Images of the last flight of *Challenger* have been published many times, but this view of children in class watching a school teacher's strive for orbit tragically ending in front of them is especially poignant and heartfelt. (bottom) In MCC Houston, TFNG Dick Covey [right] and Fred Gregory [on his right] watch events in Florida unfold on TV screens, trying to come to terms with what had just happened to their friends and colleagues while still following serious mishap procedures.

A major malfunction

The incomprehensible loss of Shuttle *Challenger* and her crew sent NASA reeling, with the badly wounded space agency desperate to find answers as a shocked America mourned the seven astronauts and demanded to know how such a tragedy could have happened. Once one of the most respected administrations in the world – with dedicated men and women whose expertise and talents had saved the crew of Apollo 13 against enormous odds – NASA would now suffer the twin hammer blows of blame and shame. Chronic complacency, arrogance, systemic smugness, a failure to listen and a badly rushed launch schedule lay undeniably at the heart of the accident. The finger of blame was pointed directly at these factors and the administrators were held responsible.

Lessons about these very issues had been painfully learned as a result of the Apollo 1 pad fire in January 1967, but over the intervening 20 years NASA had ineptly slipped back into many of its former bad habits, ignoring vital, urgent warnings from concerned engineers and contractors as the Shuttle program progressed and coming perilously close to disaster on more than one occasion.

The Shuttle program would eventually resume with the successful orbital flight of STS-26 in September 1988, but a reinvented NASA had learned some tough lessons about the catastrophic effects of ignorance and an egoistic closed-doors policy. Safety, a less driven mission schedule, openness and extreme caution became the new mandate and yardstick for the space agency, and this would be extended to the ongoing flight manifest for many succeeding years.

Sad to say, that lethal enemy of NASA – complacency – would once again creep into the agency's corridors of power and result in the loss of yet another Shuttle crew of seven astronauts just 15 years later.

Mir means Peace

Twenty-two days after the loss of *Challenger*, the Soviets launched their new space station. Thought by many in the west to be named Salyut 8, continuing the series begun in 1971, it was soon identified instead as 'Mir', meaning 'Peace' or 'World'. While it had a new name, it remained essentially an improved Salyut structure with a major redesign of its front docking pod (with one longitudinal and four radial docking ports) that could accommodate five separate spacecraft, and a sixth port at the rear. The first crew to occupy the new station was launched on March 13, docking with Mir two days later. Then, on May 5, they left Mir and transferred across to the still active Salyut 7, returning to Mir on June 25 and becoming the first crew to fly to two separate space stations on the same mission. The crew was recovered on July 16 after 125 days in space. Mir was open for business, while the American Shuttle remained firmly on the ground.

Unlike the previous Salyut stations, the hardware that Mir was based upon, the new station was not the entire structure, but was instead just the core of a much larger modular space station that expanded over the next few years. Mir became the mainstay of Soviet (and, from 1992, Russian) human space flight activities until 2000, re-entering the atmosphere in 2001 after an impressive (for that time) 15 years in orbit. Though unaware of the fact at the time, Mir would feature very prominently in the Space Shuttle program and in the careers of three TFNG during their later years in the Astronaut Office.

THE SPACE SHUTTLE RETURNS TO FLIGHT

In the wake of the loss of *Challenger*, NASA, its astronauts, its workers and contractors conducted a lot of soul searching and painful examination of the causes and effects of the accident. The formal inquiry and recovery to ensure the Shuttle could fly again safely would take time, and the original manifest of 1986 was naturally suspended. On February 10, less than two weeks after *Challenger* was lost, NASA officially announced that it was cancelling the next mission, 61E (Astro 1), and that all assigned crews were to be stood down pending the investigation and recovery effort. **[7]** Though some generic training would be completed, there would be no formal crew announcements for the next year. This was followed on February 20 by confirmation that the two Centaur deployments were officially removed from the manifest, as the tight launch windows could not be met and the next opportunities would not present themselves for several months. As the *Challenger* investigation continued, the concerns of the astronauts and engineers on the issue of flying Centaur on Shuttle were presented to senior management at NASA headquarters in Washington D.C. After reviewing the data, the whole Shuttle-Centaur program was officially cancelled on June 18, delaying the launch of the Ulysses and Galileo probes until an alternative method of deployment could be found and new trajectories mapped out. Back in Houston, the assigned astronauts and flight controllers celebrated with a 'Centaur's Been Cancelled' beer bust in the Outpost Tavern, off the NASA Parkway just over from the main gate at JSC. **[8]**

The mission that never was

In March 1985, Rick Hauck had been named as Astronaut Office project officer for integrating the liquid-fueled Centaur upper stage rocket into the Shuttle. Two months later, he was appointed CDR of the STS-61F Centaur-boosted Ulysses solar probe mission scheduled for a launch in April 1986. Following the *Challenger* accident, this mission was among many postponed, and the Shuttle-Centaur project was terminated. As STS-61F MS John Fabian said in his 2006 Oral History, "There were several things with Centaur: sloppiness more than anything else. Sloppiness [and] not following prescribed procedures. A [worker] climbing on the tank with a wrench… sticking out of his back pocket that's not tethered, and it fell against the [upper stage fuel] tank and dented it. Then another guy was trying to smooth out a burr in a weld. Instead of following the normal procedure, where you put a tool down in it, firmly attach it to the side of the stage and then put a drill in with a little tip to remove the burr, but it's all [kept] very stationary, he decided that wasn't really all necessary… 'I'll just hold the drill up there and go 'brrp', and I'll take it right off a lot faster'. And guess what happened. The drill slipped and went 'brrp' across the surface of the upper stage and made this scar right across it. And they didn't learn their lesson. They did that twice." In spite of all this, Fabian was convinced that Centaur would have flown if not for the *Challenger* accident. "It's the *Challenger* accident and the re-evaluation of risk after the accident that is really responsible for removing it from the cargo manifest." [9]

In early to mid-January 1986, Rick Hauck and his group had been working an issue with the redundancy in the helium actuation system for the liquid oxygen and liquid hydrogen dump valves on Centaur. As they explored further, it became clear that there were signs of a compromise with the margins in the propulsive force being provided by the pressurized helium. "We were very concerned about it," he explained in 2003. "We had discussions about it with the technical people. So I and other members of the Astronaut Office went to a board to argue why this was not a good idea to compromise on this feature, and the board turned down the request. I went back to the Crew Office and I said to my crew, in essence, 'NASA is doing business different from the way it has in the past. Safety is being compromised, and if any of you want to take yourself off this flight, I will support you'. And of course, it was two or three weeks later that *Challenger* blew up. Now, there is no direct correlation between my experience and *Challenger,* but it seemed to me that there was willingness to compromise on some of the things that we shouldn't compromise on."

Asked if he had ever thought about stepping down from the mission, Hauck admitted that "I probably had an ego tied up with it so much that [I thought] 'I can do this. Heck, I've flown off of aircraft carriers and I've flown in combat, and I've put myself at risk in more ways than this, and I'm willing to do it'. So I didn't ever think of saying, 'Well, I'm not going to fly this mission'. Knowing what I know now, with *Challenger* and *Columbia,* maybe I would. But NASA was a lot different back there when we'd never killed anybody in space flight up to that point. I mean, there was a certain amount of sense that it wouldn't happen. So anyhow, the *Challenger* accident happened. I was assigned to be part of a Johnson [Space] Center review of requirements, and it wasn't terribly exciting, but it was something that needed to be done. Meanwhile, other people were down at the Cape [Canaveral, Florida] sifting through wreckage and other people were assigned to the Solid Rocket Booster problem. So mine was kind of a nondescript assignment." [10]

Hauck also felt, in retrospect, that getting ready for the 'mission that never was' (STS-61F), was the most challenging personal assignment during his years at NASA, "because it's clear that that was going to be a very risky flight. As with any flight, if everything goes well, it's not risky. It's when things start to go wrong that you wonder how close you are to the edge of disaster. And with Shuttle-Centaur you're much closer to the edge than in most flights, because this was going to carry the Centaur upper-stage [filled with] liquid oxygen, liquid hydrogen, in a very fragile booster in the cargo bay of the Shuttle. If there were to be a requirement for a return-to-launch-site abort, all of the liquid oxygen and hydrogen would have had to be automatically dumped out of the booster. It's clear that that was, in my view, probably going to be one of the riskiest missions NASA had flown on the Shuttle. It won't ever compare with STS-1, which, in my view, with all the unknowns, was the riskiest one, but this was going to be one of them. So that was as demanding a time on me, I think, because the crew, along with George Abbey and the Astronaut Office, were almost like a lone voice in the agency, raising concerns about the risks involved in this mission. And, of course, when the *Challenger* accident happened, eventually Shuttle-Centaur was cancelled, which I think we're very fortunate that it was. So there was some real soul-searching. Would it have gotten to the point where I would have stood up and said, 'This is too unsafe. I'm not going to do it'? I don't know, but we were certainly approaching levels of risk that I had not seen before." **[11]**

ASTRONAUT DEPARTURES

The whole of 1986 was a difficult time for NASA, its contractors, and of course the astronauts and their families (see sidebar: *An Office in Mourning*). The loss of their colleagues, classmates and friends was particularly hard on the close-knit astronaut community, and came at a time of decision making for the military contingent of the TFNG. They were within their second secondment to NASA, with an option to return to their parent service, remain at NASA, or leave to seek new goals in the civilian world[1]. Many had planned to retire anyway, but others had to make the choice between remaining at NASA for what could be several years before they flew again, or returning to their parent service to continue their military careers, with no guarantee that such a move would enhance their chances of promotion or be rewarded with advantageous assignments. Indeed, some who had investigated that option were told they had been separated from the military process for so long that they would have to be "re-greened" or "re-blued," while some of the civilians were told that if they returned to their academic life they would have to be "re-nerded" again. The other concern was in being seen to leave NASA at a very difficult time, so close to losing *Challenger* and her crew, thereby giving the impression that the loss had triggered their decision, or that they had lost faith in NASA. Such an impression was not desirable, though not entirely inaccurate. Despite these concerns, a number of veteran astronauts, including a few from the Class of 1978, decided over the next couple of years that with the Shuttle fleet grounded it was time to move on.

---

[1] Astronauts who were serving military officers were considered to be on a seven-year tour of duty (with possible extensions) at NASA, in an understanding between the space agency and the DOD. Civilian astronauts were expected to remain at NASA for at least five years.

AN OFFICE IN MOURNING

Between the creation of the NASA astronaut team in April 1959 and its frequent expansion up to the end of 1985, the agency had lost eight astronauts in training or off-duty accidents. These all occurred in the relatively short period between October 1964 and October 1967. The eight were: Ted Freeman (Group 3, 1963), killed in a T-38 crash in October 1964; Elliot See (Group 2, 1962) and Charles Bassett (Group 3) killed together in a T-38 crash in February 1966; Gus Grissom (Group 1, 1959), Ed White (Group 2) and Roger Chaffee (Group 3), killed in the Apollo 1 pad fire in January 1967; Ed Givens (Group 5, 1966) killed in an off-duty auto accident in June 1967; and C.C. Williams (Group 3) also killed in a T-38 crash in October 1967.

For over 18 years, the Astronaut Office had not had to address the painful and difficult issue of the loss of a serving member of the Corps, until *Challenger* in January 1986. But just four months later, tragedy struck again when Group 11 Ascan Stephen Thorne was killed in an off-duty private plane crash in Houston, Texas.

Civilian MS James 'Ox' van Hoften announced in June 1986 that he would be the next member of the Class of 1978 to leave the agency, joining Bechtel National Inc., of San Francisco, California, the following month. [12] Then, on July 24, it was announced that the first representative of the United States Army, Robert 'Bob' Stewart, was to return to the service effective August 1, to take a position with the U.S. Space Command at Colorado Springs, Colorado. [13] Space Command is a joint service organization that both manages and operates the nation's military space program. Stewart, who was also in line for promotion to Brigadier General, was to work in the organization's Plans Division.

During the hiatus following the loss of *Challenger*, and the subsequent cancellation of his 61F mission to deploy Centaur, Rick Hauck was asked by NASA Administrator James Fletcher to serve as the NASA Associate Administrator for Congressional, Public, and International Relations. He was appointed to the role in August 1986 but would still retain his astronaut status, and was to resume his astronaut duties at JSC in early February 1987. [14]

After Sally Ride had completed her work on the Rogers Commission investigating the *Challenger* accident in August 1986, she was named the Special Assistant for Strategic Planning at NASA Headquarters, Washington D.C. [15] At the time, Ride had already been planning to leave the agency after her third mission, a decision brought forward in part by the cancellation of that flight.

Then, in September, Don Williams become Chief of the Mission Support Branch in the Astronaut Office, a position he held until December 1988 when he was named to his next mission. In October, Dale Gardner, who had been assigned to STS-62A, the first mission from Vandenberg, returned to the U.S. Navy (USN) and was assigned to U.S. Space Command, Colorado Springs, Colorado, as Deputy Director for Space Control.

Praising the workers

Perhaps one of the least headline-grabbing assignments which all astronauts undertake during their years at the agency is to travel across the United States to meet the thousands of workers, in hundreds of companies large and small, who build and prepare the hardware that will fly on a mission. Reaching out in person to the workforce was seen as especially important during difficult times, such as experienced in the aftermath of the *Challenger* accident, when it seemed that the Shuttle might never fly again and the task to get the vehicle back into orbit was monumental. The responsibility felt by those who worked on space hardware to ensure the astronauts were safe was acknowledged and appreciated during visits by members of the Astronaut Office in both good and bad times in the program, and such visits were intended to encourage and support the workforce as well as to congratulate and appreciate their efforts. The efforts of such workers tended to go under the radar whenever a mission was successful, but could be thrust into the spotlight when things went wrong.

Early in September 1986, STS-51A pilot Dave Walker found himself participating in one of these outreach assignments at KSC. He was visiting the Pad Operations and Mobile Launch Platform teams to brief them on the current status of the astronaut training program, as part of the Return-to-Flight initiative. "On behalf of the astronaut corps," he told them, "I want to extend my sincere appreciation for the outstanding job you've done and are still doing under some difficult circumstances. You have not been forgotten and we want you to know we still have the highest confidence in the jobs you do." He added, "This is not the time to bury our heads in the sand, but to look up and go forward with the tasks that will lead us back to flying the Shuttle again." [16]

Ground simulation 'crews'

Though the Shuttle had been grounded, the work to return the system to flight continued as the investigation and recovery from the loss of *Challenger* progressed. Over the next two years, a number of astronauts continued generic training, covering aspects of their previously assigned payloads or missions pending a return to flight. They also worked in 'crews', participating in milestone ground simulations and major training exercises as part of the Return-to-Flight Program. These were not official flight crews but were formed from teams originally assigned pre-*Challenger* or from groups assigned to fulfill the role of a formal Shuttle crew. On October 29, 1986, the former STS-61H crew (Mike Coats, John Blaha, Anna Fisher, Robert Springer and James Buchli) participated in a fully-integrated simulation at JSC.

In early November, it was revealed that the revised numbering system that had first been assigned to Shuttle missions in September 1983 had been scrapped. When the launch schedule had to be abandoned following *Challenger*, the numbering system was deemed to be both cumbersome and confusing. In future, the system that had been used up to STS-9 would be reinstated. STS-51L was the 25th time the Shuttle had launched and, despite the tragic outcome just seconds later, NASA classed the flight as a 'mission in progress.' For the resumption of flights, it was decided that the next mission, the 26th launch, would officially be designated STS-26, followed by STS-27, STS-28 and so on.

Though there were no plans to launch a Shuttle for some time, over at KSC the Orbiter *Atlantis* was moved from the Vehicle Assembly Building (VAB) to Pad 39B to support planned evaluations of the new weather protection systems installed at the pad. Taking full advantage of the opportunity to have a Shuttle on the pad again, it was also decided to conduct a number of concurrent tests on countdown and emergency procedures on and around the pad area. On November 2, an Emergency Egress Test Crew used *Atlantis* sitting on Pad 39B to develop their experiences of pad operations and emergency egress procedures. This crew consisted of rookie astronauts Frank Culbertson (CDR), Steve Oswald (PLT), Carl Meade, Kathy Thornton and David Low (all MS), and Pierre Thuot and Jay Apt (taking the role of PS for the tests). Their participation reflected the type of support roles conducted by members of the TFNG a few years earlier in the build up to STS-1. Culbertson, Low and Thornton were members of the Group 10 selection (1984), while the other four had been selected as Ascans in 1985 (Group 11).

On November 18, the former STS-61C crew (Hoot Gibson, Charlie Bolden, Steve Hawley, Pinky Nelson and Franklin Chang Díaz) participated in a partial countdown, again using *Atlantis* on Pad 39B, to give the launch team a chance to maintain their skills and procedures during the period of flight inactivity. The mock countdown had begun at 19:40 pm the previous day and continued through to 11:00 am on November 18, ending with a simulated firing of the Orbiter's three main engines. For the final two and a half hours of the test, the Gibson crew lay in *Atlantis* simulating the final moments to lift off. Despite the crew encountering a number of technical problems which could have scrubbed a real launch, the test brought back fond memories for the participants of their own experiences in trying to get STS-61C off the ground earlier in the year. "We felt it was a real good test, everything went pretty well," reported Hoot Gibson after the simulation ended. However, seeing an Orbiter on Pad 39B stirred less happy memories in the astronaut than those of his launch on 61C from Pad 39A eleven months earlier. "I think it's going to be very difficult to look at an Orbiter or look at the launch pad [without recalling *Challenger* and her crew]. They were our companions. They were our close friends, and I don't think I'm ever going to quite get over that completely. I decided a long time ago [that] I was going to keep going and that I want to fly [the Shuttle again]." Later that same day, after the pad simulation had been completed, the seven astronauts comprising the Emergency Egress Test Crew participated in another three-hour crew escape test, before *Atlantis* was finally returned to the VAB. **[17]**

ANNOUNCING A RETURN-TO-FLIGHT CREW

It had been over a year since the previous crew announcement, but on January 9, 1987, the identity of the long-awaited 26th Shuttle crew, also termed the Return-to-Flight Crew, was revealed by NASA. The CDR would be TFNG Rick Hauck, who would be returning to JSC from his role as Acting Associate Administrator for External Relations in Washington D.C. during the first week of February to train for STS-26. The mission was to deploy the much-delayed second TDRS (TDRS-C), the next in the series after the one lost on *Challenger* (TDRS-B), and was planned for launch with a crew of five no earlier than February 1988. Joining Hauck would be fellow TFNG Dick Covey as PLT (formerly PLT of STS-51I), and MS Pinky Nelson (MS STS-41C and 61C), together with John M. ('Mike') Lounge (MS STS-51I) and David C. Hilmers (MS STS-51J) from the Class of 1980. In announcing the crew, Richard H. Truly said, "I am particularly pleased to assemble a group of such experienced individuals led by one of our senor space flight veterans, and I am very proud of them." With seven flights between them, they were indeed an experienced crew, and labeling Hauck as a "senior space flight veteran" indicated just how far the TFNG had journeyed since arriving at NASA as rookies nine years before.

The selection of these astronauts for this key mission was spearheaded by George Abbey. Due to the importance of the Return-to-Flight mission, former astronaut Dick Truly (who had become the agency's new Associate Administrator for Spaceflight at NASA Headquarters in February) specified early on that the next crew would be made up of five veteran astronauts who were not susceptible to Space Adaptation Syndrome (SAS), an important consideration for such a short flight.

At the time of the *Challenger* accident, the next crew in line was the Astro-1 science mission crew headed by TFNG Jon McBride, followed by Rick Hauck's crew who were due to fly the first Shuttle Centaur mission on STS-61F. Of the two astronauts, Hauck had flown two missions, commanding the second of them (STS-51A), while McBride had only flown once as PLT (STS-41G). Hauck had also been the lead contender to be the first pilot of the 1978 group to fly, on STS-7, and was considered one of the most experienced in the group, and thus received the CDR position on STS-26. McBride retained his unconfirmed position as CDR for the re-manifested Astro-1 mission for the time being. Mike Lounge and Dave Hilmers from the 1980 selection had previously been assigned to Hauck's 61F crew, so it seemed sensible to retain them as a trained unit for this new, important mission. The question was who would fulfill the role of PLT. Hauck had lost his original PLT during the hiatus following *Challenger*. Roy Bridges, from the 1980 selection, had been assigned to 61F, but had already decided to leave NASA and return to the USAF after that mission. However, after *Challenger* and with the Shuttle fleet likely to be grounded for at least two years, he had decided to bring his departure forward. Bridges had left the Astronaut Office in May 1986 to take command of the 6510th Test Wing at Edwards AFB in California.

In his 2018 book about George Abbey, author Mike Cassutt explained that while there were other candidates for the PLT seat on STS-26, it was important to try to keep the 'crew' as intact as possible. Dick Covey had been preparing for the role of launch and landing Capcom for STS-61F (commanded by Hauck), and after the departure of Roy Bridges he had joined the rest of the crew in a series of generic simulations and was therefore integrating into the crew. While it seemed sensible to assign Covey as PLT for the mission, there was one issue which might have prevented this. Normally, a pilot astronaut would fly their first mission in the right seat, under a veteran CDR, and would then be expected to take the command seat themselves on their second mission. If Covey stayed on the STS-26 crew, he would become the first Shuttle PLT to fly in that position twice, delaying the opportunity for his own command. When informed of the plan, however, it did not take Covey long to accept a second PLT slot, reasoning that while he might not be in command, he would be flying a second time much earlier than he expected to, and on a very important mission.

To make the crew up to the minimum complement of five, Pinky Nelson, a veteran of STS-41C and 61C (and with the added experience of two EVAs behind him), had already been assigned. At the time, Nelson was on leave of absence at the University of Washington. "In terms of the technical work that was going on there [at the time], things were just being pulled together and teams were being made to work on various aspects of this and that," Nelson recalled in 2004. "We all got little assignments and went off and got back to work, which was good therapy for the folks in the office, to dig in and try and really find out what happened and see if we can get back flying again. It became obvious after a few months that not much was happening. I'm not very good at sitting around, so I actually asked to take a leave and I spent six months up in Seattle, working at the University of Washington in the Astronomy Department just thinking about astronomy, because I didn't feel like I had much of a contribution to make down there. Then, after being here for six months, George Abbey must have figured out that I was having too good of a time, so he assigned me to the STS-26 crew and made me come back." **[18]**

A new Chief Astronaut and Deputy

For the previous 13 years, since America's first astronaut Alan Shepard had retired in 1974, Gemini, Apollo and Shuttle veteran John Young had been in the coveted post of Chief of the Astronaut Office in Building 4 at JSC. Young had flown a record six missions into space; two on Gemini, two on Apollo and two on the Shuttle. If his launch from the Moon during Apollo 16 was included, he had experienced seven rocket launches in his 25-year career at NASA. In the wake of the *Challenger* accident, having written some direct and critical memos, Young was removed from the Chief Astronaut position and reassigned to an administrative role as Special Assistant to the JSC Center Director. He was replaced in the short-term by Acting Chief Hank Hartsfield (Class of 1969, USAF). [19]

According to author Michael Cassutt, there was now a desire to have a Chief Astronaut from the 'new generation' of astronauts, and the most senior group of this 'new era' were the TFNG. [20] The chosen candidate had to be a pilot astronaut and an experienced commander, while both of the previous incumbents had come from the USN (Shepard 1963−1974; Young 1974−1987). With these criteria in mind, George Abbey shortlisted four candidates who he knew would make good office leaders at some point in their careers: Dan Brandenstein, Hoot Gibson, Rick Hauck (all USN) and Brewster Shaw (USAF).

Of the four, Hauck was already in training for STS-26, while Shaw was also lined up, but not yet formally announced, for an important DOD mission, leaving only Brandenstein and Gibson available. Brandenstein had been in the post of Deputy Director for Flight Crew Operations for the past year. After deliberating, and with other plans in mind for Gibson, Abbey called Brandenstein on the evening of Sunday, April 22 and asked him to come into JSC for a meeting, where he was told he was to be the new Chief Astronaut. The Deputy Chief would be Steve Hawley, the first civilian non-pilot to hold such a position in the Office, making it clear that the rookies from less than a decade earlier were now not just veteran astronauts but showing talents for leadership and mentoring. Such characteristics would be demonstrated through their future careers, not only during their years at NASA, but also once they left the agency.

Brandenstein chaired his first Monday Morning Pilots' Meeting the following day, April 23, but the formal announcement of the new Chief Astronaut was made a few days later, on April 27. The official announcement also confirmed that Hank Hartsfield would replace Brandenstein as Deputy for Flight Crew Operations. [21]

Brandenstein explained the role of Chief of the Astronaut Office over a decade later. "You're responsible for the office. You assign the crews. You ensure that they're getting trained. You sign off that they're trained and ready to fly when the time comes. You work all the issues. Once again, the crew has input into a lot of the technical decisions that have to be made, and you get involved in those. You have to develop an 'office position.' Trying to get 100 people to agree to a position… well, you never do, but ultimately it's your job that you take the inputs from the folks in the office and establish a position. Somebody's got to make the 'Okay, this is the way we're going forward [decision]'. So you let everybody have their say, then you use your best judgment and say, 'Okay, this is the way we'll carry the office position forward. This is what it's going to be and here's why'. I made the decision. Like I say, you'll never get everybody to agree, or very seldom.

Every once in a while you might find something that's so outrageous that they'll all agree on it, but that rarely happens.

"So that was it, and taking care of the care and feeding of 100 [or so] astronauts and stuff. So that's another full-time job. They're human like anybody else. They come in and they have problems here, and this and that, and they go out and every once in a while… do some dumb things, and you have to try and keep them on the straight and narrow. If you have to have a heart to heart with them, every once in a while you have to do that. If they get blindsided with something, you try and protect them from that. And there's a lot of people tugging and pulling at folks in that position from a lot of different directions. You try and let them go off and do their job and not be tugged and pulled by extraneous sources (see sidebar: *The Max Q Band*). That's part of your job to help with some of that. It was altogether a pretty interesting job." **[22]**

"The Chief [of the Astronaut Office] and the Deputy are responsible for a lot of the technical issues the astronauts get involved with; all of the individual astronauts in one way or another are involved in their various program issues," explained Steve Hawley in 2002. "Usually there comes a time when the office itself as a group needs to take a position on some technical issue or program issue, and we would orchestrate that. Of course, we would make recommendations for job assignments for people: who would get to go work as a Capcom [Capsule Communicator], who would get to work down in Florida, who would go work at SAIL [Shuttle Avionics Integration Laboratory]. We made recommendations about crew assignments, and basically just running the administration of the office. Now, when I came in as Deputy to Dan Brandenstein, who was the Chief, I felt that it was my job to be the guy that was available for people to come talk to, because the Chief is pretty busy, and he's involved in lots of stuff that takes him away from the office, and maybe he doesn't have as much time to spend listening to the people in the office. And so I felt like it was my job to sit there with the door open and I actually did that. I sat at the end of the hall, and I'd always leave my door open, and I sat there with my desk facing the hallway, so people could see if I was in. Painful as that was, I felt it was my job to be the guy that was there, that if people wanted to come and say something or vent or ask for something, that they had somebody to go to. So I remember, for whatever reason, I remember that being something that I thought was very important in the job." **[23]**

Seventeen years after his Oral History contribution, Steve Hawley was asked about his move from an operational position in the Astronaut Office to a management role, and commented on managing the people he used to work with. "Some people were able to do that well and others not so much. I felt that there were a lot of advantages, I think, to having very experienced astronauts in some of these positions because of their experience and perspective, but not everybody is well suited to be a manager. Flying in space doesn't make you a good manager, but I know George [Abbey] felt that the experience the flown astronauts had was advantageous for the program in a variety of different ways. **[24]**

Fig. 11.3: (top) The former STS-61H, now STS-61M (T) crew take breakfast in advance of their April 1987 ground simulation as part of the preparations for the Return-to-Flight mission of STS-26. An ongoing joke played on the crew by the trainers involved carrots and their symbolic role as a reward ("carrot and stick"), including the carrot-related decorated cake shown here. The banner above Fisher and Coats proclaims the room as the 61MT Dining Hall, with the 'H' crossed out and 'MT' added. (bottom) Inside the motion-based version of the STS, the 61M (T) crew enjoys more lighthearted banter with the bunch of carrots suspended between the CDR and PLT station. [l-r] Coats, Blaha, Fisher, Springer and Buchli (standing).

A week prior to the announcement of Steve Hawley's appointment, a new simulation 'crew', STS-61M (T – for training), was named. In command of the former STS-61H five-person crew was Mike Coats, with his colleagues Anna Fisher and Jim Buchli from the 1978 selection as MS. They were joined by PLT John Blaha and MS Robert Springer, both from the 1980 selection. [25] Designed to support the efforts to return the Shuttle to flight, this 61M (T) team would participate in a simulation of the first 56 hours of the planned STS-26 mission, while remaining firmly on the ground at JSC. The mission timeline and payload would be used during the sim, held between April 28 and 30, with the Coats crew replicating the roles that Hauck's crew would perform during the real mission. The primary objective of this sim was to "exercise the people and the processes necessary" to support the next mission. Termed an integrated simulation, a team of flight controllers supported the test around-the-clock from their consoles in MCC Building 30 as if it was a real mission, although the flight crew did not remain in the Shuttle Mission Simulator (SMS) over in Building 5 at JSC during sleep periods. The simulation began at 09:00 am Central Daylight Time (CDT) on Tuesday April 28, picking up at the T-9 minutes mark and followed by the simulated launch nine minutes later. Six hours and six minutes later, the crew initiated the simulated deployment of the TDRS. The simulation ended by 17:00 pm on Thursday April 30.

"We were down for two and a half years between *Challenger* and Rick Hauck's flight [STS-26], so they wanted to keep the crew trainers and instructors sharp," recalled Mike Coats. "The crew goes through a long syllabus getting ready for a flight. We had just been through the syllabus [and] were ready to fly. So when *Challenger* happened, they said, 'Well, we need a couple of crews to go through the training syllabus again', more to keep the trainers sharp than anything else and knowing full well that we'd have to go through it again when we were finally assigned to another mission. So that was what we did. I think Brewster [Shaw] did it with his crew, and I did it. He had a classified syllabus and we had an unclassified syllabus to go through. So we went through the whole training syllabus again, which was good. It's fun to get as good as you can get in a simulator. We had the opportunity to do that. Then we were assigned another mission and essentially went through that again. So essentially for three missions I went through the training syllabus five times. We'd flown a mission. We'd gone through the training syllabus. We were about ready to fly when *Challenger* happened. Then we went through just the exercise syllabus. Then we went through again to fly the second mission. Then we went through a fifth time to fly my last mission. So it's a lot of simulations if you will. But you feel like you get pretty good in the simulator when you've done all those things. Kind of fun, but after my third mission I thought 'Man, I don't know if I can go through that syllabus again'." [26]

THE MAX Q BAND

The post-*Challenger* period was tough for NASA, the contractors, the astronauts and their families, and it was difficult to come to terms with the loss of the crew, proceed through the enquiry and implement its recommendations, and look forward to resuming the program. To help get through these dark days, a variety of diversions were explored, one of which was to turn to music.

Formed in early 1987 by three of the TFNG (Robert Gibson, George Nelson and Brewster Shaw), the band was named by Gibson after the engineering term Max Q, used to denote the maximum dynamic pressure experienced by an ascending spacecraft, although the joke description was that, like the Shuttle, the band "makes a lot of noise but no music." Over the years, the members of the band rotated between active duty astronauts, due to training commitments, crew assignments and retirements. The band was established during the down period following the loss of *Challenger*, when morale in the office was pretty low. As Shaw explained years later, "We weren't flying again yet [and] we weren't really sure when we were going to be able to fly again because they were redesigning the Solid Rocket Boosters. [So we thought] we ought to come up with some way to pick people up a little bit. So I got talking with Hoot Gibson and we decided 'let's have a sock hop'. And not only let's have a [50s style] Saturday night sock hop, but let's put together a little band that could play at this sock hop, and the band would be all astronauts. Hoot declared [that] he could play lead guitar. Hoot and I had played guitars a little bit together before. I could play rhythm guitar. And we got Pinky [Nelson] to play bass guitar, which he didn't play, really, but he learned. He learns everything in a heartbeat, so he picked that up overnight. And it turns out that [Jim] Wetherbee had been a drummer. So we got the four of us together and we practiced a bunch and learned, I don't know, three or four songs, and we played at this sock hop down in the pavilion down at the park in League City on Highway 3… it was a great time. Everybody had a lot of fun, and it absolutely did what we'd hoped it would do. It gave people a diversion and they quit thinking about not flying and how we were grounded. So I was real glad we did it, and then the band just kept going, and the band is still going today [with different members]." **[27]**

Steve Hawley was another of the TFNG who joined the band, as a keyboard player, shortly after its formation. As they were doing this on their own time away from NASA and the Astronaut Office, they did not need permission or approval to carry on, although it did help that the sponsor of the Fajita Festival they played at in the early days was one George Abbey, their boss at JSC. As Hoot Gibson observed, "We weren't good, but we weren't bad."

On July 30, Jon McBride was named as Assistant Administrator for Congressional Relations at NASA Headquarters, a role he occupied from September 1987 until March 1989. The following month, Sally Ride finally left NASA to join the Stanford University Center for International Security and Arms Control.

On September 15, NASA named the STS-27 crew, which would be commanded by Robert 'Hoot' Gibson. He would be joined by fellow TFNG Mike Mullane as MS, alongside Group 9 astronauts Guy Gardner (PLT) and Jerry Ross (MS), as well as Group 10 MS William 'Bill' Shepherd. [28] The classified DOD mission was targeted for early fall 1988 on the Orbiter *Atlantis*. This crew was mainly derived from the former STS-62A crew, which was due to be commanded by Bob Crippen and to fly *Discovery* on the first launch out of Vandenberg. However, following *Challenger*, both the mission and any operations from the West Coast were cancelled. In April 1987, Crippen was reassigned as Deputy Director for Shuttle Operations at NASA KSC in Florida, effective July that year. [29] When Crippen was moved to KSC, Hoot Gibson was assigned as the crew's new CDR, with Bill Shepherd added to make up the five-person crew requirement. George Abbey had lined up Gibson, who had been shortlisted for Chief Astronaut earlier, to command the mission as he was the only one of the seven flown commanders still active and available. Gibson was told of the assignment over a beer in a bar by Abbey one evening and was surprised when first informed of the decision as it was not his 'turn' to command, having recently commanded the last flight prior to *Challenger*, STS-61C. Gibson thought that there were plenty of other TFNG pilots to command the mission in front of him, but there was another reason for the decision. The payload was a new imaging radar system, for which the customer required an experienced veteran commander to lead the crew on this important mission of national security. As all of the pre-1978 veterans who had commanded a mission in 1984 and 1985 had now been reassigned, retired or moved to management, that left Brandenstein (now Chief Astronaut), Hauck (in training for STS-26), Shaw (in training for STS-28) and Gibson, who got the seat[2]. [30]

The year had ended positively with the assignment of two Shuttle crews and the completion of several important simulations, together with step-by-step progress towards resuming flight operations, hopefully within the next year. On the down side, this became the first year since 1980 that American astronauts had not flown in space, and all hopes rested on preparations to try to ensure that 1988 would also not be devoid of American missions.

---

[2]At one point, Manned Spaceflight Engineer (MSE) Kathy Roberts was being considered to fly on STS-27 with this crew. Her payload specialty was the Lacrosse Imaging Radar Satellite, which was the prime payload for this mission. However, when NASA instigated a rule that its astronauts would be the only ones to fly the first five post-*Challenger* missions (STS-26, 27, 28, 29 and 30), Roberts lost her place on the crew and the chance to fly into space.

With the Shuttle still grounded, only a few crews beginning to resume mission training, and new astronauts from the 1984 and 1985 selections available, the planned 1986 selection was deferred for a year. This was also the time that a number of veteran astronauts left the office, while others chose to remain and perform a wide variety of assignments or collateral duties, hopefully waiting for the call to resume mission training. One of those associated duties, which most astronauts perform at some point in their careers, was recorded in one of the final news releases of the year. **[31]** Dave Griggs was appointed as Chairman of an incident investigation committee, charged with examining an incident that had occurred on December 17, and due to report its findings to JSC Director Aaron Cohen by January 31, 1988. The incident under investigation centered upon one of NASA's high-altitude research aircraft, a General-Dynamics WB-57, which had run off the runway at Ellington Field, Houston. It had been piloted by Michael Corbett and Albert Crews, a former X-20 and MOL astronaut who had missed the age limit for selection to NASA's Group 7 in 1969 and had joined the JSC aircraft division instead. No one was injured and minimal damage was inflicted on the aircraft, which had been operating in conjunction with Project Airstream, a long-term NASA/Department of Energy multi-disciplinary study of the upper atmosphere, and was also used to collect 'cosmic dust' samples for the Solar Exploration Division at JSC. Griggs's TFNG colleagues Kathy Sullivan and Pinky Nelson both had experience as crewmembers on previous WB-57 flights.

A DECADE ON

On January 16, 1988, the Class of 1978 celebrated the tenth anniversary of their selection as astronauts. By now, they had all flown in space at least once, with eleven of them having flown a successful second mission, though another four who had embarked on that feat had perished in the attempt. Of the 35 who began the first Ascan training program back in July 1978, six had now departed the Astronaut Office, and four had died on *Challenger*. At the start of their astronaut careers they were naturally space rookies, mostly out of their depth in the hallowed halls of the Astronaut Office, but that soon changed, and ten years later those who remained now numbered among the most experienced and senior of the astronaut detachment, with very few of the pre-1970 astronauts left on the active list. So much had changed, not only in the group, but also in the office, in the program and in NASA itself. The NASA they now worked for was very different as the requirements of the program changed, particularly in the preparation of the next phase of human space flight in America, the creation of a large space station. But this would only proceed if the Shuttle could be certified safe to fly again.

Critical in ensuring that there were astronauts ready to fly the missions once the program resumed was the selection of crews. Initially at least, these would still be formed from those who had been in training prior to the loss of *Challenger*. On February 4, 1988, therefore, TFNG Brewster Shaw was finally named to command the STS-28 DOD mission aboard *Columbia,* targeted for a late 1988 launch. [32] Two of the remaining four members of the crew, PLT Richard Richards and MS Dave Leestma, came from the class of 1980, with the remaining two, MS James Adamson and Mark Brown, from the Class of 1984. This was the former STS-61N crew, apart from Richard Richards replacing Mike McCulley, with the latter having been reassigned during the hiatus to serve as technical assistant to Don Puddy, the Director of Flight Crew Operations.

Astronaut Science Support Group

A NASA news release in February 1988 reported that "Astronauts at NASA's Johnson Space Center have created an Astronaut Science Support Group to provide direct interaction with prospective experimenters on Space Shuttle and Space Station missions." [33] This group included MS members of Group 8 and 9, with each focusing on a specialty area. From the TFNG, Jeff Hoffman focused upon astrophysics and remote sensing and Rhea Seddon on life sciences. They were joined by Group 9 (Class of 1980) astronauts Franklin Chang Díaz (plasma and space physics), Mary Cleave (biological materials processing), Bonnie Dunbar (material processing), Jerry Ross (Extra-Vehicular Activity (EVA), satellite servicing and space construction).

Using the experience gained from their Shuttle missions, the group strongly believed that increasing the involvement of crewmembers in the design, development and operation of experiments would both improve the return of data and simplify the repair of faulty equipment. The primary goal of the group was to transmit their operational experience to the science and technology user community, in order to make best use of the Orbiter as a testbed for scientific and engineering research by utilizing the crew as a critical element both on the ground and during the mission. This would provide more efficient experiment operations and flexibility in real-time repairs and fine tuning of equipment. They also acted as advisors to the Shuttle and Space Station programs for issues of science and technology. Both Hoffman and Seddon commented on the group in their respective Oral Histories.

"As scientist astronauts, several of us were often asked to work with scientists who were preparing experiments for space," Hoffman explained. "For instance, I got assigned at one point [to] a combustion science working group at the Glenn [Research Center, Cleveland, Ohio]. At that point it was the Lewis Research Center, so I was basically their astronaut contact. I went up there periodically when they would have their investigators meetings and tried to give them advice: 'You're trying to design this, but you're asking the crew to do this. In weightlessness, that's going to be very difficult. We should change things around…' and so on. It was nice, actually. On my last space flight we had a glove box, and we actually performed some of the combustion experiments that I had been working on. This was early on. This may have even been in the early '80s while I was still getting ready in the early days of my training. I don't exactly remember, but it was kind of typical.

"All of us would have similar experiences, where we would be presented with experiments that people thought were ready for flight, but clearly there were aspects of the space environment that they didn't understand. We could often see ways in which things could fail or just ways to make them more efficient, so we came up with the idea [that] maybe the scientists in the [Astronaut] Office should form an advisory group so that we could work with people who were designing experiments. At the same time we thought, 'They're always having lectures to the Astronaut Office about various aspects of aircraft safety and space operations. Maybe we can invite some science people in to give maybe a monthly lecture to the office about something of interest'. So those were the two motivations.

"Franklin [Chang Díaz], Bonnie Dunbar, I, Rhea Seddon, a couple of other people were the founding members. Actually we had a fair bit of success in the first few years, but as often happens then we all got assigned to flights. You get so busy that we couldn't really continue it, but I think some of the payload people took over that kind of philosophy with the idea that we really have to look at these experiments from a user's point of view. The only difficulty is that when it's not the users or people who are really scientifically knowledgeable, but instead they tend to be the engineering contractors who do this, they tend to go very strictly by the book. That again often ends up making life more difficult for the experimenters rather than less difficult, but that's the way things evolved. We were active for a couple of years. Then after Return-to-Flight, everybody got busy doing lots of different things, and we really couldn't keep it up just as astronaut scientists." [34]

"As we got into a more mature phase of the Shuttle Program, we came to realize that scientists who either wanted to propose experiments, or had proposed experiments, frequently designed both the hardware and the experiment itself before they ever had crewmembers that they could work with," added Rhea Seddon. "We likened it to saying, 'Well, they've designed this experiment to be a 'flying brick'.' In other words, the only switch on there is the on/off switch, and the only indicator on there is a light that comes on when it's on. If you take it to space and you flip the switch and the light doesn't come on, there's nothing you can do. It's sealed up. They didn't design it to be repaired. There's no way to understand what went wrong, no insight into the mechanisms. We felt that was unfortunate, because crewmembers are more than willing to learn everything there is to know about an experiment. They take the responsibility for it. For instance, scientists I knew were designing life sciences experiments, and they were told, 'We can't guarantee that you'll have any physicians on the flight or anyone that knows anything about life sciences'. So again, they would dumb-down the experiment or they would worry about it, or they wouldn't propose it. Or they would do something that they ordinarily wouldn't do, in order that any person could do it.

"So we wanted to go out and tell them the capabilities that people had. Even if you were an astronomer, you could learn to draw blood. Even if you didn't understand a lot of the complexity of the payload that you were operating, you had the opportunity to talk to the ground. A lot of times you could do the mechanics of it and do it very well, and understand how it could go awry without understanding how the telescope was built or what the internal workings were. So being a resource to scientists was something that we felt was useful for us to be doing. I think in the time after *Challenger*, all of us were trying to look at not only how to recover from the mechanical problem that happened on *Challenger*, but how we could use this time to improve the product of our office. We could make ourselves available to scientists who didn't understand, for instance, how to work in zero gravity.

"Scientists would say, 'All this experiment requires is that you draw blood and that you collect urine'. We would remind them that we didn't have a lab centrifuge on board, and we frequently didn't have a lab refrigerator on board. So something that sounded pretty simple, [and] would have been very simple here on the ground, like collecting urine, was hard to do with the things that we had in the space environment. But we would let them know there are flights that you could stow the urine-monitoring system on board. They might want to group their experiment with some other people's so that they could assure that that equipment could go on board to support several experiments. So it was that kind of interaction, and we all enjoyed it. Most of us that were working on the support group had flown, so we understood what weightlessness was like, what you could and couldn't do, what equipment you had or could have. So it was earning our keep as good scientists to go and share some of what we had learned and what we knew about the Shuttle Program." [35] This was another example of how members of the TFNG participated in wide-ranging issues which would have later application in future programs, in this case in developing the coordination between the Astronaut Office and the science programs conducted aboard the ISS.

The last before the first
The pace of the Return-to-Flight program increased as the months rolled on, and with growing confidence that the Shuttle would indeed return to orbit that year, in March 1988 NASA issued the names of three new crews, which were all scheduled to fly during 1989. [36] It was also made clear in the release that these would be the last crew selections announced prior to the first mission in the resumption of Shuttle flights later that year (STS-26). Among the crews announced, there were seven members of the TFNG.

STS-29, commanded by Mike Coats, included MS Jim Buchli and was set for launch in January 1989 on *Discovery* carrying the third TDRS (TDRS-D) to be deployed. The crew also included three members of the 1980 selection, with John Blaha as PLT and Robert Springer and James Bagian as MS. Due to the changes in the manifest following *Challenger* and the cancellation of previously named crews in the down time, Anna Fisher decided to have a second child and stood down from the former STS-61H crew to be replaced by Bagian. At the time she was on maternity leave, but following the birth of her second child she would return to the Astronaut Office in 1991, initially in a part-time role, to work on Space Station *Freedom* training issues.

STS-30, commanded by Dave Walker, included Norman Thagard as MS. They were joined by Group 9 PLT Ron Grabe and MS Mary Cleave, with Group 10 MS Mark Lee rounding out the crew. The mission was planned for launch in April 1989 on board *Atlantis*, a four-day flight to deploy the Magellan Venus radar mapping orbiter.

STS-31 would be commanded by Loren Shriver (replacing John Young, the original CDR of what was then called STS-61J). This mission was to fly in June 1989 to deploy HST. Steve Hawley and Kathryn Sullivan had been named to the original mission and now resumed their training, along with Group 6 veteran Bruce McCandless. At the time of the announcement, Hawley was serving as Deputy Chief of the Astronaut Office, a position he would stand down from in the spring of 1989 to begin active mission training.

As the news release pointed out, STS-29 (the 28th STS mission) would now fly after STS-27 (the 27th STS mission) but before STS-28, based on the current Shuttle flight manifest, reflecting the still-confusing sequence of Shuttle missions and the delays in preparing *Columbia* for its return to operational status.

"Gooooooood Morning Discovery!"
Excitement built as the countdown entered the final hours on September 29, 1988, with more than 250,000 people and 2,400 news media gathered in the KSC area. *Discovery* lifted off flawlessly, and President Ronald Reagan declared, "America is back in space." Six hours into the flight, the astronauts deployed the TDRS-C satellite, similar to the one destroyed in the *Challenger* accident, and they also operated 11 middeck science experiments. During their four days in orbit, the crew tested more than 200 design changes made to the Shuttle. A day before returning, they paid an emotional tribute to the *Challenger* astronauts.

On Flight Day 2 (FD2, September 30) the crew were woken up, not by the expected call from the duty Capcom Kathy Sullivan, but by the unmistakable voice of actor Robin Williams with a variation of his 1987 "*Good Morning Vietnam*" radio wake-up call from the comedy-drama war film of the same name. Producer and radio KKBQ DJ Mike Cahill had approached Williams about creating the tape. Apparently, NASA knew nothing about it until Cahill offered it to the agency as a gift. He had heard previous wake-up calls and found some of them awful, so he thought it would be nice to write, produce and design songs for the astronauts. He sent the tapes to Pat Mattingly who worked in Mission Control, who in turn passed them on to Kathy Sullivan on the Capcom Console. His desire was to show that the people at NASA had a sense of humor. **[37]**

The crew were also the first American astronauts to wear pressure suits for launch and landing since STS-4 in 1982, and the first all-veteran American crew since Apollo 12 back in 1969.

STS-26 (September 29 – October 3, 1988)
Flight Crew: Frederick H. HAUCK (CDR), Richard O. COVEY (PLT), J. Michael Lounge (MS-1), David C. Hilmers (MS-2), George D. NELSON (MS-3)
Spacecraft: Discovery (OV-103) 7th mission
Objective: 26th Shuttle mission; Return-to-Flight mission; TDRS-C deployment
Duration: 4 days 1 hour 0 minutes 11 seconds
Support Assignments: For this mission, John CREIGHTON served on the ascent and entry 'Gray' team Capcom console, while Kathy SULLIVAN was on console for the 'Indigo' planning team. [38] Assigned to SPAN were Brewster SHAW (ascent/entry) and Don WILLIAMS (planning). Hoot GIBSON took the first shift in the SMS, with Mike COATS taking the second. Fred GREGORY was sent to Edwards AFB for any Abort Once Around (AOA) situation, then went to the Dryden Contingency Action Center. Chief Astronaut Dan BRANDENSTEIN was Weather Pilot (WX) and Steve NAGEL was one of the Family Escorts. Jeff HOFFMAN worked as a media representative of the Astronaut Office for the TV networks. [39]

Four new crews were announced in late November 1988, scheduled to fly in late 1989 or early 1990, with each commanded by pilots from Group 8. [40] Fred Gregory would command the STS-33 crew. He was returning to flight duty in December after completing a number of administrative and managerial appointments in recent years[3]. Gregory would be joined by Group 8 astronaut Dave Griggs as PLT, together with MS Story Musgrave (Class of 1967), Kathy Thornton and Sonny Carter (both from the Class of 1984). STS-33 was planned for an August 1989 launch using *Discovery* and carrying a classified DOD payload.

Don Williams was named as CDR of STS-34, which was planned for October 1989 on *Atlantis* and featured the deployment of the Jupiter probe Galileo, using an Inertial Upper Stage (IUS) rather than the cancelled Centaur, and with an increased transition time to reach Jupiter. Mike McCulley (Class of 1984) was assigned as PLT following his temporary managerial assignment at JSC, together with three MS: TFNG Shannon Lucid, Franklin Chang Díaz (Class of 1980) and Ellen Baker (Class of 1984).

The STS-32 mission in November 1989 was to be commanded by TFNG and Chief of the Astronaut Office Dan Brandenstein, with PLT Jim Wetherbee (Class of 1984) and MS Bonnie Dunbar (Class of 1980), David Low and Marsha Ivins (both from the 1984 selection). This mission would use *Columbia* and featured the

---

[3] Gregory's administrative assignments included: Chief of Operational Safety at NASA HQ; Chief of Astronaut Training; and member of both the Orbiter Configuration Control Board and of the Space Shuttle Program Control Board.

deployment of the Syncom IV-5 satellite and the much-delayed retrieval of the LDEF that had been deployed from STS-41C in 1984. Brandenstein would not begin full time training for this mission until the spring of 1989, when he temporarily stood down from the position of Chief Astronaut.

The CDR for STS-35 was confirmed as TFNG Jon McBride, with Guy Gardner (Class of 1980) as PLT and TFNG Jeff Hoffman, Robert Parker (Class of 1967) and Mike Lounge (Class of 1980) as MS. Due to launch in March 1990 on *Columbia*, the mission was manifested to carry the Astro-1 astronomy payload. This crew had originally been assigned to the STS-61E mission which would have been the 26th Shuttle mission had *Challenger* succeeded. In this new manifest, it was to be the 35th.

This brought the total number of crews in training to nine and reflected a number of changes due to crew reassignments and departures.

STS-27 (December 2 – 6, 1988)
Flight Crew: Robert L. GIBSON (CDR), Guy S. Gardner (PLT), R. Michael MULLANE (MS-1), Jerry L. Ross (MS-2), William M. Shepherd (MS-3)
Spacecraft: Atlantis (OV-104) 3rd mission
Objective: 27th Shuttle mission; 3rd classified DOD satellite deployment
Duration: 4 days 9 hours 5 minutes 35 seconds
Support Assignments: Capcoms on this mission included John CREIGHTON ('Gray' ascent and 'Aquila' entry teams) and Kathy SULLIVAN (Orbit 1). Three of the four SPAN shifts were covered by Group 8 astronauts Brewster SHAW (08:00–14:00), Loren SHRIVER (14:00–20:00) and Don WILLIAMS (02:00–08:00). The remaining shift (20:00–02:00) was fulfilled by members of Group 9. Steve NAGEL was originally intended to be at Edwards AFB covering any AOA incident and as a member of the Dryden Flight Research Facility (DFRF) Contingency Action Center, but was replaced in both positions by one of the newer astronauts. Mike COATS served on the first shift in the SMS and Fred GREGORY replaced the originally assigned Dave WALKER. Chief Astronaut Dan BRANDENSTEIN again served as WX pilot, both at KSC for launch and at Edwards for landing, while Anna FISHER was at the JSC Contingency Action Center. **[41]**

During this flight, an initially secret DOD payload, later identified as a radar imaging satellite called Lacrosse (also known as Onyx) was successfully deployed for the National Reconnaissance Office (NRO). According to some sources, this required a rendezvous with the deployed payload to release the solar arrays when they failed to open as planned. The nature of this flight meant that most of the mission and the activities of the crew remained secret, so suggestions that Mike Mullane operated the RMS to save the payload, or that Jerry Ross and Bill Shepherd performed an unannounced contingency EVA have been neither acknowledged nor denied, even more than 30 years after the mission. What is known is that the crew operated a number of secondary payloads and a range of experiments aimed at defining the human role as a military observer in space. As most details of the flight of STS-27 remain classified, the mission is remembered for almost becoming the second fatal Shuttle mission in less than three years, though the classified nature of the mission meant that it took some time for the details of this to emerge.

A FRIGHTENING CLOSE CALL

Eighty-five seconds after *Atlantis* launched, insulation material from the nose cap of the right-hand SRB struck the Orbiter, inflicting extensive damage to the Thermal Protection System (TPS) tiles. The crew had reported seeing white material on the windows during the eight-minute ascent to orbit, but they had no idea of the extent of the damage until the next day when, unusually, the flight controllers asked them to use the RMS cameras to inspect the starboard side of *Atlantis*.

When the crew positioned the arm over the right side of the Orbiter, they were surprised and shocked to discover the worst tile damage yet recorded during a Shuttle mission. It was so severe that Gibson thought the crew were going to die during re-entry. Mike Mullane wrote about the incident in his 2006 autobiography (devoting two chapters to his account of the events) noting that, from the RMS inspection, it looked as though hundreds of tiles had been damaged and at least one tile had been completely ripped off, with scars extending right across the leading edge of the right wing. Mullane feared that if the damage had penetrated the carbon composite panels at that point, then they were "dead men floating." **[42]** Gibson contacted Houston through the closed [to the public] channel, giving Mission Control a detailed verbal description of the damage they had seen. He was asked to send images down on the secure TV channel, and here is where the fate of the crew hung in the balance.

Because STS-27 was a classified mission, the crew were not allowed to use the standard methods of downlink and instead could only rely on the much slower encrypted transmission – one frame at a time – in which each single frame of the TV 'movie' took three seconds to send. When received at JSC and reassembled into the TV images the crew had seen on orbit, they were of such poor quality that the NASA engineers misinterpreted the damage as just "lights and shadows," commenting that the TPS did not appear to be any worse than on previous missions and that poor lighting conditions had led the astronauts to draw the wrong conclusions. Astonished, to say the least, Gibson and his crew were well aware of what they had seen in the crystal clear images from the RMS cameras – though there was little they could do about it – and Gibson was fuming. Despite stressing what they were seeing on orbit, Gibson was told that the engineering assessment of the damage was that it was not that severe. The crew could not believe what they were hearing. Between them, Gibson, Mullane and Ross had flown four previous missions in space and had been around the Shuttle program since well before it had started flying, so they had plenty of knowledge of all the missions to date, and the effects of space flight on the tiles. They naturally disagreed with the evaluation on the ground, but there was nothing to be done except fly the mission. Though it looked bad to them, they had to defer to the 'experts' on the ground, but as Gibson later admitted while up on orbit, it looked as though "we're in deep doo-doo." They were aware that there was no immediate problem, but without tools to repair the damage, no methods in place or trained for to get an EVA astronaut to the damaged area, no space station to use as a safe haven, and no second Shuttle sitting on the pad that could come and rescue them, their re-entry would be tense to put it mildly. In fact, mounting a rescue using a second Shuttle would have been challenging in the time available, as the other two orbiters, *Discovery* and *Columbia*, were in the Orbiter Processing Facility (OPF) at the time of the STS-27 mission.

As the time for entry approached on December 6, Gibson and Gardner initiated the re-entry to commence the fiery descent to Edwards AFB. On the flight deck, Gibson knew exactly how *Atlantis* would behave if the damage was indeed terminal, his eyes glued to the gauges to look for any deflections in the wing elevon. Any excess drag on the right wing would record right elevon trim, and the left and right wing elevon positions would differ as the automatic system tried to trim out the variances. Even a half degree of trim would indicate that something was wrong, and if he saw that, Gibson knew that he would have about 60 seconds to tell Mission Control what he thought of the assessment of the damage before *Atlantis* broke apart in the upper atmosphere.

Fortunately, *Atlantis* did not suffer a burn through and Gibson skillfully guided the Orbiter to a safe entry and landing on Runaway 17 at Edwards, much to the relief of all onboard, and no doubt the controllers as well. As this was a 'secret' mission, nothing was mentioned publicly for some time after the flight. As soon as the crew exited the Orbiter, they hurried to the right side of *Atlantis* to see the damage for themselves. Post-flight examinations by a review team set up to investigate the incident reported over 700 damaged tiles, with the missing tile being over the steel mounting plate for the L-band antenna. Fortunately, this was a less critical area, which probably prevented a complete burn through similar to the one which sealed the fate of *Columbia* and her crew 14 years later in 2003.

The official STS-27 Mission Report stated: "Initial post-flight inspections of the exterior surface of the Orbiter revealed significant tile damage, with 298 damage sites greater than 1 inch [2.54 cm] in area, and a total of 707 damage sites on the lower surface of the vehicle. The area of major damage was concentrated outboard of a line from the bi-pod attachment to the External Tank liquid oxygen umbilical. One tile was missing on the right side slightly forward of the L-band antenna. Also, there were many damage sites consisting of long narrow streaks with deep gouges. *The damage noted is the most severe of any mission flown yet* [authors' italics]." The report also recorded that "the tin plating on the [L-band antenna] aluminum door was melted, with aluminum appearing to be halfway between hardened and annealed. The door also had a small buckle." [43]

In a 2009 interview with *Spaceflight Now*, former Shuttle Program Manager Wayne Hale stated that the engineers were "caught off guard" when the severity of the damage was seen on *Atlantis* standing on the runway at Edwards. He recalled that back in 1988 it was a struggle to maintain security, and that the controllers had to ask the 'payload customer' if they could view the images because TV from classified missions was forbidden. This was only agreed providing the slow-scan system was used. [44]

The STS-27 mission has been recorded as having the most damaged launch-entry vehicle to return to Earth successfully. Had *Atlantis* been lost, re-entering over the Pacific Ocean, most of the debris would have sunk without trace, and without debris the exact cause may never have been determined. Gibson thought that Congress would have cancelled the whole program if there had been another disaster just two missions after *Challenger*. He and his crew had been lucky, but that luck ran out for the *Columbia* crew on February 1, 2003.

Playing catch-up
As it was a classified mission, the near-miss of STS-27 was not initially revealed, and so the new year 1989 came with an apparently renewed confidence in the Shuttle system after the difficulties of the previous three years. Though there still remained much to accomplish before the whole fleet returned to flight and the recommendations and improvements made to the hardware, systems and infrastructure were fully implemented, there was a feeling of catching up with delayed missions while progressing steadily and, publicly at least, ensuring safety with each step. But within NASA, the seriousness of the damage sustained by *Atlantis* on STS-27 was a stark warning of how close they had come to a second disaster so soon after recovering from the first. The mission announcements during this year revealed a shift towards space science missions rather than the commercial satellite deployments that had been a feature of previous years. Of course, many planned commercial satellite payloads had been shifted to ELV in the down time following the loss of *Challenger*, removing some of the revenue from those planned missions but allowing NASA to launch the delayed science missions and to look towards a new era of Shuttle orbital operations, hopefully focusing upon building the Space Station.

On February 24, NASA announced that John Creighton would command STS-36, a dedicated DOD mission aboard *Atlantis* scheduled for February 1990. Creighton would continue to serve as head of the Mission Support Branch in the Astronaut Office prior to starting full-time training for his upcoming flight. Fellow TFNG Mike Mullane was assigned as an MS on the flight, together with PLT John Casper (Class of 1984) and MS David Hilmers (Class of 1980) and Pierre Thuot (Class of 1985). In the same release, Rhea Seddon was named as an MS on STS-40, the first dedicated Spacelab Life Sciences (SLS, formerly Spacelab 4) mission, together with James Bagian (Class of 1980). *Columbia* was manifested to carry the Spacelab Long Module on a mission scheduled for June 1990. The two MS joined the two PS named to the mission back in April 1985 and this partial crew assignment enabled long-lead crew participation for payload training and experiment integrations, with the rest of the flight crew being assigned at a later date. [45]

Fig. 11.4: (top) A much-relieved STS-27 crew descends the steps after exiting *Atlantis* at the end of their eventful mission. Leading is CDR Hoot Gibson, followed by PLT Guy Gardner and then MS Mike Mullane, Jerry Ross and Bill Shepherd. This image of the left side of *Atlantis* does not reveal the extensive damage to the Thermal Protection System on the right-hand side, which came close to costing the crew their lives and the probable termination of the Shuttle program. (bottom) The experienced TFNG continued to support the Return-to-Flight program. Here, John Creighton, in shirt sleeves, and Frank Culbertson (Class of 1984 next to him), are at the Capcom console in MCC during the STS-29 mission.

STS-29 (March 13 – 18, 1989)
Flight Crew: Michael L. COATS (CDR), John L. Blaha (PLT), Robert C. Springer (MS-1), James F. BUCHLI (MS-2), James P. Bagian (MS-3)
Spacecraft: Discovery (OV-103) 8th mission
Objective: 28th Shuttle mission; TDRS-D deployment
Duration: 4 days 23 hours 38 minutes 50 seconds
Support Assignments: For this mission, John CREIGHTON served as Capcom on the 'Aquila' team for ascent and the 'Phoenix' team for entry, while Kathy SULLIVAN was on console for the 'Altair' Orbit 2 team. [46] Assigned to SPAN were Loren SHRIVER (08:00-14:00), Hoot GIBSON (20:00-02:00) and Dick COVEY (02:00-08:00). Dave WALKER worked the first shift in SMS and Fred GREGORY the second. Dan BRANDENSTEIN returned to his WX pilot role at both KSC and Edwards. Steve NAGEL was contingency landing support for AOA and Alternative End of Mission (EOM) support, both at Northrup, though neither were required. Pinky NELSON provided PAO support. [47]

Six hours into this mission, the crew deployed the TDRS-D satellite, while later on the middeck they focused upon an array of science investigations and photography using the large format IMAX camera. STS-29 was one of the smoothest missions of the program.

During March 1989, Jon McBride officially returned to the Astronaut Office from Washington to resume training as CDR of the Astro-1 mission, while Jim Buchli was assigned as Deputy Chief of the Astronaut Office after flying STS-29, replacing Steve Hawley who was reassigned to active mission training as MS for STS-31. At the same time, Mike Coats, (Buchli's CDR on STS-29) became the Acting Chief of the Astronaut Office while Dan Brandenstein returned to active mission training as CDR of STS-32. Following his success in commanding the important Return-to-Flight mission six months earlier, Rick Hauck, a veteran of three Shuttle missions and one of the leading figures in the Class of 1978, announced on March 24 that he would leave NASA on April 3 to assume the post of Director of Navy Space Programs Division, Staff of Chief Naval Operations, The Pentagon, Washington D.C., effective the end of May. "My eleven years with NASA have been extremely rewarding," Hauck said at the time. "I'll miss the challenging environment and the people. I am looking forward to continuing my career in the Navy and to the new challenges it provides." [48]

Just two days after the departure of Hauck from the Astronaut Office came the news that Steve Nagel was to take command of STS-37, with a crew comprising PLT Ken Cameron (Class of 1984), and MS Jerry Ross (Class of 1980), Jay Apt and Linda Godwin (both from the Class of 1985). This mission was to be flown on *Discovery* in April 1990, deploying the Gamma Ray Observatory, one of NASA's Great Observatories, into orbit. In the same announcement, NASA assigned three other crewmembers to STS-40 (SLS-1) from the Classes of 1980 (Bryan O'Connor and John Blaha) and 1985 (Tammy Jernigan, who replaced the retired TFNG John Fabian), joining Rhea Seddon and Jim Bagian who were already in training. **[49]**

STS-30 (May 4 – 8, 1989)
Flight Crew: David M. WALKER (CDR), Ronald J. Grabe (PLT), Mark C. Lee (MS-1), Norman E. THAGARD (MS-2), Mary L. Cleave (MS-3)
Spacecraft: Atlantis (OV-104) 4th mission
Objective: 29th Shuttle mission; Magellan Venus orbiter probe deployment
Duration: 4 days 0 hours 56 minutes 27 seconds
Support Assignments: John CREIGHTON was the sole Group 8 representative in the MCC, assigned to the 'Phoenix' team for ascent and entry. This was the last Capcom assignment for Group 8 astronauts for some time, as other astronauts from more recent selections were given the chance to gain experience of working in MCC Houston (MCC-H). **[50]** Continuing their roles in SPAN were Loren SHRIVER (02:00-08:00), Hoot GIBSON (08:00-14:00) and Dick COVEY (20:00-02:00). Fred GREGORY and Don WILLIAMS took the two shifts in SMS. Dan BRANDENSTEIN provided weather support for launch and landing and Pinky NELSON was on call at the Weightless Environment Training Facility (WETF) to support any contingency EVA. Mike MULLANE was one of the Family Escorts. **[51]**

During this four-day mission, the crewmembers deployed the radar-mapping Magellan Venus exploration spacecraft (the first U.S. planetary science mission in nine years). This was the first planetary probe to be deployed from a Shuttle Orbiter, demonstrating another capability of the Shuttle system. In addition, the crew worked on secondary payloads involving fluid research in chemistry and electrical storm studies. This mission should have used the more powerful Centaur upper stage to boost Magellan towards the shrouded plant in a shorter timescale, and using the lesser powerful IUS meant that Magellan would not reach Venus for 16 months. However, when it did finally enter the orbit of Venus on August 10, it not only completed its planned one-year primary mission but had its mission extended through October 1994, over five years after being deployed from *Atlantis*. In that time, Magellan radar-mapped 98 percent of the surface and recorded 95 percent of the planet's gravity data.

The deployment of Magellan from *Atlantis* was seen as a significant achievement, so much so that President George H. W. Bush called first-time Shuttle commander Dave Walker onboard *Atlantis* to congratulate him and his crew and invite them to the White House after the mission. While flying his T-38 jet trainer to those ceremonies on May 5, 1989, Walker inadvertently came within 100 feet (30.5 meters) of a Pan Am jetliner just outside Washington, D.C., an error that would cost him his next command.

Assignments, retirements and another tragedy

May and June 1989 was a busy period in the Astronaut Office, with Mike Coats named as Acting Chief of the Astronaut Office until March 1990, standing in for Dan Brandenstein while the latter trained as CDR for STS-32. On May 11, Dick Covey was named as CDR of STS-38, another classified DOD mission aboard *Atlantis* in May 1990. He would lead a crew consisting of PLT Frank Culbertson (Class of 1984) and MS Robert Springer (Class of 1980), Carl Meade and Charles 'Sam' Gemar (both from the Class of 1985). The same announcement listed Guion Bluford as one of three MS for STS-39, also a DOD mission, scheduled for July 1990. He was named together with members of the 1984 (Charles Lacy Veach) and 1985 (Richard Hieb) selections. Once again, the early naming of this partial crew would provide them with plenty of lead time to participate in payload training and integration. [52]

The following day, Jon McBride became the latest member of the TFNG to leave NASA. He also announced his intention to retire from the USN with the rank of Captain and pursue a career in politics. His intention to retire had been announced the previous month (April 24), shortly after he had returned from his assignment at NASA Headquarters in Washington D.C. to resume training as CDR for STS-35 (Astro-1). He would be replaced on that mission by veteran astronaut Vance Brand. "I've spent an extremely rewarding 25 years with NASA and the Navy," McBride commented at the time. "This move has been a very difficult decision for me, but in the final analysis, I felt it was time to make a career change and return to West Virginia. I'll continue to follow developments in the space program with keen interest." [53]

On June 9, it was announced that George 'Pinky' Nelson would also be leaving the agency, effective June 30, to accept academic and administrative posts at the University of Washington in Seattle, Washington State. He was named both Assistant Provost of Astronomy and Associate Professor of Astronomy at the institute. "I am excited with the prospects of a new challenge at the University," Nelson said. "At the same time, I know that I will miss NASA and the Johnson Space Center, especially the people. I don't think there is a more dedicated, motivated and skilled group around. Thanks to everyone for making the past 11 years so enjoyable. I hope to continue to promote the space program in my new career, because I believe that the exploration of space and the development of new technology is key to the future success of our civilization." [54] At just 27 in 1978, Nelson was one of the youngest candidates NASA had selected for astronaut training. Now, turning 39 the following month, he was one of the most experienced astronauts from the TFNG to leave the office, having amassed over 411 hours on three space flights, flown on three different Orbiters, and accumulated over ten hours of EVA time, as well as experience in a range of ground support and technical assignments.

Just over a week after the news that Nelson was leaving the agency, the TFNG remaining in the Astronaut Office reeled from the shock of losing another of their brethren when news came in that former naval aviator and test pilot S. David Griggs had been killed on June 17 in the crash of the North American AT-6 vintage trainer he was flying near Earle, Arkansas, while practicing for an upcoming air show. [55] He was just 39 years old and had been at NASA in Houston since 1970, initially as a research pilot in the Aircraft Division at JSC and later as Chief of the Shuttle Training Aircraft Operations Office before being selected as an astronaut in 1978. Having flown in space aboard STS-51D in 1985, he was in training as PLT for the STS-33 classified DOD mission aboard *Discovery* that was scheduled for November 1989.

The Astronaut Office had strict rules to prevent astronauts who were assigned to flight crews from participating in potentially dangerous extracurricular activities, with extreme sports and risky recreational activities, and even softball and skiing, deemed off limits to reduce the risk of accidents so close to launch. Questions were asked as to why Griggs, an experienced and talented pilot, would spend the weekend doing acrobatics so close to his scheduled launch date, clearly against these rules. As it turned out, Griggs had not sought permission for the out-of-hours flying, which for a while reflected badly on the management at JSC, who clearly were unaware of the event. While the enforcement of these rules was often criticized, there were clear reasons behind them, but as this case demonstrated, sometimes members of the office simply overlooked the rules they were expected to adhere to.

With the STS-33 mission just five months away, a recently-flown replacement had to be assigned to the mission quickly to prevent a break in the training flow or delay to the flight. The seat was given to Group 9 astronaut John Blaha (Class of 1980), who had flown his first mission as PLT on STS-29 just three months earlier and was just beginning training as PLT on STS-40. By now, he would have recovered from his first flight and his performance level would still be at a peak, and his assignment was not expected to impact the planned launch date. Blaha's place on STS-40 would be fulfilled by Sydney M. Gutierrez (Class of 1984), and a further flight crew assignment was announced in the same news release. TFNG Norman Thagard and Mary Cleave (Class of 1980) were named as MS for the first International Microgravity Laboratory (IML-1) mission, again enabling long term crew participation in payload training and experiment integration. IML-1 was planned as a nine-day flight on *Columbia* in December 1990. [56]

In a packed program

By the close of June 1989, NASA had flown four missions in nine months and had a further dozen crews, or partial crews, in various stages of training for missions planned to fly over the coming 18 months (see Table 11.2).

TABLE 11.2: NAMED SPACE SHUTTLE CREWS IN TRAINING AS OF JUNE 30, 1989 [NASA ASSIGNMENTS ONLY]

STS	Primary Payload	Launch date	Commander	Pilot	Mission Specialists					
SHUTTLE MISSIONS MANIFESTED FOR CALENDAR YEAR 1989										
28	DOD	Jul 31	SHAW	Richards R.	Adamson	Leestma	Brown M.	-	-	-
34	Galileo Jupiter probe	Oct 12	WILLIAMS D.	McCulley	LUCID	Chang-Diaz	Baker E.	-	-	-
33	DOD	Nov 19	GREGORY F.	Blaha	Carter	Musgrave	Thornton K.	-	-	-
32	Intelsat; LDEF retrieval	Dec 18	BRANDENSTEIN	Wetherbee	Dunbar	Low	Ivins	-	-	-
SHUTTLE MISSIONS MANIFESTED FOR CALENDAR YEAR 1990										
36	DOD	Feb 1	CREIGHTON	Casper	Hilmers	MULLANE	Thuot	-	-	-
31	Hubble Space Telescope deployment	Mar 26	SHRIVER	Bolden	HAWLEY	SULLIVAN	McCandless	-	-	-
35	ASTRO 1	Apr 26	Brand	Gardner G.	Lounge	HOFFMAN	Parker	-	-	-
37	Gamma Ray Observatory deployment	Jun 4	NAGEL	Cameron	Ross	Apt	Godwin	-	-	-
38	DOD	Jul 9	COVEY	Culbertson	Springer	Meade	Gemar	-	-	-
40	Spacelab Life Sciences 1	Aug 16	O'Connor	Gutierrez	SEDDON	Bagian	Jernigan	-	-	-
39	Infrared Background Signature Survey	Nov 1	TBD	TBD	BLUFORD	Hieb	Veach	TBD	TBD	-
42	International Microgravity Laboratory 1	Dec 6	TBD	TBD	THAGARD	TBD	Cleave	-	-	-

Of the 25 members of the Class of 1978 in the Astronaut Office at the start of the year (14 PLT and 11 MS), two pilots (Hauck and McBride) and one MS (Nelson) had retired, while a third pilot (Dave Griggs) had been killed, all within three months. That left 21 TFNG in the office (11 PLT and 10 MS). Of these, 16 were in training for new missions as the group embarked on their second decade at NASA. Despite all the departures and fatalities, at least one member of the 1978 selection was present on every named crew through to the end of 1990… at least for now.

STS-28 (August 8 – 13, 1989)
Flight Crew: Brewster H. SHAW (CDR), Richard N. Richards (PLT), James C. Adamson (MS-1), David C. Leestma (MS-2), Mark N. Brown (MS-3)
Spacecraft: Columbia (OV-102) 8th mission
Objective: 30th Shuttle mission; 4th classified DOD satellite deployment
Duration: 5 days 1 hour 0 minutes 9 seconds
Support Assignments: Flight control assignments were not announced for this mission, but SPAN was occupied by Hoot GIBSON (08:00-14:00), Dave WALKER (20:00-02:00) and Jim BUCHLI (02:00-08:00). Don WILLIAMS worked the first shift in SMS and Fred GREGORY the second. Mike COATS was the WX pilot for both launch and landing. Loren SHRIVER was at Ben Guerir Air Base in Morocco for Transoceanic Abort Landing (TAL) support, while Jim BUCHLI was dual assigned as WETF support. **[57]**

This fourth fully-classified Shuttle flight was also the Return-to-Flight mission for *Columbia*, completing the post-*Challenger* program to re-qualify all three remaining Orbiters for operational flights. The crew deployed the KH-12 advanced reconnaissance satellite.

On September 29, six weeks after the return of STS-28, NASA announced astronauts to no less than five Shuttle missions expected to fly in late 1990 and early 1991. **[58]** What was evident was the changing make-up of the Astronaut Office, as many of the TFNG were assigned to what would become their final missions and members of the later groups began to take over the majority of assignments. What was not mentioned was that the emphasis on academic qualifications over more traditional piloting skills, the inclusion of female and minority selectees, the selection of representatives of other military services and even foreign-born astronauts trained as fully qualified MS, a process which began with the selection of the Group 8 astronauts in 1978, was now becoming commonplace in recent astronaut selections and assignments. It was all pointing towards preparing the first international crews for the Space Station program, providing its funding and designs could be sorted out. The assignments announced in the September news release included the first astronauts from the Class of 1987 (Group 12), the first U.S. Coast Guard astronaut to fly (Bruce Melnick), the first European Space Agency (ESA) astronaut to be named as an MS (Claude Nicollier, Switzerland) and the first black female (Mae Jemison) to be selected for space flight.

Most notable was the crew originally named to STS-41, the first not to include any members of the 1978 selection since 1985, although five TFNG were named to other crews in the same announcement. In a partial crew assignment, Mike Coats was named to command *Discovery* for STS-39, a now unclassified DOD mission scheduled for November 1990. Also named were PLT Blaine Hammond (Class of 1984) and MS Gregory Harbaugh and Donald McMonagle (both from the Class of 1987), to join the previously named MS Hieb, Veach and TFNG Guion Bluford. Kathy Sullivan was named to the STS-45 Astro-1 mission scheduled for March 1991 aboard *Columbia*, together with C. Michael Foale (Class of 1987). In another partial crew assignment, Hoot Gibson was named to command STS-46, along with TFNG Jeff Hoffman, Franklin Chang Díaz (Class of 1980) and ESA astronaut Claude Nicollier as MS. This mission on *Atlantis* would deploy the Eureka ESA-sponsored free-flying satellite and demonstrate the Tethered Satellite System (TSS). The final assignment was for the science MS for STS-47, flying the Spacelab J (Japanese) life science and material processing experiments. Though no Group 8 astronaut was initially assigned to this mission, it would feature later in the story of one of the group.

STS-34 (October 18 – 23, 1989)
Flight Crew: Donald E. WILLIAMS (CDR), Michael McCulley (PLT), Shannon W. LUCID (MS-1), Franklin R. Chang Díaz (MS-2), Ellen S. Baker (MS-3)
Spacecraft: Atlantis (OV-104) 5th mission
Objective: 31st Shuttle mission; Galileo Jupiter entry probe/orbiter deployment
Duration: 4 days 23 hours 39 minutes 21 seconds
Support Assignments: For this mission, Dave WALKER was assigned to SPAN (08:00-14:00), along with Steve NAGEL (20:00-02:00). Fred GREGORY worked the first shift in SMS and John CREIGHTON the second. Mike COATS again served as weather coordinator for launch and landing and Jeff HOFFMAN was one of the immediate Family Escorts. Dick COVEY was on hand at Ben Guerir, Morocco, for any TAL situation during launch. [59]

During this four-day mission, the crew successfully deployed the Galileo spacecraft, the second planetary probe from a Shuttle Orbiter, on its long journey to explore Jupiter. On board *Atlantis*, the crew also operated the Shuttle Solar Backscatter Ultraviolet Instrument (SSBUV) to map atmospheric ozone, and performed numerous secondary experiments involving radiation measurements, polymer morphology, lightning research, microgravity effects on plants and a student experiment on ice crystal growth in space. STS-34 lasted less than five days and remained one of the shortest Shuttle missions in the series, but the mission of Galileo lasted far longer. Originally planned to be carried on a Centaur upper stage, it was also reassigned to an IUS upper stage and therefore, like Magellan and Ulysses, would require a far longer trajectory to its target planet. The journey took six years to complete, arriving on December 7, 1995, and the results from Galileo re-wrote the textbooks on Jupiter. Once again, Group 8 astronauts played their part in a significant milestone in American space flight and planetary exploration.

Also in October, having completed his STS-28 post-flight obligations, Brewster Shaw became the latest member of the 1978 selection to leave the Astronaut Office at JSC, to take a position as Deputy Director of Space Shuttle Operations at KSC.

STS-33 (November 22 – 27, 1989)
Flight Crew: Frederick D. GREGORY (CDR), John L. Blaha (PLT), Manley L. 'Sonny' Carter (MS-1), F. Story Musgrave (MS-2), Kathryn C. Thornton (MS-3)
Spacecraft: Discovery (OV-103) 9th mission
Objective: 32nd Shuttle mission; 5th classified DOD satellite deployment
Duration: 5 days 0 hours 6 minutes 48 seconds
Support Assignments: Very few Group 8 astronauts were assigned to support roles on this flight. Chief Astronaut Dan BRANDENSTEIN worked the first shift in SMS, with John CREIGHTON taking the second shift. Mike COATS again served as WX for launch and landing. **[60]**

Compared to the previous three years, 1989 had seen a significant increase in Shuttle missions to five, with *Columbia* returning to flight to join the other two surviving Orbiters, the deployment of a third TDRS and two planetary probes, and two DOD missions including the deployment from STS-33 of a SIGINT electronics signals intelligence satellite.

SCHEDULES, SCIENCE COMMANDERS AND THE SOVIETS

The new year of 1990 opened with the assignment of the Orbiter flight crew (CDR, PLT and MS-2) for STS-42 carrying the IML-1 payload. While none of these were from the 1978 selection, they did join the previously named Norman Thagard who was already in training for the mission. It was expected to be a busy year for Shuttle operations, but the gremlins were gathering once again to cause havoc with the schedules and scupper even the best laid plans.

There were ten Shuttle missions manifested at the beginning of 1990, but early in the year a 24-hour weather delay in getting *Columbia* off the ground for STS-32, together with problems with the SRB assigned to the STS-31 HST deployment mission, necessitated the latter launch being moved from March to April. STS-37 therefore slipped from July to November and STS-39 was rescheduled from its November slot to no earlier than January 1991.

STS-32 (January 9 – 20, 1990)
Flight Crew: Daniel C. BRANDENSTEIN (CDR), James D. Wetherbee (PLT), Bonnie J, Dunbar (MS-1), Marsha S. Ivins (MS-2), G. David Low (MS-3)
Spacecraft: Columbia (OV-102) 9th mission
Objective: 33rd Shuttle mission; Comsat deployment; LDEF retrieval
Duration: 10 days 21 hours 0 minutes 36 seconds
Support Assignments: In SPAN, Don WILLIAMS worked on the 21:00−03:00 shift, while Kathy SULLIVAN and Dave WALKER both supported the 09:00−15:00 shift. Guion BLUFORD was in SAIL, while SMS support was provided by John CREIGHTON on the first shift and Loren SHRIVER on the second. Mike COATS continued his role as WX pilot at both KSC and Edwards, while Jim BUCHLI fulfilled the role of WETF support. **[61]**

STS-32 was the first mission since STS-61A in October 1985 to log a duration of over seven days and the first since STS-9/Spacelab 1 in 1983 to surpass 10 days, demonstrating that the Shuttle was beginning to return to its pre-1986 flight capabilities. Most of the Shuttle flights between 1988 and 1990 were planned to catch up with the delayed manifest following the loss of *Challenger*, and STS-32 played its part in this plan by retrieving the LDEF that had been deployed by STS-41C. Originally intended to be in orbit for a year, LDEF was finally retrieved after six years, or 2,093 days in free flight. Its retrieval came too late for some of the experiments, but was useful for analyzing how materials and samples deteriorated in space over time, an important source of knowledge for designing a large space station that was planned to be in orbit for up to 30 years. On a more personal note, Dan Brandenstein became the first of the TFNG to celebrate the anniversary of their selection in space (January 16) and followed that up with his 47th birthday the very next day, which he celebrated with an inflatable cake complete with imitation candles. His piloting skills were called for at the end of the mission, as he landed *Columbia*, with the LDEF secured on board, at an overall mass of 228,376 lbs. (103,572 kg), the heaviest landing weight in the program to date.

On January 25, less than a week after the return of STS-32, NASA announced its first Science Payload Commanders (PC) as Norman Thagard (STS-42/ IML-1), Kathy Sullivan (STS-45/Atlas-1), Jeff Hoffman (STS-46/ EURECA/

TSS-1) and Group 10 astronaut Mark Lee (STS-47/Spacelab J). **[62]** The intention was to provide long term leadership in the development and planning of payload crew science activities. According to the news release, "the Payload Commander will have overall responsibility for the planning, integration and on-orbit coordination of payload/Space Shuttle activities on their mission, though the crew commander retains overall responsibility for mission success and safety of flight."

A visit to the Soviet Union
On February 1, 1990, NASA announced it had accepted an invitation extended earlier in the year by Soviet cosmonaut Alexei Leonov, through the cosmonaut chapter of the Association of Space Explorers (ASE), for a group of Shuttle astronauts to attend the launch of the next Soviet manned space launch and to tour facilities in the Soviet Union. Members of the U.S. delegation included JSC Deputy Director and astronaut Paul J. Weitz (Class of 1966), and Chief Astronaut and TFNG Dan Brandenstein, together with Ron Grabe and Jerry Ross (both from the Class of 1980). The group travelled to Moscow on February 9, and the following day were taken to the Baikonur Cosmodrome in Kazakhstan to witness the launch of Soyuz TM-9 carrying Mir EO-6 cosmonauts Anatoly Y. Solovyov and Aleksandr N. Balandin to the Mir space station at the start of their 180-day mission. They also took the opportunity to view the Buran Space Shuttle before returning to Moscow to visit the nearby Cosmonaut Training Center named for Yuri A. Gagarin. Here, the American pilot astronauts were invited to try out the Soyuz simulator, while Jerry Ross evaluated the Soviet version of the Manned Maneuvering Unit (MMU), the UPMK (Cosmonaut Transference and Maneuvering Unit – 21KS) flight tested by cosmonauts on Mir earlier that month. The group also visited the Soviet Manned Spaceflight Control Center in Kaliningrad before returning to the United States on February 14. **[63]**

Relations with the Soviets had remained good since the Apollo-Soyuz Test Project (ASTP) in 1975, despite the adverse political situation in the late 1970s which stifled future plans. This visit was to prove fortuitous in its timing, firstly because NASA was struggling to make progress in its complicated and expensive *Freedom* space station program, and secondly because serious domestic issues were developing within the Soviet Union. In fact, the following year would see the collapse of the Soviet Union and the near-cancellation of the *Freedom* project. This visit was a stepping stone which would later allow the new Russia to become a partner in the emerging International Space Station (ISS) program in 1993.

On February 20, three days before he was scheduled to fly on his third mission as an MS on STS-36, NASA announced that Richard 'Mike' Mullane would retire from NASA and the USAF effective July 1, 1990, to return to new opportunities in Albuquerque in his home state of New Mexico. However, it turned out that his last space flight would not be as soon as he or anyone else expected. **[64]** On the very next day, February 21, NASA reported that STS-36 CDR John Creighton had developed an upper respiratory tract infection and was being treated with antibiotics. NASA Spokesman Kyle Herring stated: "It's basically a sore throat with a little head congestion," but with the astronaut medically grounded, the launch was postponed for at least 24 hours. Meteorologists also warned that the weather could further delay the launch. This was the first time since Apollo 9 in 1969 that a U.S. mission had been delayed by the illness of a prime crewmember, and 20 years since health had affected a prime crew so close to launch, when Command Module Pilot Ken Mattingly was replaced by Jack Swigert five days prior to the launch of Apollo 13 due to Mattingly's exposure to German Measles. There were no plans to replace Creighton, however. While there had been no formal back-up crews since 1982, it could have been done, though it would have been difficult to choose someone from the pool of astronauts this close to launch. With the flight rate increasing, there were now a number of astronauts who were capable, if it was critical, of taking on the role. **[65]** The expected adverse weather did indeed delay the launch until the end of the month, allowing the astronaut to recover fully without needing to replace him.

On February 26, two days prior to the launch of STS-36, it was revealed that veteran of two Shuttle flights Don Williams would become the next member of the 1978 selection to leave the Astronaut Office, effective March 1, retiring from NASA and the USN to pursue a career in private industry. He would become a Senior Systems Engineer for Science Applications International Corporation in Houston. "I reached my goal as a pilot, which was to command a mission. Now it's time to go on to other challenges. JSC and NASA have been a wonderful place to work and I'm proud to have been part of the team," he said at the time. **[66]**

STS-36 (February 28 – March 4, 1990)
Flight Crew: John O. CREIGHTON (CDR), John H. Casper (PLT), R. Michael MULLANE (MS-1), David C. Hilmers (MS-2), Pierre J. Thuot (MS-3)
Spacecraft: Atlantis (OV-104) 6th mission
Objective: 34th Shuttle mission; 6th classified DOD satellite deployment
Duration: 4 days 10 hours 18 minutes 22 seconds
Support Assignments: Shannon LUCID worked in SPAN (12:00–18:00), while Guion BLUFORD was assigned to SAIL and Loren SHRIVER to the first shift in SMS. Mike COATS was again assigned as weather support at KSC and Edwards, and Hoot GIBSON served as one of the Immediate Family Escorts **[67]**

STS-36 carried a DOD payload, which included a digital imaging and electronics signals intelligence satellite and a number of secondary payloads, to a very high inclination orbit of 62 degrees, about five degrees outside the prescribed safety limits for Shuttle launches from the Cape.

With Dan Brandenstein having completed his STS-32 post-flight duties, Mike Coats stepped down as Acting Chief of the Astronaut Office during March to allow Brandenstein to resume his position as Chief of the Astronaut Office. Coats began full time mission training for STS-39.

STS-31 (April 24 – 29, 1990)
Flight Crew: Loren J. SHRIVER (CDR), Charles F. Bolden (PLT), Bruce McCandless (MS-1), Steven A. HAWLEY (MS-2), Kathryn D. SULLIVAN (MS-3)
Spacecraft: Discovery (OV-103) 10th mission
Objective: 35th Shuttle mission; Hubble Space Telescope deployment
Duration: 5 days 1 hour 16 minutes 6 seconds
Support Assignments: Fred GREGORY continued his tour in SPAN (12:00−18:00) and was joined by Shannon LUCID (18:00−24:00). Guion BLUFORD was back in SAIL, while SMS support was provided by Dick COVEY on the second shift. Having completed his STS-32 assignments, Chief Astronaut Dan BRANDENSTEIN returned to the role of WX pilot for both KSC and Edwards. Hoot GIBSON again served as one of the Immediate Family Escorts. **[68]**

On this five-day mission, the crew deployed HST and conducted a variety of mid-deck experiments involving the study of protein crystal growth, polymer membrane processing, and the effects of weightlessness and magnetic fields on an ion arc. They also operated a range of cameras, including both the IMAX in-cabin and cargo bay cameras, for Earth observations from their record-setting altitude of 380 miles (600 km), the highest recorded on an Earth-orbital flight since the Gemini missions of 1966. As this was highest apogee 'officially' attained by any Shuttle mission, it meant that the crew, including TFNG Shriver, Hawley and Sullivan, had travelled further from Earth than any space explorer since Apollo 17. However, in 2009 Kathy Sullivan suggested that this may not have been the case: "The deployment altitude for Hubble was quite high for a Space Shuttle. I'm pretty sure it's the highest altitude civilian flight. Every time I say that, KT [Kathryn Thornton] squints at me as if she went higher than that at some other point in time. Her first flight [STS-33] was a Defense Department flight, though, so if she did, she can't say anything." **[69]** Though the crew had successfully deployed Hubble, tests carried out after the crew had returned to Earth revealed the infamous error in the telescope's optics, rendering HST relatively useless – much to the chagrin of NASA – until a solution to repair the distortion could be found, a task which would take the next three years to resolve and mount a mission to service and fix the telescope's vision. **[70]**

Fig. 11.5: (top) Chief Astronaut and STS-32 CDR Dan Brandenstein celebrates his 47th birthday (January 17) in space with an inflatable cake, the very day after the 12th anniversary of the selection of the Class of 1978. (bottom) Dressed to impress. Education has always been a key element of the TFNG outreach program. On the aft flight deck of *Columbia*, while sporting a teacher's apparel of shirt and tie rather than an astronaut's spacesuit, STS-35 MS Jeff Hoffman prepares to conduct a lesson from space to school-children during the Astro-1 mission.

A fifth Orbiter at last

On May 7, 1991, a new Shuttle Orbiter arrived at KSC on top of the Shuttle Carrier Aircraft (SCA), 12 days after its rollout from Rockwell's Palmdale facility in California. Formally named OV-105, the new vehicle was more commonly known as *Endeavour*, named, via a national school competition, after the 18th century British ship *HMS Endeavour* which was captained by James Cook on his first voyage of discovery (1768–1771), hence the English spelling rather than the American version of the name (Endeavor). The U.S. Congress had approved the construction of OV-105 in 1987 as a replacement for *Challenger*, which would be fabricated from the structural spares built during the construction of *Discovery* and *Atlantis*, as it proved much cheaper to build a new vehicle from the spares rather than refit *Enterprise* or construct a brand new Orbiter from scratch. Of course, this meant that there were no structural spares left for what would have been OV-106, nor were additional spares requested to guard against a second Orbiter being seriously damaged or lost in another accident. This decision contributed to the end of the Shuttle program 11 years later, following the loss of *Columbia* and the STS-107 crew and the decision to retire the fleet rather than build another replacement Orbiter.

Due to its late construction, in addition to utilizing the available structural spares, *Endeavour* was assembled using new hardware designed to upgrade and improve its capabilities. Many of these upgrades were later incorporated into the other vehicles during their planned out-of-service maintenance periods. This included the capability to support Extended Duration Orbiter (EDO) missions of up to 28 days, though this duration was never attempted. *Columbia* flew the longest mission, of nearly 18 days, in 1996 (STS-80).

A mouthwatering mission

One year earlier, on March 14, 1990, the $150 million Intelsat VI (603) satellite had been launched into orbit from Cape Canaveral using a Titan III. Intended to reach a geosynchronous orbit of 22,500 miles (36,200 km) above the Earth, at just 25 minutes into the mission the Orbus-21S upper stage that was designed to place the satellite in its transfer orbit failed to separate from the Titan's second stage and was thus unable to fire. Though controllers were eventually able to separate the satellite and upper stage, Intelsat remained stuck in an elliptical orbit of 96 by 214 miles (154.49 by 344.4 km). The failure was subsequently traced to faulty wiring in the launch vehicle, which had been programmed for two satellites instead of one. The onboard computer was programmed to launch the phantom first satellite while the electronics were wired to launch the second and as a result no signal reached the separation device to trigger the release of the upper stage. Rescue options were evaluated almost immediately, although some reports suggested that despite previous satellite retrievals by Shuttle astronauts, the potential rescue of Intelsat by a Shuttle crew was unlikely.

However, on March 22, Intelsat's owners met with NASA officials at JSC to discuss whether it was indeed possible to mount a rescue flight. The positive response that it would be possible to deliver and install a new upper stage motor on the satellite came the following day. By May 15, NASA had agreed to mount the rescue mission and announced on June 13 that the new Orbiter *Endeavour* would attempt the rescue for its first mission, planned for February 1992. "It's an exciting mission for NASA because it gives us a chance to exercise spacewalking capabilities that will be needed for the Space Station *Freedom* project," said William Green, Payload Manager at NASA HQ in Washington D.C. Jeff Carr, NASA spokesman at JSC, added, "To have the first flight of a new Orbiter is a plum, but to have that and a sexy flight like going to pick up the Intelsat, fix it and re-boost it will just make any experienced crew commander's mouth water." [71] Exactly whose mouths would be watering back in the Astronaut Office was yet to be made clear.

New crew assignments
On May 24, 1990, NASA revealed the latest crew assignments for three missions in 1991, raising the number of crews in training to 12. The DOD mission of STS-44 was to be flown on *Atlantis* in March 1991 under the command of Dave Walker, along with PLT Tom Henricks (Class of 1985) and MS Story Musgrave (Class of 1967), Mario Runco and James Voss (both from the Class of 1987). The same announcement named Shannon Lucid as MS to STS-43 planned for May 1991 to deploy TDRS-E. The mission would be commanded by John Blaha (Class of 1980) and also included PLT Michael Baker (Class of 1985) and MS David Low and James Adamson (both from the Class of 1984). The final assignment was for the STS-45 Atlas 1 Orbiter crew, to join already-assigned PC Kathy Sullivan and the rest of the science crew on a mission projected to launch in April 1991. [72]

On June 7, just over a month after returning from STS-31, Steve Hawley left the Astronaut Office at JSC to join NASA's Ames Research Center, Moffett Field, Palo Alto, California, as Associate Director. [73]

"I got asked to go to Ames [because] at the time there was sort of a dispute between life sciences at Ames and life sciences at JSC," Hawley explained in 2019. "[Though] I wasn't directly involved in this, Headquarters felt that somebody from JSC going out to Ames might be able to figure out what's going on and get those different groups working together, and that is why they asked me to do it. So the guys at Ames had some ideas, but they didn't have a lot of real knowledge of the astronauts' experience." Expecting his experience as an astronaut/manager would help, it was actually a challenge to suggest that Ames might want to follow a particular route. "When I was first out there I started to listen to them, saying, 'Ok, here's the problem, here's what you guys need to do'. And that didn't go over very well. I learned that it's a little more complicated, with different groups pulling in the same direction, so that was very useful. Ames [is] a research center and, at the time at least, it was both an aeronautical research center and a science research center, I think it was a 60/40 split, if I remember when I was there. It was totally different from JSC, in the sense that [at] JSC, it really felt like, regardless of what we did, we all had one mission and it was flying people in space successfully. At Ames, there was no common mission at all, and that was a real eye opener for me. Everybody was doing their own thing and [had been for several years].

"I learned a lot about how different parts of NASA work, how maybe they tried to bring different group perspectives together, I sort of had mixed success of that at Ames, but I think it was helpful when I returned to JSC. So one of the things that I tell people – and I learned this when I was Deputy Chief of the Astronaut Office – was that the job that Dan [Brandenstein] and I had for most part didn't really have authority over anybody in the program. We had some authority over the astronauts, but if you wanted to get your way, you had to use influence, you had to make good arguments. You had to try the influence of the point of view, because you couldn't just say 'We're not going to do [such and such] unless we do ['X' number of] tests of this nature'. You didn't have that authority. You could try to influence someone by [suggesting that something] more than they were planning to do would be a good idea, so that was an important lesson too, as a manager." **[24]**

LEAKING ORBITERS

In early May, problems were encountered with *Columbia* out on the launch pad during preparations for STS-35, when an old fuel line in the Main Propulsion System (MPS) leaked and required replacement. Already delayed due to the slips in the STS-31 launch, another issue arose with the flow of Freon™ through one of the Orbiter's two coolant loops. Managers faced several options: Fly as is and hope everything worked for the duration of the mission, with the option of an early landing if trouble occurred; repair the problem on the pad, which had never been done before and would require about a week's work; or roll the stack back to the VAB for de-stacking and repair the Orbiter horizontally, meaning a further delay to the launch. The decision was made to attempt the repair on the pad as it was felt that not enough was known about the problem to fly with it, though the two or three weeks repair estimation was a guess. The launch was set for May 30.

By May 27, the issue was solved and May 30 was confirmed as the launch date for STS-35. However, a hydrogen leak was detected as fuel was being loaded into the Orbiter on the day before launch (May 29). Over the next few days, the problem remained elusive and threatened the chance to fly STS-35/Astro 1 before the end of the year. Workers continued to investigate the problem hoping that a rollback to the VAB would not be necessary, but by June 6 this seemed likely. This immediately had a knock-on effect on other missions planned for 1990, pushing STS-40/SLS (with Rhea Seddon) into 1991, and both STS-42/IML-1 (with Norman Thagard) and STS-45/ATLAS-1 (with Kathy Sullivan) back by four months each.

By June 12, *Columbia* was indeed back in the VAB to investigate the leak further. Two weeks later, on June 28, another blow to the manifest occurred when *Atlantis* suffered its own leak of liquid hydrogen during a test. As a result, the whole Shuttle fleet was grounded until the problem was understood and fixed, creating a ripple effect of delays extending into 1993. Investigations suggested that the Teflon™ seal used in the connections was the culprit, but when further examination revealed that the leaks had come from two different places on *Columbia* and *Atlantis*, further tests and delays became inevitable.

As a result, no Shuttles flew between April and October 1990, and months of detailed investigations continued throughout the summer, with test after test conducted. Having all the Orbiters at the Cape created a problem as there was a shortage of storage space to park them, resulting in a game of Shuttle chess as the vehicles were moved between the VAB, the pads and the OPF. Other problems continued to plague the fleet, pushing flights into 1991, with payload mishaps and the turbulent weather around the Cape delaying both STS-38 and STS-35 further into the year and allowing STS-41 on *Discovery* to fly before them. The sequence of launches in 1990 therefore looked like this: STS-36, STS-31, STS-41, STS-38, then STS-35. Although the confusing three-digit designations had been dropped after *Challenger*, it remained just as confusing to recall which mission flew in what order in this new system. **[71]**

Grounded

On July 9, 1990, two astronauts were temporarily removed from flight status due to violations of Flight Crew Operations Directorate (FCOD) guidelines. Hoot Gibson was removed from the command of STS-46 and from T-38 jet trainer flight status for one year due to a mid-air collision with a second racing aircraft on Saturday, July 7. The release stated that this was "in violation of a policy which restricts high risk recreational activities for astronauts named to flight crews." The action was in response to Gibson's participation in an airplane race at a civilian air show in central Texas.

Gibson's stunt plane accidently collided with another airplane during a race the previous Saturday, sending the second craft hurtling into a cornfield and killing its pilot, Henry W. Jones Jr., a retired fighter pilot from Virginia. Gibson was able to land his plane safely at the New Braunfels Municipal Airport, 25 miles northeast of San Antonio, Texas. Donald R. Puddy, Director of Flight Crew Operations, said "Our high risk activity policy defines plain and simple guidelines for astronauts assigned to flight crews. They are intended to preserve our crews as assigned and apply regardless of the time prior to launch. There was a clear violation of the policy." The high risk recreational activities defined by the policy included those which involve exposure to major or fatal injury. **[74]** A new CDR for STS-46, manifested to deploy the ESA Eureka free-flyers and evaluate the Italian TSS, would be announced "in the near future." In the same release, it was reported that Fred Gregory would replace Dave Walker as CDR of the STS-44 DOD mission. Walker had also been suspended from T-38 flight status for 60 days for his own infraction of JSC aircraft operating guidelines in May 1989. Both would be eligible for Shuttle crew assignment once their T-38 flying status was reinstated.

Chief Astronaut and fellow TFNG Dan Brandenstein added at the time: "Dave [Walker] and Hoot [Gibson] have each made substantial contributions to the Shuttle and Space Station programs, and have performed in an outstanding manner on their respective Shuttle flights. The actions are unfortunate, but they are in the best interests of us all. These policies are vitally important and are to be taken seriously."

Vandenberg makeover

On September 29, 1990, it was reported by *Florida Today* that the $3.5 billion launch facilities at Vandenberg Air Force Station, California, intended to support Space Shuttle launches into polar orbit and mothballed after the loss of *Challenger*, were to be given a $300−500 million makeover to convert the Space Launch Complex 6 ('Slick 6') into a launch pad for Titan 4 expendable rockets with the liquid-fueled Centaur upper stage. Four years after STS-62A (which originally included TFNG Mike Mullane) should have become the first mission to carry American astronauts from a launch pad outside of Florida, the facility and capabilities to accommodate this would no longer be available. **[75]**

Supporting but not flying

Sandwiched between STS-31 and STS 38 and flown from October 6−10, STS-41 was the 11th mission for Discovery and the 36th flight of the program. On the five-day flight, the crew deployed the Ulysses solar orbital probe. For the first time since STS-51F in July 1985, there was no representative of the TFNG on the flight crew, although several were assigned to bring their experience to the numerous support roles required for each mission. Loren Shriver worked the 12:00−06:00 shift and Shannon Lucid the 18:00−24:00 shift in SPAN, while Dick Covey was assigned to the first SMS shift. Chief Astronaut Dan Brandenstein was WX pilot at the Cape and Fred Gregory went to Ben Guerir in Morocco to support any TAL situation. **[76]**

It was around November 1990 that Jim Buchli returned to active mission training as MS for STS-48. He was replaced as Deputy Chief Astronaut, temporarily, by Dick Covey.

STS-38 (November 15 – 20, 1990)

Flight Crew: Richard O. COVEY (CDR), Frank L. Culbertson (PLT), Carl J. Meade (MS-1), Robert C. Springer (MS-2), Charles D. 'Sam' Gemar (MS-3)
Spacecraft: Atlantis (OV-104) 7th mission
Objective: 37th Shuttle mission; 7th classified DOD satellite deployment
Duration: 4 days 21 hours 54 minutes 31 seconds
Support Assignments: For the next mission, Fred GREGORY was back in Houston to work with Steve NAGEL on the 18:00–24:00 shift in SPAN, while John CREIGHTON covered the 24:00–08:00 shift. Mike COATS was in SMS during the second shift period, while Dan BRANDENSTEIN again fulfilled the role of WX pilot at KSC and Edwards. Loren SHRIVER was assigned as Mishap Representative at the Massachusetts Institute of Technology (MIT). **[77]**

During their five-day mission the crew conducted DOD operations, deploying an advanced data relay satellite. This was the seventh and final fully-classified Shuttle mission of the program.

STS-35 (December 2 – 10, 1990)

Flight Crew: Vance D. Brand (CDR), Guy S. Gardner (PLT), Jeffrey A. HOFFMAN (MS-1), J. Michael Lounge (MS-2), Robert A.R. Parker (MS-3), Samuel T. Durrance (PS-1, astronomer, John Hopkins University), Ronald A. Parise (PS-2, astronomer, Computer Sciences Corp.)
Spacecraft: Columbia (OV-102) 10th mission
Objective: 38th Shuttle mission; Astro-1 astronomical research program
Duration: 8 days 23 hours 5 minutes 8 seconds
Support Assignments: Hoot GIBSON and Loren SHRIVER worked the 24:00–06:00 shift in SPAN, while Kathy SULLIVAN covered the 06:00–12:00 shift. Mike COATS was again in SMS, this time on the first shift, with Steve NAGEL on the second. Dan BRANDENSTEIN once again fulfilled the role of WX pilot at KSC and Edwards. **[78]**

This Spacelab mission featured the pallet-only ASTRO-1 ultraviolet astronomy laboratory, a project on which Jeff Hoffman had worked since 1982. This mission would have flown in March 1986 to observe Comet Halley if the STS-51L mission had been successful, but it was not to be. The mission remained on the manifest over the next four years, and *Columbia* finally made it to orbit after its pad refueling problems at the end of the year. Almost immediately, the mission suffered a primary controller failure, requiring the crew to revert to manual procedures to allow the astronomical observation to continue. It became clear that if this had been an unmanned mission it would have failed. The presence of the crew saved the mission and provided enough data to keep the scientists busy for years. Hoffman, one of four astronomers on board, conducted live classroom lessons from space with students in Alabama and Maryland during the flight.

On December 19, 1990, and with STS-35 safely on the ground before the approaching Christmas holidays, NASA announced its final crew assignments of the year for four missions: STS-48, carrying the Upper Atmosphere Research Satellite (UARS); STS-46, manifested for the deployment of Eureka and the TSS; STS-49, intended to capture, replace a failed engine and redeploy the stranded Intelsat satellite; and STS-50, flying the United States Microgravity Laboratory (USML-1). Of the 19 new NASA astronaut assignments announced, just four were from the 1978 selection, highlighting the significant changes in the make-up not only of the crewing of missions in a new decade, but also within the Astronaut Office itself. [79]

John Creighton was named to command the STS-48 UARS deployment mission and was joined on that crew by fellow TFNG Jim Buchli. Also named were PLT Ken Reightler (Class of 1987) and MS Mark Brown (Class of 1984) and Sam Gemar (Class of 1985).

Loren Shriver replaced the grounded Hoot Gibson as CDR of the STS-46 Eureka and TSS deployment mission, with PLT Jim Wetherbee (Class of 1984) and MS Andrew Allen (Class of 1987), joining the previously named Franklin Chang Díaz, TFNG Jeff Hoffman and ESA's Claude Nicollier.

Finally, the plum assignment as CDR of the maiden flight of the new Orbiter *Endeavour* (OV-105), scheduled to rescue and redeploy the stranded Intelsat, went to Chief Astronaut Dan Brandenstein, who assigned himself together with a hand-picked crew of PLT Kevin Chilton (Class of 1987) and MS Pierre Thuot (Class of 1985), Kathy Thornton (Class of 1984), Tom Akers and Bruce Melnick (both from the 1987 selection).

A NEW DECADE DAWNS

By the end of 1990, only half of the Group 8 astronaut selection (nine PLT and eight MS) remained active in the Astronaut Office. The other 18 had retired, were reassigned to managerial positions and out of crews, or were deceased.

The spring of 1991 would see a decade of Shuttle orbital operations completed, and hopefully the start of a more rigorous flight manifest, including plans to begin the construction of the long-awaited Space Station. Whether any of the TFNG would remain on the active flight list, visit a space station, or even remain in the program during the closing years of the 20th century, remained to be seen.

References

1. Astronaut Office Memo January 13, 1986.
2. **Voyage Into History**, Ch.4 *Preparation*, William Harwood, 1986, www.cbsnews.com
3. Donald Peterson, NASA Oral History, 14 November 2002.
4. *An Oral History of the Space Shuttle Challenger Disaster*, Margaret Lazarus Dean, <u>Popular Mechanics</u> magazine, February 2016.
5. spaceflight.nasa.gov//outreach/SiginificantIncidents/assets/rogers_commission_report.pdf. Last accessed November 30, 2019.
6. Sally Ride, NASA Oral History, October 22, 2002.
7. NASA News, 86-011, February 10, 1986.
8. **The Astronaut Maker, How One Mysterious Engineer Ran Human Spaceflight for a Generation**, Michael Cassutt, Chicago Review Press, 2018, p. 299.
9. John Fabian, NASA Oral History, February 10, 2006.
10. Rick Hauck, NASA Oral History, November 20, 2003.
11. Rick Hauck, NASA Oral History, March 17, 2004 (second interview).
12. NASA Headquarters News Release 86-078, June 17, 1986.
13. *Astronaut Robert L. Stewart Reassigned*, NASA JSC News Release, 86-020, July 24, 1986.
14. NASA News 86-107, August 6, 1986.
15. NASA Headquarters News Release, 86-114, August 18, 1986.
16. *Astronaut Walker praises workers 'Outstanding job'*. <u>Star Gazer</u> September 4, 1986, p. 1, in **Chronology of KSC and KSC-Related Events for 1986**, NASA KSC, KHR-11, Ken Nail, March 1987, p. 87.
17. <u>Florida Today</u>, November 19, 1986, pp. 1A-2A, in **Chronology of KSC and KSC-Related Events for 1986**, NASA KSC, KHR-11, Ken Nail, March 1987, p. 106.
18. George Nelson, NASA Oral History, May 6, 2004.
19. NASA News 87-018, April 15, 1987.
20. Reference 8, pp. 306−307.
21. *Key Astronaut Assignments*, NASA News 87-021, April 27, 1987.
22. Dan Brandenstein, NASA Oral History, January 19, 1999.
23. Steve Hawley, NASA Oral History, December 17, 2002.
24. AIS interview with Steve Hawley, October 2019.
25. *JSC Schedules Long-Duration Simulation*, NASA News, 87-020, April 21, 1987.
26. Michael Coats, NASA Oral History, November 9, 2012
27. Brewster Shaw, NASA Oral History, April 19, 2002.
28. *STS-27 crew named*, NASA News 87-043, September 15, 1987.
29. **The Last of NASA's Original Pilot Astronauts, Expanding the Space Frontier in the Late Sixties**, David J. Shayler and Colin Burgess, Springer Praxis, 2017, p. 369.
30. Reference 8, pp. 309−310.
31. *Investigation Committee to Examine Ellington Runway Incident*, NASA News 87-056, December 23, 1987.
32. *NASA Announces Shuttle Crew*, NASA JSC News, 88-001, February 4, 1988.
33. *Astronaut Group Provides Interface with Space Shuttle customers*, NASA JSC News 88-004, February 29, 1988.
34. Jeff Hoffman, NASA Oral History, November 3, 2009.
35. Rhea Seddon, NASA Oral History, May 9, 2011.
36. *NASA Announces Three New Crews*, NASA JSC News, 88-008, March 17, 1988.
37. **JSC Query Book, STS-26, Chronology of Wake up Calls**, Colin Fries NASA History Division updated March 13, 2015: [https://history.nasa.gov/wakup%20calls.pdf] last accessed November 8, 2019].

38. *Flight Control of STS-26*, NASA JSC News Release 88-037, September 22, 1988.
39. An undated FCOD Mission Support Assignments memo. Copy on file AIS Archives.
40. *Four New Shuttle Crews Named*, NASA News Release 88-049, November 30, 1989.
41. An undated FCOD Mission Support Assignments memo. Copy on file AIS Archives.
42. **Riding Rockets, The Outrageous Tales of a Space Shuttle Astronaut**, Mike Mullane, Scribner New York, 2006, pp. 273−290.
43. **STS-27 NSTS Mission Report**, NSTS-23370, NASA JSC Houston, Texas, February 1989, p. 2 and pp. 9−10. Copy on file AIS archives.
44. *Legendary Commander Tells Story of Shuttle's Close Call*, William Horwood, from an original story for CBS News "Space Place", posted March 27, 2009, Spaceflight Now. https://spaceflightnow.com/shuttle/sts119/090327sts27/ Last accessed November 12, 2019.
45. *Space Shuttle Crew Members Named to DOD, Life Sciences Missions*, NASA JSC News 89-005, February 24, 1989.
46. *Flight Control of STS-29*, NASA JSC News 89-007, March 7, 1989.
47. An undated FCOD Mission Support Assignments memo. Copy on file AIS Archives.
48. *Hauck to Assume Navy Post at The Pentagon*, NASA JSC News 89-016, March 24, 1989.
49. *Astronauts Named to Space Science Missions*, NASA JSC News 89-018, April 5, 1989.
50. *Flight Control of STS-30*, 89-206, April 27, 1989.
51. An undated FCOD Mission Support Assignments memo. Copy on file AIS Archives.
52. *Astronauts Named to DOD Missions in 1990 (STS-38, STS-39)*, NASA JSC News 89-028, May 11, 1989.
53. *McBride to Leave NASA; Brand Named Commander of STS-35*, NASA News 89-025, April 24, 1989.
54. *Astronaut 'Pinky' Nelson to Leave NASA*, NASA JSC News 89-034, June 9, 1989.
55. *Astronaut S. David Griggs Killed in Air Crash*, NASA News 89-036, June 17, 1989.
56. *Partial Crew Assignments Announced; Blaha to replace Griggs; Gutierrez, Cleave, Thagard assigned*, NASA News 89-040, June 29, 1989.
57. FCOD Mission Support Assignments memo, dated July 7, 1989.
58. *Astronauts named to Shuttle Crews: STS-39 (IBSS), STS-41 (Ulysses), STS-45 (ATLAS-01), STS-46 (TSS-1), STS-47 (SL-J)*, NASA JSC News Release 89-053, September 29, 1989.
59. An undated FCOD Mission Support Assignments memo. Copy on file AIS Archives.
60. An undated FCOD Mission Support Assignments memo. Copy on file AIS Archives.
61. An undated FCOD Mission Support Assignments memo. Copy on file AIS Archives.
62. *Science Payload Commanders Named; Carter replaces Cleave on IML-1*, NASA News 90-009, January 25, 1990.
63. *Shuttle Astronauts to Attend Soviet Space Launch*, NASA News 90-013, February 1, 1990.
64. *Mullane to Retire from NASA, Air Force*, NASA JSC News 90-019, February 20, 1990.
65. **Chronology of KSC and KSC Related Events for 1990, NASA KSC KHR-15**, Ken Nail, March 1991, p. 23.
66. *Veteran Shuttle Astronaut Williams to Retire from NASA, Navy*, NASA JSC News 90-020, February 26, 1990.
67. An FCOD Mission Support Assignments memo, February 5, 1990. Copy on file AIS archives.
68. An undated FCOD Mission Support Assignments memo. Copy on file AIS Archives.
69. Kathryn Sullivan, NASA Oral History, May 28, 2009.
70. **The Hubble Space Telescope, From Concept to Success**, David J. Shayler with David M. Harland, Springer-Praxis, 2016.
71. Reference 65, various entries.

72. *Shuttle Crews Named for 1991 Missions (STS-43, STS-44, STS-45)*, NASA News 90-033, May 24, 1990.
73. <u>World Spaceflight News</u>, Volume 4, No. 4, June 1990.
74. *Shuttle Crew Commanders Reassigned*, NASA News 90-037, July 9, 1990.
75. *Californian Shuttle Pad Will Be Converted for Titans*, <u>Florida Today</u>, September 29, 1990, p.5A.
76. FCOD Mission Support Assignments memo, September 19, 1990.
77. FCOD Mission Support Assignments memo, November 9, 1990.
78. FCOD Mission Support Assignments memo, November 15, 1990.
79. *NASA Announces Crew Members for Future Flights*, NASA JSC Release 90-063, December 19, 1990.

12

The Final Countdowns

"Columbia, Houston, you're go at throttle up"
Dan Brandenstein,
Ascent Capcom STS-1, April 12, 1981.

Fast forward 30 years…

'*God Bless America*' wake-up call played for all
*"the men and women who put their hearts and
souls into the Shuttle program for all these years"*
Shannon Lucid, Planning Capcom,
Final shift, STS-135 July 20/21, 2011

A total of 11,057 days (or 30 years, 3 months and 9 days) elapsed between Dan Brandenstein's calls to *Columbia* during the STS-1 launch on April 12, 1981, and the final conversations between the crew of *Atlantis* and Shannon Lucid a few hours before STS-135 landed on July 21, 2011. It was a generation that encompassed the whole operational phase of the American Space Shuttle Program, one which members of the Class of 1978 had both witnessed first-hand and actively participated in. Their contributions began well before that first departure from the launch pad at LC-39A and continued with the International Space Station (ISS) for some years after the final wheelstop of *Atlantis* on Runway 15 at the Shuttle Landing Facility (SLF).

History records, by some strange twists of fate, that the first Shuttle launch occurred on the 20th anniversary of the event which took the first [Soviet] human into space, while the final Shuttle landing was completed within hours of the 42nd anniversary of the first moonwalk by American astronauts, which was also the 50th anniversary of Gus Grissom's sub-orbital flight on the Mercury-Redstone 4 mission. Most of the participation on Shuttle missions by the Thirty-Five New Guys (TFNG) occurred during the first 15 years of the program, and

© Springer Nature Switzerland AG 2020

D. J. Shayler, C. Burgess, *NASA's First Space Shuttle Astronaut Selection*,
Springer Praxis Books, https://doi.org/10.1007/978-3-030-45742-6_12

towards the end of that period came the first missions to reunite the U.S. and Russia (formerly the Soviet Union) in space since 1975. The 1995 mission of STS-71, commanded by TFNG pilot astronaut Robert 'Hoot' Gibson, saw *Atlantis* docking with Mir and bring home the first American (TFNG Mission Specialist (MS) Norman Thagard) to work with Russian cosmonauts on a space station.

Following the retirement of the Shuttle fleet, the ISS has taken center stage, and for several years one member of the TFNG (MS Anna Fisher) remained involved both in that program and in the embryonic development of the vehicle to replace the Shuttle. Though the 1990s proved to be the final years in which the remaining members of the Class of 1978 journeyed into orbit, they were by no means pushed into the background. During the final decade of the 20th century, the Shuttle program was reinvigorated by a renewed hope for the struggling Space Station, even as it included the final countdowns for the remaining TFNG.

INTO THE TWILIGHT OF AN ERA

By the spring of 1991, the Space Shuttle program had been flying orbital missions for a decade. However, five of those years had been completed under difficult conditions, with the loss of the *Challenger* crew, the recovery from that tragedy, and in overcoming a number of hardware issues that had significantly reduced the number of missions that could be flown. The belief in the early 1980s that a Shuttle would leave the launch pad every two weeks proved enormously over optimistic. In the first decade of operations, only 37 missions had been flown successfully, and despite renewed optimism, the next decade would bring its own challenges in keeping the Shuttle flying, as the accountants, diplomats, engineers and politicians wrangled over the fate of Space Station *Freedom* and the format of its replacement.

Having lived through the highs and lows of the Shuttle program, the euphoria of their own space flights and the triumphs and tragedies within the Astronaut Office, the new decade dawned with the TFNG significantly reduced in number and thus their involvement as a group within the program somewhat lessened. By January 1991, half of the group had departed to pastures new or were now assigned to managerial positions within the agency. Once the largest contingent in the office, outnumbering even the pioneering veterans of the 1960s, the TFNG were now surpassed in number by members of more recent selections. At the start of this new decade, it was also a whole new era for NASA and a time of change within the agency, which would, in time, focus on establishing greater international partnership and dependency to create and maintain the much anticipated and long awaited Space Station Program.

Into the second decade

The 17 members of the 1978 selection who remained active in the Office in 1991 were at the forefront of this new era, with eight of them already assigned to the seven different Shuttle crews scheduled to fly over the coming 12 months. However, their numbers would gradually diminish in the months which followed, as more of them decided to move on to seek new goals away from the Astronaut Office. By the end of the decade and the start of ISS assembly, all of the TFNG were out of crew training, although five remained at the Johnson Space Center (JSC) or other NASA field sites. Another would return to NASA within a few years.

STS-37 (April 5 – 11, 1991)

Flight Crew: Steven R. NAGEL (CDR), Kenneth D. Cameron (PLT), Linda M. Godwin (MS-1), Jerry L. Ross (MS-2), Jerome Apt (MS-3)
Spacecraft: Atlantis (OV-104) 8th mission
Objective: 39th Shuttle mission; deployment of the Gamma Ray (Compton) Observatory; EVA Development Flight experiments
Duration: 5 days 23 hours 32 minutes 44 seconds
Support Assignments: In SPacecraft ANalysis (SPAN), Dave WALKER worked on the 06:00–12:00 shift, while Hoot GIBSON followed in the 12:00–18:00 slot. Mike COATS worked on the first shift in the Shuttle Mission Simulator (SMS), while Transoceanic Abort Landing (TAL) support was provided by Fred GREGORY at Morón Air Base in Spain. Loren SHRIVER returned to the Massachusetts Institute of Technology (MIT) as Mishap Representative. [1]

On this mission, the astronauts deployed the Gamma Ray Observatory (GRO), one of NASA's Great Observatories, which required the Extra-Vehicular Activity (EVA) crew of Ross and Apt to assist with the deployment of the observatory's antenna during their initial EVA, the first unscheduled EVA since STS-51D in April 1985. On their second spacewalk, the two astronauts evaluated the Crew and Equipment Translation Assembly (CETA), a motorized cart device which ran along a track and was designed to save the astronauts' energy when moving over long distances. CETA was intended for use with the Space Station. The crew evaluated other space station technology, with a middeck experiment simulating the space station's heat pipe radiator element. The GRO deployed by Nagel's crew was subsequently renamed after physicist Dr. Arthur Holly Compton, who won the Nobel Prize in Physics in 1927 for his work on the scattering of high-energy photons and electrons.

One day prior to the 10th anniversary of the launch of *Columbia* on STS-1, Steve Nagel brought *Atlantis* home to complete the STS-37 mission, neatly book-ending the first decade of Shuttle orbital flight operations. Despite some early in-flight problems, the Compton Gamma Ray Observatory was able to complete its primary mission of an all-sky survey by November 1992. It continued to gather data until it was safely de-orbited in June 2000, nine years after its deployment.

STS-39 (April 28 – May 6, 1990)
Flight Crew: Michael L. COATS (CDR), Blaine L. Hammond (PLT), Gregory J. Harbaugh (MS-1), Donald R. McMonagle (MS-2), Guion BLUFORD (MS-3), C. Lacy Veach (MS-4), Richard J. Hieb (MS-5)
Spacecraft: Discovery (OV-103) 12th mission
Objective: 40th Shuttle mission; unclassified Department of Defense (DOD) mission devoted to military scientific experiments
Duration: 8 days 7 hours 22 minutes 23 seconds
Support Assignments: For this mission, Loren SHRIVER was on call for SPAN (18:00−24:00) and was again the Mishap Representative at MIT. TAL support at Ben Guerir in Morocco was provided by Dave WALKER, and the Weather Pilot (WX) for the Kennedy Space Center (KSC) and Edwards Air Force Base (AFB) was Dan BRANDENSTEIN. [2]

The next Shuttle mission, and the first to have a full crew of seven NASA astronauts, was again commanded by a veteran member of the Class of 1978, Mike Coats. He and fellow TFNG Guion Bluford (MS) were both flying their third missions. This flight was a very complicated DOD mission but this time remained mostly unclassified. The primary objective was to gather data on the development of an advanced missile defense system, with the crew working in two teams around the clock. Bluford was assigned to the Blue Shift with Harbaugh and McMonagle, while Coats worked with the remaining crewmembers on the Red Shift.

STS-40 (June 5 – 14, 1991)
Flight Crew: Bryan D. O'Connor (CDR), Sidney M. Gutierrez (PLT), James P. Bagian (MS-1), Tamara E. Jernigan (MS-2), M. Rhea SEDDON (MS-3), Francis A. 'Drew' Gaffney (Payload Specialist (PS)-1, physician), Millie E. Hughes-Fulford (PS-2, biochemist)
Spacecraft: Columbia (OV-102) 11th mission
Objective: 41st Shuttle mission; Spacelab Life Sciences 1 program (using Long Module #1)
Duration: 9 days 2 hours 14 minutes 20 seconds
Support Assignments: While some members of the eighth selection left NASA to pursue new goals, others moved into managerial roles within the agency. For this

mission, Management Support from the Flight Control Operations Directorate (FCOD) was provided by Dick COVEY for launch. In SPAN, Dave WALKER worked on the 24:00–06:00 shift, Loren SHRIVER on the 06:00–12:00 shift (once he had returned from TAL support at Morón, Spain) and Kathy SULLIVAN on the 12:00–18:00 shift. Dick COVEY also provided weather support at KSC, while Dan BRANDENSTEIN did so at Edwards. SMS support was provided by John CREIGHTON on the second shift. [**3**]

During their nine-day dedicated space and life sciences mission, the STS-40 crew worked a single shift, performing a number of experiments which explored how humans, animals and cells responded to microgravity and then re-adapted to Earth's gravity on their return. Other payloads included experiments designed to investigate materials science, plant biology and cosmic radiation, and tests of hardware proposed for the Space Station *Freedom* Health Maintenance Facility.

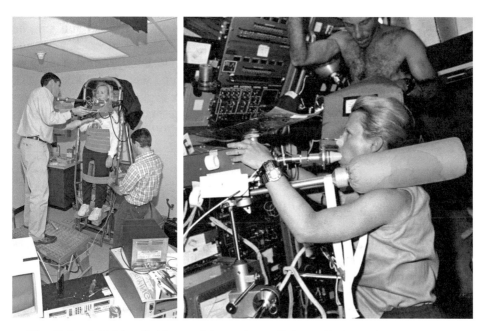

Fig. 12.1: (left) Rhea Seddon participates in pre-flight medical tests in support of her Spacelab Life Science mission and to provide baseline data for the Space Station Program. (right) During STS-40 (and the subsequent STS-58), Seddon and her colleagues gathered a wealth of life science data.

CO-OPERATING WITH THE SOVIETS

On July 31, 1991, NASA announced that the U.S. and the USSR had agreed to expand their respective civil space cooperation programs within an agreement reached by U.S. President George H. Bush and Soviet Premier Mikhail Sergeyevich Gorbachov during the U.S./USSR Summit of July 30–31 in Moscow. The plan was to conduct joint life sciences research by flying a U.S. astronaut on a long duration mission on the Soviet space station Mir, and reciprocally a Soviet cosmonaut as a crewmember on a Space Shuttle crew. [4] However, the following month saw dramatic events in the Soviet Union, which resulted in the eventual fall of the USSR and the emergence of a new Russia and a commonwealth of former Soviet states. This upheaval naturally suspended these plans while domestic issues took precedence in Russia.

Back at JSC, Mike Coats formally left NASA on August 1 and retired as a Captain from the U.S Navy (USN). He had announced his intention to leave the agency in July to assume a position as Director of Advanced Programs and Technical Planning at Loral in Houston. [5] "My years at NASA have convinced me that the finest folks in the world are attracted to the space program. I am extremely pleased to be able to change career directions and still be involved with this wonderful group of people," Coats said on leaving the agency. Director of Flight Crew Operations, Donald R. Puddy, noted at the time that he was sorry to see Coats leave but pleased that he would continue to work with NASA in his new capacity. In fact, the former astronaut's future involvement with JSC would be a lot closer than even he probably envisaged in 1991, as he would return as the Center's tenth Director in 2005, serving in that capacity for seven years.

STS-43 (August 2 – 11, 1991)
Flight Crew: John E. Blaha (CDR), Michael A. Baker (PLT), Shannon W. LUCID (MS-1), G. David Low (MS-2), James C. Adamson (MS-3)
Spacecraft: Atlantis (OV-104) 9th mission
Objective: 42nd Shuttle mission; deployment of fifth TDRS
Duration: 8 days 21 hours 21 minutes 25 seconds
Support Assignments: SPAN support was provided by Steve NAGEL and Fred GREGORY during the 21:00–05:00 shift. Weather support was fulfilled by Dick COVEY (launch and Edwards End of Mission (EOM) support) and Dan BRANDENSTEIN (KSC EOM). Dave WALKER provided TAL support at Banjul in The Gambia, West Africa. SMS support was provided by John CREIGHTON on the first shift and Fred GREGORY on the second, while Loren SHRIVER again fulfilled the role of Mishap Representative at MIT. [6]

During this flight, the crew deployed the fifth Tracking and Data Relay Satellite (TDRS-E). These TDRS deployment missions rarely captured the headlines in the media, unlike the more involved scientific research missions or those including a spacewalk, but were nevertheless important stages in developing an understanding of regular and 'routine' space flight operations, both in orbit and on the ground. In addition to deploying TDRS-E, the crew conducted 32 physical, material and life science experiments, mostly relating to the Extended Duration Orbiter (EDO) and Space Station programs. It is also noteworthy that two of the crew (John Blaha and TFNG Shannon Lucid) would later participate in long-duration residencies on the space station Mir, while a third crewmember (Mike Baker) would command one of the Shuttle docking missions with the Russian station.

Restored to flight status

On August 23, 1991, NASA named crewmembers and changes to crew assignments for no less than eight future Shuttle missions. Within those assignments, relevant to the Class of 1978, Hoot Gibson was named to command STS-47 (Spacelab J) and Dave Walker to command STS-53 (DOD), joined on that mission by fellow TFNG Guion Bluford as MS. Gibson and Walker had both been restored to T-38 flying status and were once again in line to return to flight crews, just a year after both had been grounded for flying infringements. [7]

STS-47 was planned as a joint mission with the Japanese Space Agency (then known as NASDA, later changed to JAXA) due in August 1992. Gibson was named with Pilot (PLT) Curtis Brown (Class of 1987), and joined previously assigned crewmembers. The research to be conducted on Spacelab J was dedicated to materials processing and life sciences experiments. The STS-53 DOD mission was due to be launched in October 1992, with Walker leading a crew of PLT Robert Cabana (Class of 1985), and MS Bluford, James Voss (Class of 1987) and Rich Clifford (Class of 1990).

August also saw Dick Covey taking on the challenging dual role of Acting Deputy and Acting Chief for the next three months, until Steve Nagel took on the role of Acting Chief of the Astronaut Office pending Dan Brandenstein's return after flying STS-49. At one point in late 1991, Covey was Acting Director of Flight Crew Operations as well.

Senior roles in the office

A decade into the Shuttle program, TFNG pilot astronauts were now regularly taking either the command seat on Shuttle missions, senior MS roles, or moving into managerial roles in the agency. This vindicated the potential the Astronaut Selection Board had seen in choosing the 35 new astronauts back in 1978. Over a decade after qualifying from the first Ascan training and evaluation program, the TFNG were certainly living up to expectations, having established a strong presence in the Astronaut Office, the Shuttle program and with NASA itself.

STS-48 (September 12 – 18, 1991)
Flight Crew: John O. CREIGHTON (CDR), Kenneth S. Reightler (PLT), Charles
D. 'Sam' Gemar (MS-1), James F. BUCHLI (MS-2), Mark N. Brown (MS-3)
Spacecraft: Discovery (OV-103) 12th mission
Objective: 43rd Shuttle mission; deployment of the Upper Atmosphere Research
Satellite (UARS)
Duration: 5 days 8 hours 27 minutes 38 seconds
Support Assignments: SPAN support was provided by Kathy SULLIVAN
(06:00–12:00) and Loren SHRIVER (18:00–24:00), the latter repeating his dual
assignment as Mishap Representative at MIT. Weather support was fulfilled by
Steve NAGEL for landing at KSC, while Fred GREGORY was on duty in the
SMS for the first shift. FCOD Management Support for this mission included
Dick COVEY for launch at KSC/Operations Support Room (OSR), orbit at JSC,
and landing at Dryden (he was also FCOD Management Representative for the
Crew Recovery Team), and Steve NAGEL at JSC for launch and landing. [**8**]

Following his second space flight, STS-36 in 1990, John Creighton had headed up
the Operations Development Branch within the Astronaut Office. This was a one-
year appointment prior to resuming full-time training for his next space flight as
Commander (CDR) for STS-48. During the mission, the crew deployed the
research satellite UARS, which had a design life of just three years but was still
providing data a decade after its deployment, with six of its ten instruments still
operating. In May 2005, UARS surpassed 5,000 orbits of Earth, but its batteries
began to fail over the next few months, signaling the end of its operational life in
August of that year. It was decommissioned due to budget cuts that December
after 14 years and three months of operational life. The satellite finally re-entered
Earth's atmosphere in September 2011, 20 years after it had been placed in orbit.
In addition to the UARS deployment, the STS-48 crew also conducted numerous
secondary experiments, ranging from growing protein crystals to studying how
fluids and structures react in microgravity. On the middeck of *Discovery*, Buchli
worked with Mark Brown in assembling scale models of the proposed space station truss, to test vibration characteristics on the structure's joints in the microgravity environment on orbit.

On October 23, Rhea Seddon was confirmed as Payload Commander (PC) for
Spacelab Life Science 2, manifested for STS-58 and due to launch in July 1993.
As PC, Seddon assumed overall crew responsibility for long-range planning and
integration of the numerous payloads, and provided her experience in coordinating the science activities during the mission. [**9**] According to the press release,
this mission was "dedicated to continuing research on the adaptation to microgravity in preparation for the Space Station *Freedom* program and future planetary
exploration."

Jim Buchli also completed his post-STS-48 assignments in October and resumed his position as Deputy Chief of the Astronaut Office, releasing Dick Covey from that role.

STS-44 (November 24 – December 1, 1991)
Flight Crew: Frederick D. GREGORY (CDR), Terence T. 'Tom' Hendricks (PLT), James S. Voss (MS-1), F. Story Musgrave (MS-2), Mario Runco Jr. (MS-3), Thomas J. Hennen (PS-1, U.S. Army)
Spacecraft: Atlantis (OV-104) 10th mission
Objective: 44th Shuttle mission; deployment of the Defense Support Program (DSP) satellite; Military Man-in-Space experiments (Terra Scout)
Duration: 6 days 22 hours 50 minutes 44 seconds
Support Assignments: On this flight, the system for SPAN changed, requiring the astronauts to be on 24-hour call (08:00–08:00 daily) for their particular mission days. The Group 8 astronauts fulfilling this assignment were Hoot GIBSON and Shannon LUCID for Flight Day (FD) 1, with LUCID also working FD3 and FD4. TAL support included Dave WALKER at Morón, Spain, who travelled back to California after the launch for assignment as Shuttle Training Aircraft (STA) Weather Pilot for EOM at Edwards. Steve NAGEL was STA Weather Pilot for launch and KSC EOM. Loren SHRIVER was again assigned as Mishap Representative, but this time was based at JSC. [**10**]

One of the final Shuttle DOD-dedicated flights, this mission deployed *Liberty*, one of NORAD's (North American Air Defense Command) Tactical Warning and Attack Assessment System satellites.

On December 6, NASA released the names of the Payload Crew for SLS-2/STS-58, in addition to the three PS shortlisted for the single seat available. Joining PC Seddon were fellow TFNG Shannon Lucid and David Wolf (Class of 1990). [**11**]

INTERNATIONAL SPACELABS

The new year opened with another Spacelab pressurized module flight carrying an international crew and science payload, giving an indication of what might be expected in the forthcoming Space Station Program.

STS-42 (January 22 – 30, 1992)
Flight Crew: Ronald J. Grabe (CDR), Stephen S. Oswald (PLT), Norman E. THAGARD (MS-1), William F. Readdy (MS-2), David C. Hilmers (MS-3), Roberta L. Bondar (PS-1, Canada), Ulf D. Merbold (PS-2, European Space Agency (ESA), Germany)

Spacecraft: Discovery (OV-103) 14th mission
Objective: 45th Shuttle mission; International Microgravity Laboratory-1 (IML-1) research program flying the Spacelab Long Module Unit #2
Duration: 8 days 1 hour 14 minutes 44 seconds
Support Assignments: This was the first mission for some time to include a Group 8 astronaut at the Capcom console in Mission Control Center-Houston (MCC-H), with Rhea SEDDON as one of the Orbit 2 Capcoms on the 'Antares' team. [**12**] On 24-hour callout for SPAN were Guion BLUFORD (FD3), Loren SHRIVER (FD4, and again having a dual assignment as Mishap Representative at JSC), and Shannon LUCID (FD6 and 7). Weather Support was provided by Dick COVEY at KSC and Edwards. TAL support came from Steve NAGEL at Ben Guerir in Morocco and Dave WALKER in Zaragoza, Spain. Dan BRANDENSTEIN, who was back in training for STS-49, was in the SMS for the second shift, and Fred GREGORY was one of the Immediate Family Escorts. [**13**]

With Norman Thagard as PC, the IML-1 crew investigated the effects of microgravity on materials processing and life sciences while working two shifts, with Thagard on the Blue Shift. In total, there were 55 experiments devoted to research in space medicine and manufacturing, an interesting preview of future space station operations.

This mission was flown at a time when a new Russia was slowly emerging from the collapse of the Soviet Union the previous year, with the growing prospect of even closer international relations with Russia in space exploration. A few years later, Thagard would become the first American and only TFNG to be launched and fly on a Russian spacecraft (Soyuz) from Russia (Baikonur) and serve as a resident crewmember on one of their long duration missions aboard Mir. Ironically, on January 24 during STS-42, that same station passed within 39 nautical miles (44.85 miles or 72.16 km) of *Discovery*, with the crew reporting the reflected sunlight from the solar arrays of the station "as bright as the planet Mercury."

IML-1 was another milestone in the Shuttle program. The extensive science program encompassed investigations supplied by the space agencies of the United States (NASA), Europe (ESA), Canada (CSA), France (CNES), Germany (DARA) and Japan (NASDA), as well as smaller Get Away Special (GAS) experiments provided by Australia, China, the Federal Republic of Germany, Japan, Sweden and U.S. students, plus the IMAX camera and a host of smaller middeck investigations. Over 200 scientists from 16 countries were deeply involved in this one flight and its investigations. At the time, the long-term plan was to fly four or more of these IML missions every couple of years, within a program devoted to life and material sciences over the next decade, to provide important data and experience in planning and preparing advanced experiments for Space Station *Freedom*. Despite the success of IML-1 and advanced planning for IML-2, however, *Freedom* was in serious trouble by the early 1990s due to its escalating costs and growing complexity. Ironically, when the revived and redesigned ISS program emerged along with a new partner in Russia, the limited resources available and the need to devote several Shuttle missions to assembling and supplying the station signaled the demise of the Spacelab Long Module program. This also saw the plans for IML (and other Spacelab-type missions) reduced from a ten-year multi-flight program to just two missions.

When crew assignments to three Shuttle missions were announced on February 21, 1992, they included Steve Nagel being named as CDR for STS-55/Spacelab D-2, scheduled for launch early in 1993. Spacelab D-2 was a second cooperative mission with the German Space Agency, to conduct research into robotics, materials processing and life sciences. Also selected with Nagel were PLT Tom Henricks (Class of 1985) and MS Charles Precourt (Class of 1990), who joined the previously assigned crewmembers. [**14**]

STS-45 (March 24 – April 2, 1992)

Flight Crew: Charles F. Bolden (CDR), Brian Duffy (PLT), Kathryn D. SULLIVAN (MS-1), David C. Leestma (MS-2), C. Michael Foale (MS-3), Dirk D. Frimout (PS-1, ESA, Belgium), Byron K. Lichtenberg (PS-2, United States)
Spacecraft: Atlantis (OV-104) 11th mission
Objective: 46th Shuttle mission; Atmospheric Laboratory for Application of Science-1 (ATLAS-1) research program
Duration: 8 days 22 hours 9 minutes 28 seconds
Support Assignments: Rhea SEDDON continued her tour as Capcom on Orbit 2, this time on the 'Falcon' team. [**15**] FCOD Management Support was provided by Dick COVEY at JSC for both launch and landing, while Loren SHRIVER again fulfilled the role of Mishap Representative at JSC. Weather Support was provided by Steve NAGEL at KSC and Edwards and Fred GREGORY for OSR and landing at Edwards. Dan BRANDENSTEIN was back in the SMS, this time on the first shift, honing his skills in preparation for his forthcoming flight after "flying a desk" for a couple of years. [**16**]

On her third and final mission, Dr. Sullivan served as PC and a member of the Blue Shift on STS-45, the first Spacelab mission dedicated to NASA's Mission to Planet Earth program. During this nine-day mission, the crew operated the 12 experiments that constituted the ATLAS-1 payload. ATLAS-1 obtained a vast array of detailed measurements of atmospheric chemical and physical properties, contributing significantly to improving our understanding of our climate and atmosphere. In addition, this was the first time an artificial beam of electrons had been used to stimulate a man-made aurora discharge. The mission was extended by a day in order to continue the science experiments.

STS-49 (May 7 – 16, 1992)
Flight Crew: Daniel C. BRANDENSTEIN (CDR), Kevin P. Chilton (PLT), Richard J. Hieb (MS-1), Bruce E. Melnick (MS-2), Pierre J. Thuot (MS-3), Kathryn C. Thornton (MS-4), Thomas D. Akers (MS-5)
Spacecraft: Endeavour (OV-105) 1st mission
Objective: 47th Shuttle mission; maiden flight of Endeavour; capture, repair and redeployment of Intelsat VI (F-3); Assembly of Space Station by EVA Methods (ASSEM) demonstration
Duration: 8 days 21 hours 17 minutes 38 seconds
Support Assignments: Dick COVEY again fulfilled the FCOD Management Support role at JSC during both launch and landing. Fred GREGORY was entry Capcom in MCC-H. Norman THAGARD was on 24-hour callout for SPAN during FD2. Weather Support for launch and EOM at Edwards was provided by Steve NAGEL. Loren SHRIVER was Mishap Representative again at JSC and also worked the second shift in the SMS. [17]

Chief Astronaut Dan Brandenstein commanded the newest (and final) Shuttle Orbiter on this mission, becoming the only TFNG pilot astronaut to lead a crew on the maiden flight of a new Orbiter. And what a flight it turned out to be. During the mission, the crew had to rendezvous with the stranded Intelsat VI satellite, capture it using the Remote Manipulator System (RMS), install a new Perigee Kick Motor (PKM), and then redeploy the satellite. While this sounded straightforward, it proved to be anything but. When the capture bar could not be secured automatically, an ingenious three-person EVA was staged to allow the astronauts to grab the satellite physically and then lower it down onto the capture bar. With the new PKM installed and the satellite redeployed, the new motor dispatched it to geostationary orbit the following day. The mission also included a further demonstration of constructing pyramid structures, as another evaluation of space station assembly techniques similar to those conducted on STS-61B in 1985. The demonstration was difficult and was in any case largely redundant, as by now the redesign of the space station was leaning towards deploying pre-fabricated truss structures to form its main backbone, with the least EVA input possible.

A summer of departures
With the new Shuttle flying its first mission, the fleet once again comprised four vehicles to support the manifest of the 1990s and hopefully beyond. The nature of the payloads had also changed, from mainly commercial to a more scientific orientation, including both NASA's topical Mission to Planet Earth program and preparations for the Space Station. Exactly what that Space Station would be was now mired in debates and wrangles on Capitol Hill, in the halls of NASA and at the contractors. While this was making its uncertain way through the various

political and legal stages and re-design processes, several members of the Class of 1978, now 14 years into their service with the agency and with a significant number of newer astronauts occupying the Astronaut Office, decided it was time to move on.

Fred Gregory left the Astronaut Office, but not NASA, during May. He became the Associate Administrator at the Office of Safety and Mission Assurance, a position he would hold until 2002. That same month, Jim Buchli was replaced by Dave Leestma (class of 1980) as Deputy Chief of the Astronaut Office, a position Buchli had officially held since March 1989. On May 29, it was announced that Buchli would be retiring from the U.S. Marine Corps (USMC) and NASA in August (effective September 1), to take up a position with Boeing Defense and Space Group in Huntsville, Alabama, as Manager of Station Systems Operations and Requirements. Buchli had flown four times on the Shuttle during his 14 years at NASA and had completed a number of technical assignments, including in the Astronaut Office Development Branch working on the display and controls for the Shuttle and Space Station *Freedom*. His significant contributions were described as assets for Boeing in their support of the Space Station program. On his decision to leave the Astronaut Office, Buchli said: "I'm grateful for my years of active service as a Marine and as part of the NASA team. It has allowed me to be part of two of the finest organizations in the world. I'm looking forward to changing career directions and remaining involved with the outstanding people who make up our space team." [18]

The following month, STS-50 was launched on the first U.S. Materials Laboratory mission, with *Columbia* flying the first EDO mission kit to provide the capability of extending the duration of the mission to 14 days. This kit of additional cryogenics and consumables was used on several missions in the 1990s to stretch the capabilities of the Shuttle Orbiter on science missions and acquire further baseline data, research results and experience in flying longer missions as a prelude to Space Station. STS-50 was also just the third Shuttle flight since the initial TFNG mission, STS-7 in 1983, to be launched without a member of Group 8 among the crew (the others being STS-51F in July 1985 and STS-41 in October 1990). This shows the dominance the 1978 group held in the Astronaut Office during the first decade of the Shuttle program, both before and immediately following the loss of *Challenger*.

As normal, several of the non-assigned Group 8 astronauts fulfilled support roles during this mission. FCOD Management Support was provided by the retiring Jim Buchli at JSC for launch and landing, while Shannon Lucid was on 24-hour call for SPAN during FD3. Weather Support was provided by Steve Nagel and Dick Covey for launch, with the latter also supporting EOM at Edwards. Dave Walker was at KSC for EOM. Loren Shriver was the Mishap Representative at JSC again, and was also in the SMS for the first shift. [19] Rhea Seddon was at the Capcom console for Orbit 2 on the 'Antares' team. [20]

On June 25, with her STS-45 responsibilities completed, Kathy Sullivan was formally nominated as Chief Scientist at the National Oceanic and Atmospheric Association (NOAA), succeeding Sylvia Alice Earle. Pending Senate confirmation to make this a permanent assignment, she would assume the new position effective August 17, 1992. "My 14 years at NASA have been immensely rewarding, both professionally and personally," she said. "I will take many cherished memories with me, particularly the superb people who make up the Shuttle team. I know they can craft this country an exciting spacefaring future, and I will watch their exploits with great interest and pride." [21]

On the very same day of Sullivan's appointment, John Creighton announced that he, too, was leaving the Astronaut Office and NASA, and retiring from the USN, effective July 15, to join the Commercial Airplane Group of the Boeing Co. in Seattle, Washington, where he would be working as a production test pilot and as an instructor pilot in the customer support area. "I have thoroughly enjoyed my time at NASA," Creighton said, "especially working with the outstanding people there. I feel privileged to have flown on three Shuttle missions – each unique and rewarding – but there comes a point when it's time to look for a new and different challenge. I'm looking forward to returning to Seattle where I grew up, and to beginning my new career at Boeing." [22]

On July 1, Dan Brandenstein, the Chief of the Astronaut Office for the past five years, also announced his intention to leave the agency and retire from the USN on or about October 1, to pursue other interests. "For the past 14 years I have had the opportunity to have the most challenging and interesting job in the world," Brandenstein explained. "It has been exciting, rewarding, and a pleasure to work with the many talented and motivated people who make up this country's space team. Although I have chosen to change careers, I will always be an avid supporter of the space efforts which I feel are essential to the advancement of knowledge and technology in this country." [23] With effect from December 8, 1992, Brandenstein was replaced as Chief Astronaut by Robert 'Hoot' Gibson, who had missed out on the position in 1987.

STS-46 (July 31 – August 8, 1992)
Flight Crew: Loren J. SHRIVER (CDR), Andrew M. Allen (PLT), Claude Nicollier (MS-1, ESA, Switzerland), Marsha S. Ivins (MS-2), Jeffrey A. HOFFMAN (MS-3, PC), Franklin R. Chang Díaz (MS-4), Franco Malerba (PS-1, Italian Space Agency)
Spacecraft: Atlantis (OV-104) 12th mission
Objective: 49th Shuttle mission; deployment of ESA's European Retrievable Carrier (EURECA); Tethered Satellite System-1 (TSS-1) demonstration
Duration: 7 days 23 hours 15 minutes 3 seconds
Support Assignments: Dick COVEY was back at JSC as FCOD Management Support for launch and was also Weather Support at the Cape for launch and OSR/

landing. Weather Support at Edwards was provided for EOM by Dave WALKER. Hoot GIBSON was in the SMS for the first shift, while Norman THAGARD was on 24-hour call for SPAN during FD1. [**24**]

The STS-46 mission included the deployment of the ESA-sponsored free-flying science platform EURECA and the first test flight of the TSS, a joint project between NASA and the Italian Space Agency. Jeff Hoffman had worked on the Tethered Satellite project since 1987 as a technical assignment. TSS was an experiment to generate electricity to provide spacecraft with power using tethered satellites lowered into the upper reaches of the atmosphere, at greater altitude than balloons and lower than orbiting satellites. It was intended to provide data for future operations in space broadcasting, atmospheric research and conducting microgravity experiments. Though the TSS was deployed successfully, several failures were encountered with unplanned halts of the deployment mechanics. Instead of being deployed to 20 km as planned, the TSS reached only 256 m out from the payload bay. In this two-shift mission, Hoffman was assigned to the Red Shift, while fellow TFNG and mission CDR Loren Shriver worked with either the Blue or Red Shift as required.

A Plus One

In the midst of all the departures among the Class of 1978, there was one arrival. On August 1, Steve Hawley returned to JSC from Ames Research Center – not to the Astronaut Office, but to take the position of Deputy Director of Flight Crew Operations. "This is a unique opportunity to return 'home' and play a part in running my old organization," he said. "I have enjoyed my two years at Ames and seeing a unique part of the agency. I will miss the Ames people very much." Don Puddy, the Director of FCOD, said that Hawley's experience as an "outstanding manager" made him well qualified to help lead the flight crew functions at JSC. These included recommending astronaut selections, aviation operations, domestic and international PS activities, and operational contributions to the design and development of manned spacecraft, payloads, equipment and systems. [**25**]

In an interview with the authors, Steve Hawley expanded upon his decision to return to JSC after two years at Ames. "I was lucky because when I went out there, I did it at a specific request from headquarters. It was [William B.] Bill Lenoir [Associate Administrator for Space Flight] that called me up and asked me to do it. At the time, I remember it was a tough decision, but I remember thinking [that] my plan was to spend my career at NASA and if I was going to spend my career at NASA then it might be more beneficial for me to get this experience as a manager in different parts of the agency than it would be to fly a fourth mission. They don't care if you've flown three times for four times, and so I said 'Yes, I'll do this, and if I don't like it. I will just come back'. That's more or less what happened. It wasn't that I didn't like it, I just found that [it] was really interesting and

I was learning a lot, but while I was out there I was still wanting to fly Shuttle missions. And it occurred to me that people were managing to do that without my help and that was tough to take back then. I realized that I shared this with some of my other colleagues: don't leave unless you are really prepared to watch somebody else do this. You know, if you're not really ready to go, don't leave. So I decided that I really wanted to get back into operations, I wasn't really ready to step aside, but in the time it took me to leave JSC and be at Ames for a year, or a year and a half, Bill Lenoir was gone, and two or three other people who were involved in asking me to go there had gone. [Richard H. (Dick)] Truly [former astronaut and the eighth Administrator of NASA] might have left too and gone back to the Navy, I don't remember, but anyway, all of the people that knew why I had gone out there were no longer around. Aaron Cohen [Deputy NASA Administrator] was still around, but suddenly I realized that if I tried to go back to JSC, there were going to be a lot of people who are going to go, 'who is this guy that wants to work JSC?' And so I decided that if I was going to go back, I really needed to go back now, so I got in touch with Aaron and I said, 'Hey, I would like to come back'. I think the first thing he did was offer me the job of deputy project manager for the CRV crew rescue vehicle for station. I didn't really want to do that, I wasn't sure that project was going to go anywhere, and then he said, 'Well, come back and be Don [Donald R.] Puddy's deputy' and I said 'Yes, I'll do that'. I don't remember how much longer Aaron was there afterwards, but basically all of the people that had asked me to go do this [work at Ames] had gone. I felt this window of opportunity [to return to JSC] shrinking in front of me, but the experience was great, I learned a lot, met some good friends, and learned about different parts of the agency. I learned some management skills maybe, but back in the early '90s I wasn't ready to give up flight operations." [26]

On August 27, NASA announced crew assignments for STS-58 (SLS-2) and STS-61 (Hubble Service Mission 1). There were no further TFNG assignments on STS-58 to join the previously named PC Rhea Seddon and MS Shannon Lucid, but there was one for STS-61. Three MS with spacewalking experience were named to join Story Musgrave in preparing for the first servicing mission to the Hubble Space Telescope (HST). The three were Tom Akers (Class of 1987), Kathy Thornton (Class of 1984) and TFNG Jeff Hoffman. [27]

STS-47 (September 12 – 20, 1992)
Flight Crew: Robert L. GIBSON (CDR), Curtis L. Brown (PLT), Mark Lee (MS-1, PC), Jerome Apt (MS-2), N. Jan Davis (MS-3), Mae C. Jemison (science MS), Mamoru M. Mohri (PS-1, NASDA, Japan)
Spacecraft: Endeavour (OV-105) 2nd mission
Objective: 50th Shuttle mission; Spacelab J research program flying the Spacelab Long Module Unit #2
Duration: 7 days 22 hours 30 minutes 23 seconds

Support Assignments: FCOD Management Support roles were completed by Dick COVEY at JSC for launch and landing and the recently returned Steve HAWLEY at KSC/OSR for launch, orbit and landing [authorized Crew Transfer Vehicle (CTV)]. HAWLEY also fulfilled the role of FCOD Management Representative for the Crew Recovery Team. Weather support was completed by retiring Chief Astronaut Dan BRANDENSTEIN at KSC and by Dave WALKER at Edwards, who also worked the second shift in SMS. [**28**]

The seven-person crew commanded by Hoot Gibson included the first Japanese astronaut to fly on the Space Shuttle[1]. During eight days aloft, the STS-47 astronauts focused on science and materials processing experiments in 44 investigations aboard a Spacelab module cradled in the Shuttle's cargo bay. Of these, 35 were sponsored by NASDA, seven by NASA, and two were joint investigations. The payload also included four female frogs, 30 chicken eggs, 180 oriental hornets, about 400 fruit flies, 7,200 fly lava and two Japanese carp. In the build-up to the mission, the media soon latched on to the extended 'crew' roster and started calling the mission "Hoot's Ark", referring to the commander's nickname and his extensive zoological 'crew complement'.

Missiya Spetsialist Kosmonavt

On October 6, 1992, experienced Russian spacefarers Vladimir Georgiyevich Titov and Sergei Konstantinovich Krikalev were identified as the two cosmonauts who would undergo abbreviated NASA MS training at JSC for flights on the Space Shuttle, following a recent agreement with the Russians. This closer tie between the two nations was the beginning of acceptance of the new, emerging Russia, and preceded a major review of the Space Station *Freedom* program by President Bill Clinton's administration the following year.

The next Shuttle mission, STS-52/LAGOES II/USMP-1 (October 22 – November 1) lacked any Group 8 members on the flight crew, but once again several of them served in support roles. FCOD Management Support was completed by Steve Hawley (Launch KSC/OSR orbit JSC, and landing KSC [authorized CTV]), while Dick Covey was at JSC for the landing, although he was replaced by Group 9 astronaut Dave Leestma at JSC for launch. He was also replaced in the original memo listing for Weather Support at the Cape and Edwards by Group 9 astronaut Richard Richards. On 24-hour call for SPAN were Steve Nagel (launch), Shannon Lucid (FD6) and Loren Shriver (FD9). Mishap Representative for this mission was Loren Shriver, the Crew Recovery Team FCOD Management Representative was Steve Hawley, and Dave Walker worked on the first shift in the SMS. [**29**]

---

[1] The first Japanese citizen to fly in space was Toyohiro Akiyama, a journalist who had been selected to fly a one-week mission on Soyuz TM-11 to Mir in December 1990.

In another management reshuffle in November 1992, and after only a few months in the position, Dave Leestma was replaced as Deputy Chief of the Astronaut Office by TFNG Loren Shriver. Leestma was named as the new Chief of the Flight Crew Operations Directorate at JSC.

STS-53 (December 2 – 9, 1992)

Flight Crew: David M. WALKER (CDR), Robert D. Cabana (PLT), Guion S. BLUFORD (MS-1), James S. Voss (MS-2), M. Rich Clifford (MS-3)
Spacecraft: Discovery (OV-103) 15th mission
Objective: 52nd Shuttle mission; deployment of classified DOD payload
Duration: 7 days 17 hours 19 minutes 47 seconds
Support Assignments: FCOD Management Support fell to Steve HAWLEY at KSC/OSR for launch, JSC for orbit and KSC for landing [authorized CTV], and Dick COVEY at JSC for launch and landing. HAWLEY was again the Crew Recovery Team Management Representative. Weather Support was Loren SHRIVER (KSC) and Hoot GIBSON (OSR), with SHRIVER serving as Mishap Representative at JSC and GIBSON doing so at MIT. Steve NAGEL served on the second shift in the SMS. [30]

On this flight, the tenth and final DOD Shuttle mission, the crew deployed the classified DOD-1 payload (later identified as the third Advanced Satellite Data System Intelligence Relay Satellite) and performed several Military-Man-in-Space and NASA experiments during their seven-day mission.

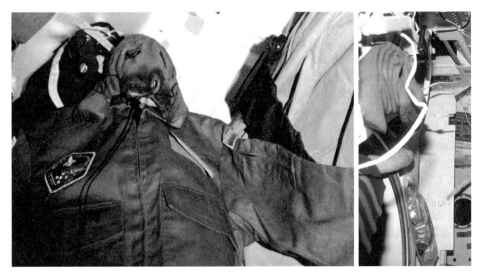

Fig. 12.2: The STS-53 'stowaway' known as "Dog Breath," an official member of the "Dogs of War" Shuttle crew. (Image courtesy Ed Hengeveld.)

The Dogs of War

Every crew takes their mission and objectives very seriously, fully aware of the huge investments in their training and in the preparations and importance each expensive mission contributes to the program. But there is usually a little time for some lightheartedness as well.

The Dogs of War is what the STS-53 crew called themselves during training, as all five were patriotic representatives of the four armed forces of the United States and active serving military officers on secondment to NASA. TFNG and mission CDR Dave Walker was a Captain in the USN and a team builder by nature, but also liked to have fun. PLT Robert Cabana was a USMC Colonel, MS Guion Bluford was a Colonel in the United States Air Force (USAF), and Lt. Colonels Jim Voss and Rich Clifford were both representatives of the U.S. Army.

Throughout his military flying career prior to arriving at NASA, red-headed Dave Walker had already acquired the nickname "*Red Flash*." But having such a military-orientated crew behind him, and with his 'dogged' determination, he decided to "dig up the bone" and adopt the 'Dog Crew' callsign. "It's a morale booster," Walker once said, "[giving] us an excuse for parties." Cabana, being a Marine, naturally became "*Mighty Dog*." Clifford, the rookie on the crew, became "*Puppy Dog*", while Bluford became "*Dog Gone*." He was also the one who gave Jim Voss the title "*Dog Face*," after the Army's 'Dogface' infantry soldiers of WWII[2]. They also adopted a mascot called "*Duty Dog*." Walker himself was called "*Red Dog*."

"In order to make training an enjoyable experience," Bluford explained, "Dave [Walker] bought a jalopy (the "*Dogmobile*") and painted our names on the side. The trainers and support people in Mission Control all had dog names. My dog name was "*Dog Gone*" [as] I had been in Europe on PR trip when Dave handed out the dog names. We even had a paper dog mascot stored in the middeck lockers, which we took into space and hung up. Though we had a lot of fun with our dog names, we took our tasks seriously when we trained." [31] The stowaway up on orbit was a rubber dog mask hung over an empty orange launch and entry suit, and was known by the name "*Dog Breath*." During preparations for the mission, the STS-53 training team became known as "*Bad Dog*," due to their tough approach with the crew in the simulations, while during the actual mission, the crew often reported that they were "working like dogs."

---

[2] The term 'Dogface' normally refers to U.S. Army foot soldiers, primarily in WWII. The origin of this is vague, but in the American Civil War, wounded soldiers had tags tied to them to indicate their wounds, similar to those used on a pet dog. During WWII, American infantrymen living in "pup tents" and "foxholes", were often cold, wet and filthy as a hunting hound, with long faces deemed to resemble a sad dog, and always being ordered around and enduring shouted commands like a half-trained dog.

On December 8, 1992, Hoot Gibson took over as Chief of the Astronaut Office at JSC. He would retain the position for the next two years until he began full-time training for his final space mission.

FREEDOM TO BECOME INTERNATIONAL

STS-54/TDRS-F was the next mission to fly with no TFNG among the crew, between January 13–19, 1993. This time, only five of the remaining Class of 1978 were assigned to support roles on the mission. FCOD Mission Support Launch (JSC) and landing (Edwards) [authorized CTV] was again Steve Hawley, with Loren Shriver fulfilling the role for landing at JSC and also serving as Mission Representative at the same field center. Weather Support for launch and KSC EOM was Hoot Gibson, while Dick Covey was over in Morón, Spain, for TAL support. Steve Nagel worked the first shift in the SMS. [**32**]

In early March, it became clear that Norman Thagard was a leading candidate for a 90-day mission to the Mir space station. At the time, he was taking Russian language lessons and had completed several trips to Russia in his role as Astronaut Office Point of Contact (POC) for Russian Manned Mission Operations. According to private sources, Thagard was the candidate preferred by the Astronaut Office and management at JSC, although Washington Headquarters preferred Bill Shepherd (Class of 1984) for the assignment.

During April 7–17, STS-56 flew the ATLAS-2 mission, again without a member of the TFNG on the flight crew. Only two of the Class of 1978 were assigned to support roles for this flight. Once again, FCOD Support for launch at JSC and landing at Edwards was provided by Steve Hawley, while Weather Support at KSC was fulfilled by Hoot Gibson. [**33**]

STS-55 (April 26 – May 6, 1993)
Flight Crew: Steven R. NAGEL (CDR), Terence T. Henricks (PLT), Jerry L. Ross (MS-1, PC), Charles J. Precourt (MS-2), Bernard A. Harris (MS-3), Ulrich Walter (PS-1, DFVLR, Germany), Hans W. Schlegel (PS-2, DFVLR, Germany)
Spacecraft: Columbia (OV-102) 14th mission
Objective: 55th Shuttle mission; Spacelab D-2 research program using Long Module Unit #1
Duration: 9 days 23 hours 39 minutes 59 seconds
Support Assignments: Steve HAWLEY again provided FCOD Management Support for launch (JSC) and landing (Edwards) [authorized CTV]. Weather Support was provided by Hoot GIBSON at KSC. [**34**]

Getting STS-55 off the ground proved to be a challenge. The original launch had slipped from February, and again several times in March, due to hardware issues. On March 22, the first two main engines had ignited at about T-3 seconds, but the launch had been aborted when the third engine failed to ignite. This type of abort was termed a Redundant Set Launch Sequencer, one in which serious problems prevented the launch after the Shuttle's onboard computers have taken over from the Ground Launch Sequencer, but just prior to the ignition of the Solid Rocket Boosters (SRB). If the SRB ignited, the launch had to occur regardless of the issue. Fortunately, this system worked as designed (as it did on the other four occasions it was needed in the program) but the STS-55 mission was delayed into April to allow the replacement of all three main engines. Frustratingly, it was delayed again on April 24 – this time for 48 hours due to a faulty Inertial Measurement Unit (IMU) – before finally leaving the pad on April 26.

This second dedicated German Spacelab mission featured a research program of 88 experiments, investigating materials and life sciences, the application of technology, Earth observations, and astronomy and atmospheric physics. The crew worked the familiar two-shift system, with Nagel working alongside Ross, Henricks and Walter on the Blue Shift and the rest of the crew on the Red Shift. During the flight, the Mission Management Team (MMT) decided that there was sufficient power onboard to extend the mission by one day, so that when Steve Nagel finally brought *Columbia* to a wheelstop on Runway 22 at Edwards AFB in California, just over 12 years since John Young had returned the same Orbiter to Runway 23 at the end of STS-1, the cumulative time logged by the five Orbiters across the 55 missions to date exceeded one year (365 days 23 hours 48 minutes).

Taking a break from training
On May 3, Rhea Seddon was working in Building 9 when she was injured in her training accident while preparing for her flight on STS-58/Spacelab Life Science 2. Seddon broke four metatarsal bones in her left foot but was not expected to be taken off the flight because of the injury. Fortunately, she missed only a couple of weeks of refresher training. As an experienced astronaut and veteran of two previous missions, most of her Space Shuttle training for this flight was mainly familiarization classes. [**35**]

The next mission, STS-57 (Spacehab 1/EURECA retrieval) flew between June 21 and July 2, 1993. The only Group 8 astronauts listed in support roles for this flight (in an Astronaut Office memo dated June 1, 1993) were Weather Pilots Hoot Gibson (launch and Edwards EOM) and Dave Walker (Edwards EOM).

Following the landing of STS-57, a number of managerial appointments came into effect that involved some of the Group 8 astronauts, moving more of the veteran TFNG away from active flight status and into management roles. Brewster

Shaw, the former Deputy Director of Space Shuttle Operations at KSC, became Director of Shuttle Operations at JSC. Loren Shriver, replaced as Deputy Chief of the Astronaut Office by Linda Godwin (Class of 1985), joined the Shuttle Program Office to assist in the management of the program as Manager for STS Launch Integration, while Dave Walker became Chief of the [Space] Station/Exploration Office at FCOD, a position he held until June 1994. [**36**]

On June 15, it was announced that Guion Bluford would be leaving NASA and retiring from the USAF in July 1993 to take a position as Vice President and General Manager of the Engineering Sciences Division, NYMA Inc., in Greenbelt, Maryland. NYMA provides engineering and software support services to the Federal Aviation Administration (FAA), the Justice Department, the Department of Defense and NASA. "I feel very honored to have served as a NASA astronaut and to have contributed to the success of the Space Shuttle program," Bluford stated at the time. "I will miss working with the people at JSC and the team spirit and *esprit de corps* that comes with flying crew members in space." [**37**]

The next mission to fly, STS-51 (ACTS deployment/ORFEUS-SPAS deployment and retrieval), flown between September 12–22, once again saw Group 8 astronauts providing only support rather than crewmembers. Steve Hawley was once more FCOD Management Support for launch and landing at JSC, and Crew Recovery Team FCOD Representative. Hoot Gibson continued his role as Weather Pilot at KSC and Edwards. [**38**]

STS-58 (October 18 – November 1, 1993)
Flight Crew: John E. Blaha (CDR), Richard A. Searfoss (PLT), M. Rhea SEDDON (MS-1, PC), William S. McArthur (MS-2), David A. Wolf (MS-3), Shannon W. LUCID (MS-4), Martin J. Fettman (PS-1, veterinarian)
Spacecraft: Columbia (OV-102) 15th mission
Objective: 58th Shuttle mission; Spacelab Life Sciences 2 research program using Long Module Unit #2
Duration: 14 days 0 hours 12 minutes 32 seconds
Support Assignments: Dick COVEY worked the second shift in the SMS. Steve NAGEL was originally assigned as STA Weather Pilot but in the memo he had been replaced for launch and EOM landing at Edwards by Jim Wetherbee. [**39**]

On this, her third and final space mission, PC Rhea Seddon and her fellow crewmembers received NASA management recognition for their contribution to the most successful and efficient Spacelab flown to date. During their 14-day flight, the second life sciences research mission, the seven-person crew worked a single shift system, similar to that followed during STS-40/SLS-1 two years earlier. The focus was upon 14 experiments across five major areas of neurovestibular, cardiovascular, cardiopulmonary, metabolic, and musculoskeletal medicine, with the crew conducting experiments on themselves and 48 rats to expand knowledge of human and animal physiology both on Earth and in space flight. The crew also performed a number of engineering tests and Extended Duration Orbiter Medical Project (EDOMP) experiments aboard *Columbia*.

STS-61 (December 2 – 12, 1993)
Flight Crew: Richard O. COVEY (CDR), Kenneth D. Bowersox (PLT), Kathryn C. Thornton (MS-1), Claude Nicollier (MS-2, ESA, Switzerland), Jeffrey A. HOFFMAN (MS-3), F. Story Musgrave (MS-4, PC), Thomas D. Akers (MS-5)
Spacecraft: Endeavour (OV-105) 5th mission
Objective: 59th Shuttle mission; first Hubble Servicing Mission (SM-1)
Duration: 10 days 19 hours 58 minutes 37 seconds
Support Assignments: Hoot GIBSON was STA Weather Pilot for launch, at Edwards, or KSC EOM, as required. [**40**]

During this 11-day flight, the HST was captured and restored to full capacity thanks to a record five EVAs by two pairs of astronauts. They replaced four gyros in the two Rate Sensing Units, two electronic control units, the solar arrays, and installed corrective optics to improve the vision of the telescope. Jeff Hoffman (EV1) logged a total of 22 hours 02 minutes on EVA and was teamed with Story Musgrave for three of the five spacewalks: December 4 (EVA #1: 7 hrs 54 min); December 6 (EVA #3 6 hrs 47 min) and December 9 (EVA #5: 7 hrs 21 min).

1994: A Barren Year
By 1994, there was a growing consistency in the number of Shuttle missions that had been launched since the return to flight (two in 1988; five in 1989; six in 1990; six in 1991; eight in 1992; and seven in 1993). There would be another seven in this year, with each mission continuing to fulfill the manifest of science and research as a precursor to Space Station. However, 1994 would also be notable as the first year since 1982 that no member of Group 8 flew in space, reflecting the changing personnel in the Astronaut Office[3]. Though most of the TFNG had retired from active flight status, there were still some important contributions to come from those in NASA management roles and from those remaining in the Astronaut Office.

---

[3] 1987 was an exception, as the whole fleet was grounded during the recovery from the *Challenger* accident. In the six decades since human space flight began in 1961, there have only been eight years in which no American has flown into space (1964, 1967, 1976–80 and 1987).

The first mission of the year was STS-60 (Spacehab 2/Wake Shield Facility 1), flown between February 3–11, which featured the first cosmonaut (Sergei Krikalev) to fly on the Space Shuttle. The only support role fulfilled by a TFNG that the authors could find for this flight was that of Chief Astronaut Hoot Gibson, who served as STS Weather Pilot for launch and KSC EOM. [41]

A THREE-PHASE PROGRAM

When the ISS Program succeeded the cancelled Space Station *Freedom*, the Russians were welcomed as full partners in what was now a 16-nation venture. As well as the redesign of the station and program, it had been decided to split the creation of ISS into several phases.

- Phase One: Following the flight of a cosmonaut on the Shuttle (Krikalev on STS-60), between 7–10 Space Shuttle-Mir docking missions would take place between 1995 and 1997, with each featuring rendezvous, docking (the first for the Americans since ASTP in 1975) and crew and logistics transfers. The Space Shuttle crew would assist in crew exchange logistics, consumable resupply and payload activities on Mir. In addition, a second Russian cosmonaut (Vladimir Titov) would fly on STS-63 in a near-rendezvous flight, to be followed by four or more NASA astronaut extended stays on Mir (all except the first launching and each landing on the Space Shuttle, to give Shuttle crews much needed rendezvous and docking practice). This would accumulate almost two-years of on-orbit experience, significantly increasing the on-orbit time of the NASA astronaut corps.
- Phase Two: Devoted to the assembly of the 'core' of the ISS, from the launch of the first Russian modules and initial elements of the truss structure, up to the attachment of the U.S. Laboratory, the first elements of the Mobile Base System (MBS) and the U.S. Airlock. This would allow the station to be occupied permanently by a core crew of three, supported by a Soyuz crew rescue vehicle and without requiring the presence of a docked Shuttle.
- Phase Three: Expansion of the station to include all the hardware, facilities and infrastructure from the international partners, to attain the status of Assembly Complete. The permanent crew could then be increased to six (occasionally increasing in the short term to nine), by crew rotation in successive overlapping teams of three.

Orbital operations after Assembly Complete were designed to expand the scientific and research capabilities of the orbital facility, as well as keeping up the maintenance and housekeeping levels to support 24/7 operation, 365 days each year, into the 2020s.

Do You speak Russian?

Before any construction or occupation of the ISS could be addressed, however, the Americans needed to gain experience in long duration space flight, beyond that pioneered on Skylab two decades before. On February 3 came the long awaited and much anticipated news that TFNG Norman Thagard would be the prime American astronaut aboard a Russian Soyuz for a three-month mission to the Mir space station. His back-up (in the first such NASA assignment since 1982) was named as Bonnie Dunbar (Class of 1980). The pair would leave for Russia later that month with a NASA support team, and each would be paired with a two-man cosmonaut team to train as a three-person resident crew. The launch would be aboard a Russian R-7 (Soyuz-U2) rocket and the flight to Mir aboard a Soyuz spacecraft. The return to Earth would be on the Space Shuttle (STS-71), with the next Mir crew flying to the station on STS-71 but returning at the end of their residency on a Soyuz spacecraft. It was also revealed that a number of NASA astronauts were undergoing Russian language training, though this did not necessarily mean that they would automatically be assigned to one of the planned Shuttle-Mir docking missions. [42]

An astronaut on Mir

Following STS-42 in early 1992, Norman Thagard had been given a technical assignment working Space Station again, and was thinking about retiring. During that summer, he was discussing his options with Dave Hilmers (Class of 1980) when he learnt of the opportunity to fly with the Russians on Mir. A few days later, Chief Astronaut Dan Brandenstein asked if he was interested in flying the Russian mission. "It was a bolt out of the blue, because I had never gone to him and volunteered," he explained, "and I said 'Absolutely', because I thought it was a neat thing to do." [43] Thagard had always wanted to learn Russian, but could not find a reason to do so until now. "It was a ride on a Soyuz rocket, it was training in Russia, and it was [a reasonably] long duration, three months, which would be unlike anything that I'd experienced in the Shuttle program, so I thought it was a great deal." Initially, he started self-studying Russian, and then classes were organized in the Astronaut Office. Shannon Lucid had been studying Russian for about a year, and in July 1993, limited funds became available for more intense lessons at a Russian language school in Monterey, California. Thagard stayed there for about four and a half months, returning to Houston a few weeks prior to the official announcement of his selection to the flight.

An astronaut's training for Mir

Thagard, his back-up Bonnie Dunbar, and their support team, moved into the Yuri Gagarin Cosmonaut Training Center in Russia (Tsentr Podgotovki Kosmanavtov, or TsPK, also known as Star City) to start Phase 1 (Group) training effective

March 1, 1994. This phase of training included technical training on the Soyuz TM spacecraft that would take Thagard and his Russian colleagues to the Mir space station. Over the next seven months, the two Americans received practical classes and numerous training sessions in the Soyuz simulator and part-task trainers; technical training on the Mir orbital complex followed by simulator and further part task training; a program of medical and biological training; simulated microgravity flights in airborne laboratories flying parabolic curves; medical examinations and physical training; and survival and wilderness training. All the training sessions and exams were conducted in Russian which, along with being so far from family and friends back in the United States, added to the challenge. Not only was a Mir assignment challenging technically and professionally, it was also a severe culture shock for the Americans. The first phase of Thagard and Dunbar's training was completed on October 7, 1994, when both astronauts were assigned to their Russian crew colleagues.

As the NASA-1 Board Engineer, Thagard was paired with the Mir-18 crew of Vladimir Nikolayevich Dezhurov, a rookie cosmonaut and CDR of the Soyuz TM-21 spacecraft they were to fly to Mir, and Flight Engineer (FE) Gennady Mikhailovich Strekalov, a veteran of four previous missions to Salyut 6, Salyut 7 and Mir, who had survived a launch pad abort on September 26, 1983. Dunbar was teamed with her CDR Anatoly Yakovlevich Solovyov, a veteran of three missions to Mir, and rookie FE Nikolai Mikhailovich Budarin[4].

The second phase of their Mir training was as part of a resident crew. This continued the practical training on the Soyuz TM simulators and further training on the Mir station, including aspects of the scientific program the crew would conduct aboard the station. There was also a program of joint training with the assigned Shuttle crew (in this case STS-71 headed by Hoot Gibson) at both TsPK and back in the United States. This phase of training was completed between October 10, 1994 and February 21, 1995, three weeks prior to the launch of Soyuz TM-21.

In total, Thagard and Dunbar accumulated 883 hours as a group for Phase 1 of cosmonaut training and 845 hours in Phase 2 Mir resident crew training, totaling an impressive 1,728 hours in the 12 months between March 1994 and March 1995. A detailed account of the training of American astronauts for flights on the Mir station is beyond the scope of this current volume, but a summary can be found in *Russia's Cosmonauts*. [**44**] A summary of the training completed by Thagard (and subsequently Shannon Lucid) for residency on Mir can be found in Table 12.1

---

[4] Solovyov and Budarin would fly to Mir on STS-71 and take over as the Mir-19 resident crew, returning on Soyuz TM-21 after a flight of 76 days having handed over in turn to the Mir-20 crew. Dunbar flew to and from Mir on STS-71, but unfortunately did not complete a residency, though she was qualified to do so.

Table 12.1: NASA GROUP 8 RUSSIAN CREW TRAINING SUMMARY

TABLE 12.1a: Scope and dates of Mir training for NASA group 8 astronauts

NASA board engineer mission	Prime astronaut	Phase 1 (Group) start/complete	Phase 2 (Crew) start/complete	Total hours in group	Total hours in a crew	Total training hours	Launch date/ mission	Mir expedition
NASA 1	THAGARD	1994 Mar 01–1994 Oct 7	1994 Oct 10–1995 Feb 21	883	845	1728	1995 Mar 16, Soyuz TM-20	EO-18
NASA 2	LUCID	1995 Jan 03–1995 Jun 24	1995 Jun 26–1996 Feb 26	795	1127	1922	1996 Mar 24, STS-76	EO-21/-22

TABLE 12.1b: NASA group 8 astronauts Mir training hours as part of a group

NASA mission	Prime astronaut	Soyuz TM training Technical	Simulators	Mir training Technical	Simulators	Medical/ biological	Independent training	Russian language	Total hours
NASA 1	THAGARD	134	173	120	50	170	86	150	883
NASA 2	LUCID	20	50	114	60	122	161	268	795

TABLE 12.1c: NASA group 8 astronauts Mir training hours as part of a crew (Prime only)

NASA mission	Prime astronaut	Soyuz TM training Technical	Simulators	Mir training Technical	Simulators	Medical/ biological	EVA train-ing	Science training	Pre-flight training	Independent training	Russian language	Total hours
NASA 1	THAGARD	35	90	128	68	94	4	311	80	11	24	845
NASA 2	LUCID	80	130	141	142	180	0	266	24	76	88	1127

Taken from: *Russia's Cosmonauts Inside the Yuri Gagarin Training Center*, Rex D. Hall, David J. Shayler & Bert Vis, Springer-Praxis, 2005, pp. 280–281
Adopted from: *Phase 1 Program Joint Report*, Editors George C. Nield and Pavel M. Vorobiev, 1999, NASA SP-1999-6108 (in English)

As Thagard and Dunbar were beginning their training in Russia, two Shuttle missions flew over the next two months. STS-62 (USMP-2/OAST-2) was flown between March 4−18 and the astronaut support roles for this mission fulfilled by members of the TFNG included: Steven Hawley as FCOD Management Representative, providing support for launch (KSC/OSR), orbit (at JSC) and landing back at the Cape; Hoot Gibson as STA Weather Pilot for launch and KSC EOM, and Dave Walker also at Edwards for the EOM; Hawley was also Management Representative for the crew recovery team. [**45**] This mission was followed by STS-59 (Space Radar Laboratory 1) between April 9−20, with the only TFNG Mission Support role completed by Hoot Gibson, as STA Weather Pilot for launch and landing EOM at KSC. [**46**]

Two months after STS-59 flew, the Shuttle crew assignments for the first mission to physically dock with the Russian Mir space station and exchange resident crews were officially announced on June 3. A seven-member crew would be led by TFNG Hoot Gibson and included PLT Charles Precourt (Class of 1990) and MS Ellen Baker (Class of 1984), Greg Harbaugh (Class of 1987) and Bonnie Dunbar (Class of 1980). Also assigned were Russian cosmonauts Anatoly Solovyov (Mir-19 CDR) and Nikolai Budarin (Mir-19 FE). The two cosmonauts would replace the Mir-18 crew and remain on the station as the next resident crew. Returning home on STS-71 at the end of the mission would be the Mir-18 crew of Vladimir Dezhurov (Mir-18 CDR), Gennady Strekalov (Mir-18 FE) and TFNG Norman Thagard (Mir-18 NASA-1 Board Engineer). The mission was scheduled for mid-1995 using the orbiter *Atlantis*, which had been specially modified to carry a docking system compatible with the Russian Mir station. As this was the first time since Skylab that an American astronaut would be returning from an extended duration mission in space (three months), *Atlantis* would also carry the Spacelab Long Module configuration, within which various life science experiments and collection of data would take place during the mission. [**47**]

On July 7, Dave Walker, who had been named as the Chief of the JSC Safety Review Board the previous month, was now named as CDR for STS-69. [**48**] Walker's new crew would include PLT Kenneth Cockrell (Class of 1990) and MS James S. Voss (Class of 1987), James Newman (Class of 1990) and Michael Gernhardt (Class of 1992). STS-69 was manifested to fly the SpaceHab-4 mission, which included the second flight of the Wake Shield Facility (WSF) experiment. The day after the STS-69 crew announcement, NASA launched the STS-65 (IML-2) mission, which flew between July 8−23. There were no signs of support roles for any of the TFNG on this flight in the undated Astronaut Office memo researched by the authors, beginning a trend over the next couple of years as the number of active members of the group reduced.

One such reduction of the TFNG in the Astronaut Office occurred on July 11, when it was announced that Dick Covey would leave NASA and retire from the USAF on August 1, 1994. He would take the position of Director of Business Development for Calspan Services Contract Division, an operating unit of Space Industries International, based in Houston. [**49**]

On September 6, Robert 'Hoot' Gibson stood down from the role of Chief Astronaut, as was the tradition, to return to full time training for the STS-71 mission. He was replaced as Chief Astronaut by Robert Cabana. In the last third of the year, the remaining members of the TFNG in the Office supported the final three missions flown. For STS-64 (LITE/SPARTAN 201; September 9–20), FCOD Management Support at JSC for launch was provided by Steve Hawley, while Dave Walker was STA Weather Pilot for Edwards EOM. [50] The Weather Pilots for STS-68 (SRL-2, September 30 – October 11) were Hoot Gibson for launch and KSC EOM and Dave Walker again for Edwards EOM. [51] For STS-66 (ATLAS-3/CRISTA-SPAS November 3–14), Steve Hawley again provided FCOD Management Support at JSC, this time for both launch and landing. [52]

To go or not to go? That is the question
On November 3, 1994, the day STS-66 launched, Shannon Lucid and John Blaha (Class of 1980) were named to training for the second of at least four scheduled astronaut residencies on Mir. The two space flight veterans, who had flown together on STS-58, journeyed to Russia in January 1995 to begin their training at TsPK for a long duration mission to the Russian station. [53]

It would be understandable to assume that an astronaut (or cosmonaut) would jump at the chance to fly if given the opportunity, and mostly this is true, but there have been instances over the years when this has not been the case. Rare instances of personality clashes or crew incompatibility aside, some of the original astronauts actually had no interest in taking a flight to the Moon, while others simply had no desire to spend weeks on board a space station, or wait years to fly the Shuttle. Now the Shuttle-era astronauts were faced with Shuttle-Mir. Here was an opportunity to do something different, but the sacrifice was to learn Russian and spend time living in Russia, initially away from family and friends for months on end. To many astronauts in the 1990s, this was not what they had joined NASA for and they had no desire to travel to Russia and train for such a mission. Therefore getting volunteers to put themselves forward for the Mir seats, as few as there were, was difficult. This was also a challenge faced by the Space Station planners early on.

There were many astronauts lining up for seats on the Space Station assembly missions, or for flights that did not visit the station, such as the Hubble Servicing Missions, but spending months training in locations around the world (USA, Europe, Canada, Japan and Russia) was not something many wanted to go through. Apart from the separation from their families, the harshness of the Russian winters, learning the Russian language and the amount of worldwide travel required for up to two years before flying a six-month mission were all just too much for some. These types of missions needed a new breed of space explorers, and from the early 2000s, with an all-time high of 149 astronauts in the office, that was what NASA looked to recruit. In 2004, the 11 astronauts selected for Group 19 were the last to undergo Shuttle mission training and fly on the vehicles before they were retired. Five years later, the nine candidates of Group 20 became the first to focus entirely on the ISS and its supporting vehicles.

Back in the mid-1990s, the Shuttle-Mir docking missions and crewing for the available seats on long duration Mir flights were the focus, with the first element of ISS still three years from launch. At the start of the new year, two of the TFNG were preparing for their historic missions to the Russian space station.

THE FIRST TFNG VISIT A (RUSSIAN) SPACE STATION

On January 3, 1995, Shannon Lucid, John Blaha and their support team began their Phase 1 (Group) training at Star City. This would last for the next six months (completed on June 24) and totaled 795 hours. "People always think that you went over there and you trained with a crew", Shannon Lucid commented in 1998. "You didn't go over there and train with a crew. All my training in Russia was with John Blaha. John Blaha and I trained together. We sat in a classroom together. It was just the two of us and the instructor for whatever classroom it was. We just talked to each other all day long and that was it." [**54**]

Just over three weeks later on January 27, NASA announced the crew for STS-75, which including Jeff Hoffman as MS-1. STS-75 was planned as the second flight of the TSS in early 1996 using *Columbia*. [**55**] The mission also included the third flight of the United States Microgravity Payload (USMP) series. Hoffman would be flying with CDR Andrew M. Allen (Class of 1987), PLT Scott Horowitz (Class of 1992), and ESA MS-trained Claude Nicollier and Maurizio Cheli. They joined Franklin Chang Díaz and the PS named earlier who were already in training.

The first Shuttle mission to fly anywhere near a space station came the following month. STS-63 (February 3–11) was officially known as the SpaceHab 3 mission, but was more commonly referred to as the 'Near-Mir' rendezvous mission. The crew, led by Jim Wetherbee (and including cosmonaut Vladimir Titov), brought *Discovery* to within 11.2 m of the Mir docking port, in practice for the full docking due on the later STS-71 mission. Though the STS-63 crew did not include anyone from Group 8, Dave Walker provided support by working the second shift in the SMS. [**56**] STS-63 was the first mission to bring American and Russian space vehicles into close proximity since the 1975 Apollo-Soyuz Test Project (ASTP) docking mission. It was a remarkable turnaround in the space programs of both countries.

On February 28, Steve Nagel retired from the USAF with rank of Colonel. He also retired from the Astronaut Office, but not NASA, the following day (March 1) to take up a position at JSC as Deputy Director of Operations, Safety, Reliability and Quality Assurance. [**57**] When the next mission flew (STS-67/Astro-2, March 2–18), Dave Walker was again the sole representative of the TFNG to support the flight, assigned to the first shift in the SMS and on 24-hour call for SPAN. [**58**]

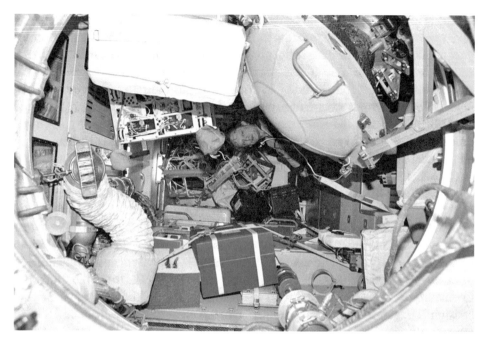

Fig. 12.3: Norman Thagard in the Spektr module during his four-month residency on the Russian station Mir.

Soyuz TM-21 (March 14 – July 7, 1995, Landing on *Atlantis*/STS-71)
Flight Crew: Vladimir N. Dezhurov (CDR), Gennady M. Strekalov (FE), Norman E. THAGARD (NASA-1 Cosmonaut Researcher, CR)
Spacecraft: Soyuz TM-21 (7K-M, spacecraft serial #70) for the launch phase and the flight to Mir. The landing would be on STS-71 (*Atlantis*)
Objective: Mir EO-18 resident crew; 1st NASA astronaut resident crewmember; Shuttle-Mir Phase 1
Duration: 115 days 8 hours 43 minutes 2 seconds
Support Assignments: There were no formal support team roles for the astronauts discovered during the authors' research, apart from Bonnie Dunbar as THAGARD's back-up. For about two months prior to his launch, Shannon LUCID and John Blaha were training at TsPK at the same time as THAGARD and Dunbar.

It had been over 15 months since a member of the 1978 selection had been in Earth orbit. There had been nine Shuttle missions in the interim, and though members of the TFNG had supported those missions on the ground, this had been the longest period of flight operations without a member of the 1978 selection flying since the Return-to-Flight. This mission certainly returned a TFNG to orbit, on a far longer mission than any of those previously flown by the group, and initially not on American hardware. As part of the agreement between the ISS partners and

Russia, Thagard's flight would be the first of a series of resident missions flown by seven veteran American astronauts to the Mir station, over what became a four-year period in preparation for residential flights on ISS. It was a chance for NASA and its astronauts to gain a feel for regular station operations, something the Americans were severely lacking in comparison to their Russian counterparts, who had over 25 years' worth of experience from their fleet of Salyut stations and now Mir. It also allowed Americans to log their first long duration flights (those exceeding one month) since the Skylab missions of two decades earlier, giving them much needed practical experience prior to embarking on ISS[5].

Chosen to fly this important first mission to Mir was TFNG Dr. Norman Thagard, on his fifth excursion into orbit. This time, he would not be a member of a Shuttle crew but a Cosmonaut Researcher (CR, or in NASA speak, Board Engineer) on the Russian Mir EO-18 crew, working aboard space station Mir along with cosmonauts Vladimir Nikolayevich Dezhurov (whose personal radio callsign was "Uragan", meaning "Hurricane") and Gennadi Mikhailovich Strekalov ("Uragan-2"). Lift off on board Soyuz TM-21, from the Baikonur Cosmodrome in Kazakhstan, occurred on March 14, 1995. Thagard (using the callsign "Uragan-3") therefore became the first American astronaut since ASTP to train on Russian soil, the first in history to enter space aboard a non-American rocket and spacecraft, and the first American occupant of a foreign space station (Mir). When Thagard and his two colleagues entered orbit, they joined the STS-67 crew who still had four more days of their Astro-2 mission to complete, raising the number of space explorers in orbit during those days to 13 (three on Soyuz TM-21, seven on STS-67 and the three EO-17 cosmonauts aboard Mir).

On March 16, once Soyuz TM-21 had docked with Mir, Thagard became the first American since Skylab 4 to enter and live on a space station. During his stay on Mir he assisted in a program of 28 experiments in the course of the 115-day flight. As the first astronaut to live longer in space than 18 days since Skylab 4 a decade earlier, Thagard found his four-month mission on Mir challenging both culturally and in terms of finding things to do. The delay in delivering the Spektr module to Mir certainly restricted his research capabilities. "I found myself with, really, too much time on my hands," he reported post-flight. At no time did he find the experience claustrophobic, or encounter any compatibility problems with his Russian crewmates as he had trained with them for some time. However, Thagard went days without speaking to controllers in Houston and missed reading newspapers. While he was proficient in Russian, sending English language reports from Earth had not been built into the daily routine on Mir. Thagard reported that boredom could be as depressing as a schedule which was too hectic to keep up with, as the Skylab 4 astronauts had experienced during the early phase of their 84-day

---

[5] Prior to 1995, the only U.S. human space flights to exceed three weeks (21 days) had been Skylab 2 (28 days in May−June 1973), Skylab 3 (59 days in July−September 1983) and Skylab 4 (84 days from November 1973−February 1974).

mission. As Marsha Freeman wrote in her 2000 book *Challenges of Human Space Exploration* (Springer-Praxis), when Thagard returned, NASA Administrator Dan Goldin offered him a public apology, stating that "We put all of our focus on the physical wellbeing of the astronauts and the success of the mission, [but] we neglected the psychological wellbeing and Dr. Thagard made it clear to us." The oversight was addressed before the next American resided on Mir.

On March 30, two weeks after Thagard left Baikonur to start his residence on Mir, Shannon Lucid was selected as the prime crewmember for a five-month stay on Mir in 1996. She was still in Russia at the time. She would be replaced at the end of her residency by Jerry Linenger (Class of 1992). [**59**] The back-up astronauts assigned were John Blaha for Lucid and Scott Parazynski for Linenger, both of whom would continue training at Star City for later flights to Mir.

On April 14, Lucid was assigned to the STS-76 crew for the launch of her mission to Mir, and to the STS-79 crew for her landing five months later. Launching with Lucid to Mir, the remainder of the STS-76 crew were CDR Kevin Chilton (named in November 1994, Class of 1987), PLT Richard A. Searfoss (Class of 1990) and MS Linda Godwin (Class of 1985), Michael R "Rich" Clifford and Ronald Sega (both Class of 1990).

Lucid would be assigned as MS-4 for the ascent, flying on the middeck and transferring to the main crew shortly after docking and hatch openings. She would remain aboard Mir as NASA Board Engineer-2 when STS-76 departed. Five months later when STS-79 docked, she would transfer to the Shuttle crew once again as MS-4 for entry and landing on the middeck, using the specially installed prone seat recliners. The remaining crew for STS-79 were CDR William Readdy (named November 1994, Class of 1987), PLT Terrence Wilcutt (Class of 1990), and MS Tom Akers (Class of 1987), Jay Apt (Class of 1985), and Carl Walz (Class of 1990). Jerry Linenger would fly as MS-4 for ascent, swapping places on the crew with Lucid. [**60**]

As the program proceeded, several factors affected this plan. Requirements for the individual missions changed, together with varying progress in training, delays with preparing hardware, and difficulties in getting some of the astronauts qualified to fly as a Mir crewmember. There was also a new requirement for Linenger to qualify in a Russian Orlan EVA suit. As Linenger wrote in his 2000 book *Off the Planet*, "John Blaha and I switched positions in the sequence in order for me to do a spacewalk." The added Russian EVA suit training meant that it would be more feasible to have the experienced Blaha fly the third residency (up on STS-79 and down on STS-81) after Lucid and delay Linenger to the fourth residency (up on STS-81 and down on STS-84).

Lucid cosmonaut training

The day before STS-71 left the launch pad destined for Mir, Shannon Lucid and John Blaha began their Phase 2 (Crew) training at Star City, Moscow. This phase alongside their cosmonaut colleagues would last eight months and be completed on

February 26, after 1,127 training hours. In total, Lucid and Blaha logged over 1,922 hours training for the second American residency on Mir (see Table 12.1, p. 437).

In February 1995, they had been paired with their Russian colleagues. Lucid was teamed with Yuri Ivanovich Onufriyenko (Callsign "Skif" meaning "Scythian") and Yuri Vladimirovich Usachev, who had been in training as a crew for two years (and who Lucid called the "two Yuris"). Blaha was initially paired with Gennadi Mikhailovich Manakov and Pavel Vladimirovich Vinogradov on the back-up crew[6]. Lucid noted that her colleagues had assumed that she would train with the cosmonauts she would fly in space with, but this was not the case, at least not at first. Day after day, she and Blaha sat in classrooms learning each specific system from a specialist, taking notes and then taking an oral exam before moving on to the next system. It was only later in their training program that they completed a very few simulations with the cosmonauts. "[Initially], we didn't interface with anybody else," Lucid recalled. "Only towards the end did we do just a few sims with the Russians, but it was very minimal. It wasn't the training with a crew like you would think," which was the way they had been taught at NASA. "You sat in the Soyuz, and my job was to be quiet and not to interfere. That's basically my job. We would do that one afternoon. We did that two or three times. We did one training session in the Mir and we did one training session over in the vacuum chamber together." [**54**] All the astronauts who participated in Mir long duration training said that it was a fascinating experience, but a big undertaking and a challenge.

STS-71 (June 27–July 7, 1995)
Flight Crew: Robert L. GIBSON (CDR), Charles J. Precourt (PLT), Ellen L. Baker (MS-1), Gregory J. Harbaugh (MS-2), Bonnie J. Dunbar (MS-3)
Mir EO-19 crew (up only): Anatoly Y. Solovyov (Russian Cosmonaut #1/Mir-19 CDR), Nikolai M. Budarin (Russian Cosmonaut #2/Mir-19 FE)
Mir EO-18 crew (down only): Vladimir N. Dezhurov (Russian Cosmonaut #1/ Mir-18 CDR), Gennady M. Strekalov (Russian Cosmonaut #2/Mir-18 FE), Norman E. THAGARD (NASA-1 CR/MS-4)
Spacecraft: Atlantis (OV-104) 14th mission
Objective: 69th Shuttle mission; 1st Shuttle-Mir docking; delivery of EO-19 and return of EO-18 resident crewmembers; Spacelab Long Module unit #2
Duration: 9 days 19 hours 22 minutes 17 seconds
Thagard Duration: 115 days 8 hours 43 minutes 2 seconds
Support Assignments: The only Group 8 support assignment was FCOD Management Support at JSC for both launch and landing provided by Steve HAWLEY. [**61**]

---

[6] During training, Manakov was grounded due to a slight heart irregularity, so the crew was replaced by Valery Grigoryevich Korzun and Aleksandr Yuryevich Kaleri. The change had minimal impact on Lucid, but Blaha later flew a 'difficult' mission with cosmonauts he hardly knew and had not trained with.

It seems appropriate that the CDR of the Shuttle destined to return Thagard from his stellar home was a fellow TFNG, Hoot Gibson, who was also privileged to be the first American to dock with a separate crewed spacecraft in Earth orbit since Tom Stafford linked up his Apollo Command and Service Module (CSM) with Alexei Leonov's Soyuz 19 during ASTP in July 1975. The modified *Atlantis* was launched from KSC on June 27, 1995 and docked with Mir two days later. There was an exchange of crewmembers, and while the Shuttle was launched with seven astronauts, (including the two EO-19 cosmonauts) it returned to Earth after ten days on July 7 with a crew of eight on board (including the three EO-18 crew). This was the 100th U.S. mission to be launched from the Cape and to commemorate the event, Gibson took with him the first American flag flown in space, carried on board with Alan Shepard on the first U.S. astronaut flight in 1961. That flag was subsequently displayed in the U.S. Astronaut Hall of Fame in Florida.

The STS-71 mission was the first to fly a profile that the Shuttle was originally designed for, a shuttling flight to and from a space station. It was a Russian station not an American one, but what was more important was that at last the astronauts could finally practice and achieve rendezvous, proximity operations, docking and undocking with a suitably massive item of space hardware. This mission was key not only to Shuttle-Mir but also to plans for the ISS, and initiated the successful run of nine dockings of the Shuttle with Mir between 1995 and 1998. They led in turn to an impressive 37 Shuttle dockings with ISS from 1998 through 2011 and the end of the Shuttle program. Gibson began this remarkable run of 46 flawless rendezvous and docking operations on June 29, 1995, when he nudged the docking system of *Atlantis* into the forward docking port on the Kristall Module on Mir, restoring the skills of docking in space that had been pioneered during Gemini and Apollo to the Astronaut Office, after a gap of two decades. While docked to Mir, the crew conducted the transfer of hardware, water and fresh supplies into the station, and stowed unwanted items and trash into *Atlantis* for the return to Earth. In the onboard Spacelab Module, over 15 biomedical and scientific investigations were completed. On July 3, which also happened to be his 52nd birthday, Thagard bid farewell to Mir after spending 109 days aboard the station, *Atlantis* was undocked from Mir on July 4 after 4 days 22 hours 9 minutes and 26 seconds together.

Ten days after Gibson brought *Atlantis* home, Shuttle *Discovery* left the launch pad for the 21st time. STS-70, whose designation for once matched its launch order as the 70th Shuttle mission, was launched on July 13, 1995, on a nine-day mission to deploy TDRS-G. From the research by the authors, none of the remaining TFNG served in a support role for this mission.

STS-69 (September 7 – 18, 1995)
Flight Crew: David M. WALKER (CDR), Kenneth D. Cockrell (PLT), James S. Voss (MS-1), James H. Newman (MS-2), Michael L. Gernhardt (MS-3)
Spacecraft: Endeavour (OV-105) 9th mission

Objective: 71st Shuttle mission; Wake Shield Facility-2 and SPARTAN-201-03 operations; EVA Development Flight Test
Duration: 10 days 20 hours 28 minutes 56 seconds
Support Assignments: FCOD Management Support at JSC for both launch and landing was provided by Steve HAWLEY. [**62**]

The original launch date for STS-69 was set for early August, but several delays pushed this back to early September. During the mission, the crew deployed and retrieved two payloads – the second Wake Shield Facility and the SPARTAN-201-03 free flying satellite – and supported the second in an important series of EVA development spacewalks. These were aimed at providing experience and developing the necessary EVA procedures for ISS assembly and future satellite servicing missions, as had been demonstrated during the STS-61 Hubble Servicing Mission in 1993. By 1995, some 12 years after the flight of the first TFNG on STS-7, the crew opportunities for those still active were diminishing, but they continued to remain at the forefront of current operations.

Fig. 12.4: The STS-69 "Dogs of Summer" crew proudly displays the "Dog Kennel" emblem on the shoulder of their flight suits. [l-r] MS Mike "Underdog" Gernhardt; PLT Ken "Cujo" Cockrell; CDR and TFNG Dave "Red Dog" Walker; MS Jim "Dog Face" Voss; and MS Jim "Pluto" Newman.

"Dogs of Summer"

STS-69 was also the flight of the second 'Dog Crew'. Although this time they were a mix of military and civilian crewmembers, they still became great buddies, encouraging Walker to revive the 'dog-tag' names he assigned to his crew as he had on STS-53. Walker remained "*Red Dog*" and flying with him again was Jim "*Dog Face*" Voss. The new "pack members" were "*Cujo*" (Cockrell), after the Steven King novel of the same name and "*Pluto*" (Newman), due to his interest in science, the fact that he was a computer wizard in the Office and, according to his crewmates, his "unique perspective on life." Newman had also told his crewmates that he would kill them if he was called "*Goofy*". The other new boy was rookie Gernhardt, who naturally became "*Underdog*," though this was also due to his previous career as a deep-sea diver prior to becoming an astronaut. Upon arriving at KSC for the mission, the crew displayed an unofficial crew patch featuring a bulldog peering out of a doghouse shaped like a Space Shuttle, with the five "Dog Tag" names listed. And there was a change to the old station wagon too. Walker's 1979 Pontiac "*Dogmobile*" acquired the name "En*dog*vour" during the lead up to STS-69. Having originally been painted flat-black to fit the classified nature of his STS-53 DOD mission, for this second flight of the Astronaut Office "guard dogs" it was changed to a peaceful T-38 training jet white… trimmed in blue but dotted with paw prints and sporting a range of surplus 'flight hardware'. [**63**]

The final two missions flown in 1995 included STS-73 (October 20–November 5), *Columbia*'s 18th mission, carrying the USML-2 LM#1 payload. Only Steve Hawley of the TFNG fulfilled support role, again providing FCOD Management Support at JSC for both launch and landing. [**64**] This mission was followed a few days later by STS-74 (November 12–20), with *Atlantis* flying the second Shuttle-Mir docking mission and delivering a new Docking Module to the station. This became the only Shuttle docking mission at Mir which did not include a NASA exchange or return. None of the TFNG supported this mission in formal roles.

THE LONGEST MISSION

Early in the new year, January 5, 1996, Norman Thagard retired from NASA to return to his alma mater, Florida State University, having accepted the position of Visiting Professor and Director of External Relations for Florida A&M University, Florida State University College of Engineering in Tallahassee. From January 16, the 18th anniversary of his selection as a NASA astronaut, he began teaching electronics, which was a long term hobby for the former astronaut. [**65**]

Between January 11−20, *Endeavour* flew STS-72 to retrieve the Japanese Space Flyer Unit and conduct the deployment, independent flight and retrieval of the OAST-Flyer. For this mission, Dave Walker was assigned as STA Weather Pilot for EOM at Edwards, even though *Endeavour* landed at KSC. [**66**] January 1996 also saw the return to the Astronaut Office of Anna Fisher, after a six-year absence to raise her family. She was assigned to Space Station issues.

More Missions to Mir
On January 30, NASA announced plans to expand the current Phase 1 Shuttle-Mir program with two additional dockings, taking the total to nine, though no further TFNG would visit the station. The emphasis in the Office was on the completion of the Shuttle-Mir missions and moving on to assembly of the ISS. It looked unlikely that any of the 1978 class would remain active long enough to fly an ISS mission, however. Indeed, there were only a handful of TFNG still active in the office at the start of the year. [**67**]

STS-75 (February 22 – March 9, 1996)
Flight Crew: Andrew M. Allen (CDR), Scott J. Horowitz (PLT), Jeffrey A. HOFFMAN (MS-1), Maurizio Cheli (MS-2, ESA, Italy), Claude Nicollier (MS-3, ESA, Switzerland); Franklin R. Chang Díaz (MS-4, PC), Umberto Guidoni (PS-1, ASI, Italy)
Spacecraft: Columbia (OV-102) 19th mission
Objective: 75th Shuttle mission; TSS re-flight; USMP-3 payload
Duration: 15 days 17 hours 40 minutes 21 seconds
Support Assignments: FCOD Management Support at JSC for launch and landing again fell to Steve HAWLEY, while Hoot GIBSON provided STA Weather Support. [**68**]

STS-75 was a 16-day mission whose principal payloads were the re-flight of the TSS and the third flight of the United States Microgravity Payload (USMP-3). The TSS successfully demonstrated the ability of tethers to produce electricity. The crew also worked around the clock performing combustion experiments and research related to USMP-3 microgravity investigations. During this mission, Hoffman, the first TFNG to enter orbit for five months, was a member of the Blue Shift with Nicollier and Chang Díaz, while the others worked the Red Shift. Interestingly, Hoffman joined Allen on the 'White Shift', allowing them to work with either of the other shifts as necessary. On this flight, Hoffman also became the first astronaut to log 1,000 hours aboard the Space Shuttle.

On February 29, Steve Hawley was named MS-2/FE for STS-82, the second Hubble Servicing Mission due in early 1997. Hawley had deployed HST using the RMS as a member of the STS-31 crew six years earlier and would be the first astronaut to return to the telescope since then. He was again assigned as primary RMS operator, supporting the EVAs. [69]

In March 1996, Hoot Gibson became Deputy Director, Flight Crew Operations, JSC, a position he held until he retired from NASA in November that year.

STS-76 (March 22 – 31, 1996)
Flight Crew: Kevin P. Chilton (CDR), Richard A. Searfoss (PLT), Ronald M. Sega (MS-1), Michael R. Clifford (MS-2), Linda M. Godwin (MS-3); Shannon W. LUCID (MS-4 up only/EO-21/22 CR/NASA-2 Board Engineer)
Spacecraft: Atlantis (OV-104) 16th mission
Objective: 76th Shuttle mission; 3rd Shuttle-Mir docking flight; transfer of NASA-2 (Lucid) to Mir EO-21 resident crew
Duration: 9 days 5 hours 15 minutes 53 seconds
Support Assignments: FCOD Management Support at JSC for launch and landing was provided this time by Hoot GIBSON, while Dave WALKER was STA Weather Pilot for EOM at Edwards. [70]

When he returned from Mir in 1995, Norman Thagard expressed relief that he had not had to endure an extended mission of six months on the aging Russian station. But this is exactly what his Group 8 colleague Shannon Lucid found herself facing when problems with the Shuttle hardware delayed her return. Lucid's journey to Mir began with a lift off aboard *Atlantis* from KSC on March 22, 1996. Following a successful docking, she transferred to the orbiting Mir station two hours after the hatches had been opened between the two spacecraft. The undocking of *Atlantis* signaled the start of a planned two-year continuous presence on Mir by successive American astronauts.

Assigned to the Mir-21 crew as NASA-2 Board Engineer, over the course of the next six months Lucid performed numerous life sciences and physical sciences experiments with two different crews of Russian cosmonauts (EO-21 and 22) during the course of her stay aboard Mir. In doing so, she set the single-mission United States space flight endurance record for a woman. Her return journey to KSC on September 26 was accomplished as part of the crew of STS-79 (*Atlantis*). Lucid was the first American astronaut to be left in space on a spacecraft they had not launched aboard.

Just over two weeks after he had provided weather pilot duties for STS-76, TFNG Dave Walker retired from NASA and the USN, effective April 15, 1996, to become a part owner and executive officer of a telecommunications company, as

Vice President for Sales and Marketing for the new startup company NDC Voice Corporation in Southern California, who were to provide integrated wireless communications and advance processing applications internationally. [71] As members of the 'old guard' departed, so new talents arrived at the Astronaut Office at JSC. On May 1, NASA announced the largest selection of Ascans since the 1978 group, with an equal amount of candidates forming the 16th Astronaut Class. [72] They began their Ascan program on August 12 and were joined by nine international astronauts, making this the largest group in NASA history to prepare for assignment as Space Shuttle and ISS crewmembers. [73]

Over May 19–29, *Endeavour* flew its 11th mission as STS-77. Onboard were the SpaceHab-4 middeck augmentation module and the SPARTAN/IAE free-flyer. For this mission, Hoot Gibson served as back-up STA Weather Pilot (to Robert Cabana) for launch and KSC EOM. [74] The following month it was *Columbia*'s turn, flying its 20th mission as STS-78. This was a 17-day Life and Microgravity Spacelab (LMS) flight using Spacelab Long Module #2. Once again, no TFNG were assigned to support this mission.

On September 15, it was announced that Rhea Seddon had been assigned to a part-time post at Vanderbilt University's Center for Space Physiology and Medicine in Nashville, Tennessee. She remained an astronaut but would act as a liaison for Vanderbilt researchers in evaluating flight equipment and developing experiment operating procedures for the 1998 Neurolab mission, which would study the adaptation of the nervous systems to microgravity during a 16-day flight on *Columbia*. Her medical background and extensive experience during two Spacelab Life Science Missions made her uniquely qualified for this position, where she represented the interests of JSC and the Astronaut Office in designing and developing protocols for this mission. [75]

STS-79 (September 16 – 26, 1996)
Flight Crew: William F. Readdy (CDR), Terrence W. Wilcutt (PLT), Jerome Apt (MS-1), Thomas D. Akers (MS-2), Carl E. Walz (MS-3); John E. Blaha (MS-4 up only/EO-22 CR/NASA-3 Board Engineer); Shannon W. LUCID (MS-4 down only/EO-21/22 CR/NASA-2 Board Engineer)
Spacecraft: Atlantis (OV-104) 17th mission
Objective: 79th Shuttle mission; 4th Shuttle-Mir docking flight; transfer of NASA-3/EO-22 crewmember (Blaha); return of NASA-2/EO-21/22 crewmember (Lucid)
Duration: 10 days 3 hours 18 minutes 26 seconds
LUCID duration: 188 days 4 hours 0 minutes 11 seconds
Support Assignments: Hoot GIBSON was STA Weather Pilot for EOM at Edwards. [76]

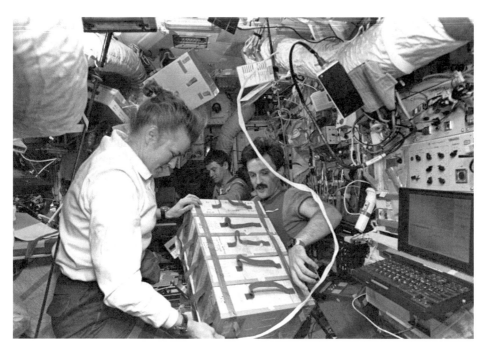

Fig. 12.5: Shannon Lucid during her six-month residency with FE Alexandr Kaleri and station Commander Valery Korzun (in the background).

Shannon Lucid was the first American female astronaut to live onboard a space station as part of a resident crew[7]. She was certainly kept occupied during her half-year on the station, conducting research into Earth sciences, fundamental biology, human life sciences, advances in technology, microgravity research and the space sciences. She adapted well to living on the Russian station, despite the setbacks and challenges, and kept track of time by wearing her pink socks each Sunday and munching on supplies of her favorite M&Ms ®™ candy. Her sleeping area on Mir was "…like living in the back of your pickup [truck] with your kids… when it's raining and no one can get out." [77] Communication with the NASA team in Moscow was not perfect, and she had not had a shower since she launched, but the best thing about doing the laundry… was not having to do the laundry and just throwing the dirty clothes away[8]. During her relaxation time she read some of the

---

[7] Shannon Lucid was only the fourth woman in history to be part of a resident station crew, after cosmonaut Svetlana Savitskaya in 1982 and again in 1984, the UK's Helen Sharman in 1991, and cosmonaut Yelena Kondakova in 1994 and 1995.

[8] Jokingly, Lucid's husband said that NASA had told him they were going to hose his wife down after landing before they gave her back to him.

books she had brought with her, although she was a little frustrated to find that the next part of a novel her daughter had given her was not included in her small library: "[I] came to the last page, and the hero, who was being chased by an angry mob, escaped… The End… Continues in Volume Two. And was there Volume Two in my bag? No. Could I dash out to the bookstore? No. Talk about a feeling of total isolation and frustration". [**78**] This shows the little problems encountered so far from home and away from normal life, even in low Earth orbit. At the end of her mission, Lucid refused the offer to help her off the Shuttle on a stretcher, preferring to walk off with some assistance to the Crew Transfer Van nearby. There, she was greeted by her family and a large box of M&Ms®™, a gift from President Bill Clinton. At 189 days, Lucid had set a new world endurance record for a female space explorer, and in doing so inspired a whole new generation of female students to follow in her footsteps.

September also saw Steve Nagel transferred to the Aircraft Operations Division at JSC to work as a research pilot. A few weeks later, towards the end of November, Hoot Gibson retired from NASA to "pursue private business interests". On leaving the office, Gibson said, "While I am looking forward to new challenges and opportunities, I will certainly miss being part of the NASA team. I am grateful for the opportunity to work with so many talented and dedicated people over the past 18 years." [**79**]

The final mission of the year was STS-80 (November 19–December 7), flown by *Columbia*. At 18 days, this was the longest Shuttle mission in the program. STS-80 carried the third Wake Shield Facility experiment and ORFEUS-SPAS II. Had Hoot Gibson remained at NASA, he would have been assigned as STA Weather Pilot for the EOM at Edwards. [**80**] He was also identified pre-flight as a member of the FCOD Management for Launch Support at KSC/OSR and for landing at JSC, as well as FCOD representative for the Crew Recovery Team. [**81**] In fact, Gibson did not remain at NASA for either of those assignments, having left after the two memos were issued and before the STS-80 mission flew. "I was not still with NASA when STS-80 flew on November 19," Gibson recalled to the authors. "My last day at JSC was November 13 and then I showed up in Dallas to begin pilot training with Southwest Airlines. I felt quite a bit depressed after leaving NASA." [**82**]

The penultimate flight

When Anna Fisher returned to the Astronaut Office in 1996, she was certain that her break to look after her family had impacted on her career, commenting later in her Oral History that she would otherwise have probably flown at least one more mission. On her return, she was assigned as Branch Chief of Operations at the Planning Branch until June 1997. She wanted to stay focused on the Shuttle, but with new classes of astronauts coming into the office at regular intervals to train to fly the Shuttle, she realized that "there weren't the people around who remembered what the early days of the Shuttle were like, so I finally gave in. [In June 1997] They made me Chief of the Space Station Branch." **[83]**

In her 2011 interview, she also noted the considerable changes to the office on her return. "By the time I came back, there were very few people left in the office that remembered the beginning of the Shuttle Program, and we were just really at the beginning of the Space Station Program at that point. I think I was able to provide an insight that they might not otherwise have had. Here we were now in the nineties, and the Shuttle was flying; it was a very proficient and experienced team. Our products, our procedures were all very good, but it wasn't like that at the beginning of the Shuttle Program. I think the expectations, particularly of some of the earlier Expedition crews, were a little unrealistic. I think I was able to provide a perspective to try to get products that were good, but to also make the other folks in the Astronaut Office realize that the Shuttle was just like this at the beginning. The simulators at the beginning didn't work, and I can't tell you the number of times I'd go over for an SMS session and it would crash and you'd go back to your office until they fixed it. That hardly ever happens now. I remember when we worked on our procedures for ascent, orbit, and entry. There were three different teams, and you could have [a] procedure for ascent, which is almost the exact same procedure for entry, and they would look totally different because different people wrote them. During the downtime for *Challenger*, that was one of the things we did; we went through the entire Shuttle Flight Data File [FDF] and tried to make it consistent. So, in a way, that downtime was valuable because I think later crews really benefited from that. The expectation at the beginning of Station was, I think, very unrealistic, and so I think I was able to provide a unique perspective. The other thing that happened at the beginning of Station – because everybody wanted to work on Shuttle – [was that] all the new people that were hired were put on working Station, rather than the experienced people. So then it was kind of like a double hit. Not only were you working a new program, which is always difficult, but it was also an international program and we were trying to work with our Russian colleagues as well as the other international partners." **[83]**

Another of the TFNG to be reassigned at this time was Jeff Hoffman, who became Lead in the Payload and Habitability Branch of Astronaut Office.

The Shuttle-Mir program continued in the new year with the fifth docking mission, STS-81 (January 12–22, 1997), which delivered Jerry Linenger to the Russian station and returned John Blaha. None of the active TFNG were listed in support roles for this mission, nor for the next mission on which Steve Hawley flew. This was the second Hubble Servicing Mission, returning the astronaut to the telescope seven years after he deployed it on STS-31.

STS-82 (February 11 – 21, 1997)
Flight Crew: Kenneth D. Bowersox (CDR), Scott J. Horowitz (PLT), Joseph P. Tanner (MS-1), Steven A. HAWLEY (MS-2), Gregory J. Harbaugh (MS-3); Mark C. Lee (MS-4), Steven L. Smith (MS-5)
Spacecraft: Discovery (OV-103) 22nd mission
Objective: 82nd Shuttle mission; 2nd Hubble Service Mission
Duration: 9 days 23 hours 37 minutes 9 seconds
Support Assignments: No listings found

By 1997, the Shuttle-Mir program was rapidly approaching the halfway point and the focus was shifting more to the assembly of the ISS, planned to begin the following year. Many scientific flights had been cut, but what remained were the servicing missions to the HST, and Hawley had been assigned to the second of these flights. His primary role was to operate the Shuttle's 50-ft (15-m) RMS robot arm to retrieve HST and then redeploy it following the completion of upgrades and repairs. He also operated the RMS during the five spacewalks, in which two teams installed two new spectrometers and eight replacement instruments. They also replaced insulation patches over three compartments containing key data processing, electronics and scientific instrument telemetry packages. HST was then redeployed having been boosted to a higher orbit by the Shuttle's OMS engines.

For the next four missions (STS-83, 84, 94 and 85), none of the Class of 1978 astronauts who remained in the Office were assigned to support roles. However, in June 1997, Anna Fisher was assigned to her new role as Branch Chief for Space Station. Then, on July 9, it was announced that Jeff Hoffman would be leaving the Astronaut Office to become NASA's European Representative in Paris, a position he would occupy until August 2001. [84] His responsibilities in Paris were in monitoring the implementation of NASA policies, and the relationships with the European space and aeronautical communities, as well as governmental, industrial and academic institutions.

During the seventh Shuttle-Mir docking mission, STS-86 (September 25-October 6), Steve Hawley fulfilled the role of FCOD Management Support for launch at KSC/OSR. For the landing, he was on call at JSC. [85] The TFNG were not involved in supporting or flying the final mission of the year, STS-87 (November 19–December 5), which carried the USMP-4 and SPARTAN-201-04 payload.

DAWN OF A NEW ERA

The 20th anniversary of the Group 8 selection occurred on January 16, 1998, but with only three members remaining on the active list, little fuss was made of this milestone. This new year would see the completion of the Shuttle-Mir docking missions, the return to orbit of 77-year-old John Glenn after 36 years, and the start of ISS assembly, but sadly it was also another year in which no member of the

Class of 1978 flew in space. Indeed, none of the group was assigned to support roles for the next six missions through to the summer of 1999, but that summer would see the final flight of one of the TFNG.

History is made
On March 5, 1998, in an historic announcement by First Lady Hillary Rodham Clinton in the Roosevelt Room at the White House in Washington D.C., the first female Shuttle CDR, Eileen Collins, was named to lead the STS-93 mission in December of that year. *Columbia* was manifested to deploy the Advanced X-Ray Astrophysical Facility Imaging System (AXAF, later called Chandra). Collins would be joined on the five-day mission by PLT Jeff Ashby (Class of 1994), and MS Catherine Coleman (Class of 1992) together with CNES astronaut Michel Tognini, who had been named to the crew in November 1997. There was one other crewman named, and their assignment was also an historic appointment, as the final member of the TFNG still in active flight training was assigned to a Shuttle mission. Steve Hawley was to be MS-2/FE, a position he had occupied on all five of his missions, and would bring to a close the assignment of the TFNG on prime crews that had begun with the announcement of the STS-7 crew 26 years earlier. [86] It was also fitting that Hawley had been selected with the first female astronauts in the NASA team, (and at one time he was married to Sally Ride, the first American female in space) and was now going to be in the crew led by the first female CDR. How much had changed since 1982.

In June 1998, following a re-organization of the Astronaut Office, Anna Fisher was reassigned as Deputy for Operations/Training, Space Station Branch, where she had oversight responsibility for inputs from the Astronaut Office to the Space Station Program. After a year in that post, Fisher was named Chief of the Space Station Branch in the Astronaut Office in June 1999. She would be responsible for coordinating the input from 40−50 astronauts and support engineers regarding the design, development and testing of station hardware, in addition to continuing to oversee station operations, procedures and training. She worked with the international partners to negotiate common designs for displays, controls and crew procedures, serving on a number of Space Station boards. With these assignments, Fisher continued the TFNG link between the end of the Apollo era, the transition to the Space Shuttle, and through to the ISS.

THE LAST FLIGHT

There had been 12 other Shuttle missions since Hawley had last flown on STS-82. The docking missions to Mir had been completed successfully, and the ISS assembly missions had begun. This was the beginning of a new era, and at the same time the end of the era in which a member of the TFNG would orbit the Earth. With most of his colleagues retired or in managerial positions, Steve Hawley provided the final milestone in the story of the Group 8 selection flying on the Space Shuttle.

He became the final member of the Class of 1978 to fly in space, 21 years after the group had arrived at JSC to begin Ascan training. Fittingly, this was also aboard *Columbia* which, as a group, they had worked so hard to support on the final push to put that vehicle into orbit back in 1981. With the last of the TFNG making their final flight under the command of the first female CDR, the Astronaut Office had come a long way since the first females and ethnic minorities arrived as part of the TFNG over 20 years earlier. The future now belonged to a new generation of space explorers who would assemble and crew the ISS.

STS-93 (July 22 – 27, 1999)

Flight Crew: Eileen M. Collins (CDR), Jeffrey S. Ashby (PLT), Catherine G. Coleman (MS-1), Steven A. HAWLEY (MS-2), Michel A. Tognini (MS-3, CNES, France)
Spacecraft: Columbia (OV-102) 26th mission
Objective: 95th Shuttle mission; Chandra X-ray Observatory deployment
Duration: 4 days 2 hours 49 minutes 37 seconds
Support Assignments: The final TFNG support assignment on a flight by one of the group went to Shannon LUCID, on 24-hour call (08:00−08:00) in SPAN on FD3

On this, his final flight, Steve Hawley again served as *Columbia*'s FE. The primary mission objective was the successful deployment of the Chandra X-ray Observatory, the third of NASA's Great Observatories after HST and the Compton Gamma Ray Observatory. He also served as the primary operator of a second telescope carried in the crew module, which was used for several days to make broadband ultraviolet observations of a variety of solar system objects. Dr. Hawley became the only astronaut to be involved in the deployment of two of NASA's great observatories (Hubble and Chandra) which, for a professional astronomer, was additionally rewarding.

Steve Hawley flew on the Shuttle five times and on each flight he served as MS-2/FE from on Seat 4 on the flight deck during launch and landing. When asked why he thought he had been assigned to the role five times he put it down to experience, and needing less in-depth training each time he was assigned because of the skills gained over many years.

"One of the tricks to being a good MS-2, I thought, was firstly you had to know your stuff, but you also had go know how to work with the pilot and the commander. Sometimes a commander may want to hear from you more often, another commander less. Sometimes, we had a commander who would only say something if it was really important, other times people wanted to know [everything that] was going on." It was not only a case of getting to know the personalities of the commander and the pilot, but also "how fast can the three of you work together to be as efficient as possible, so it's just a question of finding the right mix of tasks, in my view, whatever the commander wants." Hawley also thought that the skill of becoming a good MS-2 was all about finding a suitable balance between the requests of the pilots and the skills of the flight engineer. [26]

Fig. 12.6: The end of an era. Steve Hawley, the last of the TFNG to fly in space, stands under the nose of *Columbia* with the rest of the crew at the end of their mission. [l-r] MS Cady Coleman, TFNG MS Steve Hawley, PLT Jeff Ashby, CDR Eileen Collins and CNES French MS Michel Tognini.

THE FINAL YEARS IN THE OFFICE

As the new millennium dawned, so the era of the TFNG waned. In 2000, there were still a few flights on the manifest which were not related to assembling and supplying the space station, mainly the Hubble Servicing Missions and a few science-based missions. There was talk of retiring *Columbia*, of returning Hubble to be put on display in a museum, and of a new vehicle to replace the Orbiter fleet, but for now the Space Shuttle remained the only American vehicle to put astronauts into space. Through agreements with the Russians, space station crewmembers would rotate (in teams of three) on either Shuttle assembly or Soyuz ferry missions. The Soyuz would act as a stand-by crew rescue vehicle while the station was being built and each would be exchanged by short, visiting missions at the end of its orbital life expectancy. With Mir being decommissioned, to some regret within Russia, by early 2001 the sole operations in Earth orbit for the Americans, the Russians and their international partners in the first decade of the new millennium centered upon bringing the ISS up to full habitation standard. Waiting in the wings, however, was a new player: China. By 2001, the permanent occupancy of the ISS had begun and most of the TFNG had departed. Less than two years later, the completion of the ISS was paramount, as was deciding the future direction of U.S. human space flight following the recovery from another tragedy, the loss of *Columbia* and her crew of seven which signaled the eventual end of the Shuttle program. The first few years of the new century had become as frustrating, difficult and dark as the middle of the 1980s had been, but early in the second decade of the new millennium the station had been completed, the Shuttle retired and new goals were put forward, though a qualified replacement for the Space Shuttle remained a long way off.

But what of the remaining TFNG still with NASA at the millennium, over 20 years after their selection? Of the 35 chosen, only five were employed by NASA (although another would soon return to the agency for a few years), but none remained active for selection to a flight crew. They were all in various managerial roles, working their last assignments before finally departing the agency and leaving it to new generations of managers and astronauts.

The now-veteran 'New Guys' remaining were:

Mike Coats: After leaving NASA in 1991, Mike Coats worked in the private sector until he became the tenth Director of JSC in November 2005, a position he held until December 31, 2012, when he once again retired from NASA, just over a year after the retirement of the Space Shuttle fleet (see Chapter 13).

Anna Fisher: From January 2011 until August 2017, Fisher served as an ISS Capcom in Mission Control and became Lead Capcom for Expedition 33. In 2011, she explained her preparations for the role and how different it was from being a Capcom on an early Shuttle flight (STS-9) almost two decades earlier: "When I was a Capcom for STS-9, you just went over there and you sat with another astronaut who was there before you, and when they thought you were ready, you were ready. There was no training flow. Being a Capcom was training. That was because you didn't get in a flow until you were a crew. So getting to be a Capcom or working at SAIL [Shuttle Avionics Integration Laboratory] or any of those things was your training. Now it's the other way around. Now you have to have all this other training before you can do that. It's just kind of interesting." Towards the end of the Shuttle program she also worked Shuttle FDF and training issues, "which obviously is trailing down, and I just didn't figure it was worth giving that to someone else. I know all the people. I've been working with them for years and years, and it doesn't take that much of my time to follow that." [**83**] From 2013 until she left NASA in 2017, in addition to serving as ISS Capcom, Fisher held a concurrent assignment as a Management Astronaut working on crew display development for Orion, NASA's replacement spacecraft for the Space Shuttle that is intended to take astronauts deeper into the solar system.

Fred Gregory: After leaving the Astronaut Office in May 1992, Fred Gregory had relocated to Washington D.C. where he served in a number of senior management positions at NASA Headquarters (see Chapter 13) until his retirement from the agency in November 2005.

Steve Hawley: Following STS-93 in 1999, there were the usual debriefings and press conferences but there were no large formal post-flight activities. Shortly after completing his commitments, Steve Hawley returned to his role as Deputy Director at the Flight Crew Operations Directorate at JSC until he was asked by Roy S. Estess (JSC Acting Director February 2001–March 2002) to become Director of FCOD. From October 2001 to November 2002, Hawley served as Director, Flight Crew Operations, and from 2003 to 2004 he served as the first Chief Astronaut for the NASA Engineering and Safety Center. Up to late 2002, he was therefore still *technically* available for crew assignment, though he stated when he was Director FCOD that "I wouldn't have assigned myself…[though] I would have enjoyed going to the [ISS], I don't know that I would have wanted to take on the burden of all the training that it would have taken to have been a long duration crewmember… [perhaps on an assembly crew], that would have been more fun, I didn't really want to go to Russia to train." [26]. He then became Director of Astromaterials Research and Exploration Science Directorate between 2002 and 2008. In this role, he was responsible for directing a scientific organization conducting research in planetary and space science. The primary functions of the organization included astromaterials acquisition and curation, astromaterials research, and human exploration science. On February 27, 2008, it was announced that Steve Hawley had accepted an appointment with the faculty of the University of Kansas and would leave NASA that May.

Table 12.2: SHANNON LUCID MCC HOUSTON CAPCOM ASSIGNMENTS 2005–2012

Dates	STS Flt	MCC 'flight' name
Jul 26–Aug 9, 2005	114	Topaz
Dec 9–Dec 22, 2006	116	Pegasus
Aug 8–Aug 21, 2007	118[a]	Apex/Iron
Oct 23–Nov 7, 2007	120[a]	Kodiak/Intrepid
Feb 7–Feb 20, 2008	122	Iron
May 31–Jun 14, 2008	124[a]	Iron/Intrepid
Nov 14–Nov 30, 2008	126[a]	Iron/Defiant
Mar 15–Mar 28, 2009	119[a]	Amethyst/Onyx
Jul 15–Jul 31, 2009	127	Kodiak
Aug 28–Sep 12, 2009	128	Viper
Feb 8–Feb 21, 2010	130	Venture
May 14–May 26, 2010	132	Vega
May 16–Jun 1, 2011	134[a]	Pegasus/Defiant
Jul 8–Jul 21, 2011	135[a]	Iron/Intrepid

[a]During the STS-118, 120, 124, 126, 119, 134 and 135 missions, the 'Flight' position during Planning Shifts was split between different Flight Directors, hence the two Flight names

Fig. 12.7: (top) Between 2005 and 2011, Shannon Lucid served as shift Capcom on the Planning team for at least 14 Shuttle missions. Here, she is seen on the console for STS-126 in 2008. (bottom) Lucid on duty as Capcom during STS-135 in July 2011, ending an era begun 30 years earlier by Dan Brandenstein as launch Capcom for STS-1 in April 1981.

Shannon Lucid: On February 12, 2002, Shannon Lucid was named as NASA's Chief Scientist, replacing Dr. Kathie Olsen. In this role, Lucid would be responsible for the scientific merit of the agency's programs. She would report for that duty after completing her responsibilities in March 2002 as a Shift Capcom for STS-109 (Hubble Service Mission 3B). NASA Administrator Sean O'Keefe, who announced the appointment, said "The Chief Scientist has tremendous responsibility to develop and communicate the agency's science and research objectives to the outside world. What better selection than a NASA scientist and astronaut with extensive experience [of] living and working in the harsh environment of space." [**87**]

Following that assignment, Lucid returned to JSC in the fall of 2003 to resume technical assignments in the Astronaut Office, serving as Capcom in the Mission Control Center (MCC) for numerous Shuttle and station crews. Representing the flight crew office, she provided a familiar (and experienced) voice for dozens of her friends and colleagues in space. Starting with the Return-to-Flight Mission (STS-114) in July 2005, Lucid was at console in MCC as a shift Capcom (Planning Shift) over the next seven years (see Table 12.2). Her record of 223 days for the most days in space logged in a career by a female, set in September 1996, was surpassed in June 2007 by Sunita Williams, but Lucid's position as a pioneer in female space endurance and experience remains. On January 31, 2012, six months after the final Shuttle mission (STS-135), NASA announced that Shannon Lucid had retired from NASA after more than three decades of service to the agency. [**88**]

Loren Shriver: In June 1993, Loren Shriver became Space Shuttle Program Manager, Launch Integration, at KSC. In this position, he was responsible for final Shuttle preparation, mission execution, and the return of the Orbiter to KSC following landings at Edwards AFB in California. From 1997, he served as the Deputy Director for Launch and Payload Processing at KSC until he retired from NASA in 2000.

The last of the TFNG leaves NASA

On April 28, 2017, NASA announced that Anna Fisher had retired after nearly four decades of service to NASA. [**89**] "Flying the Shuttle was so much fun, but, boy, you work. You work very hard. Every minute you're conscious you've got a job to do and you've got to perform it, so it's not like going on vacation and kicking back and relaxing. It would be so much fun to go [back] into space and be able to do that too. I would love it…to be able to just go into space to just have fun and not have to work." [**83**]. Fisher had remained with NASA for almost 40 years but in that time flew in space just once for one week.

Since their selection on January 16, 1978, and working for the agency as an active group until Fisher departed, they had represented NASA at several field centers (JSC, KSC, HQ, Ames) for 39 years, 3 months, 12 days. Of course, their time spent actually flying in space, even including Thagard and Lucid on Mir, was just a small percentage of that total, as Steve Hawley pointed out. "Flying [in space] was almost incidental, I was there for 30 years and I flew [five missions] cumulative for about a month." [**26**].

SUMMARY

Across 30 years and 135 missions, members of NASA's Group 8 supported most of the missions either on the ground, as Capcom at Mission Control, on the flight crews, or in managerial roles. A remarkable record. They supported the program from before STS-1 and one of them was at MCC working the Capcom console for the final flight.

In April 1981, Dan Brandenstein had been launch Capcom on the very first Shuttle mission. Then Rick Hauck, John Fabian, Sally Ride, and Norman Thagard became the first members of their group to fly in space in June 1983. Two years later, all 35 had flown at least one mission into space and many would get the chance to do so again, some several times. Over the years there was inevitable attrition, and sadly losses too, and though Steve Hawley flew into space in July 1999, the final member of the group to do so, their work continued in new roles in the space program, either within or outside of NASA. It was fitting that in July 2011, one of the final Capcom shifts on the last flight day of STS-135, which ended the program, was TFNG Shannon Lucid, prior to her retirement six months later. She was followed by JSC Director Mike Coats at the end of 2012. The closure of the story of the TFNG within NASA occurred in April 2017, when Anna Fisher, one of the first six women astronauts in the Office, became the 35th and last member of the group to leave the agency.

It was over.

References

1. FCOD Mission Support Assignments memo, March 15, 1991.
2. Undated FCOD Mission Support Assignments memo.
3. FCOD Mission Support Assignments memo May 28, 1991.
4. *U.S. and USSR Expand Space Cooperation*, NASA HQ News release 91-122.
5. *Astronaut Coats to Leave NASA*, JSC News Release 91-058, July 3, 1991.
6. FCOD Mission Support Assignments memo, June 28, 1991.
7. *NASA Announces Crew Members for Future Shuttle Flights*, NASA JSC News, 91-069, August 23, 1991.
8. FCOD Mission Support Assignments memo August 14, 1991.
9. *Seddon Named Payload Commander for SLS-2*, NASA News JSC 91-077, October 23, 1991.
10. Undated FCOD Mission Support Assignments memo.
11. *Payload Crew Named for Spacelab Life Sciences-2 Mission*, NASA JSC News 91-088, December 6, 1991.
12. *Flight Control of STS-42*, NASA News 92-003, January 9, 1992.
13. FCOD Mission Support Assignments memo, December 19, 1991.
14. *Crew Assignments Announced for Future Shuttle Missions*, NASA JSC News 92-009, February 21, 1992.
15. *Flight Control of STS-45*, NASA News 92-10, March 4, 1992.
16. FCOD Mission Support Assignments memo, February 25, 1992.

17. FCOD Mission Support Assignments memo, April 29, 1992.
18. *Astronaut Buchli to Retire and Leave NASA,* NASA JSC News 92-29, May 29, 1992.
19. FCOD Mission Support Assignments memo, June 3, 1992.
20. *Flight Control of STS-50*, NASA News 92-034, June 22, 1992.
21. The White House Office of the Press Secretary press release, June 25; *Astronaut Sullivan to Become Chief Scientist at NOAA*, NASA JSC News 92-046, August 14, 1992.
22. *Astronaut Creighton to Retire and Leave NASA*, NASA JSC News 92-035, June 25, 1992.
23. *Chief Astronaut to Retire from Navy and Leave NASA*, JSC News Release 92-038, July 1, 1992.
24. FCOD Mission Support Assignments memo, July 6, 1992.
25. *Hawley to return to Johnson Space Center*, NASA News 92-037, June 30, 1992.
26. AIS interview with Steve Hawley, October 23, 2019.
27. *Crew Assignments Announced for STS-58 and STS-61*, NASA News Release 92-047, August 27, 1992.
28. FCOD Mission Support Assignments memo, August 24, 1992.
29. FCOD Mission Support Assignments memo, September 24, 1992.
30. FCOD Mission Support Assignments memo, November 9, 1992.
31. Guion Bluford, NASA Oral History, August 2, 2004.
32. FCOD Mission Support Assignments memo, December 16, 1992.
33. FCOD Mission Support Assignments memo, March 18, 1993.
34. FCOD Mission Support Assignments memo, February 3, 1993.
35. *Astronaut Seddon Injured During Training*, NASA JSC News 93-032, May 4, 1993.
36. *Peterson Announces Shuttle Program Adjustments*, NASA HQ News, March 19, 1993.
37. *Astronaut Bluford Leaves NASA*, NASA JSC News 93-045, June 15, 1993.
38. FCOD Mission Support Assignments memo, June 24, 1993.
39. Undated FCOD Mission Support Assignments memo.
40. Undated FCOD Mission Support Assignments memo.
41. Undated FCOD Mission Support Assignments memo.
42. *Astronauts Thagard and Dunbar to Train for Flight on Mir*, NASA JSC News, 94-016, February 3, 1994.
43. Norman Thagard, NASA ISS Phase 1 Oral History Project, September 16, 1998.
44. **Russia's Cosmonauts, Inside the Yuri Gagarin Cosmonaut Training Center**, Rex D. Hall, David J. Shayler and Bert Vis, Springer-Praxis, 2005, pp. 274–292.
45. FCOD Mission Support Assignments memo, February 25, 1994.
46. Undated FCOD Mission Support Assignments memo.
47. *Crew Named for First Space Shuttle-Mir Docking Mission*, NASA JSC News, 94-039, June 3, 1994.
48. *Crew Named for Second Wake Shield Facility Shuttle Flight*, NASA News Release JSC, 94-045, July 7, 1994.
49. *Shuttle Astronaut Richard Covey to Leave NASA, Air Force*, NASA News 94-044, July 6, 1994.
50. FCOD Mission Support Assignments memo, August 31, 1994.
51. Undated FCOD Mission Support Assignments memo.
52. FCOD Mission Support Assignments memo, October 28, 1994.
53. *Astronauts Blaha and Lucid to Train for Flight on Mir*, NASA JSC News 94-073, November 3, 1994.
54. Shannon Lucid, NASA ISS Phase 1 Oral History Project, June 17, 1998.
55. *Space Shuttle Crew Selected for Tethered Satellite Mission*, NASA News JSC 95-006, January 27, 1995.

56. Undated Astronaut Office Mission Support Assignments memo.
57. *Nagel Moves from Astronaut Office to SR&QA*, NASA News, 95-01, March 3, 1995.
58. Undated Astronaut Office Mission Support Assignments memo.
59. *Lucid Prime for Second Mir Stay, Linenger Selected for Third*, NASA JSC News, 95-024, March 30, 1995.
60. *Crews Selected for Third, Fourth Shuttle-Mir Docking Missions*,NASA HQ News 95-50, April 14, 1995.
61. FCOD Mission Support Assignments memo, June 13, 1995.
62. FCOD Mission Support Assignments memo, dated August 23, 1995.
63. Acknowledgement to Ken Havekotte of Space Coast Cover Service, Merrit Island, Florida, for the background to the "Dogs of War" and Dog II" stories.
64. FCOD Mission Support Assignments memo, September 11, 1995.
65. *Astronaut Thagard Leaves NASA; Returns to Alma Mater*, NASA HQ News 96-4, January 16, 1996.
66. FCOD Mission Support Assignments memo, December 4, 1995.
67. *NASA and RSA Agree to Extend Shuttle-Mir Activities*, NASA HQ News, 96-18, January 30, 1996.
68. FCOD Mission Support Assignments memo, February 6, 1996.
69. *NASA Assigns Hawley to Second Hubble Servicing Mission*, NASA HQ News 96-41, February 29, 1996.
70. FCOD Mission Support Assignments memo, February 26, 1996; FCOD Mission Support Assignments March 7, 1996.
71. *Astronauts Walker, Harris* [Bernard Harris, Class of 1990] *to Leave NASA Astronaut Corps*, NASA HQ News, 96-070, April 12, 1996.
72. *NASA Selects Astronaut Class of 1996*, NASA HQ News 96-84, May 1, 1996.
73. *International Candidates Join 1996 Astronaut Class*, NASA HQ News 96-162, August 12, 1996.
74. FCOD Mission Support Assignments memo, April 18, 1996.
75. *Seddon to Support Life Science Investigations at Vanderbilt*, NASA HQ News 96-185, September 10, 1996.
76. FCOD Mission Support Assignments memo, June 27, 1996.
77. **Shannon Lucid Space Ambassador,** Carmen Bredeson, Gateway Biographies, The Millbrook Press, Brookfield, Connecticut, 1998, p. 28.
78. Reference 77, p. 37.
79. *Veteran Shuttle Commander Retires*, NASA News 96-229, November 6, 1996.
80. STS-80 Astronaut Office Mission Support Assignments, October 2, 1996.
81. FCOD STS-80 Mission Support Assignments, October 25, 1996.
82. Email from Hoot Gibson to David J. Shayler, January 2020
83. Anna Fisher, NASA Oral History May 3, 2011.
84. *Jeff Hoffman Retires from Astronaut Corps*, NASA News 97-151, July 9, 1997.
85. FCOD Mission Support Assignments memo, September 15, 1997.
86. *Collins Named First Female Shuttle Commander*, NASA News HQ H98-37, March 5, 1998.
87. *NASA Astronaut Dr. Shannon Lucid Selected as Chief Scientist*, NASA News HQ H02-27, February 12, 2002.
88. *Legendary Astronaut Shannon Lucid Retires from NASA*, NASA News 12-038, January 31, 2012.
89. *Legendary Astronaut Retires from NASA*, NASA JSC Release J17-005, April 28, 2017.

13

Flying a desk

*"When I landed [on STS-44], within a month
I had decided that I had flown enough,
I didn't need to do that anymore,
and that I was going to have to move into
something that was more normal."*
Fred Gregory, NASA JSC Oral History

At the time of publication of this book, the 35-strong contingent of NASA's Group 8 astronauts had diminished by ten, most notably and sadly through the loss of crewmembers Dick Scobee, Ellison Onizuka, Ron McNair and Judy Resnik in the Shuttle *Challenger* tragedy of January 1986. The following are brief encapsulations of the space flights of the Thirty-Five New Guys (TFNG), and their lives subsequent to retiring from the active flight status.

HANGING UP THE SPACE SUIT

As difficult as it is to be selected as an astronaut, let alone achieve that first flight into space, it is equally difficult to give up the opportunity to experience the thrill of lift off, to witness first-hand the wonders of the universe and to explore in person the unique sensations of microgravity. For some, it was a challenge to find a new and rewarding career, having finally decided to forgo the hundreds of hours of training, the risks during each second of a mission, and putting their families and loved ones through stress and worry from the moment of lift off to final recovery (see sidebar: *All in the Family*). Not all space explorers found the transition from astronaut to more Earthbound occupations so difficult of course, having been fulfilled by what they had achieved in their NASA careers. Right from the early

© Springer Nature Switzerland AG 2020
D. J. Shayler, C. Burgess, *NASA's First Space Shuttle Astronaut Selection*,
Springer Praxis Books, https://doi.org/10.1007/978-3-030-45742-6_13

days of space flight, finding fulfillment in life after hanging up the spacesuit was another new experience, and one that NASA could not train their astronauts for, especially those who had participated in a trip to the Moon. After all, what could top the experience of standing on the dusty surface of another world and looking back at our own planet a quarter of a million miles away?

By the end of the 1980s, many of the original 73 astronauts chosen in NASA's first decade of recruitment had retired from the agency to seek new challenges and adventures away from the launch pad. When the time came for members of the first Shuttle era selection to decide to move on, the program and its opportunities had also evolved. As the first of the new generation of Space Shuttle astronauts, they helped to develop the roles of Shuttle Pilot (PLT) and Mission Specialist (MS) that continued with such success for the duration of the program. When they finally departed the Astronaut Office, they broke new ground again by taking on a variety of managerial roles within NASA, the wider space industry, or in other fields. With most aged in their mid-30s at selection, many were still young enough to forge a new career, this time well away from the media spotlight, before quietly slipping into a well-earned retirement with the families who had supported their careers for so many years.

ALL IN THE FAMILY

Being selected for space flight training, irrespective of the country or agency, means being put into the media spotlight at some point, which for most is not a natural or comfortable situation. From the obscurity of a 'normal' past, the space explorer is thrust into the unfamiliar world of the media and public attention. Some may be well prepared for the limelight, while others would prefer to remain in the background and simply get on with the job, even though they are fully aware that is no longer possible.

The focus on the first American spacefarers, who were all men, brought a level of attention that put a strain on their hopes for a 'normal' home life. The families of those astronauts also found themselves in the media spotlight, which was just as disconcerting for those used to the quieter and somewhat secluded military lifestyle. Over time, some marriages broke under the strain, with the pressures not only on the individual astronauts as they trained for and flew their missions, but also for their wives and children at home. The stories of those family upheavals, though mostly kept private, have recently been told in print and on film, in a far more honest and open way than they were originally portrayed in the 1960s, when the general reply given to the press was that the family back home was 'proud', 'thrilled' and 'happy'.

(continued)

For most of the astronauts during the 1960s and 1970s, the support of their family at home was crucial to the success of their achievements, and it continues to be so. It can be said that the most important back-up crew a space explorer can have are the "folks back home." When time came for the selection of a new group of astronauts in 1978, their families played an important role in supporting their efforts to get through the selection, complete the Ascan program, and prepare to fly. Alongside the success and fame came the setbacks and tragedies, and these, too, had to be absorbed by the families. Towards the end of the TFNG era, the upcoming long duration missions on the International Space Station (ISS), and the training required for them that took place around the globe, added to the potential stresses on home life. This became a key factor for many astronauts in the late 1990s, in deciding whether to commit to the long duration training, opt for shorter missions, or even find a new career path. Any story about the commitment and career of a space explorer should also include recognition for those who remain at home and are still there for them when they no longer don the spacesuit.

Briefly, the TFNG extended 'families' are detailed below, as of early 2020:

Guion Bluford has been married to the former Linda Tull of Philadelphia, Pennsylvania since April 7, 1964. They have two children; *Dan Brandenstein* married the former Jane A. Wade of Balsam Lake, Wisconsin, on January 2, 1966, and they have a daughter; *Jim Buchli* is married to the former Sandra Jean Oliver of Pensacola, Florida and the couple have two children; *Mike Coats* is married to the former Diane Eileen Carson of Oklahoma City, Oklahoma, and they have two children; *Dick Covey* is married to the former Kathleen Allbaugh of Emmetsburg, Iowa, and the couple have two daughters; *John Creighton* was single at selection, but subsequently married the former Terry Stanford of Little Rock, Arkansas; *John Fabian* married the former Donna Kay Buboltz of Spokane, Washington, on September 18, 1961, and together they have two children.

Anna Fisher (Tingle) was divorced from her first husband but was still using her former married name of Sims when progressing through the early stages of astronaut selection. On August 23, 1977 she married Dr. William ('Bill') Fisher, an unsuccessful candidate in 1978 who was later chosen as one of 19 Group 9 astronauts in 1980. He went on to fly one mission as an MS, STS-51I in 1985, and led a critical study of EVA tasks assigned to Space Station *Freedom* before leaving the space agency in 1992 to return to his medical career. In January 1989, the Fishers

(continued)

celebrated the birth of their second child, "There's a very small number of people whose parents have both been in space," Anna once stated when referring to their new baby daughter. She then took a rare leave of absence to raise her family, returning in January 1996. Drs. Bill and Anna Fisher were divorced in 2000.

Dale Gardner is survived by his wife Sherry and two step children. He was also the father of two children, (one of whom, his son, pre-deceased him), from his first marriage on February 19, 1977 to Sue Ticusan of Indianapolis, Indiana; *Hoot Gibson*, divorced his first wife, the former Cathy Marie Von Epps of Santa Barbara, California, shortly after selection to the astronaut program. On May 30, 1981, he married fellow TFNG astronaut Dr. M. Rhea Seddon (the first serving astronauts to marry each other) and together they have four children; *Fred Gregory* was married to the former Barbara Ann Archer of Washington D.C. for 44 years until her death in May 2008. They had two children. He married a second time to the former Annette Becke and between them they have three children; *Dave Griggs* was survived by his wife, the former Karen Frances Kreeb of Lake Ronkonkoma, Long Island, New York, and their two children; *Terry Hart* married the former Wendy Eberhardt on December 20, 1975, and they went on to have two children. The couple separated in October 1996, and were divorced on May 12, 1999; *Rick Hauck* married the former Dolly Bowman of Washington D.C. on August 27, 1962, and they had two children. They subsequently divorced and Hauck is now married to the former Susan Cameron Bruce. Together they have five children.

Steve Hawley was single at the time of selection to the astronaut program but was married to fellow Group 8 astronaut Sally Ride from July 26, 1982 until their divorce in May 1987. He subsequently married the former Eileen M. Keegan of Redondo Beach, California, a former public-affairs officer at NASA JSC who has served as the spokeswoman for Kansas Governor Sam Brownback since 2013; *Jeff Hoffman* is married to the former Barbara Catherine Attridge of Greenwich, London, England, and they have two children; *Shannon Lucid* has been married to Michael F. Lucid of Indianapolis, Indiana since 1967, and they have three children; *Jon McBride* divorced his first wife, the former Brenda Lou Stewart of Mobile, Alabama, with whom he had three children. He is now married to the former Sharon Lynne White of Nacogdoches, Texas, and together they have four children.

Ron McNair married Cheryl Moore of Jamaica, New York, on June 27, 1976 and became the father of two children; *Mike Mullane* married Donna Marie Sei of Albuquerque, New Mexico a week after his graduation from

(continued)

West Point in 1967. The couple have three children; *Steve Nagel* is survived by his wife, former astronaut Linda Godwin, and two children from his previous marriage to the former Linda Diane Penney of Los Angeles, California; *George 'Pinky' Nelson* married the former Susan (Susie) Howard of Alhambra, California, on June 19, 1971 and together they have two children; *Ellison Onizuka* is survived by his wife, the former Lorna Leiko Yoshida of Pahala, Hawaii, whom he married on June 7, 1969, and their two children; *Judy Resnik* had married Michael D. Oldak on July 14, 1970, but the couple were divorced in December 1976. She remained single up to the time of her death in 1986; *Sally Ride* was married to astronaut Steve Hawley from 1982 until 1987, and is survived by Tam O'Shaughnessy, her partner of 27 years; *Dick Scobee* is survived by his wife, the former (Virginia) June Kent of San Antonio, Texas, and their two children. His son is a rank officer (Lt-General) in the USAF. June Scobee, a NASA consultant and educator, was the founding Chairperson of the Challenger Center for Space Science, created in 1986 by the families of the *Challenger* astronauts. Today, there are 39 centers in the United States as well as international branches in Canada, South Korea and the United Kingdom, continuing the education legacy of the *Challenger* crew; *Rhea Seddon* is married to fellow Group 8 astronaut Robert (Hoot) Gibson; *Brewster Shaw*, a descendant of William Brewster of the *Mayflower*, was married to the former Kathleen Ann Mueller of Madison, Wisconsin, on May 24, 1969 and is the father of three children. Tragically their younger son was murdered, aged just 20, by carjackers in Austin, Texas, in July 1997.

Loren Shriver married the former Susan Diane Hane of Paton, Iowa. They have four children; *Bob Stewart* has been married to the former Mary Jane Murphy of La Grange, Georgia, since December 26, 1963, and they have two children; *Kathy Sullivan* remains single; *Norm Thagard* is married to Rex Kirby Johnson, formerly of Atlanta, Georgia. They have three children; *Ox van Hoften* married the former Vallarie Davis of Pasadena, California, on May 31, 1975, and they have three children: *Dave Walker* was survived by his wife Paige and two children from previous marriages to the former Patricia A. Shea of Washington D.C. and the former Stacy Randal Hall of Kilgore, Texas; *Don Williams* left behind his wife of just four years, Ann Small-Williams. They had married in 2012. He was also survived by his first wife, the former Linda Jo Grubaugh, (who he had married on September 4, 1965), two children of that marriage and their respective families.

Fig. 13.1: The support of family throughout their careers, selection, training, space flights, tragedies and retirements is always an important element in being a space explorer. (top left) Astronaut Guion Bluford with his wife Linda and two sons at home. The photo was taken in 1980 by NASA photographer Terry Slezak for a magazine article. (top right) Shuttle astronauts Rhea Seddon and Hoot Gibson appear on ABC's *Good Morning America* on August 18, 1982, together with their baby son who was born on July 26, 1982. (bottom left) Anna and William Fisher and the new addition to their family, in 1983. (bottom right) Dave Griggs enjoying a family BBQ at home in 1978.

New opportunities

By the mid-1980s, the space industry had undergone a far greater expansion than in the previous decade. Leaving the Astronaut Office presented a number of choices to the retiring TFNG. Firstly, there was the option to remain with NASA, either at the Johnson Space Center (JSC) in Houston, Texas, in an administrative role, or at one of the other field centers. The military personnel had the option to retire from service to become a civilian employee at the space agency, or to resume their military career path and finish their term of service before finally hanging up their uniform. In addition, there were opportunities in the growing space industry as an executive or as a consultant, in the academic world as a professor in a university, or perhaps in a political career in their home state.

Another avenue afforded the TFNG, and one which they helped to expand, was in senior managerial positions, either while still formally detailed to the Astronaut

Office or after leaving it. A few astronauts from the earlier selections had been able to fulfill similar roles, but not in the numbers available to the TFNG. This transition from flying in space to 'flying a desk' has evolved into the 'Management Astronaut' role, where an individual is still classed as an 'astronaut' at NASA, but is out of active flight training or major support roles and works in senior administration at JSC, Headquarters, or other NASA field centers.

When the members of the earlier selections left the program, the opportunities they were offered within the space industry or the emerging commercial space program were far fewer, often based upon their experience and achievements, and the kudos of having a 'Former NASA Astronaut' on the books. For a very few, their career after NASA was more lucrative than for some of their colleagues, but most eventually found their place in the world after space flight.

Despite eventually leaving the Astronaut Office, and in time NASA itself, an astronaut's link with the agency never completely ceases, as all former astronauts are recalled to JSC at least once a year for their annual physical as they age, adding to the growing database of medical information on the effects of space flight.

Reaching out to the public

Recent years have seen NASA increase its public outreach activities, most notably on the various social media platforms. In the *Astronaut Encounter: Meet an Astronaut* program at the Kennedy Space Center (KSC) in Florida, the public have the opportunity to meet veteran NASA astronauts and Payload Specialists (PS) at a live presentation, followed by a question and answer session and a photo opportunity. Since leaving NASA, Group 8 astronauts have participated both in this program and on the international speaker circuit as astronaut guest-speakers at various functions, sharing their experiences with the audience, hoping to inspire a new generation to continue on the journey they embarked upon as a group back in 1978.

This dedication by all space explorers, not just the members of the TFNG, to inspire others was demonstrated on January 10, 2020 by Jon McBride, who retired after 20 years of service as a 'space ambassador' overseeing the Astronaut Encounter team at the KSC Visitor Complex in Florida, over 30 years after leaving the Astronaut Office to pursue a career in business. During his time at the Visitor Complex, McBride had been instrumental in creating the *Dine with an Astronaut* program and the subsequent *Fly with an Astronaut* program. On leaving the position, McBride stated that he had been inspired by the early astronauts such as Neil Armstrong who preceded him, and that his hope was that he had "played a role in inspiring the next generation of space explorers. It's been a joyous journey."

The decision to enter this new phase in their lives had to be made by each of the surviving Group 8 astronauts, as they finally elected to "hang up their space suit" and seek new goals. To paraphrase the famous TV space series that was linked to the selection drive to recruit them in the 1970s, it was time "to boldly go where no TFNG had gone before."

WHERE ARE THEY NOW?

Six years after his selection to the NASA astronaut program, Terry Hart became the first of the 1978 group to step down from active flight status and leave the program, in June 1984. Of his remaining 34 colleagues, four (Ron McNair, Ellison Onizuka, Judy Resnik, and Dick Scobee) were lost in the 1986 *Challenger* tragedy, and a fifth (Dave Griggs) was killed in a flying accident in 1989. Over the subsequent 28 years after Hart departed, each of the remaining 29 members of the 1978 selection followed him out of the Astronaut Office for pastures new (see Table 13.1). Below, we summarize the attainments of each after stepping down from the active flight list.

Table 13.1: NASA GROUP 8 ATTRITION FROM THE ASTRONAUT OFFICE (CB) 1978–2012

Year	Leave CB	Deceased	Names	Still Active	Pilots	Mission Specialists
1978	Group 8 Ascans enter Astronaut Office at JSC			35	15	20
1984	-1		Hart	34	15	19
1986	-4		Fabian, Gardner, Stewart, van Hoften	26	14	12
		-4	McNair, Onizuka, Scobee,			
			Resnik			
1987	-1		Ride	25	14	11
1989	-4		Hauck, McBride, Nelson, Shaw	20	10	10
		-1	Griggs			
1990	-3		Hawley [1st time], Mullane, Williams	17	9	8
1991	-1		Coats	16	8	8
1992	-4		Brandenstein , Buchli, Gregory ,	12	5	7
			Creighton			
1993	-3		Bluford, Shriver, Sullivan,	9	4	5
1994	-1		Covey	8	3	5
1995	-3		Nagel , Seddon, Walker	5	1	4
1996	-2		Gibson, Thagard	4	0	4
	[+1]		[Hawley returns]			
1997	-1		Hoffman	3	0	3
1999	-1		Hawley [2nd time]	2	0	2
2006	-1		Fisher	1	0	1
2012	-1		Lucid	0	0	0

Guion (Guy) Bluford, Jr.
MS STS-8 (1983); MS STS-61A (1985); MS STS-39 (1991); MS STS-53 (1992).

As a crewmember of STS-8, Guion Bluford had the honor of being the first African-American to fly in space. By the completion of his fourth mission, he had logged over 688 hours in space. After flying STS-53, Bluford requested a break from flight assignments to allow him time to decide his future, either within NASA or outside of it.

Col. Bluford left NASA in July 1993 and retired from the U.S. Air Force (USAF) to take on the post of Vice President/General Manager, Engineering Services Division, NYMA Inc., Greenbelt, Maryland, and as Program Manager of the SETAR (Scientific, Engineering, Technical and Administrative Related service) contract at NASA's Langley Research Center. In May 1997, he became Vice President of the Aerospace Sector of Federal Data Corporation and in October 2000, he was appointed Vice President of Microgravity R&D and Operations for the Northrop Grumman Corporation. In September 2002, he became President of the Aerospace Technology Group, an engineering consulting organization based in Cleveland, Ohio. In 2003, Bluford was asked by Admiral Harold W. 'Hal' Geham to support the work of the Columbia Accident Investigation Board (CAIB). Col. Bluford was inducted into the International Space Hall of Fame in 1997, and into the U.S. Astronaut Hall of Fame in 2010.

Daniel Brandenstein
PLT STS-8 (1983); CDR STS-51G (1985); CDR STS-32 (1990); CDR STS-49 (1992); 6th Chief of the Astronaut Office (1987–1992).

After four space flights in 14 years at NASA and a 27-year career in the Navy, Dan Brandenstein realized that he was "reaching an age where if I was going to start a third career, I'd better get on with it." [1]. In October 1992, Brandenstein retired from NASA and as a captain with the U.S. Navy (USN).

Following his retirement, Brandenstein spent the next 20 years as an aerospace executive. From January 1994 to March 1996, he was appointed Director of Program Development at Loral Space Information Systems Company, later becoming their Director of Quality Assurance. Then, from March 1996 to April 1999, he became Executive Vice President, and later Program Manager, at Kistler Aerospace Corporation. In April 1999, he became Vice President (and later Program Manager) with Lockheed Martin Space Operations. At Lockheed, he was responsible for developing, and sustaining the engineering, operations and maintenance of, the Mission Control Centers and Integrated Planning System at JSC, supporting human space flight programs. He resigned from the company in September 2007 to join IBM Federal Systems Company, leading their data management system redesign team for the Space Station redesign effort. More recently, he served as Executive Vice President and Chief Operating Officer for the United Space Alliance (USA) Limited Liability Company (LLC).

Capt. Brandenstein was the 1992 winner of the Society of Experimental Test Pilots' (SETP) Iven Kincheloe Award. That same year he was inducted into the Wisconsin Aviation Hall of Fame, and in 2003 he was inducted into the U.S. Astronaut Hall of Fame. He has received numerous other medals and awards throughout his career, including the American Institute of Aeronautics and Astronautics (AIAA) Haley Space Flight Award, the Federation Aéronautique Internationale (FAI) Yuri Gagarin Gold Medal, and the American Astronautical Society (AAS) Flight Achievement Award. In 2012, he retired to a small ranch in the Texas Hill Country but continued to serve as the Chairman of the Astronaut Scholarship Foundation.

James Buchli
MS STS-51C (1985); MS STS-61A (1985); MS STS-29 (1989); MS STS-48 (1991).

James Buchli logged over 490 hours in space and, from March 1989 until May 1992, also served as Deputy Chief of the Astronaut Office.

On September 1, 1992, Buchli retired from the U.S. Marine Corps (USMC) with the rank of colonel after 25 years' service. He also resigned from NASA after 13 years in the Astronaut Office to accept a position as Manager of Space Station Systems Operations and Requirements with the Boeing Defense and Space Group, Huntsville, Alabama. In April 1993, he was reassigned as Boeing's Deputy for Payload Operations, Space Station *Freedom* Program. He then served as Operations and Utilization Manager for Space Station at the Boeing Defense and Space Group in Houston, Texas. He spent the next 15 years at the Boeing Company and USA, before joining Oceaneering International, Inc., Houston, in 2007. The following year, he received a promotion to Vice President and Program Manager. In this capacity, Buchli has led the company's efforts to design and build America's next generation space suit, the Constellation Space Suit System (CSSS), for NASA.

He is a recipient of the Defense Superior Service Medal, Legion of Merit, Purple Heart, Defense Meritorious Service Medal, Navy & Marine Corps Commendation Medal and Vietnam Gallantry Cross with Silver Star.

Michael Coats
PLT STS-41D (1984); CDR STS-29 (1989); CDR STS-39 (1991); 10th JSC Director (2005−2012)

Michael Coats logged more than 463 hours in space over his three Shuttle missions and served as Acting Chief of the Astronaut Office between May 1989 and March 1990.

Capt. Coats retired from the USN and the Astronaut Office in August 1991 and entered the corporate arena. From 1991 to 1996, he was Vice President of Avionics and Communications Operations for Loral Space Information Systems. From 1996 to 1998, he was Vice President of Civil Space Programs for Lockheed Martin Missiles and Space in Sunnyvale, California, and from 1998 to 2005, he served as Vice President of Advanced Space Transportation for Lockheed Martin Space Systems Company in Denver, Colorado. He returned to NASA in November 2005 as the tenth director of JSC in Houston, Texas. He would remain in this position until his retirement in December 2012, concluding a 44-year career that spanned 20 years with NASA, including seven as center director. At his retirement party on January 11, 2013, NASA Administrator and former astronaut Charles Bolden presented Coats with NASA's Distinguished Service Medal, calling him "an absolutely outstanding center director − an incredible gentleman."

Coats was inducted into the U.S. Astronaut Hall of Fame on May 5, 2007. He has also been recognized as a Fellow of the AIAA, and been awarded the FAI Gold Space Medal and three Distinguished Flying Crosses (DFC), in addition to a number of other honors.

Fig. 13.2: TFNG managerial roles (top) The 10th Johnson Space Center Director Mike Coats (left) and Robert Winkler of Boeing are pictured on the forward flight deck of OV-095 during final tests in the Shuttle Avionics Integration Laboratory (SAIL) at JSC in Houston, July 21, 2011, the day *Atlantis* completed the STS-135 mission and ended the Shuttle flight program. (bottom) Steve Hawley's official portrait as Director, Flight Crew Operations Directorate (FCOD) at JSC (Image from the collection of Steve Hawley, used with permission).

Richard (Dick) Covey
PLT STS-51I (1985); PLT STS-26 (1988); CDR STS-38 (1990); CDR STS-61 (1993).

A veteran of four space flights, USAF Capt. Dick Covey enjoyed a distinguished 16-year career with NASA, logging over 646 hours in space, but it was career filled with a number of highs and lows. On January 28, 1986, he was serving as Capcom during the launch of Shuttle *Challenger*, and as such became the last person to communicate with the crew when he radioed: "*Challenger*, go at throttle up," a few seconds before the Orbiter was lost in a massive fireball. In contrast, in December 1993, he was CDR for the highly successful flight of STS-61 to service and repair the Hubble Space Telescope (HST).

Throughout his astronaut career, Covey held additional technical assignments within the Astronaut Office, also serving as Acting Deputy Chief of the Astronaut Office and Acting Deputy Director of Flight Crew Operations. During 1989, he served as Chairman of NASA's Space Flight Safety Panel. Covey stated in his Oral History that he had considered whether or not he was going to remain at NASA after completing his third space flight. The offer of a number of managerial roles and the prospect of a fourth flight convinced him to stay, but he decided, after taking the management roles, that STS-61 would be his last trip into space and he would leave shortly afterwards. Reportedly, when his wife was asked whether her husband would fly again after that, she underlined that decision by saying "Not with this wife."

On August 1, 1994, Col. Richard Covey officially retired from NASA and the USAF. Between 1994 and 1996, he was a Unisys Deputy Program Manager for Space Operations in Houston. In 1996, he joined the Boeing Company as Division Director for McDonnell Douglas's Houston Operations, then later served as President of the Boeing Service Company in Colorado Springs. From 2003 to 2005, he also provided critical leadership during the exhaustive independent assessment of NASA's actions in response to the CAIB recommendations, as co-chairman of the Return-to-Flight Task Group. For this duty, he was awarded the NASA Distinguished Public Service Medal. In February 2006, he joined USA and was selected as their President and Chief Executive Officer (CEO) on September 28, 2007, becoming responsible for the direction, development and operations of the company. USA served as NASA's prime contractor for Shuttle and Space Station operations, including launch and recovery, mission planning and support, and astronaut training. He retired from USA in 2010.

Among other awards Col. Covey has received are five USAF DFCs, 16 Air Medals, the NASA Exceptional Service Medal, the AIAA Haley Space Flight Award for 1988, and the AAS Flight Achievement Award for 1988. He is a Distinguished Graduate of the USAF Academy, received the Liethen-Tittle Award as the Outstanding Graduate of USAF Test Pilot School Class 74B, and is a

Distinguished Astronaut Engineering Alumnus of Purdue University. The Arkansas Aviation Historical Society inducted Covey into the Arkansas Aviation Hall of Fame in 1995, and he was also inducted into the Astronaut Hall of Fame in May 2004.

John Creighton
PLT STS-51G (1985); CDR STS-36 (1990); CDR STS-48 (1991).

Capt. Creighton left NASA and retired from the USN in July 1992, having logged over 403 hours in space across his three missions. In his 2004 Oral History, he said that if he had wanted to, he could have remained in Houston and probably flown a couple more times. However, he and his family wanted to move back to the northwest of America, and since his wife was completing her residency program in her medical training, the timing seemed right and he reluctantly left NASA. He then became a test pilot on the 737 jetliner with Boeing, later becoming Chief Technical Pilot for Boeing Commercial Airplanes until his retirement from the company in 2007. "It wasn't space anymore, but it was still flying airplanes," he said. "I continued my flying career. [I] started out flying fighter airplanes [and] with that experience was able to fly the Space Shuttle. Then after that, [I] continued flying big commercial airplanes." **[2]**. In 1997, he ran as a Republican candidate for the Washington State Senate but was unsuccessful.

He is a member of the SETP and the Association of Space Explorers (ASE). Among other honors, he was awarded the Defense Superior Service Medal, Legion of Merit, DFC, 10 Air Medals, the Armed Forces Expeditionary Medal, the Vietnam Cross of Gallantry, the NASA Distinguished Service Medal, NASA Leadership Medal, three NASA Space Flight Medals, the French Legion of Honor and the Saudi Arabia King Fahd Medal.

John Fabian
MS STS-7 (1983); MS STS-51G (1985).

When John Fabian was assigned to his third Shuttle mission, he was delighted. His new assignment, STS-61G, was scheduled to fly on May 20, 1986, and placed him on the crew of the first Space Shuttle to carry a cargo destined for another planet. The main objective of this mission was to launch the Galileo spacecraft toward Jupiter, using the Centaur-G upper stage. The flight became one of several deferred by NASA in the wake of the *Challenger* tragedy, with the Centaur upper stage later removed from the Shuttle manifest.

In his Oral History, Fabian said that delays to the missions had pushed them into 1986, so in 1985 he began looking at opportunities outside of NASA. "My wife told me that my marriage had a two-flight limit, and I believed her, and so I was in the process of looking for a job." **[3]** Fabian subsequently left NASA on

January 1, 1986, to become Director of Space, Deputy Chief of Staff, Plans and Operations, Headquarters USAF. In June the following year, he retired from the USAF with the rank of colonel and joined Analytic Services Inc. (ANSER), a non-profit aerospace professional services firm in Arlington, Virginia, retiring as President and Chief Executive Officer in 1998. He continues to serve as an independent consultant and public speaker on the NASA space program and environmental stewardship.

Col. Fabian has been recognized as an Associate Fellow, AIAA; Fellow of the AAS; President of the ASE, and Corresponding Member of the International Academy of Astronautics. In 2010, he was named a Distinguished Member of the ASE – only the third astronaut or cosmonaut to receive the honor – joining cosmonaut Alexei Leonov and Apollo astronaut Rusty Schweickart, who were both founding members of the association. Fabian has long been involved with the association, serving for 14 years as International Co-president of the group, and for two years as President of the U.S. chapter, ASE-USA.

His many special honors include: Air Force Astronaut Wings; NASA Space Flight Medal with one Oak Leaf Cluster; FAI Komarov Diploma; Air Force Meritorious Service Medal; Defense Superior Service Medal; Legion of Merit; Defense Meritorious Service Medal; French Legion of Honor; Saudi Arabian King Abdul Aziz Medal; Air Medal with 2 Oak Leaf Clusters; Air Force Commendation Medal; Washington State University Sloan Engineering Award (1961); Air Training Command Academic Training Award (1966); Squadron Officer School Commandant's Trophy (1968); Squadron Officer School Chief of Staff Award (1968); Washington State University Distinguished Alumnus Award (1983); Washington State Service to Humanity Award (1983); Distinguished Alumnus Award (1985); Medallion of Merit (1987); Phi Sigma Kappa.

Anna Fisher
MS STS-51A (1984)

When Anna Fisher was assigned to the STS-51A crew in November 1984, it was just two weeks before giving birth to her first daughter. The mission was launched 14 months later, making her the first mother to fly in space.

Dr. Fisher was later assigned to her second mission, joining the STS-61H crew preparing to fly aboard Shuttle *Columbia* in June 1986. However, that mission was cancelled following the loss of *Challenger* in January of that year. Instead, she resumed technical roles within the Astronaut Office and served on the selection board for the 1987 class of astronauts (see sidebar: *Serving on an Astronaut Selection Board*). She then joined the Space Station Support Office, where she worked for the Operations Branch. Dr. Fisher was also the crew representative supporting space station development in the areas of training, operations concepts and the health maintenance facility.

During the early phase of building the International Space Station (ISS), through 2002, she served as Chief of the Space Station Branch, coordinating inputs to the operations of the burgeoning outpost, working closely with NASA's international partners, and supervising the assigned astronauts and engineers.

Later, from early 2011 through mid-2013, Fisher served as a space station Capcom, working in Mission Control, and was the lead Capcom for Expedition 33. In one of her final assignments at NASA, she contributed to developing the crew displays for NASA's new Orion spacecraft.

On April 28, 2017, Dr. Anna Fisher, the last member of NASA's 1978 Group 8 Shuttle astronauts still working for the space agency, retired after more than three decades of service to spend more time with her family. Among her many awards and distinctions, Dr. Fisher received the NASA Space Flight Medal; Lloyds of London Silver Medal for Meritorious Salvage Operations; University of California Los Angeles (UCLA) Professional Achievement Award; UCLA Medical Professional Achievement Award; NASA Exceptional Service Medal 1999; California Science Center Woman of the year 1986; and UCLA Alumni of the Year Award 2012. In November 2017, the UCLA presented Dr. Fisher with its first Science and Education Pioneer award.

SERVING ON AN ASTRONAUT SELECTION BOARD

At least sixteen members of the TFNG (seven pilots and nine MS) have served on the selection board for a new Astronaut group at some point in their astronaut careers. From 1984 until 2009, with the exception of the 1996 and 2004 selections, at least one member from the TFNG was appointed to the selection board.

As Steve Hawley explained in his 2002 Oral History, there are two parts to the selection process. "There's something that's called a ratings panel, and that is the group that has to go through all of the applicants' folders and make some preliminary decisions about who are the most qualified, because from the most qualified you select the ones that actually come down to Houston and go through the medical tests and actually get interviewed. And some subset of the ratings panel ends up being on the selection board itself that actually does those interviews. So fundamentally, the selection board's job is to have looked at the folders, to have reviewed the contents, to have looked at the recommendations, to have sat through the interviews, and ultimately to make some judgments as to who are the best candidates to recommend to the center director, who is the selecting official.

"I always enjoyed being on the selection board. It was kind of a humbling experience, because I thought over time we seemed to always get a more and

(continued)

more qualified bunch of people wanting to come work for us… I always thought it was kind of an honor that all of these really capable people wanted to come work in your program. Several of us used to joke that we probably wouldn't even be competitive if we were trying to get selected now. But it was also, I thought, very important to select the right kind of people to come into the program, and so being on the board was also, I thought, a great responsibility. So I was always proud to be entrusted with that responsibility. It's a real burden, because it takes a lot of time. The interviews themselves take basically six solid weeks if you interview 100, 120 individuals, which is typically what we would do, plus the time beforehand to go through hundreds or thousands of folders and try to make a determination who are the most qualified applicants. But, like I say, it's a real honor to be asked to do that, and I always enjoyed it."

Hawley did not think that the process had changed that much from when he went through it himself in 1977. "I think the quality of the people that apply has gotten better, in part, frankly, because it's a little easier now to prepare if you want… When I was a kid, people like me didn't get to be astronauts, because I didn't want to be a military test pilot, plus in those days… before my class, NASA hadn't actually picked astronauts for ten or eleven years. So the chance of actually even getting a chance to apply was very remote. Now, today, that's not true. We select more or less regularly every couple or three years, and you sort of know what the job is, and so you have a chance to tailor your development in that direction if you want to, and I think that does tend to give you more qualified applicants. I think they have a better idea what the job's like. They kind of know what the skills are that we're looking for. And we've had a number of astronauts come into the program and be successful, and if they want to, they can pattern themselves after the people that have done that.

"You still get that today. It's a little bit humbling and a little bit different now. When I first started on the selection board, people would say, 'Oh, yeah, you know, I always wanted to be an astronaut ever since I watched Apollo 11 land on the Moon', and now in later years it's been, 'Yeah, I really wanted to be an astronaut ever since I was a little kid and I watched the first Shuttles launch'. And you think, nobody ever says, 'When I watched Steve Hawley launch', but they will say, 'Hey, when I saw Sally Ride launch, then I really wanted to be an astronaut', or, 'I thought I could do that, too'." [4]

Recently, Steve Hawley explained to the authors that he had been a part of most of the Astronaut Selection process from 1984 through 2000, "except if there was a selection while I was at Ames (1990-1992). There was a

(continued)

selection in 1990 but I was part of that. So was Charlie Bolden, because I recall trying to schedule interviews around our training for STS-31." Hawley therefore served on the Astronaut Selection Boards for Groups 10 (1984), 11 (1985), 12 (1987), 13 (1990), 15 (1994), 16 (1996), 17 (1998) and 18 (2000).

The known Astronaut Rating and Selection Board Members from the Class of 1978 are: BRANDENSTEIN: Groups 11 (1985), 12 (1987) and 13 (1990); BUCHLI Group 13 (1990); COATS: Groups 11 (1985) and 13 (1990); COVEY Groups 13 (1990) and 14 (1992); FISHER Groups 11 (1985), 12 (1987) and 20 (2009); GIBSON: Groups 12 (1987), 14 (1992) and 15 (1994); Gregory: Group 15 (1994); HOFFMAN: Groups 13 (1990) and 14 (1992); NAGEL: Group 15 (1994); NELSON: Groups 11 (1985) and 12 (1987); ONIZUKA: Group 11 (1985); RIDE: Group 11 (1985); SEDDON: Groups 13 (1990) and 14 (1992); SHRIVER: Groups 11 (1985), 12 (1987), 14 (1992), 17 (1998) and 18 (2000); SULLIVAN: Group 14 (1992); WILLIAMS: Group 11 (1985).

Dale Gardner (1948–2014)
MS STS-8 (1983); MS STS-51A (1984).

Altogether, Dale Gardner logged a total of 337 hours in space on his two missions. He was then assigned as an MS on the STS-62A mission, scheduled as the first Shuttle launch from Vandenberg Air Force Base (AFB) in July 1986, but this flight was another of those cancelled following the loss of Shuttle *Challenger* that January. Gardner subsequently resigned from NASA and returned to duties with the USN, serving for over two years as the Deputy Chief of Space Control Operations Division at Cheyenne Mountain AFB, Colorado Springs. Following his promotion to the rank of captain, he became the command's Deputy Director for Space Control at Peterson AFB, also in Colorado Springs. In 1990, he retired from the USN and became Program Manager in the Engineering Operations of TRW Inc., before becoming manager of Northrup Grumman's Colorado Springs operations. He later joined the National Renewable Energy Laboratory (NREL) in Golden, Colorado, as the Associate Director for Renewable Fuels Science and Technology, retiring in January 2013.

On February 19, 2014, aged 65, Dale Gardner passed away from a brain aneurysm, and was buried at Evergreen Cemetery, Colorado Springs. Among many other honors and awards, Gardner was the recipient of the NASA Space Flight Medal, the Defense Superior Service Medal, and the DFC.

Robert (Hoot) Gibson

PLT STS-41B (1984); CDR STS-61C (1986); CDR STS-27 (1988): CDR STS-47 (1992); CDR STS-71 (1995); 7th Chief of the Astronaut Office[1].

In June 1996, Hoot Gibson retired from the USN with over 27 years' service. After 18 years with NASA, Hoot Gibson left the space agency in November 1996 and opened yet another chapter in his aviation life, flying as a captain with Southwest Airlines for the next ten years. In 2006, he was forced into mandatory retirement as a commercial pilot due to the Federal Aviation Administration's (FAA) "Age 60" rule. He made national headlines when he spoke out against the FAA rule, but there was no going back. Upon retirement from Southwest, he joined the Benson Space Company as Chief Operating Officer and Chief Test Pilot. By 2009, he was a demonstration pilot for the Hawker Beechcraft Corporation.

Capt. Gibson was inducted into the U.S. Astronaut Hall of Fame on June 21, 2003, the National Aviation Hall of Fame in October 2013, and the Tennessee Aviation Hall of Fame on November 7, 2015. He currently serves on the Astronaut Scholarship Foundation's Board of Trustees. Among his many honors and awards, he has received the FAI "Louis Bleriot Medal" (1992), and the Experimental Aircraft Association (EAA) "Freedom of Flight" Award (1989). He established world records for "Altitude in Horizontal Flight," Airplane Class C1A in 1991, and "Time to Climb to 9000 Meters" in 1994. His military awards include: the Defense Superior Service Medal; the DFC; three Air Medals; the Navy Commendation Medal with Combat "V"; a Navy Unit Commendation; Meritorious Unit Commendation; Armed Forces Expeditionary Medal; Humanitarian Service Medal; and Vietnam Campaign Medal.

Frederick (Fred) Gregory

PLT STS-51B (1985); CDR STS-33 (1989); CDR STS-44 (1991). NASA Deputy Administrator (2002−2005)

---

[1] Officially, two Group 8 astronauts (Dan Brandenstein and Hoot Gibson) have held the position of Chief of the Astronaut Office. The position was first held, unofficially, by Deke Slayton between September 1962 and November 1963. He was replaced by Alan Shepard, who took over as the first *official* Chief of the Office until 1969, with Tom Stafford temporarily replacing him while Shepard trained for Apollo 14. After his lunar mission, Shepard served a second term until he retired in 1974. He was replaced by John Young, who became the 5th Chief of the Office (4th person to hold the position). In 1987, Dan Brandenstein assumed the role as the 6th Chief (5th person), with Mike Coats serving as Acting Chief (May 1989–March 1990) and Steve Hawley serving as Deputy Chief between 1987 and 1990. When Brandenstein left NASA in October 1990, Hoot Gibson became the 7th Chief (6th person) to take on the role. He relinquished the position to Robert Cabana in September 1994 to begin training for STS-71.

In addition to his regular astronaut duties, Fred Gregory garnered extensive experience as a manager of flight safety programs and launch support operations. In his years as an active NASA pilot astronaut, he logged 455 hours in space. Gregory also served in several key positions during his years as an astronaut, including Chief of Operational Safety at NASA Headquarters, Washington, D.C., and Chief of Astronaut Training. Additionally, he served on the Orbiter Configuration Control Board and Space Shuttle Program Control Board, all of which were valuable assignments in his path to a senior NASA managerial position.

From June 1992 to December 2001, Gregory held the position of Associate Administrator, Office of Safety and Mission Assurance, at NASA Headquarters. During this period, he retired as a colonel in the USAF, in December 1993, after 29 years of service. As Associate Administrator for NASA, he was responsible for the safety, reliability, quality, and mission assurance of all NASA programs. In December 2001, he began serving – initially in an acting capacity – as Associate Administrator for the Office of Space Flight. This position became permanent in February 2002. He was now responsible for overseeing the management of the ISS; Space Shuttle operations; Space Access using Expendable Launch Vehicles (ELV) for commercial launch services; Space Communications; and Advanced Programs. On August 1, 2002, the U.S. Senate confirmed Gregory's appointment as NASA's Deputy Administrator and he was sworn in on August 12. Not only was he the first person to fill the position in ten years, he was also NASA's first African-American Deputy Administrator. He continued in this role until his retirement from NASA in 2005.

Col. Gregory is a member or past member of numerous societies, including the SETP, American Helicopter Society, Air Force Academy Association of Graduates, the National Technical Association, the Tuskegee Airmen, and the Order of the Daedalians. He is on the Board of Directors for the Young Astronaut Council, Kaiser-Permanente, the Photonics Laboratory at Fisk University, and the Engineering College at Howard University. He is also on the Trustees at the Maryland Science Center, and is a member of the Committee of the ASE.

His many awards and honors include the Defense Superior Service Medal; the Legion of Merit; the National Intelligence Medal of Achievement; 2 DFCs; 16 Air Medals; the NASA Distinguished Service Medal; 2 NASA Outstanding Leadership Medals; National Society of Black Engineers Distinguished National Scientist Award; the George Washington University Distinguished Alumni Award; and President's Medal, Charles R. Drew University of Medicine and Science. He holds Honorary Doctor of Science degrees from the College of Aeronautics and the University of the District of Columbia. He was also awarded the Air Force Association Ira Eaker Award in addition to numerous civic and community honors. On May 1, 2004, Col. Gregory was inducted into the U.S. Astronaut Hall of Fame.

S. David (Dave) Griggs (1939–1989)
MS STS-51D (1985).

On his follow-up assignment to STS-51D, Dave Griggs began training for the dedicated Department of Defense (DOD) mission STS-33 – this time as Pilot (PLT) – then scheduled for launch in August 1989.

On June 17, 1989, aged 49, Naval Reserve Rear Adm. Griggs was killed when his stunt plane slammed into a field 20 miles west of Memphis, while practicing solo aerobatic maneuvers for an air show later that day in Clarkville, Arkansas. He was flying a 1944 vintage two-seat AT-6D propeller airplane.

His many awards and honors include the DFC; Meritorious Service Medal; 15 Air Medals; Vietnamese Cross of Gallantry; the NASA Space Flight Medal; NASA Achievement Award; and NASA Sustained Superior Performance Award.

Terry (T.J.) Hart
MS STS-41C (1984).

Terry Hart's only mission in space logged a total of 168 hours. According to his Oral History, Hart was on leave of absence from Bell Telephone Laboratories (which later became AT&T) while he was at NASA. He had expected this to last for six years: "In '78, we thought we were one year from the first [Shuttle flight], and our class would start flying around the sixth mission. So I figured I'll get maybe two, maybe three [flights] and then go back. And here it was, at four years, I was put on a crew, but I wasn't going to fly until two years later, so I was gone six years." This proved to be at a difficult time at AT&T in 1984, as they were breaking up the Bell telephone system, and Hart was asked to come back after his flight to resume his career. "That all happened maybe six or eight months before I flew my mission," he explained. "George Abbey had called me in and offered me a second flight. It was a good flight. It was a science mission with the Germans [Spacelab D], and would have been interesting, but it was a three-year preparation to get ready for it. So I went back and I talked to my executive management team at Bell Labs, and was torn for a while, but then I decided I probably should settle down into a real career, because I was always an engineer at heart. I wanted to get back to that. So I felt a little bit bad not flying a couple more missions, but it would have been quite a bit longer away from my main career." **[5]**

Hart retired as an astronaut in 1984 and returned to a career in the telecommunications industry, holding several engineering management positions in the Government Data Systems Division with AT&T, and then as the Director of Engineering and Operations for the company's satellite network. During this time, he also achieved the rank of lieutenant colonel in the New Jersey Air National Guard, which he had joined in 1973. He continued flying with the Guard until 1985 and finally left that service in 1990. In 2004, he retired as President of Loral Skynet, a satellite communications company, in order to join the engineering faculty at his old alma mater, Lehigh University, teaching aerospace engineering.

Professor Hart is a member of the AIAA, the Institute of Electrical and Electronic Engineers, Tau Beta Pi, Sigma Xi, and Delta Upsilon. He has been awarded the National Defense Medal, the NASA Space Flight Medal, and was named Outstanding Officer of Undergraduate Pilot Training Class in 1970. He has also received the Rutgers Distinguished Alumnus Award, the Pride of Pennsylvania Award, and the New Jersey Distinguished Service Medal.

Frederick (Rick) Hauck
PLT STS-7 (1983); CDR STS-51A (1984); CDR STS-26 (1988).

Rick Hauck had the honor of being the first pilot from his group to fly in space, as a member of the five-person crew of STS-7. By the end of his third and final mission, Hauck had logged 436 hours in space. Like several of his colleagues, Hauck was conscious of turning 50 before he took the next step in developing his career. He had decided to leave NASA before flying STS-26, but did not announce it immediately. When he was asked at the pre-flight press conference if he was hoping for a fourth mission, he replied "No," and confirmed he had decided to retire from the Astronaut Office on one of America's morning TV shows the following day. "So I think I'm probably one of the few people who was able to announce on national television that I was on the job market. You can't buy that kind of publicity," he stated in 2004. **[6]**

In May 1989, Capt. Hauck became Director of the Navy Space Systems Division, in the Office of the Chief of Naval Operations. In this capacity, he held budgeting responsibility for the Navy's space programs. He concluded his military active duty on June 1, 1990. That October, he was appointed President and CEO of AXA Space, (formerly INTEC), a subsidiary of the international AXA insurance group, specializing in underwriting insurance for the risk of launching and operating satellites. He assumed responsibilities as CEO on January 1, 1993, and retired from this position on March 31, 2005. In May 2010, he was appointed to the board of Cianbro, a Maine-based construction company.

Capt. Hauck is a Fellow of the SETP, the AIAA, and the AAS. He was inducted into the Astronaut Hall of Fame on November 11, 2001, and serves on the boards of the Astronaut Scholarship Foundation and the U.S. Space Foundation. He is also a member of the Advisory Council of the Institute of Nuclear Power Operations. In the wake of the Space Shuttle *Columbia* tragedy and the subsequent return to flight of *Discovery* (STS-114), he featured as a news analyst on NBC and National Public Radio.

Steven Hawley

MS STS-41D (1984); MS STS-61C (1986): MS STS-31 (1990); MS STS-82 (1997); MS STS-93 (1999).

Astronomer Steven Hawley left the Astronaut Office in June 1990 in order to take up the post of Associate Director of NASA's Ames Research Center in California, before returning to JSC in August 1992 as Deputy Director of Flight Crew Operations. He resumed his astronaut flight status and training in February 1996 and went on to complete two further space flights. In completing the STS-93 mission, Hawley became the final member of the 1978 group to make a space flight, ending a remarkable series of 103 missions over a period of 15 years in which members of the 1978 group participated.

Dr. Hawley then returned to his former duties as Deputy Director of Flight Crew Operations. From October 2001 to November 2002, he was Director of Flight Crew Operations, and from 2003 to 2004, he served as the first Chief Astronaut for the NASA Engineering and Safety Center. From 2002 to 2008, he also worked as Director, Astromaterials Research and Exploration Science Directorate (ARES). In this role, he was responsible for directing a scientific organization conducting research in planetary and space science. Having logged around 32 days in space on his five space flights, Dr. Hawley retired from NASA in May 2008 to become Professor of Physics and Astronomy at the University of Kansas. In 2010, he was appointed the university's Director of Engineering Physics, and in 2012 he became their Adjunct Professor of Aerospace Engineering.

He is a member of the AAS, the Astronomical Society of the Pacific, the AIAA, Sigma Pi Sigma, and Phi Beta Kappa. Among his numerous awards and honors, he has received the Group Achievement Award for software testing at the Shuttle Avionics Integration Laboratory (SAIL), 1981; NASA Outstanding Performance Award, 1981; NASA Superior Performance Award, 1981; NASA Space Flight Medal (1984, 1986, 1990, 1997, 1999); NASA Exceptional Service Medal (1988, 1991); Exceptional Service Medal for Return to Flight, 1988; University of Kansas Distinguished Service Citation, 1998; NASA Distinguished Service Medal (1998, 2000); and the V.M. Komarov Diploma from the FAI, 1998 and 2000. He was inducted into the Kansas Aviation Hall of Fame in 1997, and the U.S. Astronaut Hall of Fame in 2007. He currently works and resides in Lawrence, Kansas.

Jeffrey Hoffman

MS STS-51D (1985); MS STS-35 (1990); MS/PC STS-46 (1992); MS STS-61 (1993); MS STS-75 (1996).

Dr. Hoffman became the first astronaut to log 1,000 hours aboard the Space Shuttle, during STS-75. At the completion of his fifth space mission, he had logged more than 1,211 hours and flown 21.5 million miles in space.

Later in 1996, Hoffman led the Payload and Habitability Branch of the Astronaut Office, but he left the astronaut program in July 1997 to become NASA's European Representative in Paris, where he served until August 2001. His principal duties were to keep NASA and its European partners informed about each other's activities, to try to resolve problems in U.S.-European cooperative space projects, to search for new areas of U.S.-European space cooperation, and to represent NASA in the European media. In August 2001, he was seconded by NASA to the Massachusetts Institute of Technology (MIT) where he became a Professor in the Department of Aeronautics and Astronautics, engaged in several research projects using the ISS and teaching courses on space operations and design. His research at MIT also focused on improving the technology of spacesuits and designing innovative space systems for human and robotic space exploration. Since 2008, he has been a Visiting Professor at the Department of Physics and Astronomy at the University of Leicester in England. He is also the director of the Massachusetts Space Grant Consortium and Deputy Principal Investigator of an experiment on NASA's Mars 2020 mission. In June 1986, his book *An Astronaut's Diary* (accompanied by a cassette tape) was released. The cassette contained excerpts of the original recordings he made with a pocket tape recorder.

Dr. Hoffman is a member of the International Academy of Astronautics, the Spanish Academy of Engineering, the AIAA, the AAS, the International Astronomical Union, Phi Beta Kappa and Sigma Xi. Among other awards, he has received five NASA Space Flight Medals, two NASA Exceptional Service Medals, two NASA Distinguished Service Medals, and the V. M. Komarov and Sergei P. Korolev Diplomas from the FAI in 1991 and 1994. In 2007, he was inducted into the Astronaut Hall of Fame.

Shannon Lucid
MS STS-51G (1985); MS STS-34 (1989); MS STS-43 (1991); MS STS-58 (1993); MS STS-76/Mir 21 & 22 NASA Board Engineer 2/MS STS-79 (1996).

Dr. Lucid became the first woman to hold an international record for the most flight hours in orbit by any non-Russian, and, until June 2007, also held the record for the most flight hours in orbit by any female astronaut. At the end of her fifth mission, Shannon Lucid had logged more than 223 days in space.

In 1993, Dr. Lucid was inducted into the Oklahoma Women's Hall of Fame, and in December 1996 she became the first woman to receive the Congressional Space Medal of Honor for her record-breaking service aboard the Mir station. She was also presented with the Order of Friendship Medal by Russian President Boris Yeltsin for that mission. This is one of the highest Russian civilian awards and the most distinguished that can be presented to a non-Russian. From February 2002 until September 2003, Dr. Lucid served as NASA's Chief Scientist, stationed at NASA Headquarters in Washington D.C., with responsibility for developing and communicating the agency's science and research objectives to the outside world. She then returned to her duties at JSC in Houston, including acting as Capcom for at least 14 Shuttle missions, starting with STS-114 in 2005 and ending with the final mission, STS-135, in 2011. On January 31, 2012, after 34 years of service with the space agency, Shannon Lucid retired from NASA.

In May 2014, during ceremonies conducted at KSC, Dr. Lucid was inducted into the United States Astronaut Hall of Fame.

Jon McBride
PLT STS-41G (1984).

Following his first flight as PLT aboard STS-41G in 1984, Jon McBride was next scheduled to fly as Commander (CDR) of STS-61E in March 1986. It had been planned for this mission to carry the ASTRO-1 observatory, which would be used to make astronomical observations including observations of Comet Halley. ASTRO-1 consisted of three ultraviolet telescopes mounted on two Spacelab pallets. However, the flight was one of those deferred by NASA in the wake of the *Challenger* accident in January 1986.

On July 30, 1987, McBride was assigned to NASA Headquarters in Washington, D.C., to serve as Assistant Administrator for Congressional Relations, with responsibility for NASA's relationship with Congress, and for providing coordination and direction to all Headquarters and Field Center communications with Congressional support organizations. He held this post from September 1987 until March 1989. On November 30, 1988, he was named to command the crew of the delayed ASTRO-1 mission, now re-designated STS-35 and scheduled for launch in March 1990. "Two or three months after that, rumors [were] floating around the halls [in Washington] that ASTRO was not going to fly, it was such a low priority," he recalled. McBride was told he could return to JSC and train for a couple of years but with no guarantee of the mission flying. He had already bought a vacation home in West Virginia, and was commuting three hours every weekend back and forth between there and Washington D.C., but the thought of commuting to and from Houston was just too much and so he decided to resign from NASA and return permanently to West Virginia. [7]

On May 12, 1989, NASA announced that Capt. McBride had officially retired from the agency and the USN in order to pursue a business career, and been replaced on the STS-35/ASTRO-1 mission by Vance Brand. Eighteen months later, Brand commanded *Columbia* on that mission. "I'm sure sorry I didn't get to fly as a commander once or twice aboard the Shuttle," McBride reflected in 2012, "but the things I got to do were just as remarkable in many ways."

He retired with an enviable record of more than 8,800 hours in over 40 different types of military and civilian aircraft. McBride subsequently became President and Chief Executive Officer of the Flying Eagle Corporation in Lewisburg, West Virginia, and President of the Constructors' Labor Council of West Virginia. In 1996, he unsuccessfully ran for the Republican nomination for Governor of West Virginia, losing to Cecil Underwood. He later left West Virginia to pursue business opportunities in Arizona, and following his retirement from business interests in 2008, moved near to Cocoa, Florida. Until January 2020, he worked at the KSC Visitor Complex as a member of the Astronaut Encounter team. In addition to being a liaison between the visiting astronauts and the Visitor Complex, he also gave motivational lectures and presentations.

His many awards and honors include the Legion of Merit; Defense Superior Service Medal; three Air Medals; Navy Commendation Medal with Combat "V"; National Defense Service Medal; Vietnam Service Medal; and the NASA Space Flight Medal. He was inducted into the West Virginia Hall of Fame in 2014.

Ronald McNair (1950–1986)
MS STS-41B (1984); MS STS-51L (1986).

While orbiting the Earth, Dr. Ronald McNair, an accomplished saxophonist, also fulfilled an ambition by becoming the first person to play the saxophone in space. He had brought his own instrument on the mission and managed to play a few numbers while circling the globe during his 191 hours in space.

In January 1985, McNair was assigned to the STS-51L mission of Shuttle *Challenger*. The mission would never be completed as *Challenger* and her crew of seven, including Ron McNair, were lost 73 seconds after lift-off from KSC on January 28, 1986, following a massive explosion in the external fuel tank.

Dr. McNair exemplified excellence and was the recipient of several honorary doctorates, fellowships and commendations. He was a member of the American Association for the Advancement of Science, the American Optical Society, the American Physical Society (APS), the APS Committee on Minorities in Physics, the North Carolina School of Science and Mathematics Board of Trustees, the MIT Corporation Visiting Committee, and Omega Psi Phi. He was a visiting lecturer in Physics at Texas Southern University. Dr. McNair was posthumously awarded the Congressional Space Medal of Honor, while a vast number of public places, institutions and programs have been renamed in his honor, including a crater on the Moon. He is now buried in his hometown at the Ronald E. McNair Memorial Park, which sits next to the old library building (now renamed the Dr. Ronald E. McNair Life History Center) in Lake City, South Carolina. Work on the Ronald E. McNair Memorial Park was begun in 2009, and when completed McNair's remains were moved from the Rest Lawn Memorial Cemetery and entombed in the new memorial park named after him.

Richard (Mike) Mullane
MS STS-41D (1984); MS STS-27 (1988); MS STS-36 (1990).

Col. Mike Mullane retired from NASA and the USAF in September 1990, having logged a total of 356 hours in space on his three missions. He then became self-employed as a professional motivational speaker and writer. His first fiction novel, *Red Sky, A Novel of Love, Space and War,* was published in June 1993. Since then, he has written an award-winning children's book, *Liftoff! An Astronaut's Dream* and a popular space-fact book, *Do Your Ears Pop In Space?* His memoir, *Riding Rockets: The Outrageous Tales of a Space Shuttle Astronaut,* was published in 2006 and was favorably reviewed in the *New York Times.*

Mullane has been inducted into the International Space Hall of Fame and is the recipient of many awards, including six Air Medals, the Air Force DFC, Meritorious Service Medal, Vietnam Campaign Medal, National Defense Service Medal, Vietnam Service Medal, Air Force Commendation Medal, and NASA Space Flight Medal. He was named a Distinguished Graduate of the USAF Navigator Training School (and recipient of its Commander's Trophy), the USAF Institute of Technology, and the USAF Test Pilot School.

Steven Nagel (1946–2014)
MS STS-51G (1985); PLT STS-61A (1985); CDR STS-37 (1991); CDR STS-55 (1993).

Steve Nagel was unique among the 1978 group in having flown in all the primary positions on a Shuttle crew. Although selected as a future Shuttle pilot by NASA, Steve Nagel flew in space as an MS on his first mission. Just four months later, he flew as PLT on his second mission and then later flew as CDR on his third and fourth missions. Altogether, he logged a total of 723 hours in space. Nagel retired from the USAF as a colonel on February 28, 1995, and formally left the Astronaut Office the next day to join the Safety, Reliability, and Quality Assurance Office at JSC. In September 1996, he moved to the Aircraft Operations Division as a research pilot, Chief of Aviation Safety and Deputy Chief of the division.

After retiring from NASA on May 31, 2011, Nagel joined the University Of Missouri College of Engineering in Columbia, Missouri, where he served as an instructor in the University's Mechanical and Aerospace Engineering Department.

His numerous awards included the Air Force DFC and the Air Medal with seven Oak Leaf Cluster. For pilot training, he received the Commander's Trophy, the Flying Trophy, the Academic Trophy and the Orville Wright Achievement Award (Order of Daedalians). He also received the Air Force Meritorious Service Medal. He earned four NASA Space Flight Medals; two Exceptional Service Medals; an Outstanding Leadership Medal; the AAS Flight Achievement Award; the

Outstanding Alumni Award of the University of Illinois; a Distinguished Service Medal; the Distinguished Alumni Award, California State University Fresno; and the Lincoln Laureate of the State of Illinois.

Col. Nagel died on August 21, 2014, after a two-year battle with an aggressive form of melanoma. He was 67 years old.

George (Pinky) Nelson
MS STS-41C (1984); MS STS-61C (1986); MS STS-26 (1988).

George "Pinky" Nelson logged over 407 hours in space on his three missions. He left NASA in 1989 and became an Assistant Provost at the University of Washington. He now directs the Science, Mathematics and Technology Education program at Western Washington University in Bellingham, and is the principal investigator of the North Cascades and Olympic Science Partnership, a mathematics and science partnership grant from the National Science Foundation.

His many awards include the NASA Exceptional Engineering Achievement Medal; NASA Exceptional Service Medal; three NASA Space Flight Medals; the AIAA Haley Space Flight Award; the FAI V. M. Komarov Diploma; and the Western Washington University Faculty Outstanding Service Award. He is also an elected member of Washington State Academy of Science and an Elected Fellow of the American Association for the Advancement of Science. On May 2, 2009, Dr. Nelson was inducted into the U.S. Astronaut Hall of Fame.

Ellison (El) Onizuka (1946–1986)
MS STS-51C (1985); MS STS-51L (1986).

On his 1985 STS-51C flight aboard *Discovery*, and newly promoted to lieutenant colonel, Onizuka became the first American of Japanese ancestry (and the first Buddhist) to fly in space. When asked about his STS-51C flight and his role as an astronaut, Lt. Col. Onizuka said that he felt very fortunate to have been given an opportunity so few people would experience. "Being out in space, you really realize the potential and you start to understand what this new frontier is all about," he once told a reporter. "It was an opportunity for me to do something I had dreamed of doing for a lifetime. And it was also an opportunity to serve our country." **[8]**

Onizuka was then assigned to serve as an MS aboard STS-51L. Among his other duties during the six-day *Challenger* flight, he was scheduled to facilitate the deployment of TDRS-B and to film Halley's Comet with a hand-held camera. Lt. Col. Ellison Shoji Onizuka, USAF, was one of the seven crewmembers killed in the loss of Shuttle *Challenger* on January 28, 1986.

Following the accident, Onizuka was laid to rest at the National Memorial Cemetery of the Pacific in Honolulu, Hawaii. He was posthumously promoted to the rank of colonel in the USAF and awarded the Congressional Space Medal of Honor. An asteroid, and a crater on the Moon, were also named after him, as well as Onizuka Air Force Station in Sunnyvale, California. The Onizuka Village family housing on Hickam AFB and the Astronaut Ellison S. Onizuka Space Center at Kona International Airport were also dedicated to him. In Hawaii, a visitor's center on Mauna Kea bears his name.

During his lifetime, Col. Onizuka became a member of the Society of Flight Test Engineers, the Air Force Association and the AIAA. Among other honors, he was presented with the Air Force Commendation Medal; Air Force Meritorious Service Medal; Air Force Outstanding Unit Award; Air Force Organizational Excellence Award; and National Defense Service Medal.

Judith (Judy) Resnik (1949–1986)
MS STS-41D (1984); MS STS-51L (1986).

In flying on STS-41D, Judy Resnik became the second American woman – following Sally Ride – to journey into space, and only the fourth since human space flight began. With the completion of that flight, Dr. Resnik had logged 144 hours and 57 minutes in space.

For her next mission, STS-51L (*Challenger*), Resnik was scheduled to operate the Canadian-built Remote Manipulator System (RMS) to deploy the SPARTAN, a network of scientific instruments which would have floated in space for several hours to study Halley's Comet. On January 28, 1986, only 73 seconds after lift-off during her second mission, Resnik died in the tragic loss of Shuttle *Challenger* nine miles above the Atlantic Ocean, along with her six colleagues, leaving a country shocked and in mourning. Judy Resnik was 36 years old.

Dr. Resnik was a member of the Institute of Electrical and Electronic Engineers (IEEE); American Association for the Advancement of Science; IEEE Committee on Professional Opportunities for Women; American Association of University Women; AIAA; Tau Beta Pi; Eta Kappa Nu; Mortarboard; and a Senior Member of the Society of Women Engineers. She had received the NASA Space Flight Medal in 1984, and was posthumously awarded the Congressional Space Medal of Honor.

Sally Ride (1951–2012)
MS STS-7 (1983); MS STS-41G (1984).

Group 8 MS Dr. Sally Ride would achieve lasting fame as the first American woman to fly in space. On June 18, 1983, Dr. Ride achieved her first (and history-making) flight on the STS-7 mission, which lifted off that morning from launch Complex 39A at KSC.

Following the loss of Shuttle *Challenger* and her crew of seven in January 1986, Dr. Ride served as a member of the Presidential Commission investigating the accident. Upon completion of the investigation, she was assigned to NASA Headquarters as Special Assistant to the Administrator for long-range and strategic planning. In 1987, she coauthored a report on the possible future options for the U.S. space program. She had always planned to retire after her third flight and return to academia in the summer of 1986, but the loss of *Challenger* and her involvement in the Commission and the report changed this. In August 1987, she resigned from NASA to take a position at the Center for International Security and Arms Control (CISAC) at Stanford University

In 1989, Dr. Ride joined the faculty at the University of California San Diego (UCSD) as a Professor of Physics and Director of the University's California Space Institute. During the 1990s, she led two public-outreach programs for NASA – the ISS EarthKAM and GRAIL MoonKAM cooperative programs with NASA's Jet Propulsion Laboratory (JPL) and UCSD – to enable middle school students to request images of the Earth and the Moon. In 2001, she founded her own company, Sally Ride Science, in order to pursue her long-time passion of motivating girls and young women to consider careers in science, math and technology. The company created entertaining science programs and publications for upper elementary and middle school students and their parents and teachers. A long-time advocate for improved science education, Dr. Ride wrote or co-wrote seven books, including five science books for children: *To Space and Back*; *Voyager: An Adventure to the End of the Solar System; The Third Planet*; *The Mystery of Mars*; and *Exploring Our Solar System*. She also initiated and directed education projects designed to fuel middle school students' fascination with science.

Sally Ride was a member of the President's Committee of Advisors on Science and Technology and the National Research Council's Space Studies Board. She also served on the boards of the Congressional Office of Technology Assessment, the Carnegie Institution of Washington and the National Collegiate Athletic Association (NCAA) Foundation. She was a Fellow of the American Physical Society, a member of the Pacific Council on International Policy and served on the boards of the Aerospace Corporation and the California Institute of Technology. She was the only person to have served on the commissions investigating both the *Challenger* and *Columbia* accidents.

The recipient of numerous honors and awards, Sally Ride was inducted into the National Women's Hall of Fame and the U.S. Astronaut Hall of Fame (2003) and received the Jefferson Award for Public Service, the von Braun Award, the Lindbergh Eagle and the NCAA's Theodore Roosevelt Award. She was twice awarded the NASA Space Flight Medal.

Until her death from pancreatic cancer on July 23, 2012, aged 61, Dr. Ride continued to help students and worked with science programs and festivals around the United States. She is survived by Tam O'Shaughnessy, her partner of 27 years, who also serves as Chief Operating Officer and Executive Vice President of Sally Ride Science, continuing the work begun by Sally.

Several books have been written about Dr. Sally Ride. The most definitive was *Sally Ride: America's First Woman in Space*, by Lynn Sherr, published two years after Ride's death.

Francis (Dick) Scobee (1939–1986)
PLT STS-41C (1984); CDR STS-51L (1986).

While waiting and training for his first assignment, and in addition to his astronaut duties, Dick Scobee served as an instructor pilot on the NASA/Boeing 747 Shuttle Carrier Aircraft (SCA), which transported the Shuttles between ground stations. His first space flight came as PLT on STS-41C in 1984 and by the end of it, Maj. Scobee had logged a total of 167 hours and 40 minutes in space.

Promoted to CDR, he was next assigned to STS-51L, a second mission for him aboard Shuttle *Challenger*. Following a number of launch delays, *Challenger* finally lifted off from KSC at 11:38:00 EST on January 28, 1986. The massive explosion in the fuel tank 73 seconds into the flight resulted in the loss of *Challenger* and her seven crewmembers. Lt. Col. Francis (Dick) Scobee was interred at Arlington National Cemetery on what would have been his 47th birthday.

Among his many special honors, Dick Scobee was awarded the Air Force DFC, the Air Medal, and two NASA Exceptional Service Medals. On July 9, 1994, the San Antonio College Planetarium was rededicated as The Scobee Planetarium. In 2004, he was posthumously awarded the Congressional Space Medal of Honor and was inducted into the Astronaut Hall of Fame. After the *Challenger* disaster, a number of schools, streets, and municipal facilities in the U.S. were renamed in his honor. North Auburn Elementary School in Auburn, Washington, was renamed Dick Scobee Elementary, and Auburn Municipal Airport became Dick Scobee Field. Dick Scobee Road in Myrtle Beach, South Carolina, also commemorates his name. In Houston's George Bush Park, there is a Radio Control Flying Field named in his honor. His many organizations included membership of the SETP, the Experimental Aircraft Association and the Air Force Association.

Margaret (Rhea) Seddon
MS STS-51D (1985); MS STS-40 (1991); MS/PC STS-58 (1993).

It would be nearly seven years after her Group 8 selection before Rhea Seddon first ventured into space, as an MS on STS-51D in 1985. In her three space flights,

Dr. Seddon logged just over 722 hours in space. Seddon was not approached to fly a fourth mission, as she and husband Hoot Gibson were planning to add to their family and move to Tennessee once Gibson had completed his fifth flight, but as he became so busy after STS-71, their plans were delayed a year.

"So we wanted to go back in '96, and it was pointed out to me that I needed one more year with NASA to qualify for a pension. They were offering early outs, so I needed to be 50 and have 20 years with the government. Luckily, I had some time in the V.A. [Veterans Affairs] Hospital system when I was a resident, so I had a little extra time, but I needed to be 50, and that was in 1997. So I thought, 'How am I going to work this?' Luckily, the folks at NASA said, 'Well, you know, if you could work for NASA at a different place, like somewhere in Tennessee, then you could fulfill that last-year requirement.' Luckily, Vanderbilt had been working with the Neurolab people, and there was a Neurolab experiment coming out of Vanderbilt, so I negotiated a part-time job there, and came to Vanderbilt.

"Drew [F. Andrew] Gaffney, who had flown with me on SLS-1, was a professor at Vanderbilt at the time. I said to him, 'I'd really like to come back to Tennessee, and one of the places that I think I'd like to work would be Vanderbilt, the Medical Center. Can you make some introductions?' And luckily, he did, with the vice chancellor at the time and his deputy, and they sent me around to talk to a bunch of people. I also talked to people at other hospitals in Nashville and looked around there. I was going to be working part-time in one of the Vanderbilt research labs two days a week for NASA. I let the Vanderbilt Hospital people know I could work for them three days a week until my NASA commitment was up in a year." **[9]**

In September 1996, Rhea Seddon was detailed by NASA to Vanderbilt University Medical School in Nashville, Tennessee. There, she assisted in the preparation of cardiovascular experiments which flew aboard Space Shuttle *Columbia* on the Neurolab Spacelab flight in April 1998. She retired from NASA in November 1997 and is now the Assistant Chief Medical Officer of the Vanderbilt Medical Group in Nashville, Tennessee.

On May 30, 2015, Rhea Seddon became the eighth woman inducted into the U.S. Astronaut Hall of Fame during a public ceremony held at the KSC Visitor Complex. A week prior to her induction, Dr. Seddon released her first book, the memoir *Go For Orbit*. In October that year, she was also inducted into the Tennessee Women's Hall of Fame.

Brewster Shaw, Jr.
PLT STS-9 (1983); CDR STS-61B (1985); CDR STS-28 (1989).

Brewster Shaw logged over 533 hours in his three space flights. Following his third mission, he was advised that it would be at least three or four years before he would fly in space again, and in the interim was asked by Robert Crippen to take over the job he had been doing down at the Cape, which he accepted. But when he asked if he could retain his T-38 flying status and still be eligible for space flight, the answer was a firm "No and No."

Brewster Shaw left JSC in October 1989 to assume the NASA Headquarters senior executive position of Deputy Director, Space Shuttle Operations, located at KSC. As operations manager, Shaw was responsible for all operational aspects of the Space Shuttle Program and had Level II authority over the Space Shuttle elements, from the time the Orbiters left the Orbiter Processing Facility (OPF) and were mated to the External Tank (ET) and Solid Rocket Boosters (SRB), through transport to the launch pad, launch and recovery, to their return to the OPF. He was the final authority for the launch decision, and chaired the Mission Management Team. Shaw then moved on to serve as the Deputy Program Manager, Space Shuttle, as a NASA Headquarters employee located at KSC. In addition to the duties he previously held, he also shared full authority and responsibility for the conduct of the Space Shuttle Program with the Space Shuttle Program Manager.

He was then appointed Director, Space Shuttle Operations, with responsibility for the development of all Space Shuttle elements, including the Orbiter, ET, SRB, Space Shuttle Main Engines (SSME), the facilities required to support mission operations and in the planning necessary to conduct Shuttle operations efficiently.

Post-NASA, Shaw joined Rockwell in 1996 after 27 years with the USAF and the space agency. The Boeing Company acquired Rockwell in December 1996. Initially, Shaw served as Director of Major Programs, Boeing Space and Defense Group. He then became Vice President and Program Manager of ISS Electrical Power Systems at Rocketdyne Propulsion and Power. The contract included the development, testing, evaluation and production of the electrical power system to be assembled in space during multiple Space Shuttle launches. Shaw's next role was to lead the consolidated Boeing teams at Huntsville, Alabama, and Canoga Park and Huntington Beach, California, in the design, development, testing, evaluation, production and flight preparation of ISS hardware and software. Boeing was NASA's prime contractor and supplier for the ISS.

In mid-2003, Brewster Shaw left Boeing and became the Chief Operating Officer of USA. In that position he had primary responsibility for the day-to-day operations and overall management of USA, the prime contractor for the Space Shuttle Program, and its 10,000 employees in Florida, Texas, Alabama and Russia. In January 2006, he returned to the Boeing Company's Houston campus, and was then acting as the Vice President and General Manager of the Space Exploration Division, which controlled Boeing's ISS and Space Shuttle programs.

Shaw has earned numerous honors and awards, including 28 medals in Vietnam. He received the Defense Superior Service Medal, the Air Force DFC with Seven Oak Leaf Cluster and the Defense Meritorious Service Medal. He was inducted into the U.S. Astronaut Hall of Fame on May 6, 2006.

Loren Shriver
PLT STS-51C (1985); CDR STS-31 (1990); CDR STS-46 (1992).

Loren Shriver, a veteran of three space flights and with more than 386 hours logged in space, was assigned as Deputy Chief of the Astronaut Office in October 1992. The following year, he accepted the position of Space Shuttle Program Manager for Launch Integration at KSC. Then, from 1997, he served as Deputy Director for Launch and Payload Processing, before resigning from NASA in March 2000 to take up the role of Deputy Program Manager for the Space Shuttle Program for USA. In 2006, he retired from this position, and also took his retirement from the USAF with the rank of colonel.

Shriver's accomplishments have earned him many notable awards. He has received the Air Force DFC; the Defense Superior Service Medal; the Defense Meritorious Service Medal; the Air Force Meritorious Service Medal; and the Air Force Commendation Medal. His NASA awards include the NASA Distinguished Service Medal, the NASA Outstanding Leadership Medal, and the NASA Space Flight Medal (three times). In 1990, he received the Flight Achievement Award from the AAS and the Haley Space Flight Award from the AIAA. He was inducted into the Astronaut Hall of Fame in 2008.

Now living in Estes Park, Colorado, he has stated that even though he has retired, he is becoming a regular on the "speaker's circuit," sharing his experiences at schools, businesses and organizations.

Robert (Bob) Stewart
MS STS-41B (1984); MS STS-51J (1985).

Bob Stewart logged a total of 289 hours in space including approximately 12 hours of Extra-Vehicular Activity (EVA) operations. Although astronauts who had served in the U.S. Army Air Force during World War II had flown previously (including Mercury astronauts Gus Grissom and Deke Slayton), Stewart was the first active-duty Army officer to fly into space.

While in training for his scheduled third flight (STS-61K, ultimately cancelled as a result of the *Challenger* disaster in January 1986), the U.S. Army promoted Stewart to brigadier general, the highest military rank attained by a member of the TFNG. Upon accepting this promotion, he was reassigned from NASA to be the Deputy Commanding General, U.S. Army Strategic Defense Command, in Huntsville, Alabama. In this capacity, he managed research efforts in developing ballistic missile defense technology. Three years later, he was reassigned as Director of Plans, U.S. Space Command, Colorado Springs, Colorado. He retired from the army in 1992 and now makes his home in Woodland Park, Colorado. He is presently employed as Director, Advanced Programs, Nichols Research Corporation in Colorado Springs.

Brig. Gen. Stewart is a member of the SETP, ASE, a past member of Phi Eta Sigma, and the Scabbard and Blade (military honor society). During his army and NASA careers, he has been awarded the Army Distinguished Service Medal; the Defense Superior Service Medal; two Legions of Merit; four DFC; a Bronze Star; a Meritorious Service Medal; 33 Air Medals; the Army Commendation Medal with Oak Leaf Cluster and "V" Device; two Purple Hearts; the National Defense Service Medal; the Armed Forces Expeditionary Medal; the U.S. and Vietnamese Vietnam Service Medals; and the Vietnamese Cross of Gallantry. He has also been the recipient of the Army Aviator of the Year 1984, AHS Feinberg Memorial Award, and AIAA Oberth Award, and was awarded the NASA Space Flight Medal (1984 and 1985). He was inducted into the U.S. Army Aviation Hall of Fame in Atlanta, Georgia, in 2007.

Kathryn (Kathy) Sullivan
MS STS-41G (1984); MS STS-31 (1990); MS/PC STS-45 (1992).

Dr. Kathy Sullivan's first space mission was on STS-41G, in 1984, during which she and fellow MS David Leestma successfully conducted a 3.5-hour spacewalk to demonstrate the feasibility of satellite refueling, making her the first U.S. woman to perform an EVA. At the completion of her third mission, Dr. Sullivan had logged more than 532 hours in space.

In the months prior to her final space flight, Sullivan had been pondering her future career after 15 years with NASA. At this time, a friend who had been serving as Chief Scientist at the National Oceanic and Atmospheric Administration (NOAA) was stepping down for personal reasons and asked if Sullivan would be interested in putting her name forward as a replacement. Sullivan had been considering whether to remain at NASA or pursue interests elsewhere, but when the NOAA offer came up it included many of the areas she was personally interested in. Shortly after the STS-45 mission, Sullivan decided this would be an interesting avenue to pursue and her name was put forward. As this was a Presidential appointment, it had to go through the nomination process and be confirmed by the Senate. In the interim, she moved to Washington D.C., as a NASA person on loan to NOAA, while awaiting the decision. Her nomination was originally under the George H.W. Bush administration, but after Bush lost his bid for a second term in the presidential election in November 1992, Sullivan had to be nominated again to the new William J. Clinton administration in early 1993. By March of that year, her appointment was finally confirmed by the Senate.

As Chief Scientist at the NOAA, she oversaw a wide array of research and technology programs, ranging from climate and global change to satellites and marine biodiversity. From 1996 to 2006, she served as President and CEO of the Center of Science & Industry (COSI) in Columbus, Ohio. Under her leadership,

COSI strengthened its impact on science teaching in the classroom and its national reputation as an innovator of hands-on, inquiry-based science learning resources. Dr. Sullivan then served as the inaugural Director of the Battelle Center for Mathematics and Science Education Policy in the John Glenn School of Public Affairs at Ohio State University.

On March 6, 2014, Dr. Sullivan was confirmed by the Senate as the Under Secretary of Commerce for Oceans and Atmosphere, and the tenth NOAA Administrator, having served as Acting NOAA Administrator since February 28, 2013. Prior to her appointment as Acting Administrator, Dr. Sullivan held the positions of Assistant Secretary of Commerce for Environmental Observation and Prediction, and Deputy Administrator. As Assistant Secretary, Dr. Sullivan played a central role in directing Administration and NOAA priority work in the areas of weather and water services, climate science and services, integrated mapping services and Earth-observation capabilities. She also provided agency-wide direction with regard to satellites, space weather, water, and ocean observations and forecasts, to best serve American communities and businesses. She is the United States co-chair of the Group on Earth Observations (GEO), an intergovernmental body that is building a Global Earth Observation System of Systems (GEOSS) to provide environmental intelligence relevant to societal needs. Her tenure as Director of NOAA ended on January 20, 2017, with the swearing in of President Donald Trump.

She then became the 2017 Charles A. Lindbergh Chair of Aerospace History at the Smithsonian Institution's National Air and Space Museum. Her 2019 book *Handprints on Hubble* recounts her experiences as an astronaut and being part of the team which launched, rescued, repaired and maintained HST.

Among her many honors and awards, in 1991 Dr. Sullivan received the Haley Space Flight Award for "distinguished performance in the deployment of the Hubble Space Telescope on Mission STS-31 during April 1990." In 2004 she was inducted into the Astronaut Hall of Fame and received the Adler Planetarium Women in Space Science Award. In 2014, she was included in *Time* magazine's 100 Most Influential People list for that year.

Norman (Norm) Thagard
MS STS-7 (1983); MS STS-51B (1985); MS STS-30 (1989); MS/PC STS-42 (1992); NASA Cosmonaut Researcher Soyuz TM-21/Mir EO18 NASA Board Engineer 2/MS STS-71 (1995).

Dr. Norman Thagard's fifth space flight was different to any of his other missions. In 1995, he became the first American cosmonaut-researcher on a Russian Mir-resident space station crew. As a member of the Mir EO-18 crew, along with cosmonauts Vladimir Nikolayevich Dezhurov and Gennadi Mikhailovich Strekalov, he became the first American astronaut to live and work on a space station since Skylab in 1974. Lift off from the Baikonur Cosmodrome in Kazakhstan occurred on March 14, 1995. Thagard was the first American to train on Russian soil, the first to enter space aboard a non-American craft, and the first American occupant of Mir. Thagard, Dezhurov and Strekalov landed at KSC aboard *Atlantis* on STS-71 on July 7, 1995. With the completion of his fifth mission, Dr. Thagard had logged over 140 days in space. Following this flight, Thagard resigned from NASA to become a tenured Professor of Electrical Engineering and Dean of Public Relations at Florida State University's FAMU-FSU College of Engineering, where he is currently an Associate Dean. He also serves as the Director of the Challenger Learning Center in Tallahassee, Florida, and is on the boards of various private corporations.

Among his many awards and honors, Dr. Thagard has earned 11 Air Medals; the Navy Commendation Medal with Combat "V" and the Marine Corps "E" Award; the Vietnam Service Medal; and the Vietnamese Cross of Gallantry with Palm. He was inducted into the U.S. Astronaut Hall of Fame on May 1, 2004. Dr. Thagard resigned from the USAF as a captain in 1971.

James ("Ox") van Hoften
MS STS-41C (1984); MS STS-51I (1985).

On his second space flight (STS-51I in 1985), James "Ox" van Hoften, PhD became the first human being to launch a satellite by hand, having spun and pushed a repaired communications satellite away from the orbiting Shuttle *Discovery* during an EVA. Over his two space missions, Dr. van Hoften logged a total 338 hours in space, including 22 hours of EVA flight time. In July 1985, he was assigned to a third Shuttle mission, STS-61G, scheduled for launch on May 22, 1986, but when the flight was cancelled due to the *Challenger* tragedy, he resigned from NASA and joined the Bechtel Corporation in 1986. He had already told his wife that he was going to retire and leave NASA after STS-61G, but the loss of *Challenger* and the down time after that tragedy changed his plans. "After the accident, everything just shut down, and everyone went off and started doing accident investigations and other things. I just knew that this was going to be a tough recovery, and I just said, 'I've got to go out and do something else'," he said in a 2007 interview. "I started interviewing and looking at lots of different options. At that time I felt I didn't want to [leave the Astronaut Office], but I just felt like it was the thing to do. I'm pretty good at making transitions, and I just figured 'this is another transition in life, and you've got to move on', and never looked back.

"I interviewed [for] all sorts of different things. I asked myself what I really wanted to do in life. I've been a college professor… but I didn't go to work there, and then I looked at going into aerospace… [but] I didn't have any interest in doing that. I was a civil engineer, which is kind of unusual, so I looked for a company that was a civil engineering-type company that I had some interest in doing work in that area, and ended up going to work for Bechtel. It was great. I had a great career there and spent 20 years with them. I was a partner and worked on huge jobs all over the world. So I had four careers, basically, and enjoyed them all." **[10]**

Over the next six years, van Hoften managed the San Francisco-based company's engineering and construction business for the defense and space markets. He later became Senior Vice President and a partner in Bechtel, responsible for airport developments in the Middle East, Japan, and North and South America. In the early 1990s, he was the program manager of the $23 billion Hong Kong Airport Core Program, including the new Hong Kong Airport. He later acted as Director of Projects for the UK National Air Traffic Services. In 2009, he was appointed a non-executive director of Gatwick Airport Ltd. Four years later, in September 2013, he joined the board of directors with construction firm Cianbro Corporation.

During his extensive career in the USN Reserve and USAF Reserve (retiring as a lieutenant colonel in that service), and as an astronaut with NASA, van Hoften received a number of awards and honors. These included the Meritorious Service Medal; two Navy Air Medals; Vietnam Service Medal; National Defense Service Medal; and two NASA Space Flight Medals. He was installed as a Jimmy Doolittle Fellow of the Aerospace Education Foundation in 1990.

David Walker (1944–2001)
PLT STS-51A (1984); CDR STS-30 (1989); CDR STS-53 (1992); CDR STS-69 (1995).

Dave Walker flew four Shuttle missions for NASA, three of them as CDR. By the end of his fourth space flight, he had logged 724 hours in space. His last technical assignments were as Chief of the Station/Exploration Support Office in the Flight Operations Directorate and as Chairman of the JSC Safety Review Board.

Capt. Walker retired from the USN and resigned from NASA in 1996, taking up a sales and marketing position with NDC Voice Communications in San Diego, California. In 1999, he joined Ultrafast, Inc. of Malvern, Pennsylvania, as Vice President of Sales. On his retirement, he moved to Boise, Idaho, although he often worked as a consultant. Until he became seriously ill with cancer, he was active as President of the Idaho Aviation Foundation, a non-profit corporation promoting general aviation in the state of Idaho.

His honors and awards included the Defense Superior Service Medal; the DFC; the National Intelligence Medal of Achievement; the Legion of Merit; two Defense Meritorious Service Medals; six Navy Air Medals; the Battle Efficiency Ribbon; the Armed Forces Expeditionary Medal; the National Defense Service Medal; two NASA Distinguished Service Medals; the NASA Outstanding Leadership Medal; four NASA Space Flight Medals; the Vietnamese Cross of Gallantry; the Vietnam Service Medal; and the Republic of Vietnam Campaign Medal.

Capt. Walker passed away on April 23, 2001, at the age of 56, in the M.D. Anderson Cancer Center in Houston, Texas. He is interred at Arlington National Cemetery.

Donald Williams (1942–2016)
PLT STS-51D (1985); CDR STS-34 (1989).

Don Williams logged a total of 287 hours and 35 minutes in space on his two Shuttle missions. In March 1990, he retired from the USN with the rank of captain and also resigned from NASA. "The goal that I set for myself when I came [to NASA] … was to successfully command a mission. And then, what's next?" he explained in a 2002 interview. **[11]** "I distinctly thought about that some time before the [STS-34] mission and [asked myself] 'What am I going to do after this? This is a lot of fun. I'd love to stay here and fly for a long time'. But once you reach that goal, then what do you do?

"I had decided that I had four alternatives. I could stay and fly again, which was an option, and that was offered to me. I could go into a NASA management job as a civil servant, which was also a possibility. I could go back to the Navy as an officer, because I was still on active duty and military officers were detailed to NASA. Or I could go into private industry and see if I [could] make a living there. I explored all four of those over the next few months, and it turned out, just by a matter of timing I guess it was, I ran into Neil B. Hutchinson [who] had been a neighbor of mine and had been a Flight Director in Houston. [He] was looking for a person with the qualifications that I had, for a potential job that he had for the company he worked for then. It turned out it was a good fit mutually for both of us, and so it was time to move on and do something else. I did explore all three of the other options, though, and decided on that one and never looked back."

That appointment was as a Division Manager with Science Applications International Corporation (SAIC), working on several projects in the Houston area, nationally and internationally. He retired in 2007 and moved to Henderson, Nevada, where he served on the board of Sun City Anthem's Veterans Club and became its Vice President Emeritus.

Increasingly suffering from the effects of dementia in the last months of his life, Capt. Williams passed away on February 23, 2016, aged 74. His awards included the Legion of Merit; the DFC; the Defense Superior Service Medal; two Navy Commendation Medals with Combat "V"; two Navy Unit Commendations; a Meritorious Unit Commendation; the National Defense Medal; and an Armed Forces Expeditionary Medal. He also received the Vietnam Service Medal (with four stars), a Vietnamese Gallantry Cross (with gold star), and the Vietnam Campaign Medal. At NASA, he earned the NASA Outstanding Leadership Medal, the NASA Space Flight Medal and the NASA Exceptional Service Medal.

MOVING ON

In January 2018, the surviving members of the 1978 group marked the 40th anniversary of their selection. Those four decades had seen the whole Shuttle flight program completed, growth in human space flight international cooperation, Americans living on Mir and the creation of the ISS, commercial space flight, and the prospect, once again, of a return to the Moon.

There had been highs and lows, tragedy and triumph, success and disappointment along the way, but those 35 rookie Ascans who sat on the podium in the glare of media spotlights in 1978 all lived up to the reasons they were selected in the first place. They took on the challenge and delivered many times over, some giving their lives in the pursuit of the quest for an affordable, safe and reliable, regular service to access space, where they could perform useful work before returning to a runway near the launching site to begin the preparations for the next mission. While the Space Shuttle Program itself may not have fully delivered on expectations, these astronauts achieved and often surpassed their goals. The legacy of the TFNG is explored in the next chapter, but one thing became clear early on: Each of them proudly and confidently carried – and redefined – the mantle of the "*Right Stuff*" from the pioneering decade of American human space flight into the closing decades of the 20th century and on into the new millennium. Now, a new generation of American space explorers, with a new era of vehicles, is ready to take up the challenge and create new pages in space history.

While the current generation of astronauts attains new milestones in space exploration history with the first crewed flight of the Dragon reusable spacecraft to the ISS, the surviving members of the TFNG continue to attain their own new goals.

On June 7, 2020, over 35 years after making history as the first American female to walk in space, Kathy Sullivan became the first woman and eighth person to dive to the lowest point on Earth, known as Challenger Deep. During a successful 12-hour expedition to the deepest point of the Pacific Ocean's Mariana Trench, over 35,000 feet (10,668 m) below the ocean surface, Sullivan accompanied Victor Vescovo, pilot of the Deep Submersible Vessel (DSV) *Limiting Factor* on an exploratory dive to almost seven miles (11 km).

Following her return to the mother ship, the DSSV *Pressure Drop*, Sullivan made contact with the crew on ISS orbiting 254 miles (408 km) above her, explaining that it had been "an extraordinary day, a once in a lifetime day, seeing the moonscape of the Challenger Deep, then comparing notes with my colleagues on ISS about our remarkable, reusable inner-space, outer-space craft."

A similar event occurred almost 55 years earlier, in August 1965, when Gemini 5 astronaut Gordon Cooper spoke from orbit with his fellow Mercury astronaut-turned-aquanaut colleague Scott Carpenter, who was 205 feet (62 m) beneath the surface of the Pacific off the coast of California, in the USN Sealab II.

References

1. Dan Brandenstein, NASA Oral History, January 19, 1999.
2. John Creighton, NASA Oral History, May 3, 2004.
3. John Fabian, NASA Oral History, February 10, 2006.
4. Steve Hawley, NASA Oral History, December 17, 2002
5. Terry Hart, NASA Oral History, April 10, 2003.
6. Rick Hauck, NASA Oral History, March 17, 2004.
7. Jon McBride, NASA Oral History, April 17, 2012.
8. *"Onizuka tried to prepare family for the worst,"* Beverley Creamer, *The Honolulu Advertiser*, issue 29, January 1986, pp. A-1, A-1A.
9. Rhea Seddon, NASA Oral History, May 10, 2011.
10. James van Hoften, NASA Oral History, December 5, 2007.
11. Donald Williams, NASA Oral History, July 19, 2002.

14

Reflections

"I hope that I see humans back on the Moon in my lifetime,
and perhaps on Mars, but I don't know now.
It's looking increasingly unlikely that that will happen,
but it's a matter of time."
NASA JSC Oral History Project, July 19, 2002
Donald E. Williams (1942-2016)

In the words of the late Don Williams, whether we will be lucky enough to be able to witness significant events that await us in the future is "a matter of time". The same could have been said for all 35 members of the Class of 1978 when they entered the NASA Astronaut Program as young, eager rookies, with their hopes and expectations for the future, as well as uncertainty and probably a little trepidation as they embarked on their great adventure together as a group. Over four decades later, the achievements of the Class of 1978 are now consigned to the history books, and though all flew in space, which is an achievement in itself, unfortunately none of them experienced the International Space Station (ISS). It is possible that some of them will be able to see the hopes of Williams fulfilled, with a return to the Moon and maybe on to Mars. After all, it's just "a matter of time."

A MATTER OF TIME

The study of history, according to the dictionary, is the "continuous, chronological record of important events," or the "study of past events," while making history is defined as "doing something memorable which influences the course of history." [1]

D. J. Shayler, C. Burgess, *NASA's First Space Shuttle Astronaut Selection*,
Springer Praxis Books, https://doi.org/10.1007/978-3-030-45742-6_14

Clearly, most achievements in the exploration of space fall under such definitions, even if those who participate in making history consider their input at the time to be simply doing the job they trained for rather than "doing something memorable." Over six decades into the 'space-age', each day in a given year can add to the roll-call of significant events in the annals of space history, and since 1961 and the pioneering one-orbit flight of cosmonaut Yuri A. Gagarin in Vostok, the human exploration of space has certainly created its own pages in the history books. Certain dates remain memorable in one's life, and for those lives that have been interwoven with human space flight, their personal memories can often be celebrated alongside milestones in space history.

In 1978, for the Thirty-Five New Guys (TFNG), the date of January 16 was one such key milestone, and the date on which they officially became known around the world as NASA's eighth group of astronauts has shared its significance with other moments in human space history across the years. For example:

- On January 16, 1969, Soviet cosmonauts Yevgeni Khrunov and Alexei Yeliseyev spacewalked from the Soyuz 5 they had launched in across to the docked Soyuz 4 they would return home in, completing the first Soviet Extra-Vehicular Activity (EVA) for nearly four years.
- In 1976, exactly two years before the TFNG were revealed to the world, a crew module arrived at Rockwell's Palmdale facility in California for assembly as Orbital Vehicle 101, later known as Orbiter *Enterprise*.
- Earlier on the very same day as the Class of 1978 were paraded in front of the press, the Soviet Soyuz 26 Descent Module landed, with cosmonauts Vladimir Dzhanibekov and Oleg Makarov returning from a one-week visit to the Salyut 6 station, having exchanged the older Soyuz 26 with the fresher vehicle Soyuz 27 they launched in.
- In 2003, as the Class of 1978 celebrated their silver jubilee, the crew of *Columbia* was launched on the ill-fated STS-107 mission.

Regardless of all the other anniversaries, celebrations and memories, however, it is January 16, 1978 that meant the most to the individuals who were named America's latest astronauts on that day, one of whom – Michael Coats – also celebrated his 32nd birthday with added pride. There would be other memorable dates for the TFNG of course, such as June 29, 1978 when they arrived at NASA's Johnson Space Center (JSC) for the Ascan training program. Or August 31, 1979, when they officially ceased being Ascans and received their Silver Pins. Then there were the individual days on which each of them entered space for the first time (or the second, third, fourth, or fifth time) and the days they each returned safely to Earth. Or the time they were awarded the NASA astronaut Gold Pin. Alongside these good days there were the bad days of course, with January 28, 1986 being foremost of the 'dark times' when the group lost seven colleagues and friends including four of their own. But amid all the highs and lows, that 16th day of January in 1978 will always remain very special to each of them.

2018: A RUBY YEAR

Fig. 14.1: The official 40th anniversary emblem designed by Tim Gagnon and Jorges Cartes. The Earth is taken from the STS-7 emblem, the first mission with TFNG crew-members, while the Shuttle stack is taken from the STS-8 launch honoring Guion Bluford's involvement in the original emblem. The Sun was updated and a spiral galaxy added to honor astronomer Steve Hawley, the last TFNG to fly in space. The ten stars honor the both the ten stars on the original emblem and the 10 TFNG who were deceased at the time the patch was designed. The number '40' is ruby red to acknowledge the milestone and the class acronym 'TFNG' is colored in silver and gold to signify the transition from silver to gold Astronaut Pins when they fly (Image courtesy of Tim Gagnon and Dr. Jorge Cartes, used with permission).

With Anna Fisher, the last remaining member of the class, leaving the agency in 2017, the next January 16 in 2018 was the first in four decades without any of the group still at NASA to mark the anniversary of their selection. Ironically, that year was also the Ruby Anniversary of the group's selection, a momentous event that just had to be celebrated. It was just done a little early, on December 14, 2017, when most of the surviving members of the selection and their spouses attended a reunion in Houston. According to Rhea Seddon, the timing was due to another event that most of the group wanted to attend: "Our 40th was done in December because the All Astro[naut]s Reunion was being held then and rather than all of us traveling twice we decided to do ours then attend the other." [2]

Time marches on for all of us, even for the youthful Class of 1978, as author and TFNG Mike Mullane eloquently observed of the reunion on his website: "As it is for all attendees at a class reunion, we see our age in the changes of others. I was certainly reminded of mine. Gravity had replaced the sculptured hardness of our youth with sags and lumps. Hairlines were galloping in retreat… or missing altogether. Faces were etched with the arroyos of age. Hearing aids were de rigueur. Reading glasses were holstered in pockets, ready for a quick draw when the menu came." [3]

After dinner, a slide show highlighted milestones in the Class of 1978 saga, with Rick Hauck taking the lead for a series of heartfelt toasts to the ten colleagues forever absent. The guests of honor at the dinner were Apollo and Skylab astronaut Alan Bean, who had been their mentor during their Ascan year, and former Director of Flight Crew Operations at JSC, George Abbey, who was also a member of the selection board that chose them and, according to Mullane, considered the "father of the TFNG." Abbey gave a reflective speech, "kind of making the point that we [as a group] had helped to shape the Shuttle program," according to Steve Hawley. [4] According to Mullane, Abbey told them how proud he was of them all and that, given a second chance, he would not change any of his selections. The next evening, the group attended the all-astronaut reunion, with many attending from other selections including the latest dozen Ascans chosen in the Class of 2017 (Group 22 "The Turtles")[1], in whom Mullane recalled the youth, passion and determination of himself and his peers.

Looking back over their careers in the Astronaut Office, many of those in the TFNG collective shared some of their most memorable moments.

---

[1] It must have been sobering to the TFNG who attended the all-astronaut reunion that all but three of the 12 "Turtles" were born *after* 1978, with the youngest born a decade after the Class of 1978 had been selected and after all of them had flown in space.

Fig. 14.2: The TFNG 40th anniversary reunion, December 2018. [Front l-r] Shaw, Fisher, Seddon, Hauck, Nelson, Sullivan, Hart, Thagard, Shriver, Lucid and Mullane. [Rear l-r] Gibson, Covey, McBride, Creighton, Brandenstein, Bluford, van Hoften, Fabian, Hoffman, Hawley and Coats. Not featured in the image: Gregory, Buchli, Stewart (Image from the collection of Rhea Seddon, used with permission).

A 'throwback mission'

In 2019, Steve Hawley was reminded he had been the final member of the selection to fly in space, two decades earlier, and was asked to reflect on how much had changed over the 16 years between the first flight of the TFNG in 1983 and the last on STS-93 in 1999. "Ironically, STS-93 in a sense was sort of a throwback mission, because if you look at what we were doing in those days, certainly over time, the missions became more and more complex, more and more difficult, [and] we took on more and more mission objectives. I always said early in the program [that] we held back a lot of margin on behalf of the crew, and as we got more experience and more mature in the program we started to give up some of that margin for the accomplishment of the mission. But 93 was a five-day mission where we deployed a satellite, [like] back in the old days, so it was a little bit of an [outsider] at the time. For me it felt kind of familiar, but I would say that in the 16 years from launching satellites to servicing Hubble Space Telescope, [and] building ISS, I think we realized a lot of the potential that we had always said that the Shuttle represented. I would like to think that our group helped that, first of all because there were enough of us to be involved in maybe more aspects of the program than could be done by a smaller astronaut core, which brought the astronaut prospective to different activities and also allowed the Astronaut Office to know what was going on maybe a little more broadly.

"I used to joke about this, although I think there is truth in it, that people would say 'Was it a culture change when your class was selected because you had women and minority astronauts for the first time?' I would usually say 'Yes'. I mean, there was certainly a culture change but I felt that the biggest culture change was the introduction of civilians into what was really a military structure to the Astronaut Office. The way I described it was that if you give a pilot a procedure, the first thing a pilot will do is learn how to execute it. You give it to a scientist and they will look at it and say, 'Yeah, well you know, I don't think that's the way I would have written it'. And that combination, I think, led to the ability to identify things that we could improve. And sometimes we were sent packing, and sometimes we weren't, while getting the improvements implemented." Hawley accepted that when they first entered the Office, there was perhaps a sense of breaking the barriers of the traditional Astronaut Office genre to establish their own credentials, which was one of the reasons they were selected in the first place. "I felt it was, but maybe not everybody felt it was that way. But I think legitimately there were questions, particularly about the civilian scientists, and the questions were of the form, 'Are these guys really capable of being operational?' In fact, when I was on the selection board, that was one of the hardest things to evaluate. I mean, if you're hiring a pilot, a military test pilot or military flight test engineers, there is no doubt that the guy is operational because he has had to do that. But if you hire a guy fresh out of grad school, somebody like Pinky [Nelson] or me or Sally [Ride], will they be able to develop a kind of operational mind set to make it successfully as a crewmember? I think people sort of assumed, until we proved we could do, or weren't capable, or at least maybe they were skeptical about it. One of the things I tried to do early in my career, when I got assigned to SAIL [Shuttle Avionics Integration Laboratory], the software verification lab, was I wanted to learn as much as I could about the software, about how it worked, and that allowed me, not only in testing but later on in training, to demonstrate that I had some operational skills." **[4]**

REFLECTIONS FROM THE CLASS OF '78

One of the most useful resources for a space historian is the chance to interview key individuals about their participation and achievements in the program, but it is often impractical to get to interview all the people you would like to speak to in person. This is where the official Oral Histories are invaluable, and the archives of these interviews are a gold mine of information for the researcher. Throughout this trio of cooperative titles, as well as all the other titles the authors have penned, first-hand interviews remain the paramount and most precious source, but the archive of Oral Histories can be a useful and informative back-up option. Fortunately, NASA has recognized that this is a valuable resource and has created an online catalogue of interviews under the NASA JSC Oral History Project. Readers are encouraged to browse the treasure trove of information within these interviews. **[5]**

JSC Oral History Project
This project was established at NASA JSC in Houston in 1996, with the goal of capturing "history from the individuals who first provided the country and the world with an avenue to space and the Moon." Oral History interviews began in 1997, and in the two decades since then almost 1,000 individuals have participated. These include managers, engineers, technicians, doctors, astronauts and other employees of NASA and aerospace contractors who served in key roles during Project Mercury, Gemini, Apollo, Skylab, Space Shuttle, and more recently in the Shuttle-Mir Phase 1 program (including a number of Russian participants) and the ISS project. Within these archives are 54 interviews by 26 of the TFNG between 1998 and 2016. Those still awaited (if they have been, or plan to be conducted) are interviews with Jim Buchli and Bob Stewart, with hopefully further sessions from Hoot Gibson, Shannon Lucid and Norman Thagard to name a few. Those opportunities lost by other "moments in time" are the *Challenger* four (Ron McNair, Ellison Onizuka, Judy Resnik and Dick Scobee), plus Dale Gardner, Dave Griggs, and Dave Walker[2].

---

[2] The TFNG Oral History interviews (number of sessions in brackets) to date (2019) are: Bluford 2004 (1); Brandenstein 1999 (1); Coats 2008–2015 (6); Covey 2006–2007 (4); Creighton 2004 (1); Fabian 2006 (1); Fisher A. 2009–2011 (3); Gibson R. 2013–2016 (2); Gregory 2004–2006 (3); Hart 2003 (1); Hauck 2003–2004 (2); Hawley 2002–2003 (3); Hoffman 2009–2010 (3) Lucid 1998 (1); McBride 2012 (1); Mullane 2003 (1); Nagel 2002 (1); Nelson G. 2004 (1); Ride 2002 (2); Seddon 2010–2011 (4); Shaw 2002 (1) Shriver 2002–2008 (3); Sullivan 2007–2009 (4); Thagard 1998 (2); van Hoften 2007 (1); Williams 2002 (1).

Within these official Oral Histories, each individual was asked about their youth, early careers, the numerous assignments and space flights during their years at NASA, and the move into management or appointments outside of the space agency. To include a personality feature, each participant was also asked about their most challenging milestone, their most significant accomplishments, and what had given them the most pride during their career at NASA. In this summary, we have extracted a selection of those replies to present their thoughts on their personal achievements and successes at NASA.

Brewster Shaw recognized that future generations will be the judge of the Shuttle Program. "Well, this is a history. I mean, we're talking about history here. We don't know what history is going to say about this program. We know what history said about Apollo: it was a great success. And then history quickly forgot about Apollo. Instead of flying several rockets we had lined up to go to the Moon, we just parked them out here. So, history is fickle and it's certainly subject to politics and the winds of the nation. So the history of the Space Shuttle Program and human space flight, as we know it now, has yet to be written, but I hope that the historians are reasonable and are kind, because I think that the Space Shuttle is a wonderful accomplishment and so is the Space Station. If we don't continue to do things like this, our whole lives will end up being virtual and not real from the aspect of human knowledge, and human exploration, and humans understanding our world and our lives, and that would be a great travesty if we virtualize our future rather than living it." Though he did not mention the TFNG as a group, their place in that history will also be judged. **[6]**

Mike Coats was asked about his perspective from his two very different experiences, as a NASA astronaut and later as a field center (JSC) manager. "I've told people for years, and I don't think anybody believes me, but it's actually tougher to watch a launch than it is to ride through a launch, and it comes down to the 'control freak' part of it. When you're in the Orbiter and you've been trained up to peak efficiency, you're very confident you can handle anything that can humanly be handled. The things that can't be handled, why are you worried about it? You're anxious not to embarrass yourself when you're in the Orbiter and make a mistake. On the other hand, you have complete confidence in your crew and in the ground controllers and yourself, and 'let's go do it'. You're just so mission focused, getting it done. You're proud of the level of proficiency the crew has achieved in the simulators, getting ready to go. It's almost like, 'Okay, throw it at me, I can handle it'. Which is what you want them to be thinking.

"When you're watching a launch as Center Director, you have no control; you have nothing you can do except hope everything goes okay. That's a horrible feeling, to have no control, and because you've been through the training, you know all the different emergencies that could happen to you. You can't help but be thinking about, 'Okay, if we lose an engine now, what happens? If we lose an engine now, what do we do? If we lose an engine now? If we lose the hydraulic system, what happens? If we lock up an engine?' That stuff goes through your mind, and you can't help it. You lived it, trained it, for years and years. So, it's tougher to watch a launch than it is to ride through it, or at least it was for me. On the other hand, boy, you're as happy as the crew when they come back safely. When you're hugging the crew out there, boy, you really mean it, 'It's so good to have you back'. That's the big difference to me. You feel responsible for the crew. The Center Director has final approval on crew assignments, and you just feel responsible for every crew when they're up there; you want them to come back safely. That's a big responsibility." **[7]**

In a subsequent interview, *Coats* expressed satisfaction in having "the opportunity to work with a lot of really good people, and I got to see what dedicated and talented people can do in some real crisis moments. We had some problems with the Shuttle and then the Station that any space program is going to have, and watching the teams work together to solve the problems and come up with some truly elegant technical solutions was really fun and really satisfying. I wish my kids and grandkids could have that kind of satisfaction. Being part of a very successful team, a motivated team, was pretty neat, a good memory. I think it was a privilege to be an astronaut and fly into space. It was even more of a privilege to be the Director of the Johnson Space Center and feel like you're part of a team, and to be able to, from that perspective, see the different parts of the team and how they work together, what they accomplished. I think being the Center Director is a unique perspective because you get to see so much. I just can't say enough about how motivated this team is. They think they're making a difference, and they are, and that's pretty neat." **[8]**

Guion Bluford is most proud of having the opportunity to become an astronaut and going on to participate in four very successful Shuttle flights. The years at NASA were "a great experience which I will always cherish. I also felt very privileged to have been a role model for many youngsters, including African-American kids, who aspired to be scientists, engineers and astronauts in this country." **[9]**

Dan Brandenstein felt that one of the "neat things about being an astronaut" was the opportunity to work with such a great team. Of course, he admitted that "Flying in space is neat, but it [the satisfaction] kind of goes back to that diverse thing. You get to know and work with the people building the vehicle; you get to work with the various people that develop the experiments and the payloads; and the flight controllers; many of the guys that process it down at the Cape; and just being able to come in contact with so many dedicated people, it's, in my mind, the best part of being at NASA."

The most memorable time for Brandenstein was the STS-49 mission (maiden flight of *Endeavour* and Intelsat re-boost), "because we had a real [challenging] time. That obviously makes it very rewarding, because you've had a problem with something, you sorted through it, and successfully accomplished what you'd originally set out to do, even though you did it differently. But, once again, that was not just the crew or just myself, that was the whole team kicked in on that. That was, I'm sure, the most challenging, and makes it probably the most rewarding, too." **[10]**

Dick Covey was sure of the most challenging milestone in his years at NASA "Well, in my mind there's no doubt that preparing for STS-61 [the first Hubble Service Mission] was the most challenging thing because of the complexity of what we wanted to do and gaining a constituency that believed that the crew could go and do all those things that had never been done before. You know, expanding from three spacewalks max [maximum] on a mission, where two was nominal and three was exceptional, and going to five; the number of complex operations that we had. It wasn't that there was resistance. It's just that we were pushing the bounds, and so the way you have to demonstrate, prove, that you're ready to go do that, not just the crew but the agency, was, I thought, a major accomplishment.

"Overcoming challenge sometimes is the most significant accomplishment. There's no doubt that having commanded STS-61 and what we did on that mission was the greatest accomplishment that I was able to make to the agency and our nation. No question about it." And one of the proudest moments in his career? "Oh yes, [it] still is. It's way up there. Other than babies and grandbabies, getting married and things like that." When it was pointed out he rated becoming an astronaut as pretty good but felt that getting married and having children was a little higher, he replied "Yes, you've got to balance those things. If you put them all together, you're going to default over to the family stuff, there's no question about it." **[11]**

John Creighton thought that the most exciting event was the day in early 1978 when he received the telephone call informing him that he had successfully made the cut and was one of America's latest astronauts … "And then launch morning on my first space flight. They're all exciting, every launch is exciting, but I think your first one is the most exciting. Sitting out there waiting, and somebody saying, 'Ten, nine, eight, seven...'" **[12]**

John Fabian said that being part of such a remarkable group of individuals was a high point of his time as an astronaut: "When you come into a group like that, a group that is so highly selective, you don't come into it with a lack of appreciation of your own talents and abilities, and you don't come into it with a lot of self-doubt about what it is that you can do or might be capable of doing. But those thoughts are challenged. They're challenged by the very nature of the people that you're now involved with. And somehow, through the process of working with these people and working within the system and so forth, you need to come to an appreciation that your original thoughts are probably right; you do have capabilities, and you do have things to contribute, and so forth. But that's not an easy decision to make when you see the greatness of the people that are around you.

I think that that adjustment hits everybody who comes into the Astronaut Corps, because no one who comes into the Astronaut Corps has ever been a slacker."

As for his most significant accomplishment in the Astronaut Office, Fabian immediately picked out his work with the Shuttle Remote Manipulator System (RMS) "We really had an opportunity with the RMS to work on the human interface, to make it something which is straightforward and easy to use, intuitive in its application. That's now followed over into the Space Station, and potentially it will go on to other applications. I think it's the most significant thing that I did in my time, and I think it's the thing I'm proudest of." **[13]**

Terry Hart, the first of the group to leave the Astronaut Office shortly after flying on STS-41C in 1984, felt that the whole process of being an astronaut was challenging, "but there's nothing that is insurmountable, because you know you have so much help. It's not like you're ever worried about failing or whatever, because there's so much support behind you. So you just need to learn how to take advantage of all that support. So I guess, in a sense, that's the challenge, to fully integrate yourself with the team at NASA to ensure that you're all doing what's necessary for the mission. You're all focused on the right thing, and when things go wrong, you can fall back on that and find ways to work around the problems."

When asked about his most significant accomplishment at NASA, Hart replied that it was his work during STS-41C. "I guess the most visible thing was the Solar Max capture, and getting the satellite onboard. I'm sure George Abbey felt that I earned my keep in that ten seconds that I reached over and grabbed the satellite. But I'd like to think more that I helped a lot with that sense of building the team. Within my class, I felt good about my relationships with all the other guys in my class, and the gals. We all worked together well. So I always kind of looked upon myself as being a bit of an instigator of teamwork and someone that helped promote those kinds of things, which, you know, [is] a very natural thing at NASA." **[14]**

Rick Hauck considered his most challenging milestone to be preparing for STS-61F, the Shuttle-Centaur mission, which he referred to as "the mission that never was" because it was clearly going to be a very risky flight. In contrast, he thought that his most significant accomplishment during his time at NASA was that "I didn't screw up. I don't know, I just worked very hard and I was very fortunate. I was rewarded by always being given good assignments and working with a good team that answered the mail, I think. I don't know, I think my strength is probably in flying ability, but also perhaps in a leadership role in bringing people together and working through issues, but I can't point to anything in particular." **[15]**

Steve Hawley found it difficult to single one thing out when asked to highlight the most challenging aspect of his career, both personally and professionally, back in 2003. "Well … it wasn't as big a leap to go from being a scientist to being an operator as perhaps it was to go from being an operator to being a manager, and having to learn how to be successful in the world of people that do business-type things for a living. At least, I hope I was successful. That was pretty challenging. I've had a few people say, in particular recently now that I've moved on from FCOD [Flight Control Operations Directorate], that I was one of the few people they dealt with that had any business sense. Part of that, frankly, I think, is we have a little bit of a history of people in those jobs who are there for a short period of time and then rotate back into flight assignment. I was maybe one of the first to come out of the Astronaut Office that actually did it for a prolonged period of time, where I was able to actually develop some experience in how to do business with the world of resources and procurement and all of that. Maybe if I was able to develop something of a reputation as being pretty good at that part of the business where I really had no training at all going into it, then that may have been among the most challenging things to do."

In response to a question about his most significant accomplishment at NASA, Hawley admitted "Well, that's another hard one. There's so little that you really get to do yourself. Everything you do is as part of a team, I think. I mean, obviously, the things we did with Hubble are things that I'll remember forever, but they're not things that I did uniquely. I did them as part of a team. And if it hadn't been me, it would have been somebody else that would have done it just as well. I guess every once in a while I had a good idea that somebody incorporated. I suppose, in general, particularly as I got into management jobs, the thing that gave me satisfaction on a day-to-day basis was 'did I do something today for somebody that actually helped them?' My experience, which I guess is not unique, is you go do battle on a hundred different fronts, and you're going to lose 95 of them, but you may win a few. And you have to be satisfied, I've always thought, with the knowledge that you're not going to win all of them, you may not even win most of them, but you may win some. And the ones you win are going to help somebody do the right thing, or they're going to help the program do the right thing. If you can take satisfaction in the few that you win, knowing that you really did some good for somebody or some program, then I think you do get job satisfaction in that. So, sitting here right now, I'm not sure I could tell you what any of those are, but I do feel like I won a few over the years that probably gave me a lot of satisfaction at the time; that we were able to win this point, or successfully fight for funding in some area that we thought was important, where they had cut our budget, or win some technical point that was going to make the program better." **[16]**

Mike Mullane clearly appreciated his time at NASA: "I feel very blessed that I got to fly in space," and, he added, working with "I would have to say, one of the best teams that I'd ever been with in my life, the best in MCC [Mission Control Center], the best in the crews and the LCC [Launch Control Center]. No question about it, they were the best, and I felt privileged to be involved with it." **[17]**

Steve Nagel recalled that taking on the running of the Astronaut Office, as the Acting Chief, was one of the hardest things during his time at NASA. "That was a hard job, rewarding but difficult. Certainly the time after *Challenger* was a real hard time, but after we got our mourning process behind us and started looking forward, it was my best time there. It was real challenging, fast moving … Training for the Shuttle flights is kind of easy compared to the other stuff. You're kind of spoon-fed a syllabus. You're going back to being a student again. I don't mean to make it sound like it's trivial, easy. There's a lot to learn. You're really drinking out of a fire hose, but your life is pretty structured there. So that's not so difficult. Where I am now [2002, in the Aircraft Operations Division] is a slower, a little less demanding pace of life than being in the Astronaut Office, even though I travel a lot. I'm on the go a lot. But it's kind of a nice change of pace out there. So I guess the hardest, most challenging time was Chief of the Astronaut Office, and secondly was the *Challenger* thing. It's all been great, though. I wouldn't trade any of it. There's been the good times and the bad times, but it was a great opportunity to be there. I always felt very fortunate to have been where I was."

As for his most significant accomplishment while at NASA, Nagel thought that his work on the crew escape system was perhaps more important than flying on the Shuttle. "The flights were great," he acknowledged, "but in a way, I got more personal satisfaction out of having worked on something that I think did some good like that [the crew escape system] than even the flights. Not that I want to trade the flights, because that's why I came here, but that was the high point for me. I've talked to a lot of these people that worked on this project, that crew escape project, not just in NASA … even Bill Chandler, the manager, the chief Navy jumper, that did the jump on the pole. They said this was the high point of their career. It was the best thing they ever did … because it was so important to everybody. I mean, the Shuttle's grounded, and we're doing something that's going to get it flying again. I think everybody, almost no matter what they worked on through that time period, had a similar feeling. Really, it must have been kind of like people felt going to the Moon, a little bit. I don't want to compare it exactly, but that kind of a feeling.

"I had that overall feeling working on the Shuttle Program, anyway, or in Station Program, no matter what capacity, even in what I'm doing now. You feel like you're a part of something that's bigger than you, that's worth doing, which is really great. Not everybody gets that opportunity to work on something like we do here. But I really had that feeling on the crew escape project for that year or two." **[18]**

George 'Pinky' Nelson came up with a different perspective. "Getting selected was certainly the most challenging thing," he felt. He then continued, "I think the job of being an astronaut is not all that hard. You're in an incredible support system. You're extremely well trained and provided with resources that you need to get trained, so if you're a motivated person who has basic, good motor control and [is] fairly smart, you can learn how to do all the procedures. It's a little different for the pilots, I think, who have to have special skills and who are just really terrific at flying the machine. I can't think of anything I had to do that was really hard. I found it all just very enjoyable. It took a lot of effort to learn how to work in the spacesuit really well, but it was an interesting challenge and fun. The rest of it just came naturally. Being a person with a fairly short attention span – working within the system of going to endless meetings, working within the NASA system, which is really the one that Chris [Christopher C.] Kraft [Jr.], Al [Alan B.] Shepard, Deke [Donald K.] Slayton, and those guys made up on how to run a space program, [and which] really functions well in terms of checks and balances and procedures – learning to work in that system has its challenges, because you have to really carefully document and account for everything you do. So just working in that system was a bit of a challenge for somebody like me." As for his biggest accomplishment while working at NASA, Nelson rated the support of his family for what he was doing the highest. "I would rank the family support plan as one. I think that's had a big impact on the quality of life of the astronaut program. Then just technically, I think the work that I did on the spacesuit made it a useful tool." **[19]**

Rhea Seddon thought that her greatest accomplishment while at NASA was to ensure she also had a normal life. "When I went into NASA, I thought, 'If I choose this important and fascinating career, am I ruining all of my chances for ever getting married? If I'm an astronaut and I decided to have a baby without being married, what would be the consequences of that? Is there something that's going to happen to me that will make it difficult for me to have kids? Will I ever meet somebody whose life will mesh with this insane life I'm about to undertake?' So it was one of those unknowns, and one of the things that I hoped to accomplish at NASA was to come out on the other end with what I considered to be a normal life, the kind of life that I wanted to have. I was able to do that. I don't know if it was fate or the Good Lord looking out for me, or the right person coming along, or a combination of all those, but I'm happy that I didn't have to give up those other parts of life in order to do my work at NASA. Was I the world's best astronaut? Probably not. Was I the world's best wife and mother? Probably not. But I got to do all of those things, and that was very important to me. I wanted to look back when I got to the age where I am now and say, 'I was able to do all of those things'. When they were handing out flight assignments and Hoot and I had just gotten married, there was the decision of whether to wait to have children or to go ahead and try. I considered [that] when I'm 60 and I look back, would I rather have an early flight and no children, or children and maybe no flights, or maybe later flights? The children were a higher priority. So I was glad to be able to look back when I turned 60 and say, 'It all worked out. It worked out just fine'. I think some other people had to choose one or the other, and I was lucky that I didn't."

As for the most challenging aspect of working at NASA, Seddon selected "probably some of the physical aspects of training and doing the work. The scuba training was hard for me. It was physically hard for me. I doubted whether I was going to be able to do it. The suit work that we had to do, there were times when I thought, 'I'm not sure that I'm going to be able to do this, and if I'm even able to do it, would I be able to do it in an emergency and do it well and quickly?' So I think the most difficult part for me was the challenge of being small, because NASA – even though they broadened the height requirements so that they could take more women, and bless them for that – I don't think they had really planned or thought about what it would be like for a really small person to do the things that larger men found pretty easy. There were things along the way that I had to accommodate to. Getting into the T-38 was always a struggle. Certainly there were the physical aspects of being pregnant and wanting to continue to work when I'd rather be home with my feet up. I was just determined to work through my pregnancies and come back healthy and ready to go again, and to have good care for my children. So that was part of it, just the strength-type things. I wish I could say something more wonderful about challenges that I overcame. You know, people say, 'Wasn't it difficult to fit into an all-men's world?' I just ignored that part of it. If they didn't want me there, that was their tough luck. And so what, if I didn't fit the mold. I wasn't a test pilot. I didn't get much credit for what I did on my flights. Everybody else got a lot of glorification for doing spacewalks and rendezvous and being commanders of things, and life sciences got kind of short shrift. I didn't care. That didn't bother me. I was there to do what I wanted to do and what needed doing. Nobody else wanted life sciences flights, but that's what I wanted. I felt like I had accomplished good things in being on those flights." [20]

James 'Ox' van Hoften, when asked about his most significant accomplishment in the years he spent in the Astronaut Office, replied "Oh, I think the EVA side of this thing [on STS-41C and STS-51I]. Obviously, I was just in the right place at the right time. I was really lucky, and anybody could have done the same thing. Nothing that I did was, I think, something that somebody else couldn't have done. I would never try to convince myself of that. But I think doing those things and making it work. That second mission was tricky. That did take a lot of strength and a lot of work at that. But the one thing I've been famous for is I'm very calm under pressure, and I don't rattle very easy. On something like that, it worked. It was nice to be able to do that. I look back on it; it was a pretty short time. I was only there eight years, and it seems awful short, but it seemed like just the perfect time for me, and it was … the experience was extraordinary. I came away with nothing but positives on that. There's a lot of people I know that stayed there for a long time and came away with some pretty mixed stories. I don't have any of those. There's a few bad ones that I don't tell anybody, but nothing that was not real positive for me."

Like many of his colleagues, van Hoften felt that his toughest challenge was getting selected for the astronaut program in the first place, and then "once you're there, trying to maintain the right attitude. I mean … the attitude you have when everyone is getting assigned to a mission and you're not is tough. That's a really tough one, and to keep going forward and doing like you're supposed to do and just wondering, 'What the heck do I do? How do I get on the ladder here?' Looking back – it seemed like forever at the time – it was just a matter of months, and once you're there, it was just wonderful. Once you were finally in the loop, it was brilliant. There were some other guys who were really good. I can't even remember who flew last, but there were a lot of guys like Covey, for instance, that didn't fly until late in the program, and he's one of my best friends, and I couldn't understand why he was not getting selected for something. He had kind of a bad run at it in the beginning, but there was a lot of them. There's a lot of stories about the Air Force getting [overlooked], you know – [George] Abbey always supposedly had favor for the Navy, and who knows? But once guys all got in the circle there, it worked great, and once everybody started flying, people stayed friends … [But] There was an awful lot of competition, although not openly; you know, there was a lot of competition to just get attention, I guess. That's hard. A lot of us aren't real good at that. You don't want to go out and kiss up to people, and some people tried, and that wouldn't work. So no one knew what the heck to do." **[21]**

Don Williams firmly believed that becoming a mission commander was by far his greatest accomplishment at NASA. "The responsibility that weighs somewhat heavily upon you and the accountability that's always there, always there sitting on your shoulder like, 'Okay, Don. This is your chance. Go do it'. That was a challenging thing to do. The second most one, from a crewmember and a pilot perspective, is learning enough about a very complex vehicle that you can fly and operate it in all of its environments from the launch pad to the landing, [and] post-landing activities. I'll give you an example. I flew A-7 [Es], the Corsair [II] airplane, a light attack airplane, off of the [carrier USS] *Enterprise* in all kinds of weather, at night and [in a] wartime environment. That airplane had inertial systems, radars, weapons systems, environmental control systems, hydraulics, electrical, heads-up displays, electronic warfare systems, and communications. It was a fairly complex airplane, and flying that by yourself at night, off a carrier, on a scale of one to ten is about a seven or eight, roughly, in complexity.

"The Shuttle on that same scale is about a 30. It's two orders of magnitude more complex. It's amazing. It's largely due to the environment it flies in, from a rocket launch to an orbiting spacecraft, to a hypersonic re-entry, to a conventional gliding kind of airplane. The systems onboard to make that all operate are very complex and they all interact with each other and they have a lot of redundancy by design. But the redundancy also adds complexities, because you have to know how to gracefully fault-tolerant them down as you have failures, which you have lots and lots of in simulators and hopefully not too many in orbit. So it's a very complex flying machine and it requires a lot of attention to detail by everybody; not just the crew, by everybody.

"The mission commander challenges, for me personally. [were] being able to … I don't know whether you ever master something as complex as a Shuttle, but you get pretty familiar with it to a point where the 800 or so switches and controls onboard there, you know what every one of them does. And when you pull one, push one, turn one, or toggle it, or whatever you do to it, you have some idea of what happens inside the spacecraft, whether it's under your foot, next to your arm, or 70 feet away in the back."

Asked if he had any other comments for his Oral History, Williams replied "Maybe two, the first one being the teamwork that it takes to pull this off. I think you may have heard a lot of other people say this, but in order to successfully pull off a Shuttle mission, or any human space flight mission, [it] takes thousands and thousands of people all doing their job correctly every single time. You cannot afford [for] anybody to fail and [still] have a successful mission. And I think every crewmember that you ever talk to will probably say that, if you ask them about the teamwork that's required. A lot of credit goes to the 'invisible people', as they call them.

"The training teams, the flight control teams, the planning teams, the contractors, and civil servants that make all this happen. The secretaries, the people that take out the trash, the people that glue the tiles on the Orbiter, the people that do the launch countdown, the people that paint the towers, and the people that make the hardware and the software that make it all work. Every single one of those people has to do their job, and there's thousands of them, and it all has to come together in the end. Those of us who get on the pointy end of the spear, if you will, maybe get a lot more credit than we deserve. You have to have trust. You have to trust them, and that's a special thing about being part of this human space flight program. It's a very special thing and it's very unique about this program and perhaps aviation in general, particularly high-performance carrier aviation, where you depend on the people that work on the airplane, that work on the catapults, that work on the arresting gear, that drive the ship, that fuel the airplane, that load the airplane, that fix everything that goes wrong and get it ready to go, so you can go fly it and do your job. That's part of the team. So, being part of a huge team like that is a really fascinating and wonderful thing. I mean, that's an enjoyable thing to do.

"Then, finally, the perspective you get from on orbit of this planet of ours, which has been described in much better words than I can put it, but you say to yourself, 'Hey, it really is round', which is good, because otherwise all this orbital mechanics and stuff wouldn't work. The perspective you get, that there are no boundaries between nations and there's no labels on anything. The only visible boundaries, quite truly, that I recall seeing as I go back and look at the video and the pictures, is between the land and the sea. Therefore, why is it that we fight over things so much and we disagree and we argue about lines that are drawn on the surface by humans, which seems to cause people to disagree about things, when, in fact, we're all in this together? I'm not necessarily an advocate of world peace and world unity, but on the other hand, getting along with each other seems like the right thing to do, because we don't have any place else to go right now. Maybe someday we will." **[22]**

SUMMARY

Over four decades after their selection, and 20 years after the final member orbited the Earth, many of the surviving members of the Class of 1978 participated in the NASA JSC Oral History Program, reflecting on their very personal contributions and achievements to the nation's space program as it matured from Apollo into the Space Shuttle and on through the creation of the ISS to the emergence of the next generation of American human spacecraft. The new vehicles are the successors to the Shuttle program that the TFNG were chosen for and flew on, one which was greatly enhanced by their dedication, skills and contributions. Exactly how their efforts will be recorded in history will be for future generations, as Brewster Shaw indicated, and after all, it is all "a matter of time." But what is their legacy in the program today, with a new generation of space explorers hoping to embark in the next era of human space flight, returning us to the Moon and venturing out deeper into space? That is explored in the final chapter in this story of the TFNG.

References

1. The Concise Oxford Dictionary, 1991 edition p. 558–559.
2. Email from Rhea Seddon to David J. Shayler, January 26, 2020.
3. *Anniversary of the Astronaut Class of 1978…The TFNGs*, by Mike Mullane, https://mike-mullane.com/40th-anniversary-astronaut-class-1978-tfngs/. Used with permission.
4. AIS interview with Steve Hawley, October 23, 2019.
5. The NASA Oral History Project can be found at http://historycollection.jsc.nasa.gov/JSCHistoryPortal/history/oral_histories/oral_histories.htm
6. Brewster Shaw, NASA Oral History, April 19, 2002.
7. Michael Coats, NASA Oral History, July 14, 2015.
8. Michael Coats, NASA Oral History, September 14, 2015.
9. Guion Bluford, NASA Oral History, August 2, 2004.
10. Dan Brandenstein, NASA Oral History, January 19, 1999.
11. Dick Covey, NASA Oral History, March 28, 2007.
12. John Creighton, NASA Oral History, May 3, 2004.
13. John Fabian, NASA Oral History, February 10, 2006.
14. Terry Hart, NASA Oral History, April 10, 2003.
15. Rick Hauck, NASA Oral History 17 March 2004.
16. Steve Hawley, NASA Oral History, January 14, 2003.
17. Mike Mullane, NASA Oral History, January 24, 2003.
18. Steve Nagel, NASA Oral History, December 20, 2002.
19. George 'Pinky' Nelson, NASA Oral History, May 6, 2004.
20. Rhea Seddon, NASA Oral History, May 10, 2011.
21. James van Hoften, NASA Oral History, December 5, 2007.
22. Don Williams, NASA Oral History, July 19, 2002.

15

The legacy

"For those who have seen the Earth from space,
and for the hundreds and perhaps thousands more who will,
the experience most certainly changes your perspective.
The things that we share in our world are
far more valuable than those which divide us."
Don Williams quote, recalling his observations from orbit:
From "Eyes Turned Skyward." [1]

Those words from Don Williams summarize the bond that all space explorers have felt over the past six decades, witnessing with their own eyes the beauty and fragility of our planet from orbit. The desire to apply for astronaut training burns strongly in many of us, but having the skills and the opportunity to put that dream into practice is a completely different matter. First there is the challenge of progressing through the selection process. Then, if successful, comes the basic preparation course to qualify for assignment to a space flight. After this is the waiting, perhaps for years, for the opportunity to fly for the first and perhaps only time, although there is always the hope for more. But leaving the Earth and arriving minutes later in orbit has to be worth all the effort and sacrifices to get to that point.

Taking a space flight has never been a solo event, even in the days of single-crew spacecraft. There is a whole team, or rather a community, behind those who fly. That team, whether national or international, designs the missions, manages the program, builds the hardware, tests the systems, prepares the vehicles, and controls the operations, then moves on to repeat their part of the sequence in time

© Springer Nature Switzerland AG 2020
D. J. Shayler, C. Burgess, *NASA's First Space Shuttle Astronaut Selection*,
Springer Praxis Books, https://doi.org/10.1007/978-3-030-45742-6_15

for the next flight. At the forefront of all this are the flight crew, the small group who are in the public (and political) spotlight to execute the objectives and bring the results home. Superbly prepared and trained, each flight crew is a group of highly talented individuals who mold together over time to become a single unit. Trust is important, between each other, with those on the ground, and with the hardware and software around them. Isolated from the home planet, the flight crew is totally dependent on all the equipment and procedures working well each and every time, or on possessing the knowledge and capability to overcome problems when it does not.

Sometimes events happen so fast that, no matter how good the preparation and training, fate demands a different outcome. In his presentations to the public, after asking his audience, by a show of hands, which of them would like to ride on a rocket to space, Skylab 4 astronaut and member of the Group 8 selection board, Ed Gibson, usually reminds them that each of the several million parts of the vehicle have to work as designed, first time, every time, and that those parts are usually provided by suppliers who came in with the lowest bid to build them! When he asks again who would volunteer to fly, the number of hands shown usually reduces considerably. Space flight is, and will always be, inherently dangerous and a risky profession. No matter how far we progress and develop our capabilities, there will always remain the elements of surprise and luck, potential failures, and unfortunately the possibility of a tragedy.

Those selected to partake in such an exciting and dangerous adventure as an astronaut will know that their colleagues have gone through a similar journey in order to reach orbit, and most will probably become close friends as well as trusted colleagues over time. The original 73 astronauts chosen between 1959 and 1969 were the American pioneers of the new frontier called space. Their dedication provided the human element of the system that demonstrated the United States could launch men (no women at that time) into space, and even send them to the Moon, support them while there and get them home alive. It was built upon their aviation and professional skills and was emblazoned across the media over the years; in particular Tom Wolfe's 1979 book *The Right Stuff* and the 1983 feature film derived from it. However, by the 1970s, this pioneering spirit of adventure had changed. The so-called race to the Moon against the Soviets was won and there was now a new focus on space exploration, mostly looking back at Earth "for all mankind," as the plaque on the Apollo 11 Lunar Module (LM) *Eagle* proclaimed. The program and its objectives had to change, opening up new opportunities for new participants over the long term. Part of that adjustment came with the Space Shuttle and the selection of a new generation of astronauts to fly it – the first of whom were the Thirty-Five New Guys (TFNG) of the NASA class of 1978.

A CLOSE-KNIT GROUP

Some of the military pilots had worked together in the past while others had met during the selection progress, but now they were a new tight knit group, not only because of the media attention but also because of what they had to go through and learn at NASA. They were the first to be classed as Ascans and the first group to train specifically for the Shuttle, and they were heavily involved in preparing for its first flight and supporting the early missions. Then there were the obvious 'firsts', with the inclusion of women and minorities in the NASA Astronaut Office, and the selection of the first Mission Specialists (MS). As with previous groups, the TFNG set up home in the suburbs around the Johnson Space Center (JSC) in Houston, Texas.

For several years the group were close-knit, and though they naturally started to drift away from the Astronaut Office as time wore on, they still kept in touch, as Jeff Hoffman recalled in 2009: "For the first couple years, we definitely had a lot of parties and really built up a big rapport. I think that our group in particular [the TFNG] kept that, and in fact we're the only group that has had regular reunions just on our own. We had a ten-year reunion and a 20-year reunion in Houston, and then last year for our 30th reunion in 2008, since we were all Shuttle astronauts, we figured that by the time our 40th reunion came around the Shuttle wasn't going to be flying anymore, so we had our 30th reunion in Florida and watched a Shuttle launch. It was great. I think about 20 people came. A large number of the group showed up. I guess there's 29 of us still alive. So yes, I think you get quite close. Over the years, as we started to get assigned to missions, people went off in different directions, and we didn't have quite so active a group social life as we did the first two or three years. But it's an important way to bond with the people you're going to be working with, and I really enjoyed that." **[2]**

For the record
The 1978 Group 8 astronaut selection was only the third group since 1959 to have *all* the members of their selection (35 in this case) achieve at least one space flight. The other two groups were the seven original Mercury astronauts chosen in 1959, and the seven former Manned Orbiting Laboratory (MOL) astronauts who transferred to NASA in 1969.

THE LEGACY OF THE TFNG

During the 1950s, a wide variety of images emerged in popular culture which depicted astronauts exploring new worlds and constructing huge structures in space, as well as vast colonies on the distant planets. These were different pioneers of space exploration. Like the American Wild West of the 19th century, these were more like homesteaders, technicians and professionals, establishing an infrastructure away from Earth with which the human race could expand safely beyond the home planet.

We may still be some way from such predictions coming true but we are moving closer. As the first step beyond *The Right Stuff* American pioneers of human space flight, the TFNG represented a shift in the perception of what an astronaut was. They were no longer required to be 'hot-shot' fighter pilots pushing the limits of their vehicles and risking their lives for the greater good. Instead, these were a far more diverse generation, breaking the molds of what had gone before and creating new models to admire and emulate. There were pilots of course, selected to fly the Space Shuttle and command the missions, but there was a new designation in NASA parlance – the Mission Specialists. They were professionals who did not have to pilot the vehicle to and from space, or even move the vehicle once there. These astronauts focused on the mission, the payload, the experiments and the objectives. They were the ones who created research studies in space as well as operating experiments; they helped American astronauts regain long-unused skills in Extra-Vehicular Activity (EVA), rendezvous and proximity operations, and docking, and acquire new ones in satellite operations in Earth orbit. The TFNG had their roots in the decades of unrest and change in the United States and around the world following WWII, as education and opportunity for betterment devolved away from a select few and gradually encompassed minorities and females too.

People, especially the young, need role models to aspire to, and the TFNG provided a greater diversity of such role models for a wider range of people, though they were often unaware of how much they did inspire a younger generation to strive for success in the class room, at work, in the military and in the world at large. There are many astronauts who followed the Class of 1978 into space who have commented that seeing the TFNG lead the way made them dream that they, too, could achieve if they worked hard and persevered. Inspiring the young to persevere and to achieve is, perhaps, the most lasting and noteworthy legacy of the Class of 1978, probably more so than the space flights each completed.

None of this had been done before in this way. The job description was different, the selection process was different, and the training program was different. Their space vehicle was certainly nothing like what had come before, but the reward was the same as for the veteran space explorers – a flight into orbit to see our planet in all its glory. Once there, the pilots looked after the Orbiter, but it was the MS, supported by the Payload Specialists (PS) and helped out at times by the pilots, who fulfilled many of the mission objectives as enthusiastic participants in science research.

The TFNG helped to develop new techniques and systems for training and flights before moving into higher management at NASA or out into the wider space industry. The remaining members from earlier selections may have helped to develop the Space Shuttle program from the drawing board to the hardware, but it was the TFNG who molded it into an operational system for others to use for over two decades until the program's closure. A few went on to work on the emerging Space Station, applying skills learned on Shuttle to that new program and those which followed it.

Right from the start, the TFNG had a hand in developing the Shuttle training program and the roles of the support team. They expanded the role of the MS-2/Flight Engineer (FE), molded the responsibilities of the Payload Commander (PC) and pushed the boundaries of EVA and robotics. Their work in small pressurized laboratories aboard the Shuttle laid the foundations for the program of research on the U.S. segment of the International Space Station (ISS).

From before the first Shuttle flight until just after the final mission flew, members of the Class of 1978 were there, in active flight or support roles, in management or in the space industry. These 35 very determined individuals created a new image of a 'group' or class, as a single unit rather than individual characters. Following NASA Group 8, each astronaut group came up with its own identity, including a nickname and emblem, to identify the 'family' of that selection in more ways than the first seven groups ever did. The NASA that had sent the first American astronauts into space had changed considerably when the TFNG arrived at JSC. It also evolved significantly through the four decades they were there, and was changing again as the final members left the Astronaut Office or the space agency, handing over the mantle to new classes for them to make their own mark on American and human space flight history.

NASA AFTER THE TFNG

Among the 73 original astronauts, there were natural leaders and characters, pioneers and heroes. They were modern-day trailblazers and iconic figures whose exploits and achievements created the myth and ethos of the Astronaut Office during the 1960s, as immortalized in *The Right Stuff*. As remarkable and inspiring as this early era was, it would not and could not continue in the same vein. Pioneers have always given way to the next, second generation of explorers that exploit and expand upon what was achieved by the vanguard and set the course for the subsequent generations of settlers, to establish an infrastructure and systems to build further. The 35 men and women of Group 8 were the first of this second generation of American astronauts. They took the Astronaut Office from the 1960s into the 1990s, and evolved the program from Apollo to the Shuttle and on to the ISS. It is at the ISS that the third generation is currently establishing the foundations for the next developments in the history of American astronauts, the settlers and homesteaders.

Fig. 15.1: At the Kennedy Space Center in Florida in 2014, Anna Fisher, the last of the TFNG remaining with NASA, casts a keen eye on the work being carried out to prepare the Orion spacecraft, one of the next generation of American spacecraft to carry astronauts. The crews will not feature members of the TFNG, whose time at the forefront of American space flight had lasted an impressive four decades. But the legacy of the group is assured by their achievements and in the changes made in the Astronaut Office.

New Roles in the Astronaut Office

Following the retirement of the Shuttle in 2011, there has perhaps been even less requirement for pure 'pilots' in the NASA Astronaut Office, with no new vehicle to 'fly' for some years. For almost a decade since the grounding of the Shuttle fleet, the use of the Russian Soyuz to take the crew to the ISS has meant that not one American has been in command, or officially allowed to control, a small crew transport space vehicle into or out of orbit, or in solo flight. Those aboard the space station as increment crewmembers are designated as Flight Engineers or single Increment Commanders, and while they can be American astronauts, the roles have not necessarily been filled by professional pilots. The fact that *all* Soyuz spacecraft have been commanded during ascent, orbital flight and descent by a Russian cosmonaut further restricts the opportunities for American pilot astronauts. Interestingly, Soyuz Commanders have not always been military pilots, with a few chosen from the cosmonaut engineer team. In saying this, over the last 25 years a handful of foreign nationals (including, since the early 2000s, a few NASA astronauts) have trained as Soyuz Flight Engineer 1 in the left seat position. That means they are qualified as center seat Soyuz Return Commanders in the event of an emergency, although such a situation has fortunately not yet arisen during a mission.

From 2020 onwards, it is planned that new American human spacecraft will begin to fly with a crew. The SpaceX Crewed Demo-2 mission will include a Commander and 'Joint Operations CDR', although the crewing roles for the Boeing CST-100 Starliner crew remain to be determined beyond that of Commander. Crew positions and roles for the additional crewmembers on each spacecraft are still to be formally announced, though over the past few years NASA has used the terms "Operator-1" or "Operator-2" for future planning purposes, not necessarily as confirmed crew designations. Any crewmembers that transfer to an expedition crew on ISS will revert to the Flight Engineer expedition designation.

Since the retirement of the Shuttle, the role of MS has demised as well, as all crewmembers who fly to the ISS are classed as 'Flight Engineer' initially. During the change of increment, one will be 'promoted' to station commander for the duration of that new phase. So, is the position of MS now as redundant as the Apollo Command or Lunar Module Pilot became after 1972? The MS role was created for the Space Shuttle program and astronauts are no longer chosen under that designation, so the position was very much of its time (1970s–2011) and those skills and experiences are changing as the program moves on. TFNG Steve Hawley, who flew five times as MS-2/FE, thinks some elements will remain the same, but the role and thereby the identification will change. "If you look at it from the ground up, I think you are still going to pick people who have the hard core group of basic requirements, things like good technical skills, good teamwork skills, good leadership skills, good communication skills and some operational capability. On top of that, I think there is a lot to be said for having a diversity of those experiences. Again, I go back to saying that the pilot culture would be 'Give me the procedure and I'll go execute', and the scientist culture might be 'Give me the procedure and I'll give you some improvement'. That's what I see in the last couple of selections, for people that are presumably going to get to fly the missions on Orion. These people could have been called Mission Specialists, if you reflected off the Shuttle, because they have a lot of the same background, but now the jobs they're going to do are going to be different, though I don't think that the skill set is all that much different. Just the name of the role." **[3]**

The first of a new generation

On January 10, 2020, a week shy of the 42nd anniversary of the Class of 1978 being announced to the world, NASA held another ceremony at JSC to state that the members of the 22nd group of astronauts (The "Turtles") had graduated from their two-year Ascan basic training program and were eligible for flight assignment. At the ceremony, NASA Administrator James Bridenstine commented that this was the first class to receive their Silver Astronaut Pins under the new Artemis program, which aims to send the next man and the first woman to the Moon by 2024. Class members from this selection would now be available for assignment to expedition crews on ISS, launched aboard the new U.S. commercial crew spacecraft. They may indeed reach the surface of the Moon, over five decades since the last men of Apollo departed, and may perhaps become one of the first humans to step onto the surface of Mars. To suggest that men and women could crew such missions together shows how far the program has evolved since the TFNG stepped onto the podium at JSC at the beginning of the Shuttle era of American human space flight. Now a new generation of astronauts has taken up that challenge to forge, together, the next steps deeper into space.

References

1. spacequotations.com available online at http://www.spacequotations.com/earth.html
2. Jeffrey Hoffman, NASA Oral History, April 2, 2009.
3. AIS interview with Steve Hawley, October 2019.

Afterword

Following the selection and naming of the Mercury astronauts in 1959, NASA was on a high. America, and indeed most of the world, admired – even adored – the seven courageous test pilots who had been chosen to tackle and conquer the new and largely unknown high frontier of space. All seven were superbly healthy, highly-skilled test pilots, drawn from their country's military services. To NASA's complete surprise, the seven men would achieve instant and enormous nationwide fame, well before any of them had even seen, let alone set foot inside, a spacecraft.

As the citizens of the United States continued their admiration of these pioneering astronauts and the space program, there were some rumblings of discontent. Why, many argued, were women not considered for inclusion in NASA's astronaut program? There were any number of women flying jet aircraft, and they began to question publicly what they perceived as the space agency's seemingly entrenched attitude of maintaining the astronaut corps as an exclusively male domain.

In 1960, the renowned aviator Jacqueline Cochran, who had been the first woman to exceed the speed of sound in 1953, teamed up with Dr. Randolph Lovelace at his clinic in Albuquerque, New Mexico, with the specific goal of testing a number of women for astronaut abilities. Twelve other women pilots, including Geraldyn 'Jerrie' Cobb, underwent 87 rigorous tests at the Lovelace Clinic, followed by further testing at a naval medical laboratory in Pensacola, Florida. Despite the successful completion of their self-set program, NASA Administrator James Webb ordered the abandonment of the test project.

In July 1962, along with fellow trainee Jane Hart, Jerrie Cobb testified before the House of Representatives Space Subcommittee, accusing NASA of unreasonable discrimination. They complained that they were excluded from applying by the

© Springer Nature Switzerland AG 2020   531
D. J. Shayler, C. Burgess, *NASA's First Space Shuttle Astronaut Selection*,
Springer Praxis Books, https://doi.org/10.1007/978-3-030-45742-6

rigid requirement that anyone chosen for astronaut training had to be a jet test pilot, and this remained a military and exclusively male domain. In response to these complaints, Mercury astronaut John Glenn defended NASA's policies, claiming at the hearing that if there were qualified women, NASA would accept them into the astronaut program. Glenn also queried the women's lack of flight training and questioned whether they should be subjected to the space environment. It is interesting to note that by then Jerrie Cobb had accumulated over 10,000 hours' flying time – double that of John Glenn. Another Mercury astronaut, Gordon Cooper, was a little more pungent and chauvinistic when asked for his feelings on the subject. "All this talk about brains [scientists] and dames in space is bunk," he retorted. "As for the ladies, to date there have been no women – and I say absolutely zero women – who are qualified to take part in our space program."

NASA, always keen to present an image of innovation and forward thinking, derived some of its worst publicity at that time over its no-women-in-space policy. One former NASA flight director, the late Christopher Kraft, once recalled that the space agency had placed itself in an invidious position regarding women astronauts. "Had we lost a woman back then because we decided to fly a woman rather than a man, we would have been castrated!" he stated.

One person who wrote to NASA around this time, requesting details on how to become an astronaut, was a feisty high school student named Hillary Rodham. Much to her chagrin, she received a reply from NASA stating that 'girls need not apply.' Hillary would later marry a promising young lawyer named Bill Clinton.

Additionally, NASA was also fending off insinuations of having an all-white policy, which meant excluding minority candidates. In September 1962, despite these lingering questions, NASA announced the selection of a further nine astronauts to train and work on the two-man Gemini program. Just 13 months later, the astronaut corps expanded even further with the addition of another 14 astronauts in preparation for both the Gemini and Apollo programs.

The fourth group, selected in June 1965, was the first to break away from the test pilot qualification, in that the six men selected were scientists. NASA had realized that once its program of human lunar exploration began, they would need qualified people such as geologists who knew exactly what to look for and examine, rather than military pilots who would receive minimal training in such sciences. NASA was also looking to the advent of Earth-orbiting laboratories in its Apollo Applications Program (later renamed Skylab) where their presence would be crucial.

But still no women were selected, although a persistent few kept applying. This position would be maintained when the fifth group of 19 astronauts was selected in April 1966, and again in August 1967 when 11 more Scientist-Astronauts were chosen for possible future flights, although as the years passed without any mission assignments, these scientists came to call themselves the XS-11 (Excess 11), and several quit to go back to their studies.

Then, in August 1969, with the Apollo lunar landing program in full swing, NASA announced that seven military astronauts, previously with the abandoned U.S. Air Force (USAF) Manned Orbiting Laboratory (MOL) program, had been seconded into the astronaut ranks. There were still no women or minority applicants in the astronaut ranks, but that was all about to change.

In 1973, NASA Administrator James Fletcher announced that with the successful Apollo lunar program now at an end, there would soon be an opportunity for women and minorities to become astronauts sometime in the late 1970s, with the advent of its next generation of crewed spacecraft – the Space Shuttle. These changing times would finally come to pass in July 1976 when NASA announced that it was recruiting astronauts to fly on the Shuttle, but this time in two different categories – pilot astronaut and Mission Specialist. For the very first time in the history of the space agency, women and minorities were not only eligible to apply in both categories, they were encouraged to do so.

NASA had at last come of age in astronaut selections in 1978, and in the years since, many women and minorities have flown (and died) alongside their astronaut colleagues.

This book ends with the exciting presumption that both male and female space explorers of all creeds, color and nationalities will continue the amazing work carried out by the 108 astronauts selected in NASA's Groups 1–8, and will take us on a whole new series of challenges and to whole new destinations.

In 2019, it was pledged that the very next person to set foot on the Moon will be a woman. That will be a long-anticipated day.

APPENDIX 1

NASA CLASS OF 1978 ASTRONAUT APPLICANTS

In a NASA News Release dated 15 July 1977, it was stated that over 8,000 individuals had applied for the 30 to 40 positions the space agency had opened for Space Shuttle astronaut candidates. To be eligible, their applications had to be postmarked prior to midnight on June 30, although many had waited until the last day and the total application numbers were still being tallied.

The astronaut selection board at the Johnson Space Center (JSC) then began the process of narrowing the selection to those best qualified in each category. Of these, approximately 150 applicants would be selected for preliminary screening and physicals at JSC. The 30 to 40 pilot and Mission Specialist (MS) candidates would be notified of their selection by December, and would then undergo a training and evaluation period in 1978, also at JSC.

A total of 24,618 inquiries were received by NASA after the selection program was announced in 1976, of which 20,440 requested and were sent application packages. At the time of the News Release publication, 8,037 men and women (later updated to 8,079) had applied for the astronaut program. The MS category had 6,735 applicants and the pilot category had 1,302 applicants.

The first 20 applicants were asked to report to JSC on Tuesday, August 2, 1977. "We are pleased with the quality of applicants," noted JSC Director Christopher Kraft, Jr. in a NASA News Release dated July 29, 1977. "It is difficult to narrow the field for interviews and paring that number will be a real challenge."

The names, ages, affiliations, places of birth and current duty stations of those applicants (with successful applicants indicated in bold type) were:

© Springer Nature Switzerland AG 2020

D. J. Shayler, C. Burgess, *NASA's First Space Shuttle Astronaut Selection*, Springer Praxis Books, https://doi.org/10.1007/978-3-030-45742-6

FIRST GROUP OF (20) ASTRONAUT APPLICANTS (AUGUST 2, 1977)
(PILOT APPLICANTS ONLY)

Maj. Gerald K. Bankus, 34, USAF; Milan, Missouri; Pentagon, Washington D.C.

Lt. Richard E. Batdorf, 30, USN; Wauseon, Ohio; NAS Patuxent River, Maryland

Maj. John E. Blaha, 34, USAF; San Antonio, Texas; Pentagon, Washington D.C. (*selected NASA Group 9, 1980*)

Capt. Gary D. Bohn, 33, USAF; Halstead, Kansas; Tyndall AFB, Florida

Capt. Claude M. Bolton, Jr., 31, USAF; Sioux City, Iowa; Edwards AFB, California

Lt. Cmdr. Daniel C. Brandenstein, 34, USN; Watertown, Wisconsin; Attack Squadron 145, FPO San Francisco, California

Maj. Roy Bridges, Jr., 34, USAF; Atlanta, Georgia; Headquarters USAF/RDPN, Washington D.C. (*selected NASA Group 9, 1980*)

Maj. Frederick T. Bryan, 35, USMC; Melrose, Massachusetts; Pacific Missile Test Center, Point Mugu, California

Capt. John Casper, 34, USAF; Greenville, South Carolina; Edwards AFB, California (*selected NASA Group 10, 1984*)

Lt. Cmdr. Michael L. Coats, 31, USN; Sacramento, California; USN Post-Graduate School, Monterey, California

Maj. Stewart E, Cranston, 33, USAF; Watertown, South Dakota; Tyndall AFB, Florida

Lt. Cmdr. John O. Creighton, 34, USN; Orange, Texas; Fighter Squadron Two, FPO San Francisco, California

Lt. Cmdr. William V. Cross II, 31, USN; Omaha, Nebraska; USS *Nimitz*, Norfolk, Virginia

Capt. Edward L. Daniel, 32, USAF; Eagle Pass, Texas; Edwards AFB, California

Capt. Michael E. Durbin, 35, USAF; Dallas, Texas; 36th Tactical Fighter Wing, Bitburg, Germany

Lt. James O. Ellis, Jr., 30, USN; Spartanburg, South Carolina; NAS Patuxent River, Maryland

Mr. James D. Erickson, 35, civilian; Spokane, Washington; Federal Aviation Administration, Fort Worth, Texas

Lt. Cmdr. Kent H. Ewing, 34, USN; San Angelo, Texas; NAS Cecil Field, Florida

Capt. Guy S. Gardner, 29, USAF; Altavista, Virginia; Edwards AFB, California (*selected NASA Group 9, 1980*)

Capt. Thomas E. Fitzpatrick, 32, USMC; Winter Haven, Florida; Naval Air Test Center, Patuxent River, Maryland

SECOND GROUP OF (20) ASTRONAUT APPLICANTS (AUGUST 16, 1977) (PILOT APPLICANTS ONLY)

Lt. Col. Leslie B. Anderson III, 36, USAF; Wooster, Ohio; 436th Tactical Fighter Training Sq., Holloman AFB, New Mexico

Capt. Richard S. Couch, 31, USAF; Hamilton, Ontario, Canada; 4950th Test Wing, Wright-Patterson AFB, Ohio

Maj. Richard O. Covey, 31, USAF; Fayetteville, Arkansas; AFFTC, Detachment 2, Eglin AFB, Florida

Capt. Dale S. Elliott, 32, USAF; Lake Charles, Louisiana; 3246th Test Wing, Eglin AFB, Florida

Lt. Robert L. Gibson, 30, USN; Cooperstown, New York; Naval Air Test Center, Patuxent River, Maryland

Capt. Ronald J. Grabe, 32, USAF; New York, New York; USAF/RAF Exchange Program, Amesbury, Wiltshire, England (*selected NASA Group 9, 1980*)

Mr. Stanley D. Griggs, 37, civilian; Portland, Oregon; CC52/Johnson Space Center, Houston, Texas

Maj. James G. Hart, 35, USMC; Minneapolis, Missouri; Air Test & Evaluation Squadron 5, China Lake, California

Lt. Cmdr. William B. Hayden, 32, USN; Oakland, California; Fighter Squadron 14, NAS Oceana, Virginia Beach, Virginia

Lt. David T. Hunter, 29, USN; Tacoma Park, Maryland; Naval Air Test Center, Code SY90, Patuxent River, Maryland

Maj. Jack M. Jannarone, 34, USAF; Fort Gordon, Georgia; 6512th Test Squadron/ DOTF, Edwards AFB, California

Maj. Don E. Kenne, 35, USAF; Baltimore, Maryland; 475th Test Squadron, ADWC, Tyndall AFB, Florida

Capt. Kerry E. Killebrew, 30, USAF; Murray, Kentucky; 6512th Test Squadron/ DOTF, Edwards AFB, California

Maj. James R. Klein, 35, USAF; Dubuque, Iowa; USAF Test Pilot School, Edwards AFB, California

Lt. Joseph F. Lucey, 31, USN; Minneapolis, Minnesota; Naval Air Test Center, Patuxent River, Maryland

Lt. Cmdr. John M. Luecke, 33, USN; Macomb, Illinois; Box 84, COM NAV ACTS UK, FPO New York

Lt. Cmdr. Jon A. McBride, 33, USN; Charlestown, West Virginia; Air Test & Evaluation Squadron 4, Point Mugu, California

Lt. Cmdr. Charles R. McRae, 33, USN; Miami, Florida; Naval Air Test Center, Patuxent River, Maryland

Capt. Michael D. Marks, 34, USAF; Salt Lake City, Utah; Air Force Flight Test Center/DOVA, Edwards AFB, California

Capt. Marvin L. Martin, 30, USAF; Nevada, Missouri; USAF Test Pilot School, Edwards AFB, California

THIRD GROUP OF (20) ASTRONAUT APPLICANTS (AUGUST 29, 1977) (MISSION SPECIALIST APPLICANTS ONLY)

James P. Bagian, 25, MD; Philadelphia, Pennsylvania; Naval Air Test Center, Patuxent River, Maryland (*selected NASA Group 9, 1980*)

Stephen C. Boone, 39, MD, PhD; Navasota, Texas; Walter Reed Army Medical Center, Washington D.C.

Nitza M. Cintron, 27, PhD; San Juan, Puerto Rico; John Hopkins University School of Medicine, Baltimore, Maryland

Lt. Mark S. Davis, 33, USN, MD; Philadelphia, Pennsylvania; Naval Regional Medical Center, Oakland, Pennsylvania

Danielle J. Goldwater, 29, MD; West Haven, Connecticut; Stanford Hospital, Stanford University, California

Lionel O. Greene, Jr., 29, PhD; Brooklyn, New York; NASA Ames Research Center, Moffett Field, California

Dale A. Harris, 31, PhD; Amarillo, Texas; Letterman Army Institute of Research, Presidio of San Francisco, California

Lt. Col. James R. Hickman, 35, USAF, MD; Elkhorn City, Kentucky; USAF School of Aerospace Medicine, Brooks AFB, Texas

Michael P. Hlastala, 33, PhD; Uniontown, Pennsylvania; University of Washington, Seattle, Washington

Lt. Cmdr. Harry P. Hoffman, 34, USN; MD; New Bern, North Carolina; Air Test & Evaluation Sq. 4, NAS Point Mugu, California

Bruce A. Houtchens, 39, MD; Olympia, Washington; University of Utah, Salt Lake City, Utah

Capt. Michael D. Kastello, 32, US Army, PhD, DVM; LaSalle, Illinois; USA Medical Institute of Infectious Disease, Fredrick, Maryland

Maj. Wayne F. Kendall, 39, USAF, MD; Harrison, Arkansas; Wright-Patterson Aerospace Medical Research Laboratory, Ohio

Shannon W. Lucid, 34, PhD; Shanghai, China; Oklahoma Medical Research Foundation, Oklahoma City, Oklahoma

B. Tracey Sauerland, 29, MD, PhD; New Britain, Connecticut; JSC Space and Life Sciences Directorate, Houston, Texas

M. Rhea Seddon, 29, MD; Murfreesboro, Tennessee; City of Memphis Hospital, Memphis, Tennessee

Anna L. Sims (Fisher), 28, MD; Albany, New York; Harbor General Hospital, Torrance, California

Stephen C. Textor, 29, MD; Denver, Colorado; Boston University Hospital, Boston, Massachusetts

Lt. Cmdr. Victoria M. Voge, 34, USN, MD; Minneapolis, Minnesota; Naval Aerospace Medical Institute, Pensacola, Florida

Millie H. Wiley (married name Hughes-Fulford), 31, PhD; Mineral Wells, Texas; Veterans Administration Hospital, San Francisco, California (*1991 Payload Specialist, STS-40 Spacelab Life Sciences 1*)

FOURTH GROUP OF (20) ASTRONAUT CANDIDATES (SEPTEMBER 19, 1977)
(PILOT APPLICANTS ONLY)

Cmdr. Frederick H. Hauck, 36, USN; Long Beach, California; Naval Air Station, Whidbey Island, Washington

Capt. Ralph J. Luczak, 31, USAF; St. Louis, Missouri; USAF 4950th Test Wing, Wright-Patterson AFB, Ohio

Maj. Donald L. Marx, 35, USAF; Gary, Indiana; Air Command & Staff College, Maxwell AFB, Alabama

Mr. Edward D. Mendenhall, 41, civilian; Orange, New Jersey; NASA JSC, Aircraft Operations, Houston, Texas

Maj. Edward T. Meschko, 33, USAF; Trenton, New Jersey; USAF Flight Test Center, Edwards AFB, California

Capt. Alfred P. Metz, 31, USAF; Springfield, Ohio; USAF Flight Test Center, Edwards AFB, California

Maj. David W. Milam, 37, USAF; Tucson, Arizona; USAF Test Pilot School, Edwards AFB, California

Capt. Stephen J. Monagan, 33, USAF; Waterbury, Connecticut; 3246th Test Wing, Eglin AFB, Florida

Maj. Roger A. Moseley, 31, USAF; St. George, Utah; 3246th Test Wing, Eglin AFB, Florida

Capt. Steven Nagel, 30, USAF; Canton, Illinois; USAF Flight Test Center, Edwards AFB, California

Capt. Bryan D. O'Connor, 31, USMC; Orange, California; Naval Air Test Center, Patuxent River, Maryland (*selected NASA Group 9, 1980*)

Maj. Alva E. Peet, Jr., 38, USMC; Jacksonville, Florida; Naval Air Station, Point Mugu, California

Lt. Cmdr. Larry G. Pearson, 34, USN; Redlands, California; Pacific Missile Test Center, Point Mugu, California

Maj. Gary L. Post, 37, USMC; Trilla, Illinois; Naval War College, Newport, Rhode Island

Lt. Cmdr. Kenneth N. Rauch, 33, USN; Syracuse, New York; Student, Canadian Forces College, Toronto, Ontario, Canada

Lt. Richard N. Richards, 31, USN; Key West, Florida; Naval Air Test Center, Patuxent River, Maryland (*selected NASA Group 9, 1980*)

Maj. Francis R. Scobee, 38, USAF; Cle Elum, Washington; USAF Flight Test Center, Edwards AFB, California

Capt. Brewster H. Shaw, Jr., 32, USAF; Cass City, Michigan; USAF Test Pilot School, Edwards AFB, California

Capt. Loren J. Shriver, 32, USAF; Jefferson, Iowa; USAF Flight Test Center, Edwards AFB, California

Maj. Ivan J. Singleton, 38, USAF; Tulsa, Oklahoma; USAF Flight Test Center, Edwards AFB, California

FIFTH GROUP OF (20) ASTRONAUT CANDIDATES (SEPTEMBER 26, 1977) (PILOT AND MISSION SPECIALIST APPLICANTS)

Lt. Col. Norman L. Suits, 39, USAF; Chebanse, Illinois; F-15 Joint Test Force, Edwards AFB, California

Maj. Bronson W. Sweeney, 33, USMC; Fall River, Massachusetts; Naval Air Test Center, Patuxent River, Maryland

Capt. Paul D. Tackabury, 32, USAF; Canastota, New York; USAF Flight Test Center, Edwards AFB, California

Capt. James W. Tilley II, 34, USAF; Milwaukee, Wisconsin; 3246th Test Wing, Eglin AFB, Florida

Lt. Cmdr. David M. Walker, 33, USN; Columbus, Georgia; NAS Oceana, Virginia Beach, Virginia

Lt. Cmdr. George J. Webb, Jr., 34, USN; Jacksonville, Florida; Cecil Field, Florida

Lt. Cmdr. Donald E. Williams, 35, USN; Lafayette, Indiana; NASA Leemore, California

Lt. Cmdr. Robert C. Williamson, Jr., 31, USN; West Chester, Pennsylvania; NAS Oceana, Virginia Beach, Virginia

Maj. Paul D. Young, 34, USMC; Ada, Oklahoma; Fleet Marine Force Pacific (Okinawa)

Steven A. Hawley, 25, PhD; Ottawa, Kansas; Cerro Tololo Inter-American Observatory, La Serena, Chile

Maj. Gary W. Matthes, 35, USAF St. Louis, Missouri; USAF Flight Test Center, Edwards AFB, California

Maj. Robert L. Oetting, 37, USAF; Alton, Illinois; USMC Air Station, Cherry Point, North Carolina

Maj. Isaac S. Payne IV, 37, USAF; Malakoff, Texas; HQ USAF; The Pentagon, Washington D.C.

Wilton T. Sanders III, 29, PhD; Greenwood, Mississippi; Dept. of Physics, University of Wisconsin, Madison, Wisconsin

Maj. Michael E. Sexton, 36, USAF; Pendleton, Oregon; 36th TAC Fighter Wing F-15 Squadron, Bitburg, Germany

Lt. Cmdr. David M. Sjuggerud, 35, USN; Blair, Wisconsin; Naval Air Systems Command, Washington D.C. Agency, Atlanta

Paul S. Skabo, 35, USN; Dixon, Illinois; Federal Aviation Agency, Atlanta, Georgia

Lt. Cmdr. James L. Spencer III, 35, USN; Charleston, South Carolina; Naval Air Test Center, Patuxent River, Maryland

Capt. Will R. Stewart, 32, USAF; Montclair, New Jersey; USAF Flight Test Center, Edwards AFB, California

Joseph J. C. Degioanni, 31, MD, PhD; Resident, Aerospace Medicine, JSC, Houston, Texas

SIXTH GROUP OF (20) ASTRONAUT APPLICANTS (OCTOBER 3, 1977) (MISSION SPECIALIST APPLICANTS ONLY)

George R. Carruthers, 38, civilian; Cincinnati, Ohio; Naval Research Laboratory, Washington D.C.

Capt. Douglas L. Dowd, 34, U.S. Army; Miami, Florida; Mission Planning & Analysis Division, NASA/JSC, Houston, Texas

Michael J. Frankston, 26, civilian; New York, New York; Remote Sensing Laboratory, MIT, Cambridge, Massachusetts

Maj. Robert B. Giffen, 35, USAF; Princeton, New Jersey; USAF Academy, Colorado

Alan M. Goldberg, 28, civilian; Providence, Rhode Island; Dept. of Earth & Planetary Sciences, MIT, Cambridge, Massachusetts

Douglas R. Hansmann, 32, civilian; Olympia, Washington; Cardio-Dynamics Laboratories, Los Angeles, California

Frank R. Harnden, Jr., 31, civilian; Pittsfield, Massachusetts; Smithsonian Astrophysical Observatory, Cambridge, Massachusetts

William D. Heacox, 35, civilian; Pipestone, Minnesota; NRC Research Associate, MIT, Cambridge, Massachusetts

Jeffrey A. Hoffman, 32, PhD, civilian; New York, New York; Center for Space Research, MIT, Cambridge, Massachusetts

R. Jerry Jost, 30, civilian; Portland, Oregon; Space Physics Dept., Rice University, Houston, Texas

Donald W. McCarthy, Jr., 29, civilian; Minneapolis, Minnesota; Lunar & Planetary Lab, University of Arizona, Tucson

Maj. Roger P. Neeland, 35, USAF; Milwaukee, Wisconsin; USAF Academy, Colorado

George D. Nelson, 27, PhD, civilian; Charles City, Iowa; Astronomy Dept., University of Washington, Seattle, Washington

Capt. Arthur L. Pavel, 30, USAF; Downey, California; USAF Test Pilot School, Edwards AFB, California

Charles J. Peterson, 31, civilian; Seattle, Washington; Cerro Tololo Inter-American Observatory, LaSerena, Chile

Larry D. Petro, 29, civilian; Lansing, Michigan; Hale Observatories, Carnegie Institute, Pasadena, California

Lawrence S. Pinsky, 31, civilian; New York, New York; Physics Dept., Stanford University, Stanford, California

Sally K. Ride, 26, PhD, civilian; Los Angeles, California; Physics Dept., Stanford University, Stanford, California

Richard J. Terrile, 26, PhD, civilian; New York, NY; Division of Geological & Planetary Science, CalTech, Pasadena, California (*Spacelab 1 Payload Specialist semi-finalist, not selected*)

Bobby L. Urich, 30, civilian; Bryan, Texas; National Radio Astronomy Observatory, Tucson, Arizona

SEVENTH GROUP OF (20) ASTRONAUT APPLICANTS (OCTOBER 17, 1977) (MISSION SPECIALIST APPLICANTS ONLY)

Jack L. Burton, 32, PhD; Kenmore, New York; NASA/Goddard Space Flight Center, Greenbelt, Maryland

Samuel H. Clarke, Jr., PhD; Bristol, Virginia; U.S. Geological Survey, Office of Marine Biology, Menlo Park, California

Kathleen Crane, 26, PhD; Washington D.C.; Scripps Institute of Oceanography, La Jolla, California

Bonnie J. Dunbar, 28, civilian; Sunnyside, Washington; Rockwell International Space Division, Downey, California (*selected NASA Group 9, 1980*)

Brady A. Elliott, 30, civilian; Columbus, Ohio; Texas A&M University (Research Assistant), College Station, Texas

Joan J. Fitzpatrick, 27, PhD; Bayonne, New Jersey; Colorado School of Mines Research Inst., Golden, Colorado

Salvatore Giardina, Jr., 34, civilian; Hoboken, New Jersey; State of Arizona, Oil & Gas Commission, Tempe, Arizona

David S. Ginley, Jr., 27, PhD; Denver, Colorado; Sandia Laboratories, Albuquerque, New Mexico

Carolyn S. Griner, 32, civilian; Granite City, Illinois; NASA/Marshall Space Flight Center, Huntsville, Alabama (*Spacelab candidate but not selected*)

Evelyn L. Hu, 30, PhD; New York, NY; Bell Laboratories, Holmdel, New Jersey

Carol B. Jenner, 27, PhD; Washington D.C.; University of Wisconsin, Madison, Wisconsin

Mary Helen Johnston, 32, PhD; West Palm Beach, Florida; NASA/Marshall Space Flight Center, Huntsville, Alabama (*1985 Spacelab 3 back-up Payload Specialist*)

H. Louise Kirkbride, 24, civilian; Philadelphia, Pennsylvania; Jet Propulsion Laboratory, Pasadena, California

Larry A. Mayer, 25, civilian; New York, New York; Scripps Institute of Oceanography, La Jolla, California

Harry Y. McSween, Jr., 32, PhD; Charlotte, North Carolina; University of Tennessee, Knoxville, Tennessee

Richard W. Newton, 29, PhD; Baytown, Texas; Texas A&M University, College Station, Texas

William H. Peterson, 36, civilian; Brooklyn, New York; University of Miami, Graduate Research Assistant, Miami, Florida

Wayne R. Sand, 36, civilian; Conrad, Montana; University of Wyoming, Laramie, Wyoming

Cmdr. Brian H. Shoemaker, 40, USN; Noranda, Quebec, Canada; NAS North Island, San Diego, California

Ritchie S. Straff, 24, civilian; Philadelphia, Pennsylvania; MIT Graduate Student, Cambridge, Massachusetts

EIGHTH GROUP OF (20) ASTRONAUT APPLICANTS (OCTOBER 25, 1977) (MISSION SPECIALIST APPLICANTS ONLY)

Maj. Thomas N. Almojuela, 34, U.S. Army; Seattle, Washington; NASA Ames Research Center, Moffett Field, California

Capt. Robert F. Behler, 29, USAF; Rome, New York; 6512 Test Squadron, Edwards AFB, California

Maj. Donald C. Bulloch, 32, USAF; Alexandria, Louisiana; Air Command & Staff College, Maxwell AFB, Alabama

Maj. William J. Fields, 35, USAF; Baltimore, Maryland; Armament Development Test Center, Eglin AFB, Florida

Lt. Cmdr. William F. Harrison, 32, USN; Charleston, South Carolina; NAS Whidbey Island, Oak Harbor, Washington

Capt. Jane L. Holley, 30, USAF; Shreveport, Louisiana; USAF Tactical Fighter Weapons Center, Nellis AFB, Nevada

Capt. Robert A. Lancaster, 30, USAF; Washington D.C.; Aeronautical Systems Division, Wright-Patterson AFB, Ohio

Capt. Johnnie B. Ligon, 35, USAF; Henderson, Kentucky; 3246th Test Wing, Eglin AFB, Florida

Lt. John M. Lounge, 31, USN; Denver, Colorado; Naval Electronics Systems Command, Washington D.C. (*selected NASA Group 9, 1980*)

Capt. Richard M. Mullane, 32, USAF; Wichita Falls, Texas; 3246th Test Wing, Eglin AFB, Florida

Capt. George C. Nield IV, 27, USAF; Washington D.C.; USAF Flight Test Center, Edwards AFB, California

Capt. Frederick K. Olafson, 30, USAF; Seattle, Washington; 3246th Test Wing, Eglin AFB, Florida

Capt. Ellison S. Onizuka, 31, USAF; Kealakekua, Hawaii; USAF Test Pilot School, Edwards AFB, California

1st Lt. Michael T. Probasco, 26, USAF; Houston, Texas; SA-ALC/MMSRE, Kelly AFB, Texas

Capt. Jerry L. Ross, 29, USAF; Gary, Indiana; USAF Flight Test Center, Edwards AFB, California (*selected NASA Group 9, 1980*)

Capt. Vernon P. Saxon, Jr., 32, USAF; Birmingham, Alabama; USAF Test Pilot School, Edwards, AFB, California

Capt. Charles W. Schillinger, 34, USMC; Chicago, Illinois; NAS Whidbey Island, Oak Harbor, Washington

Maj. Robert L. Stewart, 35, U.S. Army; Washington D.C.; U.S. Army Aviation Eng. Flt. Activity, Edwards AFB, California

Erik M. Stolle, 29, civilian; Pensacola, Florida; USAF Test & Evaluation Center, Kirkland AFB, New Mexico

Capt. Charles A. Vehlow, 31, U.S. Army; Waukesha, Wisconsin; Naval War College of Command & Staff, Newport, Rhode Island

NINTH GROUP OF (23) ASTRONAUT APPLICANTS (NOVEMBER 7, 1977) (PILOT AND MISSION SPECIALIST APPLICANTS)

Lt. Franklin S. Achille, 29, USN; Doylestown, Pennsylvania; Naval Air Test Center, Patuxent River, Maryland

Lt. David W. Anderson, 29, USN; Lincoln, Illinois; Naval Air Test Center, Patuxent River, Maryland

Maj. Guion S. Bluford, Jr., 34, USAF; Philadelphia, Pennsylvania; Wright-Patterson AFB, Ohio

Lt. Joseph C. Boudreaux III, 30, USN; New Orleans, Louisiana; Cruiser Destroyer Group 5, FPO San Francisco, California

David R. Dougherty, 32, PhD; Enid, Oklahoma; Louisiana State University, Baton Rouge, Louisiana

Capt. Thomas E. Edwards, 35, PhD, U.S. Army; Starkville, Mississippi; NASA Langley Research Center, Hampton, Virginia

Maj. John M. Fabian, 38, PhD; USAF; Goosecreek, Texas; USAF Academy, Colorado

William F. Fisher, 31, MD; Dallas, Texas; UCLA Harbor General Hospital, Los Angeles, California (*selected NASA Group 9, 1980*)

Cmdr. Stuart J. Fitrell, 38, USN; Cleveland, Ohio; Comm. Officer, Attack Squadron 66, NAS Cecil Field, Florida

Lt. Dale A. Gardner, 28, USN; Fairmont, Minnesota; NAS Point Mugu, California

Robert L. Golden, 37, PhD; Alameda, California; NASA Johnson Space Center, Houston, Texas

Terry J. Hart, 31, civilian; Pittsburgh, Pennsylvania; Bell Telephone Laboratories, Whippany, New Jersey

Barbara J. Holden, 32, PhD; Los Angeles, California; Naval Weapons Center, China Lake, California

Gary R. Jackman, 32, PhD; Waterbury, Connecticut; University of Florida, Gainesville, Florida

Lawrence W. Lay, 32, DO; Kansas City, Missouri; Flint Osteopathic Hospital, Flint, Michigan

Samuel E. Logan, 30, MD; Los Angeles, California; UCLA Medical School, Los Angeles, California

Gregory B. McKenna, 28, PhD; Pittsburgh, Pennsylvania; National Bureau of Standards, Washington D.C.

Judith A. Resnik, 28, PhD; Akron, Ohio; Xerox Corporation, El Segundo, California

Capt. Eugene A. Smith, 32, USAF; Utica, New York; Office of the Secretary of the Air Force, Los Angeles, California

Maj. Robert C, Springer, 35, USMC; St. Louis, Missouri; Naval Air Test Center, Patuxent River, Maryland (*selected NASA Group 9, 1980*)

Norman E. Thagard, 34, MD; Marianna, Florida; Medical University of South Carolina, Charleston, South Carolina

James van Hoften, 33, PhD; Fresno, California; University of Houston, Houston, Texas

Capt. Robert C. Ward, 35, USAF, MD; Homestead, Florida; Hill AFB, Utah

TENTH GROUP OF (25) ASTRONAUT APPLICANTS (NOVEMBER 14, 1977) (PILOT AND MISSION SPECIALIST APPLICANTS)

William B. Atwood, 30, PhD; Nashua, New Hampshire; CERN, EP Division, Geneva, Switzerland

Robert C. Belcher, 28, civilian; Del Rio, Texas; University of Texas (graduate student), Austin, Texas

Lt. Cmdr. Ronald S. Bird, 35, USN, PhD; Ann Arbor, Michigan; Pacific Missile Test Center, Point Mugu, California

Capt. James F. Buchli, 32, USMC; New Rockford, North Dakota; Naval Air Test Center, Patuxent River, Maryland

John T. Cox, 33, PhD; New York, New York; NASA Johnson Space Center, Houston, Texas

Andrew C. Cruce, 34, PhD; Fresno, California; Naval Air Test Center, Patuxent River, Maryland

David J. Diner, 24, PhD; New York, New York; CalTech, Pasadena, California

Ayre R. Ephrath, 35, PhD; Czechoslovakia; University of Connecticut, Storrs, Connecticut

Richard S. Galik, 26, PhD; Hackensack, New Jersey; Rittenhouse Labs, University of Pennsylvania, Philadelphia, Pennsylvania

Maj. Frederick D. Gregory, 36, USAF; Washington D.C.; Armed Forces Staff College, Norfolk, Virginia

Hamilton Hagar, Jr., 37, PhD; New York, New York; Jet Propulsion Laboratory, Pasadena, California

John F. Jones, Jr., 31, PhD; Detroit, Michigan; Sandia Laboratories, Livermore, California

Byron K. Lichtenberg, 29, civilian; Stroudsburg, Pennsylvania; MIT, Cambridge, Massachusetts (*1983 Payload Specialist, STS-9/Spacelab 1; 1992 PS, STS-45/Atlas 1*)

Richard E. Maine, 26, civilian; Louisville, Kentucky; NASA Dryden Flight Research Center, Edwards AFB, California

Ronald E. McNair, 27, PhD; Lake City, South Carolina; Hughes Research Laboratories, Malibu, California

Joseph K.E. Ortega, 31, PhD; Trinidad, Colorado; University of Colorado, Boulder, Colorado

Capt. Harold S. Rhoads, 31, USAF, PhD; Lexington, Kentucky; 4950th Test Wing, Kirtland AFB, New Mexico

David W. Richards, 34, MD, PhD; San Pedro, California; North Broward Emergency Physician, Fort Lauderdale, Florida

Lt. Cmdr. Paul B. Schlein, USN, PhD; Stockton, California; NAVELEX, Washington D.C.

Alan L. Sessoms, 30, PhD; New York, New York; Harvard University, Cambridge, Massachusetts

Lt. Cmdr. Joseph A. Strada, 32, USN, PhD; Philadelphia, Pennsylvania; SAMSO, Los Angeles Air Force Station, California

Kathryn D. Sullivan, PhD; Paterson, New Jersey; Dalhousie University (graduate student), Halifax, Nova Scotia, Canada

David J. Vieira, 27, PhD; Oakland, California; Lawrence Berkeley Laboratory, University of California, Berkeley, California

Capt. James R. Walton, 30, USAF; PhD; Ithica, New York; 366th Tactical Fighter Wing, Mountain Home AFB, Idaho

Lt. Charles R. Weir, 29, USCG; Sidney, Nebraska; Oceanographic Unit, U.S. Coast Guard, Washington D.C.

APPENDIX 2

THE CLASS OF 1978

NASA GROUP 8 (1978) CAREER SUMMARY

Mission data can vary depending upon the source and method of calculation.

For this table, we have extracted Shuttle flight durations from the NASA online Shuttle archives

https:www.nasa.gov/mission_pages/shuttle/shuttlemissions/index.html

Name	Year Born	Military Service	Category	Total flights	1st mission	2nd mission	3rd mission	4th mission	5th mission	Cumulative time in space dd:hh:mm:ss	Left CB	Deceased
Bluford Jr. Guion S.	1942	USAF	MS	4	1983 (STS-8)	1985 (STS-61A)	1991 (STS-39)	1992 (STS-53)		28:16:35:46	1993 July	
Brandenstein. Daniel C.	1943	USN	Pilot	4	1983 (STS-8)	1985 (STS-51G)	1990 (STS-32)	1992 (STS-49)		32:21:05:49	1992 September	
Buchli. James F.	1945	USMC	MS	4	1985 (STS-51C)	1985 (STS-61A)	1989 (STS-29)	1991 (STS-48)		20:10:14:34	1992 August	
Coats. Michael L.	1946	USN	Pilot	3	1984 (STS-41D)	1989 (STS-29)	1991 (STS-39)			19:07:57:17	1991 August	
Covey. Richard O.	1946	USAF	Pilot	4	1985 (STS-51I)	1988 (STS-26)	1990 (STS-38)	1993 (STS-61)		26:21:11:01	1994 August	
Creighton. John O.	1943	USN	Pilot	3	1985 (STS-51G)	1990 (STS-36)	1991 (STS-48)			16:20:24:52	1992 July	
Fabian. John M.	1939	USAF	MS	2	1982 (STS-7)	1985 (STS-51G)				13:04:02:51	1986 January	
Fisher. Anna Lee	1946	Civilian	MS	1	1984 (STS-51A)					7:23:44:56	2006 June	
Gardner. Dale A.	1948	USN	MS	2	1983 (STS-8)	1984 (STS-51A)				14:00:53:39	1986 October	2014 February
Gibson. Robert L.	1946	USN	Pilot	5	1984 (STS-41B)	1986 (STS-61C)	1988 (STS-27)	1992 (STS-47)	1995 (STS-71)	36:04:18:01	1996 November	

Name	Year	Branch	Position	#	Flight 1	Flight 2	Flight 3	Flight 4	Flight 5	Duration	Status	Notes
Gregory. Frederick D.	1941	USAF	Pilot	3	1985 (STS-51B)	1989 (STS-33)	1991 (STS-44)			18:23:06:18	1992 May	
Griggs. S. David	1939	USNR	Pilot	1	1985 (STS-51D)					6:23:55:23	Deceased 1989 June	Vintage WWII training plane crash
Hart. Terry J.	1946	USAF	MS	1	1984 (STS-41C)					6:23:40:07	1984 June	
Hauck. Frederick H.	1941	USN	Pilot	3	1983 (STS-7)	1984 (STS-51A)	1988 (STS-26)			18:03:09:06	1989 April	
Hawley. Steven A.	1951	Civilian	MS	5	1984 (STS-41D)	1986 (STS-61C)	1990 (STS-31)	1997 (STS-82)	1999 (STS-93)	32:02:42:47	1999 August	
Hoffman. Jeffrey A.	1944	Civilian	MS	5	1985 (STS-51D)	1990 (STS-35)	1992 (STS-46)	1993 (STS-61)	1995 (STS-75)	50:11:55:36	1997 July	
Lucid. Shannon M. W.	1943	Civilian	MS	5	1985 (STS-51G)	1989 (STS-34)	1991 (STS-43)	1993 (STS-58)	1996 (STS-76/Mir EO 21/22/STS-79)	223:02:52:21	2012 January	
McBride. Jon A.	1943	USN	Pilot	1	1984 (STS-41G)					8:05:23:38	1989 May	
McNair. Ronald E.	1950	Civilian	MS	2	1984 (STS-41B)	1986 (STS-51L)				7:23:17:08	Deceased 1986 January	STS-51L accident
Mullane. Richard M.	1945	USN	MS	3	1984 (STS-41D)	1988 (STS-27)	1990 (STS-36)			14:20:20:01	1990 July	
Nagel. Steven R.	1946	USAF	Pilot	4	1985 (STS-51G)	1985 (STS-61A)	1991 (STS-37)	1993 (STS-55)		30:01:36:28	1995 March	2014 August
Nelson. George D.	1950	Civilian	MS	3	1984 (STS-41C)	1986 (STS-61C)	1988 (STS-26)			17:02:44:09	1989 June	

Name	Birth	Affiliation	Position	Flights	Flight 1	Flight 2	Flight 3	Flight 4	Flight 5	Duration	Status
Onizuka. Ellison S.	1946	USAF	MS	2	1984 (STS-51)	1986 (STS-51L)				3:01:24:36	Deceased 1986 January / STS-51L accident
Resnik. Judith A.	1949	Civilian	MS	2	1984 (STS-41D)	1986 (STS-51L)				6:00:57:17	Deceased 1986 January / STS-51L accident
Ride. Sally K.	1951	Civilian	MS	2	1983 (STS-7)	1984 (STS-41G)				14:07:47:37	2012 July
Scobee. Francis R.	1939	USAF	Pilot	2	1984 (STS-41C)	1986 (STS-51L)				6:23:41:20	Deceased 1986 January / STS-51L accident
Seddon. Margaret Rhea	1947	Civilian	MS	3	1985 (STS-51D)	1991 (STS-40)	1993 (STS-58)			30:02:22:15	1995 September
Shaw. Brewster H.	1945	USAF	Pilot	3	1983 (STS-9)	1985 (STS-61B)	1989 (STS-28)			22:05:52:21	1989 October
Shriver. Loren J.	1944	USAF	Pilot	3	1985 (STS-51C)	1990 (STS-31)	1992 (STS-46)			16:01:54:32	1993 June
Stewart. Robert L.	1942	USA	MS	2	1984 (STS-41B)	1985 (STS-51J)				12:01:00:33	1986 August
Sullivan. Kathryn D.	1951	USN	MS	3	1984 (STS-41G)	1990 (STS-31)	1992 (STS-45)			22:04:49:12	1993 December
Thagard. Norman E.	1943	USMC	MS	5	1983 (STS-7)	1985 (STS-51B)	1989 (STS-30)	1992 (STS-42)	1995 (Soyuz TM-21/ Mir EO-18/ STS-71)	140:13:26:58	1996 January
Van Hoften. James D. A.	1944	Civilian	MS	2	1984 (STS-41C)	1985 (STS-51I)				14:01:57:49	1986 July
Walker. David M.	1944	USN	Pilot	4	1984 (STS-51A)	1989 (STS-30)	1992 (STS-53)	1995 (STS-69)		30:04:30:06	1995 April
Williams. Donald E.	1942	USN	Pilot	2	1985 (STS-51D)	1989 (STS-34)				11:23:34:44	1990 March

Group 8 Order of most experience

Position	Astronaut	Total Flights	Total duration DD:HH:MM:SS
1st	Lucid	5	223:02:52:21
2nd	Thagard	5	140:13:26:58
3rd	Hoffman	5	50:11:55:36
4th	Gibson	5	36:04:18:01
5th	Brandenstein	4	32:21:05:49
6th	Hawley	5	32:02:42:47
7th	Walker	4	30:04:30:06
8th	Seddon	3	30:02:25:12
9th	Nagel	4	30:02:22:15
10th	Bluford	4	28:16:35:46
11th	Covey	4	26:21:11:01
12th	Shaw	3	22:05:52:21
13th	Sullivan	3	22:04:49:12
14th	Buchli	4	20:10:14:34
15th	Coats	3	19:07:57:17
16th	Gregory	3	18:23:06:18
17th	Hauck	3	18:03:09:06
18th	Nelson	3	17:02:44:09
19th	Creighton	3	16:20:24:52
20th	Shriver	3	16:01:54:32
21st	Mullane	3	14:20:20:01
22nd	Ride	2	14:07:47:37
23rd	Van Hoften	2	14:01:57:49
24th	Gardner	2	14:00:53:39
25th	Fabian	2	13:04:02:51
26th	Stewart	2	12:01:00:33
27th	Williams	2	11:23:34:44
28th	McBride	1	8:05:23:38
29th	Fisher	1	7:23:44:56
30th	McNair	2	7:23:17:08
31st	Griggs	1	6:23:55:23
32nd	Scobee	2	6:23:41:20
33rd	Hart	1	6:23:40:07
34th	Resnik	2	6:00:57:17
35th	Onizuka	2	3:01:24:36
Totals		**103**	**981:08:31:08**

This equates to **over 2 years, 8 months, 8 days** flight time by the 35 Group 8 members, flying a total of 103 separate missions.

NASA GROUP 8 INDIVIDUAL SPACE MISSION DURATIONS

NAME	NO. OF FLIGHTS	MISSION	MISSION DURATION DD:HH:MM:SS	CULMULATIVE EXPERIENCE DD:HH:MM:SS
Bluford, Jr., Guion S,	4	STS-8 STS-61A STS-39 STS-53	6:01:08:43 7:00:44:53 8:07:22:23 7:07:19:47	28:16:35:46
Brandenstein, Daniel C.	4	STS-8 STS-51G STS-32 STS-49	6:01:08:43 7:01:38:52 10:21:00:36 8:21:17:38	32:21:05:49
Buchli, James F.	4	STS-51C STS-61A STS-29 STS-48	3:01:23:23 7:00:44:43 4:23:38:50 5:08:27:38	20:10:14:34
Coats, Michael L.	3	STS-41D STS-29 STS-39	6:00:56:04 4:23:38:50 8:07:22:23	19:07:57:17
Covey, Richard O.	4	STS-51I STS-26 STS-38 STS-61	7:02:17:42 4:01:00:11 4:21:54:31 10:19:58:37	26:21:11:01
Creighton, John O.	3	STS-51G STS-36 STS-48	7:01:38:52 4:10:18:22 5:08:27:38	16:20:24:52
Fabian, John M.	2	STS-7 STS-51G	6:02:23:59 7:01:38:52	13:04:02:51
Fisher, Anna Lee	1	STS-51A	7:23:44:56	7:23:44:56
Gardner, Dale A.	2	STS-8 STS-51A	6:01:08:43 7:23:44:56	14:00:53:39
Gibson, Robert L.	5	STS-41B STS-61C STS-27 STS-47 STS-71	7:23:15:55 6:02:03:51 4:09:05:35 7:22:30:23 9:19:22:17	36:04:18:01
Gregory, Frederick D.	3	STS-51B STS-33 STS-44	7:00:08:46 5:00:06:48 6:22:50:44	18:23:06:18

(continued)

Griggs, S. David	1	STS-51D	6:23:55:23	6:23:55:23
Hart, Terry J.	1	STS-41C	6:23:40:07	6:23:40:07
Hauck, Frederick H.	3	STS-7 STS-51A STS-26	6:02:23:59 7:23:44:56 4:01:00:11	18:03:09:06
Hawley, Steven A.	5	STS-41D STS-61C STS-31 STS-82 STS-93	6:00:56:04 6:02:03:51 5:01:16:06 9:23:37:09 4:22:49:37	32:02:42:47
Hoffman, Jeffrey A.	5	STS-51D STS-35 STS-46 STS-61 STS-75	6:23:55:23 8:23:05:08 7:23:15:03 10:19:58:37 15:17:41:25	50:11:55:36
Lucid, Shannon M.W.	5	STS-51G STS-34 STS-43 STS-58 STS-76/ Mir NASA-1/ STS-79	7:01:38:52 4:23:39:21 8:21:21:25 14:00:12:32 188:04:00:11	223:02:52:21
McBride, Jon A.	1	STS-41G	8:05:23:38	8:05:23:38
McNair, Ronald E.	2	STS-41B STS-51L	7:23:15:55 1:13	7:23:17:08
Mullane, Richard M.	3	STS-41D STS-27 STS-36	6:00:56:04 4:09:05:35 4:10:18:22	14:20:20:01
Nagel, Steven R.	4	STS-51G STS-61A STS-37 STS-55	7:01:38:52 7:00:44:53 5:23:32:44 9:23:39:59	30:01:36:28
Nelson, George D.	3	STS-41C STS-61C STS-26	6:23:40:07 6:02:03:51 4:01:00:11	17:02:44:09
Onizuka, Ellison S.	2	STS-51C STS-51L	3:01:23:23 1:13	3:01:24:36

(continued)

Resnik, Judith A.	2	STS-41D STS-51L	6:00:56:04 1:13	6:00:57:17
Ride, Sally K.	2	STS-7 STS-41G	6:02:23:59 8:05:23:38	14:07:47:37
Scobee, Francis R.	2	STS-41C STS-51L	6:23:40:07 1:13	6:23:41:20
Seddon, Margaret Rhea	3	STS-51D STS-40 STS-58	6:23:55:23 9:02:14:20 14:00:12:32	30:02:22:15
Shaw, Brewster H.	3	STS-9 STS-61B STS-28	10:07:47:24 6:21:04:49 5:01:00:08	22:05:52:21
Shriver, Loren J.	3	STS-51C STS-31 STS-46	3:01:23:23 5:01:16:06 7:23:15:03	16:01:54:32
Stewart, Robert L.	2	STS-41B STS-51J	7:23:15:55 4:01:44:38	12:01:00:33
Sullivan, Kathryn D.	3	STS-41G STS-31 STS-45	8:05:23:38 5:01:16:06 8:22:09:28	22:04:49:12
Thagard, Norman E.	5	STS-7 STS-51B STS-30 STS-42 Soyuz TM-21/Mir EO-18/ STS-71	6:02:23:59 7:00:08:46 4:00:56:27 8:01:14:44 115:08:43:02	140:13:26:58
Van Hoften, James D.A.	2	STS-41C STS-51I	6:23:40:07 7:02:17:42	14:01:57:49
Walker, David M.	4	STS-51A STS-30 STS-53 STS-69	7:23:44:56 4:00:56:27 7:07:19:47 10:20:28:56	30:04:30:06
Williams, Donald E.	2	STS-51D STS-34	6:23:55:23 4:23:39:21	11:23:34:44

APPENDIX 3

GROUP 8 SPACE SHUTTLE EXPERIENCE

Space Shuttle Flight Crew Positions By Mission.

STS	Commander	Pilot	Mission Specialist 1	Mission Specialist 2/ Flight Engineer	Mission Specialist 3	Mission Specialist 4
7		Hauck	Fabian	Ride	Thagard	
8		Brandenstein	Gardner D	Bluford		
9		Shaw				
41B		Gibson R	McNair	Stewart		
41C		Scobee	Hart	Nelson G	Van Hoften	
41D		Coats	Mullane	Hawley	Resnik	
41G		McBride	Ride [2]	Sullivan		
51A	Hauck [2]	Walker D		Fisher A	Gardner D	
51C		Shriver	Onizuka	Buchli		
51D		Williams D	Griggs	Hoffman	Seddon	
51B		Gregory F		Thagard [2]		
51G	Brandenstein [2]	Creighton	Fabian	Nagel	Lucid	
51I		Covey	Van Hoften [2]			
51J				Stewart [2]		
61A		Nagel [2]		Buchli [2]	Bluford [2]	
61B	Shaw [2]					
61C	Gibson R [2]		Nelson G [2]	Hawley [2]		
51L	Scobee [2]		Onizuka [2]	Resnik [2]	McNair [2]	
26	Hauck [3]	Covey [2]			Nelson [3]	
27	Gibson R [3]		Mullane [2]			
29	Coats [2]			Buchli [3]		
30	Walker D [2]			Thagard [3]		
28	Shaw [3]					

© Springer Nature Switzerland AG 2020
D. J. Shayler, C. Burgess, *NASA's First Space Shuttle Astronaut Selection*,
Springer Praxis Books, https://doi.org/10.1007/978-3-030-45742-6

34	Williams D [2]	Lucid [2]		
33	Gregory F [2]			
32	Brandenstein [3]			
36	Creighton [2]	Mullane [3]		
31	Shiver [2]		Hawley [3]	Sullivan [2]
38	Covey [3]			
35		Hoffman [2]		
37	Nagel [3]			
39	Coats [3]			Bluford [3]
40				Seddon [2]
43		Lucid [3]		
48	Creighton [3]		Buchli [4]	
44	Gregory F [3]			
42		Thagard* [4]		
45		Sullivan* [3]		
49	Brandenstein [4]			
46	Shriver [3]			Hoffman* [3]
47	Gibson R [4]			
53	Walker D [3]	Bluford [4]		
55	Nagel [4]			
58		Seddon* [3]		Lucid [4]
61	Covey [4]			Hoffman [4]
71	Gibson R [5]			Thagard [5, down]
69	Walker D [4]			
75		Hoffman [5]		
76				Lucid [5, up]
79				Lucid [5, down]
82			Hawley [4]	
93			Hawley [5]	

[#] Denotes total number of space flights (2nd, 3rd, 4th & 5th)

*= Payload Commander (PC)

WHO FLEW ON WHICH ORBITER

[OV flights *not* missions]

Astronaut Name	Total Flights	Sequence	Columbia OV-102	Challenger OV-099	Discovery OV-103	Atlantis OV-104	Endeavour OV-105
Bluford	4	1st		8	39		
		2nd		61A	53		
Brandenstein	4	1st	32	8	51G		49
Buchli	4	1st		61A	51C		
		2nd			29		
		3rd			48		
Coats	3	1st			41D		
		2nd			29		
		3rd			39		
Covey	4	1st			51I	38	61
		2nd			26		
Creighton	3	1st			51G	36	
		2nd			48		
Fabian	2	1st		7	51G		
Fisher A.	1	1st			51A		
Gibson R.	5	1st	61C	41B		27	47
		2nd				71	
Gardner D.	2	1st		8	51A		
Gregory	3	1st		51B	33	44	
Griggs	1	1st			51D		
Hart	1	1st			51A		
Hauck	3	1st		7	51A		
		2nd			26		
Hawley	5	1st	61C		41D		
		2nd	93		31		
		3rd			82		
Hoffman	5	1st	35		51D	46	61
		2nd	75				
Lucid	6[1]	1st	58		51G	34	
		2nd				43	
		3rd				76[2]	
		4th				79[3]	
McBride	1	1st		41G			
McNair	2[4]	1st		41B			
		2nd		51L[5]			
Mullane	3	1st			41D	27	
		2nd			36		
Nagel	4	1st	55	61A	51G	37	
Nelson G.	3	1st	61-C	41C	26		
Onizuka	2[4]	1st		51L[5]	51C		
Resnik	2[4]	1st		51L[5]	41D		
Ride	2	1st		7			
		2nd		41G			
Scobee	2[4]	1st		41C			
		2nd		51L[5]			
Seddon	3	1st	40		51D		
		2nd	58				

(continued)

Shaw	3	1st	9			61B	
		2nd	28				
Shriver	3	1st			51C	46	
		2nd			31		
Stewart	2	1st	41B			51J	
Sullivan	3	1st	41G		31	45	
Thagard	5[6]	1st	7		42	30	
		2nd		51B		71[7]	
Van Hoften	2	1st	41C		51I		
Walker	4	1st			51A	30	69
		2nd			53		
Williams	2	1st			51D	34	

Notes:

1 Lucid complete *five* missions into space but logged *six* separate flights on an Orbiter.

2 Lucid was launched on STS-76 (Atlantis) but remained on Mir when it undocked for landing, ending that Shuttle mission.

3 Lucid landed on STS-79 (Atlantis) but had not been launched on that mission.

4 McNair, Onizuka, Resnik and Scobee were killed in the STS-51L (*Challenger*) tragedy before reaching orbit. This was their 2nd mission and should have resulted in their 2nd space flight.

5 STS-51L exploded just 73 seconds after launch and had not attained the official space flight threshold altitude, so it does not count as a 'space mission.' But it was in 'flight' when the vehicle exploded and therefore is counted here as a mission in progress.

6 On his 5th mission, Thagard was launched to Mir on the Russian R-7 launch vehicle (Soyuz TM-21) but returned on *Atlantis*, thus completing 5th flight (but landing only) on an Orbiter.

7 Thagard landed on STS-71, so it counts as a Shuttle 'flight' and part of his 5th space flight.

Between 1983 and 1999, Group 8 astronauts flew missions on all five operational Orbiters

- *9 Group 8 astronauts flew on COLUMBIA once; of these, 4 flew her twice.*
- *20 Group 8 astronauts flew on CHALLENGER once; of these, 5 flew her twice.*
- *The greatest number of Group 8 astronauts, 28, flew DISCOVERY at least once; 7 went on to fly her a 2nd time; and 3 a 3rd time.*
- *15 Group 8 astronauts flew ATLANTIS once; 2 of these flew a 2nd mission and 1 flew on her four separate times.*
- *Only 5 flew the newest orbiter ENDEAVOUR and only for one flight each.*

APPENDIX 4

GROUP 8 EVA EXPERIENCE

EXTRA-VEHICULAR ACTIVITY EXPERIENCE (order 1st EVA)

Name	EVAs	Mission	Orbiter	EVA Designation	Date of EVA	Duration H:M	Notes on EVA
Stewart	1st	STS-41B	Challenger	EV2	1984 Feb 7	5:55	1st Group 8 astronaut EVA; tested MMU.
	2nd				1984 Feb 9	6:17	MMU flight; evaluated procedures for planned satellite repairs.
Nelson Van Hoften	1st 1st	STS-41C	Challenger	EV1 EV2	1984 Apr 8	2:57	Nelson attempted to capture Solar Max satellite by using the MMU but was unsuccessful; captured by RMS.
Nelson Van Hoften	2nd 2nd	STS-41C	Challenger	EV1 EV2	1984 Apr 11	6:16	Repair of Solar Max in payload bay; redeployed by RMS; van Hoften test flew MMU in payload bay
Sullivan	1st	STS-41G	Challenger	EV2	1984 Oct 11	3:27	First U.S. female to perform EVA; completed satellite refueling demonstration in payload bay

© Springer Nature Switzerland AG 2020
D. J. Shayler, C. Burgess, *NASA's First Space Shuttle Astronaut Selection*,
Springer Praxis Books, https://doi.org/10.1007/978-3-030-45742-6

Gardner D.	1st	STS-51A	Discovery	EV2	1984 Nov 12	6:00	Assisted in recovery of rogue Palapa comsat
	2nd				1984 Nov 14	5:42	Flew MMU to recover rogue Westar comsat
Hoffman Griggs	1st 1st	STS-51D	Discovery	EV1 EV2	1985 Apr 16	3:00	1st unscheduled (contingency) U.S. EVA; crew attached "flyswatter" device to RMS for a later (unsuccessful) attempt to activate Leasat
Van Hoften	3rd	STS-51I	Discovery	EV1	1985 Aug 31	7:08	Manual capture of Leasat deployed from STS-51D; commenced repairs in payload bay
	4th				1985 Sep 1	4:26	Completed Leasat repairs and redeployed from end of RMS by hand
Hoffman	2nd	STS-61	Endeavour	EV1	1993 Dec 4	7:54	1st HST Service Mission EVA; replaced malfunctioning gyroscopes
	3rd				1993 Dec 6	6:47	3rd HST EVA; Installed new camera in Hubble
	4th				1993 Dec 8	7:21	5th HST EVA; installed new control system; FINAL EVA by Group 8 astronaut

NOTES:

EVA partners [Group 8 members in **Bold**]

STS	EV1	EV2
41B	Bruce McCandless	**Robert Stewart**
41C	**George Nelson**	**James van Hoften**
41G	David Leestma	**Kathryn Sullivan**
51A	Joe Allen	**Dale Gardner**
51D	**Jeffrey Hoffman**	**David Griggs**
51I	**James van Hoften**	William Fisher
61	**Jeffrey Hoffman**	Story Musgrave

EVA ORDER OF MOST EXPERIENCE

Overall Position	Name	Order 1st EVA by Group	Total STS EVAs	Space flights	Total EVA Time
1st	Jeffrey Hoffman	6th	4	2 (STS-51D; STS-61)	25h 02m
2nd	James van Hoften	3rd	4	2 (STS-41C; STS-51I)	20h 45m
3rd	Robert Stewart	1st	2	1 (STS-41B)	12h 12m
4th	Dale Gardner	5th	2	1 (STS-51A)	11h 42m
5th	George Nelson	2nd	2	1 (STS-41C)	9h 13m
6th	David Griggs	7th	1	1 (STS-51D	3h 00m
7th	Kathryn Sullivan	4th	1	1 (STS-41G)	3h 27m

EVA SEQUENCE
[Order of 1st EVA]

Position	1st EVA	2nd EVA	3rd EVA	4th EVA
1	Stewart*	Stewart*	Hoffman	Hoffman
2	Nelson. G*	Nelson. G		
3	Van Hoften	Van Hoften*		
4	Sullivan	Gardner. D*		
5	Gardner. D	Hoffman		
6	Hoffman			
7	Griggs			

* Includes MMU Flight

MMU EXPERIENCE
(Order of 1st MMU Flight)

Date	MMU Flight Sequence Number	Astronaut	MMU Unit Flown	MMU Time Hr:Min	Mission Flown	Mission EVA	Astronaut Total MMU Time Hr:Min
1984 Feb 2	2	Stewart	#3 Flt 2	1:09	STS-41B	1	1:53
1984 Feb 9	4		#2 Flt 2	0:44		2	
1984 Apr 8	6	Nelson G.	#2 Flt 3	0:42	STS-41C	1	0:42
1984 Apr 11	7	Van Hoften	#3 Flt 4	0:28		2	0:28
1984 Nov 14	9	Gardner D.	#2 Flt 4	1:40	STS-51A	2	1:40

Bob Stewart was the only Group 8 astronaut to fly an MMU twice, logging the most time (1 hr 53 min); Dale Gardner had the second-most MMU flight experience (1 hr 40 min) on one flight. Pinky Nelson was third (42 min) and Ox van Hoften fourth (28 min), each with only a single MMU flight.

APPENDIX 5

NASA ASTRONAUT GROUP 8 (1978)

A SELECTED CHRONOLOGY 1976-2019

YEAR	DATE	EVENT
1976	Jul 8	NASA announces the call to select its first Shuttle astronauts
1977	Jul 15	Over 8,000 apply for Shuttle astronaut program
	Jul 19	1st group of astronaut applicants report to JSC
	Aug 9	2nd group of applicants report to JSC
	Aug 25	3rd group of applicants, including 8 women, report to JSC
	Sep 15	4th group of applicants report to JSC
	Sep 26	5th group of applicants report to JSC
	Oct 3	6th group of applicants report to JSC
	Oct 12	7th group of applicants report to JSC
	Oct 20	8th group of applicants report to JSC
	Nov 7	9th group of applicants report to JSC
	Nov 11	10th and final group of applicants reports to JSC
1978	Jan 16	NASA names 35 Group 8 astronaut candidates in the first new selection since 1967
	Jun 29	Ascans arrive at JSC
	Jul 10	1st day of Ascan training
	Jul	Ascans learn water survival techniques at Holmstead AFB
	Aug	Ascans parasail training Vance AFB
1979	Apr 8	Ascans assigned individual technical assignments
	Aug 31	The two-year training program for Group 8 Ascans is cut to just one year on evaluation of the group's performance and they are designated as astronauts having completed basic training. They begin an advanced training program prior to assignment to a Space Shuttle flight crew.
1981	Sep 13	McNair is seriously injured in an automobile accident; his recovery took several months
1982	Sep	Williams assigned as Deputy Manager, Operations Integration, NSTS Program Office, JSC, until Jul 1983

© Springer Nature Switzerland AG 2020

D. J. Shayler, C. Burgess, *NASA's First Space Shuttle Astronaut Selection*, Springer Praxis Books, https://doi.org/10.1007/978-3-030-45742-6

1983	Jun 18	*STS-7 launched; landed Jun 24 (Hauck; Fabian; Ride; Thagard).* FIRST SPACE FLIGHTS BY GROUP 8 ASTRONAUTS; RIDE BECOMES FIRST AMERICAN FEMALE TO FLY IN SPACE
	Aug 30	*STS-8 launched; landed Sep 5 (Brandenstein; Bluford; Gardner D.).* BLUFORD BECOMES FIRST AFRICAN-AMERICAN TO FLY IN SPACE
	Nov 28	*STS-9/Spacelab 1 launched; landed Dec 8 (Shaw)*
1984	Feb 3	*STS-41B launched; landed Feb 11 (Gibson R.; McNair; Stewart)*
	Apr 6	*STS-41C launched; landed Apr 13 (Scobee; Hart; van Hoften; Nelson G.)*
	Jun 15	Hart resigns from NASA
	Aug 30	*STS-41D launched; landed Sep 5 (Coats; Mullane; Hawley; Resnik)*
	Oct 5	*STS-41G launched; landed Oct 13 (McBride; Sullivan; Ride).* SULLIVAN BECOMES FIRST U.S. FEMALE AND FIRST GROUP 8 ASTRONAUT TO PERFORM EVA
	Nov 8	*STS-51A launched; landed Nov 16 (Hauck; Walker D.; Fisher A; Gardner D.).* HAUCK BECOMES FIRST GROUP 8 MEMBER TO COMMAND A MISSION
1985	Jan 24	*STS-51C launched; landed Jan 27 (Shriver; Onizuka; Buchli).* ONIZUKA BECOMES FIRST ASIAN-AMERICAN IN SPACE
	Apr 12	*STS-51D launched; landed Apr 19 (Williams D.; Griggs; Hoffman; Seddon)*
	Apr 29	*STS-51B/Spacelab 3 launched; landed May 6 (Gregory F., Thagard)*
	Jun 17	*STS-51G launched; landed Jun 24 (Brandenstein; Creighton; Fabian; Nagel; Lucid)*
	Jul	Williams becomes Deputy Chief, Aircraft Operations Division, JSC, until Aug 1986
	Aug 27	*STS-51I launched; landed Sep 3 (Covey; van Hoften)*
	Sep	Fabian resigns NASA to join USAF as Director, Space Programs, USAF HQ, The Pentagon. Effective Jan 1, 1986
	Oct 3	*STS-51J launched; landed Oct 7 (Stewart)*
	Oct 30	*STS-61A/Spacelab D1 launched; landed Nov 6 (Nagel; Buchli; Bluford)*
	Nov 26	*STS-61B launched; landed Dec 3 (Shaw)*
		Walker, Lead Cape Crusaders, KSC
1986	Jan 1	Fabian leaves NASA to become Director of Space, Deputy Chief of Staff, Plans and Operations, Headquarters USAF. He was in training as MS STS-61G and SLS-1
	Jan 12	*STS-61C launched; landed Jan 18 (Gibson R.; Nelson G.; Hawley)*
	Jan 28	*STS-51L launched; vehicle and crew lost after an explosion at T+73 seconds (Scobee; Onizuka; Resnik; McNair)*
	Jul	Van Hoften resigns NASA to join Bechtel National, San Francisco, California
	Aug 1	Stewart, selected for promotion to Brigadier General, was reassigned by US Army to join SDI/Redstone Arsenal, Huntsville, Alabama
	Aug 11	Hauck appointed NASA Associate Administrator for External Relations until Feb 1987
	Aug 18	Ride named Special Assistant, Strategic Planning, NASA HQ
	Sep	Williams assigned as Chief, Mission Support Branch, Astronaut Office, until Dec 1988
	Oct	Gardner D. returns to US Navy and assigned to US Space Command, Colorado Springs, Colorado. Deputy Director Space Control. Had been assigned to STS-62A
1987	Apr 27	Brandenstein becomes Chief, Astronaut Office (until Sep 1992) replacing John Young
		Hawley named Deputy, FCOD
	Jul 30	McBride assigned to NASA HQ, Assistant Administrator for Congressional Relations from Sep 1987 to Mar 1989
	Aug	Ride resigns NASA to join Stanford University Center for International Security and Arms Control

1988	Jan 16	Group 8 selection 10th anniversary
	Feb 29	Astronaut Science Support Group formed. Hoffman supports research efforts in astrophysics and remote sensing sciences; Seddon supports research in life sciences
	Sep 29	*STS-26 launched; landed Oct 3 (Hauck; Covey; Nelson G.)*
	Dec 2	*STS-27 launched; landed Dec 6 (Gibson R.; Mullane)*
		Fisher A. on extended leave of absence until 1996
1989	Mar 13	*STS-29 launched; landed Mar 18 (Coats; Buchli)*
	Mar	McBride returns to Astronaut Office
		Buchli serves as Deputy Chief, Astronaut Office until May 1992; Hawley returns to active mission training (STS-31)
		Coats becomes Acting Chief, Astronaut Office until Mar 1990 while Brandenstein trains for STS-32
	Apr 3	Hauck leaves NASA, reassigned Pentagon, Washington D.C., Director of Navy Space Programs Division, Staff of Chief Naval Operations. Effective late May
	May 4	*STS-30 launched; landed May 8 (Walker D.; Thagard)*
	May 12	McBride retires from NASA and USN to pursue career in politics; he is replaced as CDR Astro-1 by Vance Brand
		Walker flies within 100ft of a Pan Am jet, resulting in him being grounded Jul–Sep 1990 and losing command of STS-44
	Jun 17	Griggs is killed in aircraft crash, Earle, Arkansas, rehearsing aerobatics for off-duty airshow appearance. He was in training as MS STS-33 (and was replaced on that crew by John Blaha Class of 1980)
	Jun 30	Nelson G. resigns NASA to become Assistant Provost and Associate Professor of Astronomy, University of Washington, Seattle, Washington
	Aug 8	*STS-28 launched; landed Aug 13 (Shaw)*
	Oct 18	*STS-34 launched; landed Oct 23 (Williams D.; Lucid)*
	Oct	Shaw leaves JSC to become Deputy Director, Space Shuttle Operations, KSC; later Deputy Program Manager, Space Shuttle at KSC, then Director Space Shuttle Operations
	Nov 22	*STS-33 launched; landed Nov 27 (Gregory F.)*
1990	Jan 9	*STS-32 launched; landed Jan 20 (Brandenstein)*
	Feb 21	Three days prior to launch of STS-36, Commander Creighton is medically grounded to recover from an upper respiratory tract infection
	Feb 28	*STS-36 launched; landed Mar 4 (Creighton; Mullane)*
	Mar 1	Williams retires NASA and USN to join Science applications International Corporation (SAIC), Houston, Texas, as Senior Systems Engineer,
	Mar	Coats steps down as Acting Chief, Astronaut Office
		Buchli becomes Deputy Chief, Astronaut Office
		Mullane retires from NASA and USAF
	Apr 24	*STS-31/HST deploy launched; landed Apr 29 (Shriver; Hawley; Sullivan)*
	Jun 7	Hawley leaves Astronaut Office to join Ames Research Center, Moffett Field, Palo Alto, California, as Associate Director
	Jul 1	Mullane resigns from NASA and retires from USAF
	Jul 9	Shuttle crews reassigned due to two veteran astronauts violating JSC FCOD flight guidelines:
		Gibson removed from STS-46 for off-duty midair collision with 2nd racing plane Jul 7
		Walker removed from STS-44 and suspended T-38 flying for 60 days for post-STS-30 near-miss incident, May 1989
		Gregory named replacement for Walker
		Shriver subsequently named replacement for Gibson R.
	Nov 15	*STS-38 launched; landed Nov 20 (Covey)*
	Nov	Buchli resumes active mission training for STS-48 and is replaced temporarily by Covey as Deputy Chief, Astronaut Office
	Dec 2	*STS-35/Astro-1 launched; landed Dec 10 (Hoffman)*

1991	Apr 5	*STS-37 launched; landed Apr 11 (Nagel)*
	Apr 28	*STS-39 launched; landed May 6 (Coats; Bluford)*
	Jun 5	*STS-40/SLS-1 launched; landed Jun 14 (Seddon)*
	Aug 1	Coats resigns from NASA and retires from USN to become Director of Advance Programs and Technical Planning for the Loral Corporation, Houston, Texas
	Aug 2	*STS-43 launched; landed Aug 11 (Lucid)*
	Aug	Covey becomes both Acting Deputy and Acting Chief, Astronaut Office, until Nov 1991
	Sep 12	*STS-48/UARS launched; landed Sep 18 (Creighton; Buchli)*
	Oct	Buchli resumes position of Deputy Chief, Astronaut Office
	Nov 24	*STS-44 launched; landed Dec 1 (Gregory F.)*
	Nov	Nagel becomes Acting Chief, Astronaut Office until Brandenstein returns
	Late 1991	Covey serving as Acting Deputy/Acting Chief, Astronaut Office *and* Acting Director, Flight Crew Operations
1992	Jan 22	*STS-42/IML-1 launched; landed Jan 30 (Thagard)*
	Mar 24	*STS-45/Atlas-1 launched; landed Apr 2 (Sullivan)*
	May 7	*STS-49 launched; landed May 16 (Brandenstein)*
	May	Buchli replaced as Deputy Chief, Astronaut Office
	May	Gregory becomes Associate Administrator, Office of Safety and Mission Assurance (until 2002)
	Jun 25	Sullivan nominated Chief Scientist of NOAA, succeeding Sylvia Alice Earle
	Jul 15	Creighton retires NASA and USN with rank of Captain to become a production test pilot and customer support pilot for Boeing Commercial Airplane Group, Seattle, Washington.
	Jul 31	*STS-46 launched; landed Aug 8 (Shriver; Hoffman)*
	Aug 1	Hawley, former Associate Director of Ames Research Center, returns to JSC to become Deputy Director, Flight Crew Operations
	Aug	Dave Leestma (Class of 1980) replaces Buchli as Deputy Chief, Astronaut Office
	Sep 1	Buchli retires from NASA and USMC to accept position as Manager, Space Station Systems Operations and Requirements, Boeing Defense and Space Group, Huntsville, Alabama
	Sep 12	*STS-47/Spacelab J launched; landed Sep 20 (Gibson R.)*
	Sep	Brandenstein steps down as Chief, Astronaut Office
	Oct	Brandenstein retires with rank of Captain, USN
		Shriver becomes Deputy Chief, Astronaut Office when Leestma becomes Director, FCOD.
	Dec 2	*STS-53 launched; landed Dec 9 (Walker D.; Bluford)*
1993	Mar	US Senate confirms appointment of Sullivan as Chief Scientist, NOAA
	Apr 26	*STS-55/Spacelab D2 launched; landed May 6 (Nagel)*
	May 3	Seddon breaks four metatarsal bones in left foot during egress training, JSC
	May	Shriver is replaced as Deputy Chief, Astronaut Office by Linda Godwin (Class of 1985)
	Jun	Shaw – former Deputy Director, Space Shuttle Operations, KSC, becomes Director, Shuttle Operations, JSC
		Shiver becomes Space Shuttle Program Manager, Launch Integration at KSC
		Walker becomes Chief, Station/Exploration Office, FCOD until Jun 1994
	Jul	Bluford resigns from NASA and retires from USAF with rank of Colonel to become Vice President/General Manager, Engineering Services Division, NYMA, Greenbelt, Maryland.
	Oct 18	*STS-58/SLS-2 launched; landed Nov 1 (Seddon; Lucid)*
	Dec 2	*STS-61/HSM-1 launched; landed Dec 12 (Covey; Hoffman)*
		Sullivan retires from NASA

1994	Mar 1	Thagard and Dunbar (Class of 1980) formally begin cosmonaut training at the Cosmonaut Training Center named for Yu. Gagarin, Russia (TsPK)
	Jun	Walker becomes Chair, JSC Safety Review Board
	Aug 1	Covey retires from NASA and USAF with rank of Colonel to join CalSpan Service Contract Division SII as Director, Business Development, Houston
1995	Jan 3	Lucid and John Blaha (Class of 1980) formally begin their cosmonaut training at TsPK, Russia
	Feb 28	Nagel retires from USAF with rank of Colonel
	Mar 1	Nagel retires from Astronaut Office to become Deputy Director, Operations, Safety, Reliability and Quality Assurance, JSC
	Mar 14	*Soyuz TM-21 launched; crew land on STS-71 Jul 7; (Thagard up only).* THAGARD BECOMES FIRST AMERICAN ASTRONAUT TO LAUNCH AND FLY ON FOREIGN SPACECRAFT (Soyuz and Mir)
	Jun 27	*STS-71 (SMM-1) launched; landed Jul 7; (Gibson R; Thagard down only)*
	Sep 7	*STS-69 launched; landed Sep 18 (Walker D.)*
1996	Jan 5	Thagard retires from NASA and returns to his alma mater Florida State University as Visiting Professor and Director of External Relations for Florida A&M University- Florida State University College of Engineering Tallassee.
	Feb 22	*STS-75 launched; landed Mar 9 (Hoffman)*
	Mar 22	*STS-76 launched; landed Mar 31 (Lucid up only)*
	Apr 15	Walker leaves NASA and retires USN with rank of Captain to be part owner and officer of a telecommunications company as VP, Sales and Marketing, for NDC Voice Corporation Southern California
	Sep 15	Seddon assigned part-time post, Vanderbilt University for Space Physiology and Medicine Nashville Tennessee – remains member of Astronaut Office
	Sep 26	*STS-79 landed. (Lucid down only)*
	Sep	Nagel transfers to Aircraft Operations Division, JSC, as a research pilot, Chief of Aviation Safety and Deputy Chief of the division
	Mid Nov	Gibson retires from NASA and USAF with rank of Colonel to join Rockwell; private business opportunities
		Shaw retires from NASA and USAF to join Rockwell
		Fisher ends extended leave of absence and returns to Astronaut Office; assigned to Operations Planning Branch
		Hoffman leads the Payload and Habitability Branch of Astronaut Office
1997	Feb 11	*STS-82 (HSM-2) launched; lands Feb 21 (Hawley)*
	Jun	Fisher assigned as Branch Chief, Operations Planning Branch until Jun 1998
	Jul 9	Hoffman leaves Astronaut Office to become NASA European Representative in Paris until Aug 2001
	Aug 15	Shriver becomes Deputy Director, KSC Launch and Payload Processing
	Nov	Seddon retires from NASA
1998	Jan 16	Group 8 20th anniversary
	Jun	Fisher A. reassigned Deputy for Operations/Training, Space Station Branch until Jun 1999
1999	Jun	Fisher A. becomes Chief, Space Station Branch Astronaut Office, CB Rep on numerous Space Station boards and Multi-lateral boards; later assigned Space Shuttle Branch
	Jul 22	*STS-93 (Chandra) launched; landed Jul 27 (Hawley)* FINAL SPACEFLIGHT BY A GROUP 8 ASTRONAUT
	Aug	Hawley resumes role as Deputy Director, Flight Crew Operations, JSC
2000	Mar	Shriver resigns from NASA to take up the role of Deputy Program Manager, United Space Alliance, for the Space Shuttle Program

2001	Apr 23	Walker dies from cancer aged 56
	Aug	Hoffman seconded by NASA to MIT as Professor, Department of Aeronautics and Astronautics
	Jun-Dec	Gregory F., Associate Administrator, Office of Safety and Mission Assurance, becomes Acting Associate Administrator for the Office of Space Flight
	Oct	Hawley assigned as Director of Flight Crew Operations until Nov 2002
2002	Feb 12	Lucid serves as Chief Scientist NASA for the next 18 months
	Feb	Gregory F., appointment as Associate Administrator, Office of Space Flight, becomes a permanent position
	Aug 12	Gregory F., former Associate Administrator, Office Space Flight, becomes NASA Deputy Administrator
	Late	Hawley becomes Director Astromaterials Research and Exploration Science Directorate (ARES), JSC.
2003	Jan 16	Group 8 25th anniversary
	May	Hawley retires from NASA to accept a position with the Faculty of University of Kansas
	Sep	Lucid returns to Astronaut Office JSC; receives technical assignments and is assigned Capcom duties in MCC
2004		
2005	Feb 20	Gregory F. becomes NASA Acting Administrator upon departure of Sean O'Keefe until Apr 2005
	Apr 14	Gregory F. resumes role as Deputy Administrator upon swearing in of Michael Griffin
	Dec	Coats becomes the 10th Director of JSC (until end of Dec 2012)
		Gregory F. retires from NASA
2008	Jan 16	Group 8 30th anniversary
	May	Hawley retires from NASA to become Professor, Physics and Astronomy, University of Kansas
2011	Jan	Fisher begins first tour as ISS Capcom
	May 31	Nagel retires from NASA
	Jul	Lucid serves on her final shift as Lead Capcom on Planning (overnight team) for final mission of the program (STS-135)
2012	Jan 31	Lucid announces retirement from NASA
	Jul 23	Ride dies of pancreatic cancer aged 61
	Dec 31	Coats steps down as Director, JSC
2013		Fisher has concurrent assignment as ISS Capcom and technical assignments as Management Astronaut in support of Orion development until 2017
2014	Feb 19	Dale Gardner dies following a stroke aged 65
	Aug 21	Steven Nagel dies aged 67
2016	Feb 23	Williams D. dies aged 74
2017	Apr 27	Fisher A. retires from NASA as a [management] 'astronaut'. Had recently been working on Orion (display development) and as an ISS Capcom. LAST GROUP 8 ASTRONAUT TO LEAVE NASA
2018	Jan 16	Group 8 40th anniversary of selection
2019	Jul	20 years since last spaceflight of Group 8 astronaut [Hawley on STS-93] is observed in the same month as the 50th anniversary of Apollo 11
	Aug 31	40th anniversary of Group 8 graduation of Ascan training

Bibliography

This title follows on from the authors' series of works on the earlier NASA astronaut selections between April 1959 and August 1969. The research of this and the earlier works, both singularly and cooperatively, was conducted over many years, involving archival searching, first-hand interviews, written and electronic correspondence, telephone conversations, official documentation and other contemporary sources.

A complete list of references and sources, from research of this magnitude, is beyond the scope of these few pages, but listed below are the primary interviews, major references and also suggested titles for further reading about the NASA astronaut Class of 1978, their training, the space flights in which they participated and about the program in which they were active.

PERSONAL INTERVIEWS

Coats: Sep 9, 1994 (Shayler)
Covey: Sep 1994 (Shayler)
Hawley: Mar 1, 2012 (Shayler)
Hoffman: Sep 8, 1994 (Shayler); Aug 1996 (Shayler); Jun 3, 2004 (Shayler)
Nelson: Jul 26, 2013 (Shayler)
Seddon: Apr 21, 2004 (Shayler); Feb 19, 2019 (Shayler)
Sullivan: Aug 1988 (Shayler)
Thagard   Aug 1988 (Shayler)

© Springer Nature Switzerland AG 2020   568
D. J. Shayler, C. Burgess, *NASA's First Space Shuttle Astronaut Selection*,
Springer Praxis Books, https://doi.org/10.1007/978-3-030-45742-6

NASA JOHNSON SPACE CENTER ORAL HISTORY PROJECT

Established in 1996 with the goal of capturing history from the individuals who participated in America's quest for space, the program began with its first interviews during the summer of 1997 and has expanded since then to include nearly 1,000 interviews with managers, engineers, technicians, doctors, astronauts, and other employees of NASA. Here, from that program, we list the Oral Histories conducted with members of the Class of 1978 and the date the interview took place.

Bluford (1)	Aug 2, 2004
Brandenstein (1)	Jan 18, 1999
Buchli (0)	*No formal interviews released to date (2019)*
Coats (6)	Jan 4, 2008, Apr 16, 2008, Nov 9, 2012, Jul 14, 2015, Aug 5, 2015, Sep 14, 2015
Covey (4)	Nov 1, 2006, Nov 15, 2006, Feb 7, 2007, Mar 28, 2007
Creighton (1)	May 3, 2004
Fabian (1)	Feb 10, 2006
Fisher, A. (3)	Feb 17, 2009, Mar 3, 2011, May 3, 2011
Gardner, D. (0)	*No formal interviews released to date (2019)*
Gibson, R. (2)	Nov 1, 2013, Jan 22, 2016
Gregory, F. (3)	Apr 29, 2004, Mar 14, 2005, Apr 18, 2006
Griggs (0)	*Deceased prior to commencement of NASA Oral History Program*
Hart (1)	Apr 10, 2003
Hauck (2)	Nov 20, 2003, Mar 17, 2004
Hawley (3)	Dec 4, 2002, Dec 17, 2002, Jan 14, 2003
Hoffman (4)	Apr 2, 2009, Nov 3, 2009, Nov 12, 2009, Nov 17, 2010
Lucid (1)	Jun 17, 1998
McBride (1)	Apr 17, 2012
McNair (0)	*Lost in the 1986 Challenger accident*
Mullane (1)	Jan 23, 2003
Nagel (1)	Dec 20, 2002
Nelson, G. (1)	May 6, 2004
Onizuka (0)	*Lost in the 1986 Challenger accident*
Resnik (0)	*Lost in the 1986 Challenger accident*
Ride (2)	Oct 22, 2002, Dec 6, 2002
Scobee (0)	*Lost in the 1986 Challenger accident*
Seddon (4)	May 20, 2010, May 21, 2010, May 9, 2011, May 10, 2011
Shaw (1)	Apr 19, 2002
Shriver (3)	Dec 16, 2002, Dec 18, 2002, Apr 23, 2008
Stewart (0)	*No formal interviews released to date (2019)*
Sullivan (4)	May 10, 2007, Sep 11, 2007, Mar 12, 2008, May 28, 2009

Thagard (2)   Apr 23, 1998, Sep 16, 1998
Van Hoften (1)   Dec 5, 2007
Walker, D. (0)   No formal interviews released to date (2019)
Williams, D. (1) Jul 19, 2002

UNRECORDED DISCUSSIONS [1985–2019]
Gibson, R.; Hawley; Hoffman; Seddon

WRITTEN/ELECTRONIC CORRESPONDENCE [2004–2020]
Gibson, R.; Hawley; Hoffman; Nelson; Seddon; Sullivan

BIOGRAPHICAL/SPACEFLIGHT RECORDS
Since 1978, both authors have amassed their own personal collection of material for each of the Group 8 astronauts, including official documentation, media reports, flight documentation and new articles for each astronaut. This in-depth research has been supplemented by face-to-face discussions, personal correspondence, email and telephone calls, over many years.

It is supported by the extensive Space Shuttle mission archive held by AIS, including program background files, hardware information, the ALT series and every flight between 1981 and 2011.

PERIODICALS
Aviation Week and Space Technology
JSC News Roundup
NASA Activities
Orbiter (Astro Info Service Ltd)
Spaceflight (British Interplanetary Society)
Spaceflight News
Quest
World Spaceflight News

NEWSPAPERS
Florida Today
Houston Chronicle
Houston Post
The Daily Telegraph
The Times (London)
Washington Post

NASA REPORTS

1988   A Shuttle Chronology 1969-1973: Abstract Concepts to Letter Contracts, [5 volumes], JSC-23309, NASA LBJ Space Center

1997   Walking to Olympus: An EVA Chronology, David S. F. Portree and Robert C. Treviño, NASA Monographs in Aerospace History Series #7 (October 1997)

NASA PUBLICATIONS

Astronautics and Aeronautics: 1976 (SP-4021, 1984); 1977 (SP-4002, 1986); 1978 (SP-4023, 1986); 1979-1984 (SP-2024, 1990); 1985 (SP-4025, 1988); 1986-1990 (SP-4027, 1997) 1991–1995 (SP-2000-4028, 2000); 1996-2000 (SP-209-4030, 2009)

1988   NASA Historical Data Book, Volume III, Programs and Projects 1969-1978, Linda N. Ezell, NASA SP-4012

1999   NASA Historical Data Book, Volume V, NASA Launch Systems, Space Transportation, Human Spaceflight and Space Station, 1979-1988

OTHER REPORTS

1972   *Space Shuttle – Skylab: Manned Space Flight in the 1970s.* Status Report for the Subcommittee on NASA Oversight of the Committee on Science and Astronautics, U.S. House of Representatives, 92nd Congress, 2nd Session, Serial N. (January 1972)

1974   *Space Shuttle, Space Tug, Apollo-Soyuz Test Project – 1974.* Status Report for the Committee on Science and Astronautics, U.S. House of Representatives, 93rd Congress, 2nd Session, Serial K. (February 1974)

1980   *Space Shuttle 1980.* Status Report for the Committee on Science and Astronautics, U.S. House of Representatives, 96th Congress, 2nd Session, Serial AA. (January 1980)

BOOKS AUTHORED BY GROUP 8 ASTRONAUTS

In contrast to the size of the selection, there have been relatively few books directly authored by members of the 1978 selection, even several decades after their selection.

1986   *To Space and Back: U.S. Astronaut Sally Ride Shares the Adventure of Outer Space,* Sally Ride with Susan Okie, Lothrop Lee & Shepard Books, New York, NY [Juvenile literature]

1995   *Liftoff! An Astronaut's Dream,* R. Michael Mullane Silver Burdett Press, Parsippany, NJ. [Juvenile literature]

1986   *Astronaut's Diary,* Jeffrey A., Hoffman, Caliban Press, Montclair, NJ
1993   *Red Sky: A Novel of Love, Space & War* (fiction), R. Michael Mullane, Northwest Publishing, Waukesha, WI.
2006   *Riding Rockets: The Outrageous Tales of a Space Shuttle Astronaut,* Mike Mullane, Scribner Books, New York, NY.
2015   *Go for Orbit: One of America's First Women Astronauts Finds Her Space,* Rhea Seddon, Your Space Press, Murfreesboro, Tennessee.
2019   *Handprints on Hubble: An Astronaut's Story of Invention*, Kathryn D. Sullivan, The MIT Press.

JUVENILE BOOKS

Though only a few from the 1978 selection have penned their own books, the composition of the group inspired a significant number of what is normally categorized as 'juvenile literature.' At first glance, these seem to be insignificant to the avid space book collector or reader, but they can frequently offer a greater, and in some cases, only insight into the person. In these titles, members of the eighth NASA astronaut group were portrayed as national role models of a post-Apollo generation, to be looked up to by the young generation of America, hoping to inspire them to follow in their footsteps.

A brief selection is listed below:

1984   *Space Challenger: The Story of Guion Bluford,* Jim Haskins and Kathleen Benson Carolrhoda Books, Inc., Minneapolis
1986   *Silver Linings: Triumph of the Challenger 7*, June Scobee Rodgers Peake Road, Macon, Georgia.
1990   *Judith Resnik,: Challenger Astronaut,* Joanne E Bernstein & Rose Blue with Alan Jay Gerber, Lodestar Books, New York, NY.
1991   *Ronald McNair: Astronaut,* Black Americans of Achievement Series, Corinne Naden, Chelsea House Publishers, New York, NY
1995   *Black Stars in Orbit: NASA's African American Astronauts,* Khephra Burns and William Miles, Gulliver Books, San Diego, California.
1998   *Shannon Lucid: Space Ambassador,* Bredeson, Carmen, The Millbrook Press, Brookfield, Connecticut.

OTHER BOOKS

1983   *United States Test Pilot School: Historical Narrative and Class Information, 1945 to 1983*, U.S. Naval Test Pilot School Alumni Association, Fishergate Publishing Company, Annapolis

1985   *The Real Stuff: A History of NASA's Astronaut Recruitment Program,* Joseph D. Atkinson, Jr., and Jay M. Shafritz, Praeger Publishers, New York, NY

1986   *Challengers: The Inspiring Life Stories of the Seven Brave Astronauts of Shuttle Mission 51-L,* Staff of the Washington Post Pocket Books, New York, NY

1986   *Ellison S. Onizuka: A Remembrance,* Dennis M. Ogawa and Grant Glen, Signature Publishing/Mutual Publishing, Kailua-Kona, Hawaii,

1987   *Before Lift-Off: The Making of a Space Shuttle Crew*, Henry S. F. Cooper, Jr., John Hopkins University Press, Baltimore, Maryland.

1987   *NASA Space Shuttle (From the Flight Deck 2),* Harry S. Siepmann and David J. Shayler, Ian Allan Ltd, Shepperton, UK

1987   *Challenger, Aviation Fact File*, David J. Shayler, Salamander Books, London, UK

1992   *Men and Women of Space,* Douglas B., Hawthorne, Univelt Inc., San Diego, California,

1992   *The Shuttlenauts 1981-1992: The First 50 Missions, Volume 1 Mission Data,* David J. Shayler, AIS Publications.

1992   *The Shuttlenauts 1981-1992: The First 50 Missions, Volume 2 Shuttle Flight Crew Assignments,* David J. Shayler, AIS Publications.

1993   *Who's Who in Space (International Space Year Edition),* Michael Cassutt, Macmillan Publishing Company, New York NY,

1994   *U.S. Air Force Test Pilot School: 50 Years and Beyond, 1944-1994*, Staff of the U.S. Air Force Test Pilot School, U.S. Air Force Publication, Edwards, California,

1994   *Deke! U.S. Manned Space: From Mercury to the Shuttle,* Deke Slayton and Michael Cassutt, Forge Books, New York, NY,

1994   *They Had a Dream: The Story of African-American Astronauts,* J. Alfred Phelps, Presidion Press, Novato, CA.

1996   *The Shuttlenauts 1981-1992: The First 50 Missions, Volume 3 Flight Crew Biography Profiles,* David J. Shayler, AIS Publications.

2000   *Teacher in Space: Christa McAuliffe and the Challenger Legacy*, Colin Burgess, University of Nebraska Press, Lincoln, Nebraska,

2000   *Disasters and Accidents in Manned Spaceflight*, David J. Shayler, Springer-Praxis Books, New York, NY/Chichester, UK,

2004   *Walking in Space,* David J. Shayler, Springer-Praxis Books, New York, NY/Chichester, UK.

2005   *Women in Space,* David J. Shayler and Ian Moule, Springer-Praxis Books, New York, NY/Chichester, UK.

2005   *Space Shuttle Columbia, Her Missions and Crews,* Ben Evans, Springer-Praxis Books, New York, NY/Chichester, UK

2005   *Russia's Cosmonauts: Inside the Yuri Gagarin Training Center,* Rex D., Hall, David J., Shayler, Bert Vis, Springer-Praxis Books, New York, NY/ Chichester, UK

2007   *NASA's Scientist-Astronauts,* David J. Shayler, and Colin Burgess, Springer-Praxis Books, New York, NY/Chichester, UK.

2007   *Space Shuttle Challenger, Ten Journeys into the Unknown,* Ben Evans, Springer-Praxis Books, New York, NY/Chichester, UK

2007   *Praxis Manned Spaceflight Log 1961-2006,* Tim Furniss and David J. Shayler with Michael D. Shayler, Springer-Praxis Books, New York, NY/Chichester, UK.

2012   *Forever Young: A Life of Adventure in Air and Space*, John W. Young with James R. Hansen, University Press of Florida, Gainesville, FL

2012   *Inventing the American Astronaut,* Matthew H. Hersch, Palgrave Macmillan

2012   *Tragedy and Triumph in Orbit: The Eighties and Early Nineties,* Ben Evans, Springer-Praxis Books, New York, NY/Chichester, UK.

2013   *Spacewalker: My Journey in Space and Faith as NASA's Record-Setting Frequent Flyer,* Jerry Ross with John Norberg, Purdue University Press, West Lafayette, Indiana.

2013   *Wheels Stop: The Tragedies and Triumphs of the Space Shuttle Program, 1986-2011,* Rick Houston, University of Nebraska Press, Lincoln, NE

2014   *Sally Ride: America's First Woman in Space,* Lynn Sherr, Simon & Schuster, New York, NY

2014   *Bold They Rise, The Space Shuttle Early Years, 1972-1986,* David Hitt and Heather R. Smith, University of Nebraska Press

2014   *Partnership in Space: The Mid to Late Nineties,* Ben Evans, Springer-Praxis Books, New York, NY/Chichester, UK.

2015   *The Twenty-First Century in Space* Ben Evans,, Springer-Praxis Books, New York, NY/Chichester, UK.

2016   *The Hubble Space Telescope: From Concept to Success,* David J Shayler with David M. Harland, Springer-Praxis Books, New York, NY/ Chichester, UK,

2016   *Enhancing Hubble's vision: Service Missions That Expanded Our View of the Universe,* David J Shayler with David M. Harland, Springer-Praxis Books, New York, NY/Chichester, UK.

2016   *Space Shuttle: Developing an Icon 1972-2013,* (3 Volumes), Dennis R. Jenkins, Specialty Press, Forest Lake, Minnesota.

2017   *The Last of NASA's Original Pilot Astronauts,* David J. Shayler, and Colin Burgess Springer-Praxis Books, New York, NY/Chichester, UK,

2018   *The Astronaut Maker: How One Mysterious Engineer Ran Human Spaceflight for a Generation,* Michael Cassutt, Chicago Review Press, Chicago, Illinois

About the authors

DAVID J. SHAYLER

Space flight historian David J. Shayler, F.B.I.S. (Fellow of the British Interplanetary Society or – as Dave likes to call it – Future Briton In Space!), was born in England in 1955. After leaving school, Dave began training as an engineering draughtsman prior to serving in HM Forces Royal Marines. Following his return to civilian life, he worked in a variety of roles within the retail industry for almost 30 years before becoming a full-time writer.

His lifelong interest in space exploration began by drawing rockets at the age of 5, but it was not until the launch of Apollo 8 to the Moon in December 1968 that his interest in human space exploration became a passion. He fondly recalls staying up all night with his grandfather to watch the Apollo 11 moonwalk. He joined the British Interplanetary Society (BIS) as a Member in January 1976, becoming an Associate Fellow in 1983, and Fellow in 1984. He served on the Council of the BIS from 2013 to 2019 and, from 2020, became the third Editor of the BIS *Space Chronicle*. The BIS published his first articles in the late 1970s, and in 1982 he created Astro Info Service (www.astroinfoservice.co.uk) to focus his research efforts.

Dave's first book was published in 1987 and has been followed by over 20 other titles, featuring works on the American and Russian space programs, spacewalking, women in space, and the human exploration of Mars. His authorized biography of Skylab 4 astronaut Jerry Carr was published in 2008.

In 1989, Dave applied as a cosmonaut candidate for the UK Project Juno cooperative program with the Soviet Union (now Russia). The mission was to spend seven days in space aboard the space station Mir. He did not reach the final selection but progressed further than he expected. The mission was flown by Helen Sharman in May 1991. In support of his research, Dave has visited NASA field

© Springer Nature Switzerland AG 2020
D. J. Shayler, C. Burgess, *NASA's First Space Shuttle Astronaut Selection*,
Springer Praxis Books, https://doi.org/10.1007/978-3-030-45742-6

centers in Houston and Florida in the United States and the Yuri Gagarin Cosmonaut Training Center in Russia. During these trips, he was able to conduct in-depth research, interview many space explorers and workers, tour training facilities, and handle real space hardware. He also gained valuable insights into the activities of a space explorer and the realities of not only flying and living in space, but also what goes into preparing for a mission and planning future programs.

Over many years, Dave has befriended a large number of former and current astronauts and cosmonauts, some of whom have accompanied him on visits to schools across the country. For over 30 years, he has delivered space-themed presentations and workshops to children and social groups across the UK. This program is intended to help the younger generation develop an interest in science and technology and the world around them, in addition to informing the general public and interested individuals about the history and development of human space exploration.

These days, Dave lives in the West Midlands region of the UK and enjoys spending time with his wife Bel and their rather large, young, white German Shepherd called Shado, while indulging in his love of cooking, fine wines and classical music. His other interests are in reading about military history, visiting historical sites and landmarks, and following Formula 1 motor racing.

Authors Colin Burgess and Dave Shayler, 2018 (Image provided by D. Shayler).

COLIN BURGESS

Australian author Colin Burgess grew up in Sydney's southern suburbs, where he and his wife Patricia still live. They have two grown sons, two grandsons and a granddaughter.

His working life began in the wages department of a major Sydney afternoon newspaper, where he first picked up the writing bug, and later as a sales representative for a precious metals company. He subsequently joined Qantas Airways as a passenger handling agent in 1970 and two years later, he transferred to the airline's cabin crew. He retired from Qantas as an onboard Flight Service Director/Customer Service Manager in 2002, after 32 years' service.

During that period, several of his books on the Australian prisoner-of-war experience were published, as well as the first of his biographical books on space explorers such as Australian payload specialist Dr. Paul Scully-Power and *Challenger* teacher Christa McAuliffe. He has also written extensively on space flight subjects for astronomy and space-related magazines in Australia, the United Kingdom and the United States.

Colin admits that his interest in human space exploration was sparked by the orbital flight of Mercury astronaut John Glenn in 1962, and 36 years later he was invited as a VIP guest to the launch of the former astronaut on his second space mission aboard Shuttle Discovery in 1998. He travelled to Florida from Australia and witnessed the only launch he has attended to this time, but declares it was the one mission launch he was determined would not pass him by.

In 2003, the University of Nebraska Press appointed Colin as Series Editor for their ongoing Outward Odyssey series of books detailing the entire social history of space exploration. Fifteen years on, he still retains that position and continues to appoint and assist what he refers to as "a whole new breed of space authors." He was also involved in co-writing three of the Outward Odyssey volumes and one under sole authorship. His first Springer-Praxis book, *NASA's Scientist-Astronauts*, also co-written with British-based space historian David J. Shayler, was released in 2007.

NASA's First Space Shuttle Astronaut Selection will be Colin's 14th title in a decade-long association with Springer-Praxis, for whom he continues to research further books for future publication. He regularly attends astronaut functions in the United States and is well known to many of the pioneering space explorers, allowing him to conduct personal interviews for his publications.

Other works by the authors

By David J. Shayler and Colin Burgess in this series
NASA's Scientist-Astronauts (2006), ISBN 0-387-21897-1
The Last of NASA's Original Pilot Astronauts (2017), ISBN 978-3-319-51012-5

Other space exploration books by David J. Shayler
Challenger Fact File (1987), ISBN 0-86101-272-0
Apollo 11 Moon landing (1989), ISBN 0-7110-1844-8
Exploring Space (1994), ISBN 0-600-58199-3
All About Space (1999), ISBN 0-7497-4005-X
Around the World in 84 Days: The Authorized Biography of Skylab Astronaut Jerry Carr (2008), ISBN 9781-894959-70-0

With Harry Siepmann
NASA Space Shuttle (1987), ISBN 0-7110-1681-X

With Robert Godwin
Outpost in Orbit: A Pictorial & Verbal History of the Space Station (2018), ISBN 978-1989044-03-2

Other books by David J. Shayler in this series
Disasters and Accidents in Manned Spaceflight (2000), ISBN 1-85233-225-5
Skylab: America's Space Station (2001), ISBN 1-85233-407-X
Gemini: Steps to the Moon (2001), 1-85233-405-3
Apollo: the Lost and Forgotten Missions (2002), ISBN 1-85233-575-0
Walking in Space (2004), ISBN 1-85233-710-9
Space Rescue (2007), ISBN 978-0-387-69905-9
Linking the Space Shuttle and Space Stations: Early Docking Technologies From Concept to Implementation (2017), ISBN978-3319497686

© Springer Nature Switzerland AG 2020
D. J. Shayler, C. Burgess, *NASA's First Space Shuttle Astronaut Selection*,
Springer Praxis Books, https://doi.org/10.1007/978-3-030-45742-6

Assembling and Supplying the ISS: The Space Shuttle Fulfils its Mission (2017), ISBN 978-3319404417

Gemini Flies! Unmanned Flights and the First Manned Mission (2018), ISBN 978-3319681412

Gemini 4: An Astronaut Steps Into the Void (2018), ISBN 978-3319766744

With Rex Hall, M.B.E.
The Rocket Men (2001), ISBN 1-85233-391-X
Soyuz: A Universal Spacecraft (2003), 1-85233-657-9

With Rex Hall, M.B.E. and Bert Vis
Russia's Cosmonauts (2005), ISBN 0-38721-894-7

With David M. Harland
Hubble Space Telescope: From Concept to Success (2016), ISBN 978-1-4939-2826-2

Enhancing Hubble's Vision: Service Missions That Expanded Our View of the Universe (2016), ISBN 978-3-319-22643-9

With Ian Moule
Women in Space: Following Valentina (2005), ISBN 1-85233-744-3

Other books by David J. Shayler and Michael D. Shayler in this series
Manned Spacecraft Log II: 2006-2012 (2013), ISBN 978-1-4614-4576-0

With Andrew Salmon
Marswalk One: First Steps on a New Planet (2005), ISBN 1-85233-792-3

With Tim Furniss
Praxis Manned Spaceflight Log: 1961-2006 (2007), ISBN 0-387-34175-7

Other space exploration books by Colin Burgess
Space: The New Frontier (1987), ISBN 978-0868963617

Oceans to Orbit: The Story of Australia's First Man in Space, Paul Scully-Power (1995), ISBN 978-0949853330

Australia's Astronauts: Three Men and a Spaceflight Dream (1999), ISBN 978-0731808311

Teacher in Space: Christa McAuliffe and the Challenger Legacy (2000), ISBN 978-0803261822

Australia's Astronauts: Countdown to a Spaceflight Dream (2009), ISBN 978-0975228616

Footprints in the Dust: The Epic Voyages of Apollo, 1969-1975 (2010), Ed. By Colin Burgess, ISBN 978-803226654

Shattered Dreams: The Lost and Canceled Space Missions (2019), ISBN 978-1496206756

With Kate Doolan
Fallen Astronauts: Heroes Who Died Reaching for the Moon (2003, revised 2016), ISBN 9778-0803262126

With Francis French
Into That Silent Sea: Trailblazers of the Space Era, 1961-1965 (2007), ISBN 978-0803211469
In the Shadow of the Moon: A Challenging Journey to Tranquility, 1965-1969 (2007), ISBN 978-0803211285978

Other books by Colin Burgess in this series
Selecting the Mercury Seven: The Search for America's First Astronauts (2011), ISBN 978-1-4419-8404-3
Moon Bound: Choosing and Preparing NASA's Lunar Astronauts (2013), ISBN 9787-1-4614-3854-0
Freedom 7: The Historic Flight of Alan B. Shepard, Jr. (2014), ISBN 978-3-3190-1155-4
Liberty Bell 7: The Suborbital Flight of Virgil I. Grissom (2014), ISBN 978-3-319-04390-6
Friendship 7: The Epic Orbital Flight of John H. Glenn, Jr. (2015), ISBN 978-3-319-15653-8
Aurora 7: The Mercury Space Flight of M. Scott Carpenter (2015), ISBN 978-3-319-20438-3
Sigma 7: The Six Mercury Orbits of Walter M. Schirra, Jr. (2016), ISBN 978-3-319-27982-4
Faith 7: L. Gordon Cooper, Jr., The Final Mercury Mission (2016), ISBN 978-3-319-30562-2

With Chris Dubbs
Animals in Space: From Research Rockets to the Space Shuttle (2007), ISBN 978-0-387-36053-9

With Rex Hall, M.B.E.
The First Cosmonaut Team: Their Lives, Legacies and Historical Impact (2009), ISBN 928-0-387-84823-5

With Bert Vis
Interkosmos: The Eastern Bloc's Early Space Program (2016), ISBN 978-3-319-24161-6

Index

Printed in the United States
By Bookmasters